RC–ACTIVE CIRCUITS
Theory and Design

PRENTICE-HALL SERIES IN ELECTRICAL AND COMPUTER ENGINEERING

Leon O. Chua, Editor

Bruton, RC-ACTIVE CIRCUITS
Chua and Lin, COMPUTER-AIDED ANALYSIS OF ELECTRONIC CIRCUITS: ALGORITHMS AND COMPUTATIONAL TECHNIQUES
Lam, ANALOG AND DIGITAL FILTERS: DESIGN AND REALIZATION

RC-ACTIVE CIRCUITS
Theory and Design

LEONARD T. BRUTON

Head, Dept. of Electrical Engineering
The University of Calgary
Calgary, Alberta, Canada

PRENTICE-HALL, INC., *Englewood Cliffs, New Jersey 07632*

Library of Congress Cataloging in Publication Data

Bruton, Leonard T.
RC-active circuits

Bibliography: p.
Includes index.
1. Electronic circuits. 2. Electric networks, Active. I. Title.
TK7867.B78 621.3815'3 79-20212
ISBN 0-13-753467-1

Editional production supervision
and interior design by: JAMES M. CHEGE

Cover design by: Edsal Enterprises

Manufacturing buyer: GORDON OSBOURNE

Printed in the United States of America

10 9 8 7 6 5 4 3 2 1

PRENTICE-HALL INTERNATIONAL, INC., *London*
PRENTICE-HALL OF AUSTRALIA PTY. LIMITED, *Sydney*
PRENTICE-HALL OF CANADA, LTD., *Toronto*
PRENTICE-HALL OF INDIA PRIVATE LIMITED, *New Delhi*
PRENTICE-HALL OF JAPAN, INC., *Tokyo*
PRENTICE-HALL OF SOUTHEAST ASIA PTE. LTD., *Singapore*
WHITEHALL BOOKS LIMITED, *Wellington, New Zealand*

This book is dedicated to

Avis

Alex, Michelle, Nicole

Archie and ***Marie***

PREFACE

The text describes the modern approach to the theory and design of RC-active circuits. It has been written as a teaching text with worked examples and problems accompanying each chapter.

A basic difficulty that usually arises in teaching this subject is that undergraduate students have not generally developed significant expertise in passive network theory, approximation theory and semiconductor circuit design and analysis. In this text the reliance on passive network theory, approximation theory and semiconductor theory are minimized to the point that it is not essential that the reader be familiar with these topics. The concept of the *nullor element* is used to unify and classify a variety of concepts, including controlled-sources, immittance converters, transistor circuit analysis, and the classification and design of RC-active filters.

A brief review of prerequisite material is given in Chapter 1 with emphasis on the systems approach to the characterization of linear networks. The concepts of linearity, causality, time-invariance, transform methods, poles, zeros, stability, frequency response and transfer functions are briefly reviewed.

In Chapter 2, a treatment of passivity, losslessness, and activity is used as the basis for classifying the building-blocks or elements of RC-active circuits. The $\pm$R, $\pm$L, $\pm$C elements, the FDNR elements, the gyrator, the transformer, the PICs, the PIIs, the GICs, the NICs, the controlled-sources, the independent sources, the nullator, the norator and, most importantly, the nullor are described and classified as active or passive one- or two-port building blocks.

Techniques of network analysis are applied to RC-nullor networks in Chapter 3 where the elementary methods of loop and nodal analysis are found to be directly applicable to the systematic analysis of RC-nullor networks. Furthermore, a variety of GIC networks are classified according to their elementary R-nullor models. The material in the third chapter is *not* prerequisite to an understanding of subsequent chapters and may be omitted if necessary.

Chapters 4 and 5 are concerned with linear circuit applications of the bipolar transistor and the operational amplifier, respectively. The modeling, nonideal performance, and applications of these devices are described. The physical semiconductor principles underlying the device behavior are not mentioned and no attempt is made to describe *nonlinear* circuits such as modulators, oscillators, trigger circuits, logic circuits, logarithmic amplifiers, analog multipliers, etc. The objective in these two chapters is to establish a unified approach to the modeling of the bipolar transistor and the operational amplifier that allows linear circuits to be analyzed, understood and modeled in a straightforward way. Both chapters contain a large number of worked examples and problems that directly refer to *practical* transistor circuits or practical operational amplifier circuits. Emphasis is placed on direct coupled (capacitorless) circuits that may be made in microelectronic form. The reader learns to replace bipolar transistors by three-terminal nullors and operational amplifiers by four-terminal nullors in order to obtain a rapid first order approximation of circuit behavior which is often more useful in understanding the behavior of complex multi-device circuits. The transistor and operational amplifier circuits that are used widely in RC-active filter synthesis, such as controlled sources, gyrators, simulated-inductance and FDNR circuits are described and their nonideal performance is established for use in the subsequent chapters (7, 8, 9) that describe RC-active filter design and performance.

The topic of network sensitivity is introduced in Chapter 6 with emphasis on the classical algebraic sensitivity function and its use in establishing worstcase design and element tolerances. Some invariance properties are discussed as a means of deriving lower-bound *benchmarks* for the estimation of the tolerance performance of RC-active circuits. In Chapter 7, the major RC-active circuit structure are introduced and their principal sensitivity characteristics are described.

Chapter 8 is devoted to the important topic of biquadratic circuit synthesis and design. The various biquadratic circuits are classified and compared. Chapters 9 and 10 describe in detail the theory and design of high-quality RC-active filters. Emphasis is on *ladder* filter design and the estimation of nonideal performance. Concepts of immittance-simulation, signal flow graph simulation, FDNR elements, etc. are described. Some relevant design case studies are included.

It is hoped that the book will be useful to several categories of readers. First, it may be used as an undergraduate textbook where students are adequately prepared in linear system theory and Laplace Transform techniques but not necessarily familiar with semiconductor physics, circuit theory, and passive network synthesis. For this application, Chapters 6 and 7 may be omitted. Secondly, the text may find application at the postgraduate level in which case Chapters 6 and 7 provide pertinent material on the sensitivity performance of electrical networks. Finally, Chapters 8, 9 and 10 are applications-oriented and they contain specific design information that is important to the designer of high-quality RC-active filters. Therefore, the industrial designer and user of RC-active filters should find that the design techniques and modeling equations are useful guidelines for practical design. From the viewpoint of teaching, the worked examples and the problems are integral parts of the text and an effort has been made to choose problems that involve practical and useful circuits.

For the benefit of the graduate level reader and current researchers in this field, a rather selective but useful bibliography is given at the end of the book, in addition to the recommended reading lists to be found in each chapter.

I believe this book will be useful to the practising active circuit designer. The design-orientation of the material and the extensive tabulations of key formulae should permit a rapid appreciation of the more important and useful design approaches. The emphasis on nonideal performance and sensitivity limitations will perhaps be most appreciated by those readers who have already built and tested their first microelectronic active circuit and found that it did not meet specifications!

I particularly want to express my thanks to some of those who have helped to make this book possible. First, to my wife Avis for continual encouragement and tolerance during the many hours that have been given to the task. I am grateful to Professors S. K. Mitra and L. O. Chua for their encouragement and interest and to the many students who have studied the drafts of the manuscript, especially M. K. N. Rao, P. A. Ramamoorthy and R. Wedding. Also, I thank Mrs. Norean Kowalchuk for typing the manuscript and the University of Calgary for the privilege of a Sabbatical Leave during which much of the book was written.

LEONARD T. BRUTON

Calgary, Alberta
Canada

RC–ACTIVE CIRCUITS
Theory and Design

1

FUNDAMENTAL CONCEPTS

1-1 INTRODUCTION

The material in this book is concerned primarily with the analysis and design of *RC*-active electronic circuits. The purpose of this first chapter is to provide a brief review of the fundamental theoretical concepts on which the topic is founded. The concepts of linearity, time-invariance, differential equations, the Laplace transform, poles, zeros, stability and steady-state sinusoidal frequency response are briefly reviewed. The second-order lowpass, bandpass, highpass, notch and allpass transfer functions are defined.

1-2 LINEARITY AND TIME-INVARIANCE

Consider the system in Fig. 1.1 with input $x(t)$ and output $y(t)$. Now, let a specific input waveform $x_1(t)$ be applied and let the corresponding output waveform be defined as $y_1(t)$. Further, given some other input waveform $x_2(t)$, let the corresponding output waveform be defined as $y_2(t)$. This is

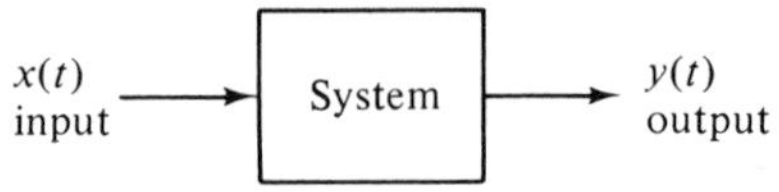

FIGURE 1.1

written as follows:

$$\left.\begin{aligned} x_1(t) &\rightarrow y_1(t) \\ x_2(t) &\rightarrow y_2(t) \end{aligned}\right\} \tag{1.1}$$

Consider now the application of the waveform $[x_1(t) + x_2(t)]$ to the system. The condition for the system to be a *linear system* is that for all possible $x_1(t)$ and $x_2(t)$ the output waveform corresponding to the input $[x_1(t) + x_2(t)]$ is $[y_1(t) + y_2(t)]$; that is

$$\boxed{[x_1(t) + x_2(t)] \rightarrow [y_1(t) + y_2(t)]} \quad \text{LINEARITY CONDITION} \tag{1.2}$$

for all possible $x_1(t)$, $x_2(t)$, $y_1(t)$, and $y_2(t)$.

The system in Fig. 1.1 is a *time-invariant system* if, for all possible $x_1(t)$ and $y_1(t)$ and all real T,

$$\boxed{x_1(t - T) \rightarrow y_1(t - T)} \quad \text{TIME-INVARIANT CONDITION} \tag{1.3}$$

A system that satisfies equations (1.1), (1.2), and (1.3) is a *linear time-invariant system*. All of the small-signal circuit models considered in subsequent chapters of this book are examples of linear time-invariant systems.

EXAMPLE 1.1: Prove that the *ideal* capacitance element C is an example of a linear time-invariant network by associating the voltage $v(t)$ with the output waveform and the current $i(t)$ with the input waveform of Fig. 1.1.

SOLUTION: By definition of the ideal capacitance element

$$i(t) = C\frac{dv(t)}{dt} \tag{1.4}$$

for all possible $v(t)$ and $i(t)$. Thus, if the excitation $i_1(t)$ is applied, the response $v_1(t)$ is

$$v_1(t) = \frac{1}{C}\int_{-\infty}^{t} i_1(t)\, dt \tag{1.5}$$

and, if excitation $i_2(t)$ is applied then the response $v_2(t)$ is

$$v_2(t) = \frac{1}{C}\int_{-\infty}^{t} i_2(t)\,dt \tag{1.6}$$

If an excitation $[i_1(t) + i_2(t)]$ is applied, it follows from equation (1.4) that the response $i_3(t)$ is given by

$$v_3(t) = \frac{1}{C}\int_{-\infty}^{t} [i_1(t) + i_2(t)]\,dt = \frac{1}{C}\int_{-\infty}^{t} i_1(t)\,dt + \frac{1}{C}\int_{-\infty}^{t} i_2(t)\,dt$$

which, by comparison with equations (1.5) and (1.6), implies that

$$v_3(t) = v_1(t) + v_2(t) \tag{1.7}$$

The response $v_3(t)$ therefore equals the superposition (or sum) of the individual responses $v_1(t)$ and $v_2(t)$ and the ideal capacitance element C is therefore linear. It is easily shown to be time-invariant; substituting $(t - T)$ for t in the left side of equation (1.4) gives the corresponding response $v(t - T)$.

EXAMPLE 1.2: The output $y(t)$ of a system is related to the input $x(t)$ by

$$y(t) = x(t) \cos \omega_0 t, \qquad \omega_0 \text{ constant} \tag{1.8}$$

Is this system linear? Is this system time-invariant?

SOLUTION: From equation (1.8),

$$y_1(t) = x_1(t) \cos \omega_0 t \tag{1.9}$$

$$y_2(t) = x_2(t) \cos \omega_0 t \tag{1.10}$$

Defining the response to $[x_1(t) + x_2(t)]$ as $y_3(t)$ and using equation (1.8), we find that

$$y_3(t) = [x_1(t) + x_2(t)] \cos \omega_0 t \tag{1.11}$$

Now, adding the separate responses in equations (1.9) and (1.10) *does* give the response $y_3(t)$ in equation (1.11); therefore, the system *is* linear. However, replacing $x(t)$ by $x(t - T)$ in equation (1.8) provides the response $y_4(t)$, where

$$y_4(t) = x(t - T) \cos \omega_0 t \tag{1.12}$$

However, from equation (1.8), $y(t - T) = x(t - T) \cos[\omega_0(t - T)]$, and this is *not* equal to equation (1.12) for all T. The system is therefore time-varying (it is not time-invariant).

EXAMPLE 1.3: A system with input $x(t)$ has an output $y(t)$, as follows:

$$y(t) = x^2(t) \tag{1.13}$$

Is this system linear?

SOLUTION: We have the individual responses

$$y_1(t) = x_1^2(t) \tag{1.14}$$

and

$$y_2(t) = x_2^2(t) \tag{1.15}$$

It follows from equation (1.13) that the response to the excitation $[x_1(t) + x_2(t)]$ is

$$y_3(t) = [x_1(t) + x_2(t)]^2 \tag{1.16}$$

which is *not* equal to the sum of equations (1.14) and (1.15) for all $x_1(t)$ and $x_2(t)$. The system is therefore *not* linear (nonlinear).

From a practical point of view, a real physical system cannot be *exactly* a linear time-invariant system. For example, the region of approximate linearity of a practical system is strictly limited by the *finite* signal amplitude capability of all practical systems.

Active devices such as transistors are only approximately linear over a particular region of the input-output signal characteristics, which limits the application of modern microelectronic operational amplifiers to situations where the output voltage and current waveforms are typically restricted to the regions $|v(t)| < 30$ V and $|i(t)| < 100$ mA.

1-3 TIME-DOMAIN AND *s*-DOMAIN BEHAVIOR OF LINEAR TIME-INVARIANT SYSTEMS

All of the circuits described in this book have an input waveform $x(t)$ and an output waveform $y(t)$ that are *ideally* related by an ordinary time-domain differential equation of the form

$$\sum_{i=0}^{N} b_i \frac{d^i y(t)}{dt^i} = \sum_{i=0}^{M} a_i \frac{d^i x(t)}{dt^i} \tag{1.17}$$

where the $(N + 1)$ coefficients b_i and $(M + 1)$ coefficients a_i are real constants numbers that may be determined by routine analysis of the circuit. (The reader is asked to prove in Problem 1.3 that equation (1.17) describes a linear time-invariant network.) The integer N is defined as the order of the circuit; for example,

$$b_2 \frac{d^2y(t)}{dt^2} + b_1 \frac{dy(t)}{dt} + b_0 y(t) = a_1 \frac{dx(t)}{dt} + a_0 x(t)$$

is an example of an input-output equation for a second-order circuit ($N = 2$, $M = 1$).

Much of the material in this book is concerned with analyzing and designing RC-active circuits that realize *specific* input-output differential equations of the type given in equation (1.17). Thus, we are concerned with circuits that implement the desired coefficients a_i and b_i to within some acceptable accuracies. In fact, some of the major problem areas in the design of RC-active circuits are concerned with determining the accuracies with which the coefficients a_i and b_i must be realized and, more importantly, with finding suitable circuits (interconnections of resistors, capacitors, and transistors) that are capable of achieving the necessary accuracies at the lowest overall cost. This might involve the use of thin film and thick film microelectronic resistor-capacitor networks in conjunction with microelectronic operational amplifiers and other special-purpose transistor devices.

The most extensively adopted approach to the specification and design of RC-active circuits is to characterize equation (1.17) in terms of its Laplace transform, as follows:

$$\sum_{i=0}^{N} b_i s^i Y(s) = \sum_{i=0}^{M} a_i s^i X(s) \qquad (1.18)$$

where it is assumed that $y(t)$, $x(t)$, the first $(M - 1)$ derivatives of $y(t)$, and the first $(N - 1)$ derivatives of $x(t)$ are zero at time $t = 0$, and where

$$X(s) \equiv \mathcal{L}[x(t)], \qquad Y(s) \equiv \mathcal{L}[y(t)]$$

The transform transfer function $H(s)$ of the circuit is defined as

$$H(s) \equiv \frac{Y(s)}{X(s)} \qquad (1.19)$$

so that, from equations (1.18) and (1.19),

$$\boxed{H(s) = \frac{a_M s^M + a_{M-1} s^{M-1} + \ldots + a_2 s^2 + a_1 s + a_0}{b_N s^N + b_{N-1} s^{N-1} + \ldots + b_2 s^2 + b_1 s + b_0}}$$

TRANSFORM TRANSFER FUNCTION (1.20)

It is conventional to write $H(s)$ in the form

$$H(s) = K\frac{(s - z_M)(s - z_{M-1}) \ldots (s - z_2)(s - z_1)}{(s - p_N)(s - p_{N-1}) \ldots (s - p_2)(s - p_1)} \tag{1.21}$$

where, by comparison of equations (1.20) and (1.21), it follows that

$$K = \frac{a_M}{b_N} \tag{1.22}$$

and $z_1, z_2, \ldots, z_{M-1}, z_M$ are the (generally complex) roots of the polynomial

$$P(s) = \frac{1}{a_M}\left(a_M s^M + a_{M-1}s^{M-1} + \ldots + a_2 s^2 + a_1 s + a_0\right) \tag{1.23}$$

and $p_1, p_2, \ldots, p_{N-1}, p_N$ are the (generally complex) roots of the polynomial

$$Q(s) = \frac{1}{b_N}\left(b_N s^N + b_{N-1}s^{N-1} + \ldots + b_2 s^2 + b_1 s + b_0\right) \tag{1.24}$$

so that

$$H(s) = K\frac{P(s)}{Q(s)} \tag{1.25}$$

The numbers $z_1, z_2, \ldots, z_{M-1}, z_M$ are known as the *zeros* of the function $H(s)$; note that $H(s)$ is zero wherever s is equal in value to $z_1, z_2, \ldots, z_{M-1}$ or z_M. The numbers $p_1, p_2, \ldots, p_{N-1}, p_N$ are known as the *poles* of the function $H(s)$; note that $H(s)$ is infinite wherever s is equal in value to $p_1, p_2, \ldots, p_{N-1}$ or p_N. In the complex s-plane there are, therefore, M finite points at which $H(s)$ is zero and N finite points at which $H(s)$ is infinite.

EXAMPLE 1.4: Find the poles and zeros of the function

$$H(s) \equiv \frac{3s^2 + 9}{s^2 + 2s + 101} \tag{1.26}$$

SOLUTION: Equation (1.26) corresponds to $M = 2$, $N = 2$, $a_2 = 3$, $a_1 = 0$, $a_0 = 9$, $b_2 = 1$, $b_1 = 2$, $b_0 = 101$. In terms of equation (1.21),

$$H(s) = 3\frac{(s + j\sqrt{3})(s - j\sqrt{3})}{(s + 1 + j10)(s + 1 - j10)} \tag{1.27}$$

corresponding to

$$\left.\begin{array}{lll} K = 3, & z_1 = j\sqrt{3}\,, & z_2 = -j\sqrt{3} \\ p_1 = -1 - j10, & p_2 = -1 + j10 & \end{array}\right\} \tag{1.28}$$

1-3.1 Stability

A general definition of bounded-input bounded-output (BIBO) stability is as follows:

DEFINITION: *A system is BIBO stable if, for all possible bounded-amplitude input waveforms $x(t)$, the output waveform $y(t)$ is bounded in amplitude.*

This leads to some straightforward conclusions in the case of a *linear time-invariant system*; from equation (1.19), the transform of the output $y(t)$ is given by

$$Y(s) = X(s)H(s)$$

which may be expressed as a rational function of s containing the denominator terms

$$(s - p_N)(s - p_{N-1}) \ldots (s - p_2)(s - p_1)$$

It may be shown that partial fraction expansion of $Y(s)$ allows it to be written in the form

$$Y(s) = \sum_{i=1}^{N} \left(\frac{K_i}{s - p_i} \right) + Y_1(s)$$

where $Y_1(s)$ is a transform function and K_i are constants. Thus,

$$y(t) = \underbrace{\sum_{i=1}^{N} K_i e^{p_i t} u(t)}_{\text{transient response}} + \mathcal{L}^{-1}[\,Y_1(s)] \tag{1.29}$$

The transient response term is composed of a sum of N exponential terms, each of which is associated with a pole p_i. Now, BIBO stability requires by definition that $|y(t)| < \infty$ for all possible $Y_1(s)$, K_i, and p_i in equation

(1.29), implying that

$$\left| \lim_{t\to\infty} \underbrace{\left[\sum_{i=1}^{N} K_i e^{p_i t} u(t) \right]}_{\text{transient response term}} \right| < \infty \quad \text{BIBO STABILITY CONDITION} \tag{1.30}$$

Writing

$$p_i = \sigma_i + j\omega_i, \qquad i = 1, 2, 3, \ldots, N$$

and using equation (1.30) leads to the equivalent necessary BIBO stability condition that

$$\sigma_i \leqslant 0, \qquad i = 1, 2, 3, \ldots, N \quad \text{BIBO STABILITY CONDITION} \tag{1.31}$$

That is,

A LINEAR TIME-INVARIANT SYSTEM IS BIBO STABLE IF AND ONLY IF NONE OF THE POLES p_i IS IN THE RIGHT-HALF s-PLANE.	(1.32)

Marginal and asymptotic stability: The preceding definition of a BIBO stable system permits at least one of the poles to be exactly on the imaginary s-plane axis (that is, $\sigma_i = 0$ for some i). Consider a pole $p_i = j\omega_i$ on the imaginary axis; this leads to a *periodic* term $K_i e^{j\omega_i t} u(t)$ in equation (1.29) for the output waveform $y(t)$. This term, and the summation of all such terms corresponding to purely imaginary poles, is clearly bounded in amplitude (as required by BIBO stability), but the transient part of equation (1.29) does not decay to zero as $t \to \infty$ due to these periodic terms. Such a linear time-invariant system, containing at least one purely imaginary pole, is known as a *marginally stable system.*

If *all* the poles are in the left-half s-plane, *all* the terms in the transient response of equation (1.29) tend to zero as $t \to \infty$; thus, the transient response itself decays to zero as $t \to \infty$, and the system is said to be *asymptotically stable*. Thus,

$$\boxed{\lim_{t\to\infty} \left[\sum_{i=1}^{N} K_i e^{p_i t} u(t) \right] = 0} \quad \text{ASYMPTOTIC STABILITY} \tag{1.33}$$

implies that $\sigma_i < 0$ for $i = 1, 2, 3, \ldots, N$. All of the *RC*-active circuits considered in this book are *ideally* asymptotically stable to zero and, therefore, all of the poles of the circuit are ideally in the left-half s-plane. It occasionally happens that the nonideal characteristics of the active devices cause one or more of the poles to be in the right-half plane and thereby cause the circuit to exhibit spurious nonideal oscillatory behavior.

1-4 STEADY-STATE SINUSOIDAL (PHASOR) FREQUENCY RESPONSE $H(j\omega)$

The input-output transfer function of an *RC*-active linear network is often characterized under conditions of *steady-state sinusoidal excitation*. Thus, the input waveform $x(t)$ is defined as a complex phasor quantity, as follows:

$$x(t) \equiv X(j\omega) \equiv e^{j\omega t} \tag{1.34}$$

Thus, $x(t)$ is a phasor of unity magnitude rotating anticlockwise at the uniform angular speed ω rad/sec, as shown in Fig. 1.2(a). It may be shown (see Problem 1.6) that the corresponding output complex phasor quantity is

$$y(t) \equiv Y(j\omega) = H(j\omega)e^{j\omega t} \tag{1.35}$$

where $H(j\omega)$ is obtained from $H(s)$ in equation (1.20) by substituting $s = j\omega$. Now, for a constant input frequency ω it follows that $H(j\omega)$ is a

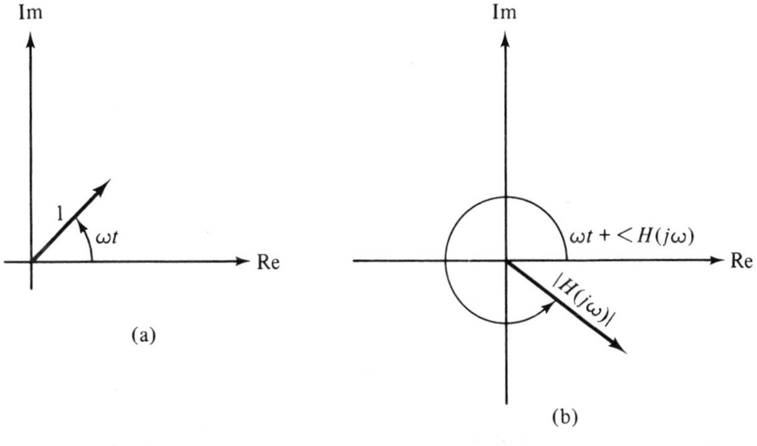

FIGURE 1.2 *(a) Input phasor $e^{j\omega t}$; (b) Output phasor $H(j\omega)e^{j\omega t}$.*

constant and generally complex number. Thus, it follows from equation (1.35) that $y(t)$ is a phasor rotating anticlockwise at ω rad/sec but with magnitude $|H(j\omega)|$ and phase $[\omega t + \angle H(j\omega)]$, as shown in Fig. 1.2(b). Thus, the gain frequency response $M(\omega)$ and phase frequency response $\theta(\omega)$ are defined as follows:

$$\begin{aligned} M(\omega) &\equiv |H(j\omega)| \quad \text{Gain Frequency Response} \\ \theta(\omega) &\equiv \angle H(j\omega) \quad \text{Phase Frequency Response} \end{aligned} \tag{1.36}$$

EXAMPLE 1.5: Find the gain frequency response and phase frequency response of a linear network with input $v_1(t)$, output $v_2(t)$, and an input-output differential equation given by

$$b_2 \frac{d^2 v_2(t)}{dt^2} + b_1 \frac{dv_2(t)}{dt} + b_0 v_2(t) = a_1 \frac{dv_1(t)}{dt} \tag{1.37}$$

SOLUTION: Taking the Laplace transform of equation (1.37), we obtain

$$b_2 s^2 V(s) + b_1 s V_2(s) + b_0 V_2(s) = a_1 s V_1(s) \tag{1.38}$$

Rearranging gives the transform transfer function $H(s)$ as

$$H(s) \equiv \frac{V_2(s)}{V_1(s)} = \frac{a_1 s}{b_2 s^2 + b_1 s + b_0} \tag{1.39}$$

so that, by substituting $s = j\omega$, we have

$$H(j\omega) = \frac{j a_1 \omega}{(b_0 - \omega^2 b_2) + j\omega b_1} \tag{1.40}$$

as the steady-state (phasor) sinusoidal frequency response function. Taking the magnitude and phase of $H(j\omega)$, we find that

$$M(\omega) = \frac{a_1 \omega}{\left\{(b_0 - \omega^2 b_2)^2 + \omega^2 b_1^2\right\}^{1/2}} \quad \text{Gain Frequency Response} \tag{1.41}$$

and

$$\theta(\omega) = \frac{\pi}{2} - \arctan\left(\frac{\omega b_1}{b_0 - \omega^2 b_2}\right) \quad \text{Phase Frequency Response} \tag{1.42}$$

1-4.1 Estimation of the Frequency Response from the Pole-Zero Locations

It is noted from equation (1.21) that $H(s)$ may be defined completely by specifying the locations, in the s-plane, of all the finite poles and zeros and the multiplicative scalar constant K. For each and every value of s in the complex s-plane ($s \equiv \sigma + j\omega$, σ real, ω real) it is clearly possible to calculate the (complex) number $H(s)$. The reader is asked to imagine that this calculation is performed for all points in the s-plane and that the magnitude $|H(s)|$ of the resultant *surface* is plotted in a third dimension above the s-plane, as shown in Fig. 1.3(a). At every zero z_j the surface "touches" the s-plane and at every pole p_i the surface is of infinite height above the s-plane, as indicated in Fig. 1.3(a). Circuit designers find it useful to visualize the general shape of this surface for different pole-zero patterns in the s-plane.

The gain response $M(\omega)$ is defined in equation (1.36) as the function $|H(j\omega)|$. It is, therefore, given by the *curve* on the surface $|H(s)|$ where σ is zero; that is, by the intersection of the plane $\sigma = 0$ with the surface $|H(s)|$ in Fig. 1.3(a). Obviously, if a pole exists "close to" the imaginary axis, then the curve $|H(j\omega)|$ on the surface $|H(s)|$ will "peak" in the vicinity of the pole. Similarly, if a zero exists close to the imaginary axis, the curve $|H(j\omega)|$ will approach zero in the vicinity of the zero, as shown in Fig. 1.3(b). Clearly, the frequency-selectivity and general shape of the gain response $M(\omega)$ may be inferred from the pole-zero diagram.

In quantitative terms, the complex frequency response function $H(j\omega)$ is obtained by substituting $s = j\omega$ in equation (1.21):

$$H(j\omega) = K\frac{(j\omega - z_M)(j\omega - z_{M-1}) \ldots (j\omega - z_2)(j\omega - z_1)}{(j\omega - p_N)(j\omega - p_{N-1}) \ldots (j\omega - p_2)(j\omega - p_1)} \tag{1.43}$$

It is now proposed that equation (1.43) be evaluated using *geometrical interpretation* of complex numbers in the s-plane. Consider the location of the jth zero and the ith pole, as shown in Fig. 1.4(a). Then the vectors $(j\omega - z_j)$ and $(j\omega - p_i)$ may be constructed in the s-plane, as shown in Fig. 1.4(a); the magnitude and phases of $(j\omega - z_j)$ and $(j\omega - p_i)$ are shown in Fig. 1.4(b) as $M_{pi}(\omega)$, $M_{zj}(\omega)$, $\theta_{pi}(\omega)$ and $\theta_{zj}(\omega)$. Thus,

$$\left.\begin{array}{lll} M_{pi}(\omega) \equiv |j\omega - p_i|, & \theta_{pi}(\omega) \equiv \arg[j\omega - p_i], & i = 1, 2, 3, \ldots, N \\ M_{zj}(\omega) \equiv |j\omega - z_j|, & \theta_{zj}(\omega) \equiv \arg[j\omega - z_j], & j = 1, 2, 3, \ldots, M \end{array}\right\} \tag{1.44}$$

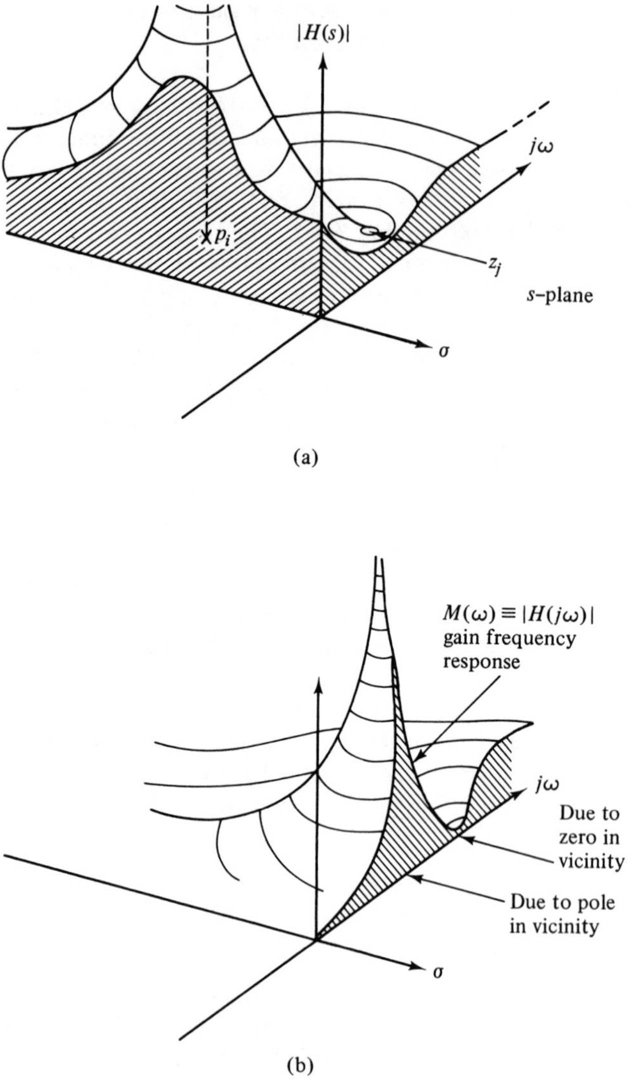

FIGURE 1.3 *(a) Behavior of |H(s)| above a pole and a zero; one quadrant shown; (b) Effects of pole and zero close to the imaginary axis.*

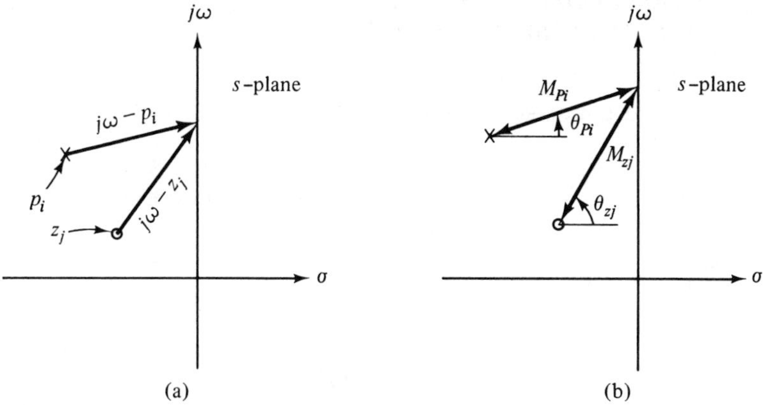

FIGURE 1.4 *(a) Vectors $\overline{j\omega - p_i}$ and $\overline{j\omega - z_j}$; (b) Definitions of M_{Pi}, M_{Zi}, θ_{Pi}, and θ_{Zi}.*

From equations (1.43) and (1.44) it follows that

$$M(\omega) \equiv |H(j\omega)| = |K| \frac{M_{Z1}(\omega) M_{Z2}(\omega) \ldots M_{ZM}(\omega)}{M_{P1}(\omega) M_{P2}(\omega) \ldots M_{PN}(\omega)} \qquad \text{GAIN}$$

$$\theta(\omega) \equiv \arg[H(j\omega)] = \arg[K] + \sum_{j=1}^{M} \theta_{Zj}(\omega) - \sum_{i=1}^{N} \theta_{pi}(\omega) \qquad \text{PHASE} \tag{1.45}$$

The two expressions above are useful for obtaining estimates of the frequency response function $H(j\omega)$ by means of geometrical construction in the s-plane. It follows from equations (1.45) that the gain frequency response $M(\omega)$ at a particular ω is determined by $|K|$ and the *products* $M_{p1}M_{p2} \ldots M_{pN}$ and $M_{z1}M_{z2} \ldots M_{zN}$. Thus if a pole p_i is "close" to the imaginary axis in some region of ω, then $M_{pi}(\omega)$ is "small" in this same region of ω, thereby tending to make $M(\omega)$ "large" in this region of ω. Similarly, a zero z_j close to the imaginary axis has the effect of suppressing $M(\omega)$ to a small value in the same region of ω.

The phase frequency response $\theta(\omega)$ in equation (1.45) depends on the sums $[\theta_{p1} + \theta_{p2} + \ldots + \theta_{pN}]$ and $[\theta_{z1} + \theta_{z2} + \ldots + \theta_{zM}]$. A pole p_i or zero z_j "close" to the imaginary axis in some region of ω implies that θ_{pi} or θ_{zj} varies rapidly with ω in this same region of ω. An example is given in the following subsection.

1-5 THE NORMALIZED BANDPASS QUADRATIC TRANSFER FUNCTION

The normalized bandpass quadratic function $t_{\mathrm{BP}}(s)$ is defined as

$$H(s) \equiv t_{\mathrm{BP}}(s) \equiv \frac{\omega_p s}{s^2 + \dfrac{\omega_p s}{Q_p} + \omega_p^2} \qquad \text{NORMALIZED BANDPASS BIQUADRATIC FUNCTION} \tag{1.46}$$

Writing $t_{\mathrm{BP}}(s)$ in the form

$$t_{\mathrm{BP}}(s) = K\frac{(s - z_1)}{(s - p_1)(s - p_2)} \tag{1.47}$$

gives, by comparison of equations (1.46) and (1.47),

$$\left.\begin{aligned} K = \omega_p, \qquad z_1 = 0, \qquad p_1 &= -\frac{\omega_p}{2Q_p} + j\omega_p\left(1 - \frac{1}{4Q_p^2}\right)^{1/2} \\ p_2 &= -\frac{\omega_p}{2Q_p} - j\omega_p\left(1 - \frac{1}{4Q_p^2}\right)^{1/2} \end{aligned}\right\} \tag{1.48}$$

The one zero and two poles are shown in Fig. 1.5(a) for $Q_p > \frac{1}{2}$; note that the poles p_1 and p_2 are complex conjugates lying on a circle of radius ω_p. For $Q_p \gg 1$, the poles $p_{1,2}$ are "close to" the imaginary axis and to the points $\pm j\omega_p$.

The expressions in equations (1.45) and the geometric properties of Fig. 1.5(a) are now used to estimate the gain frequency response $M(\omega)$ and phase frequency response $\theta(\omega)$ for the function $t_{\mathrm{BP}}(s)$, where $Q_p \gg 1$. The geometry of Fig. 1.5(a) is considered in five separate regions of ω; for $\omega \ll \omega_p$, $\omega \gg \omega_p$, $\omega = \omega_p$, and $\omega = \omega_p[1 \pm 1/2Q_p]$. It follows from equations (1.45) and Fig. 1.5(a) that

$$\left.\begin{aligned} M(\omega) &= |K|\frac{M_{z1}(\omega)}{M_{p1}(\omega)M_{p2}(\omega)} \\ \text{and} \qquad \theta(\omega) &= \arg[K] + \theta_{z1}(\omega) - \theta_{p1}(\omega) - \theta_{p2}(\omega) \end{aligned}\right\} \tag{1.49}$$

Now, if $Q_p \gg 1$, it follows from equation (1.48) that

$$p_1 \approx -\frac{\omega_p}{2Q_p} + j\omega_p, \qquad Q_p \gg 1 \tag{1.50}$$

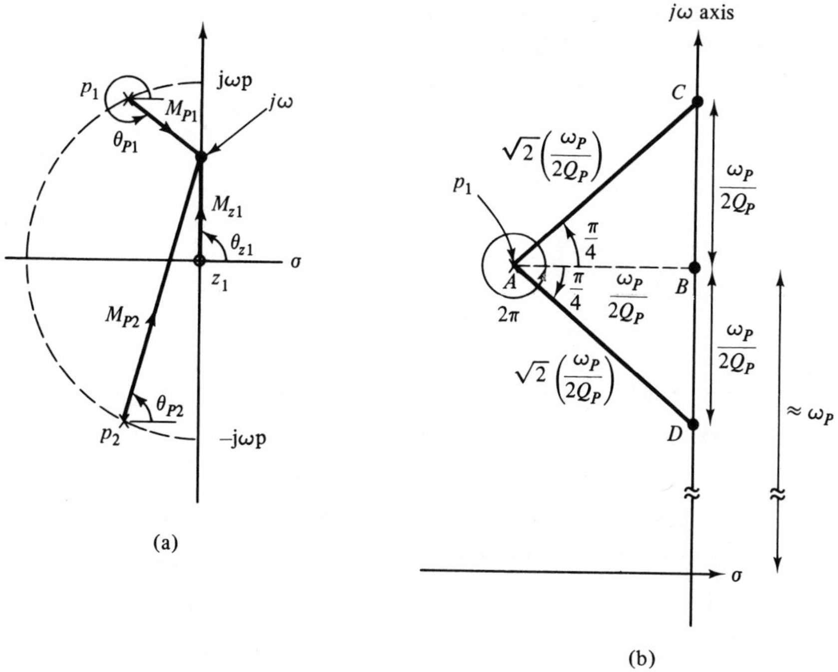

FIGURE 1.5

By inspection of Fig. 1.5(a),

(i) if $\omega \ll \omega_p$:

then

$$\left.\begin{aligned} &\theta_{p1} \approx \frac{3\pi}{2}, \qquad \theta_{p2} \approx \frac{\pi}{2}, \qquad \theta_{z1} = \frac{\pi}{2} \\ &\text{and} \\ &M_{p1}(\omega) \approx M_{p2}(\omega) \approx \omega_p, \qquad M_{z1}(\omega) = \omega \end{aligned}\right\} \tag{1.51}$$

(ii) if $\omega \gg \omega_p$:

then

$$\left.\begin{aligned} &\theta_{p1} \approx 2\pi + \frac{\pi}{2}, \qquad \theta_{p2} \approx \frac{\pi}{2}, \qquad \theta_{z1} = \frac{\pi}{2} \\ &\text{and} \\ &M_{p1} \approx M_{p2}(\omega) \approx \omega, \qquad M_{z1}(\omega) \approx \omega \end{aligned}\right\} \tag{1.52}$$

(iii) $\underline{\text{if } \omega = \omega_p\left(1 - \dfrac{1}{4Q_p^2}\right)^{1/2} \approx \omega_p\text{:}}$

This is the point B shown in Fig. 1.5(b). We have

and

$$\left.\begin{aligned} &\theta_{p1} = 2\pi, \qquad \theta_{p2} \approx \frac{\pi}{2}, \qquad \theta_{z1} = \frac{\pi}{2} \\ &M_{p1}(\omega) = \frac{\omega_p}{2Q_p}, \qquad M_{p2}(\omega) \approx 2\omega_p, \qquad M_{z1}(\omega) \approx \omega_p \end{aligned}\right\} \tag{1.53}$$

(iv) $\underline{\text{if } \omega = \omega_p\left(1 \pm \dfrac{\omega_p}{2Q_p}\right) \equiv \omega_{1,2}\text{:}}$

These are the points C and D in Fig. 1.5(b). Thus, the point C is defined as that point on the imaginary axis that subtends the angle $\left(\frac{\pi}{4} + 2\pi\right)$ to the pole p_1, as shown, and point D as the point subtending the angle $\left(-\frac{\pi}{4} + 2\pi\right)$, as shown. It follows by simple geometry that

$$\text{if } \omega \equiv \omega_{1,2} \equiv \omega_p\left(1 \pm \frac{\omega_p}{2Q_p}\right):$$

then

$$\left.\begin{aligned} &\theta_{p1} = \pm\frac{\pi}{4} + 2\pi, \qquad \theta_{p2} \approx \frac{\pi}{2}, \qquad \theta_{z1} = \frac{\pi}{2} \\ &M_{p1}(\omega) = \sqrt{2}\,\frac{\omega_p}{2Q_p}, \qquad M_{p2}(\omega) \approx 2\omega_p \\ &M_{z1}(\omega) \approx \omega_p \end{aligned}\right\} \tag{1.54}$$

Substituting equations (1.51) to (1.54) into (1.49) and using $|K| = \omega_p$, we have

$\underline{\text{if } \omega \ll \omega_p\text{:}}$

$$\left.\begin{aligned} &\theta(\omega) \approx \frac{\pi}{2} - \frac{3\pi}{2} - \frac{\pi}{2} = -\frac{3\pi}{2} \\ &M(\omega) \approx |K|\frac{\omega}{\omega_p^2} = \frac{\omega}{\omega_p} \end{aligned}\right\} \tag{1.55}$$

if $\omega \gg \omega_p$:

$$\left.\begin{aligned} \theta(\omega) &\approx \frac{\pi}{2} - \frac{5\pi}{2} - \frac{\pi}{2} = -\frac{5\pi}{2} \\ M(\omega) &\approx |K|\frac{\omega}{\omega^2} = \frac{\omega_p}{\omega} \end{aligned}\right\} \tag{1.56}$$

if $\omega = \omega_p\left(1 - \dfrac{1}{4Q_p^2}\right)^{1/2}$:

$$\left.\begin{aligned} \theta(\omega) &\approx \frac{\pi}{2} - \frac{\pi}{2} - 2\pi = -2\pi \\ M(\omega) &\approx |K|\frac{\omega_p}{(2\omega_p)\left(\dfrac{\omega_p}{2Q_p}\right)} = Q_p \end{aligned}\right\} \tag{1.57}$$

if $\omega = \omega_{1,2} \equiv \omega_p\left(1 \pm \dfrac{1}{2Q_p}\right)^{1/2}$:

$$\left.\begin{aligned} \theta(\omega) &\approx \frac{\pi}{2} - \frac{\pi}{2} \pm \frac{\pi}{4} - 2\pi = -2\pi \pm \frac{\pi}{4} \\ M(\omega) &\approx |K|\frac{\omega_p}{\sqrt{2}\left(\dfrac{\omega_p}{2Q_p}\right)(2\omega_p)} = \frac{Q_p}{\sqrt{2}} \end{aligned}\right\} \tag{1.58}$$

Equations (1.55) to (1.58) may be used to rapidly sketch the phase and gain frequency response functions shown in Fig. 1.6. The quantity 2π is added to $\theta(\omega)$ of Fig. 1.6(a) since a phase shift of one period does not alter the steady-sinusoidal output waveform. The phase response $\theta(\omega)$ decreases by π radians (180°) over the region of the resonant response of $M(\omega)$; the low-frequency and high-frequency responses of $M(\omega)$ are asymptotic to ω/ω_p and ω_p/ω, respectively, with a resonant peak of Q_p at ω_p. The value of $M(\omega)$ at $\omega_{1,2}$ is $Q_p/\sqrt{2}$, where $\omega_{1,2}$ are sometimes referred to as the -3 db frequencies because

$$20 \log_{10}\frac{M(\omega_{1,2})}{M(\omega_p)} \approx -3 \text{ db} \tag{1.59}$$

It should be noted that $\theta_{z1} = \pi/2$ and $\theta_{p2} \approx \pi/2$ over all positive frequencies ω so that, from equation (1.49),

$$\theta(\omega) \approx -\theta_{p1}(\omega) \tag{1.60}$$

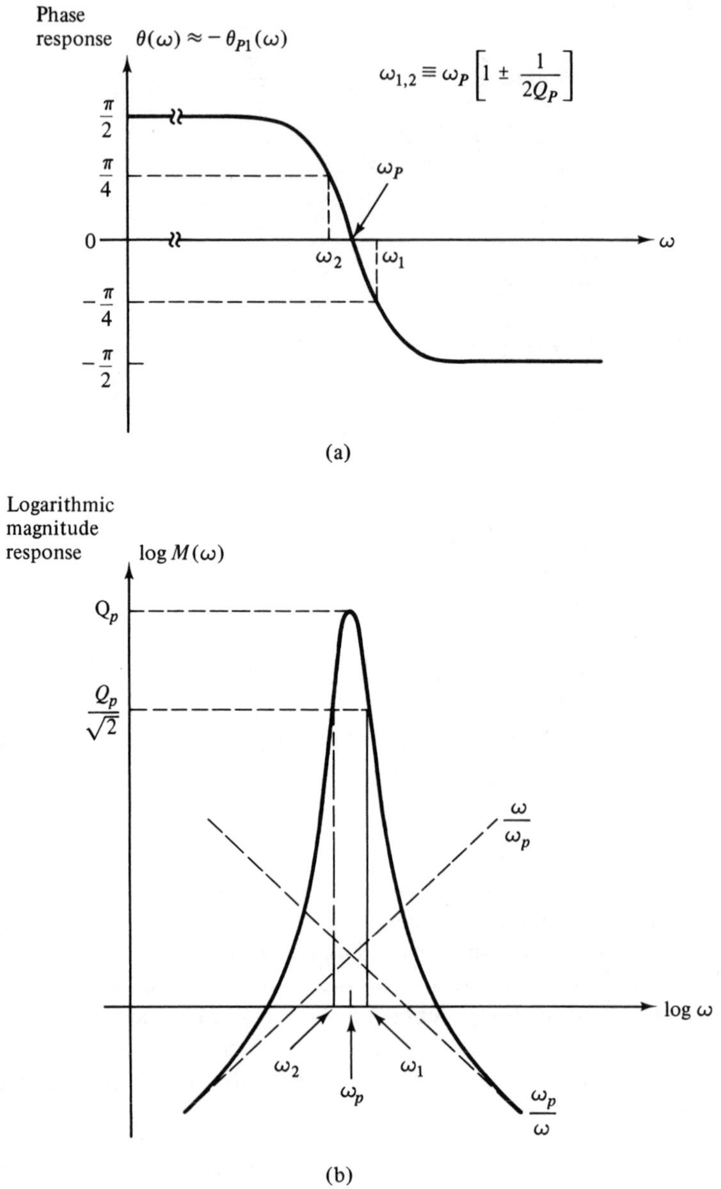

FIGURE 1.6 *(a) Phase response $\theta(\omega)$ for $t_{BP}(s)$; (b) Logarithmic magnitude response $M(\omega)$ for $t_{BP}(s)$.*

implying that the phase response $\theta(\omega)$ is primarily determined by the angle $\theta_{p1}(\omega)$ subtended to pole p_1. Further, the rate of change of phase with frequency, $\partial\theta(\omega)/\partial\omega$, in the region of ω_p increases with increasing Q-factor Q_p.

It is also noted that the -3 db bandwidth BW is

$$\mathrm{BW} \equiv \omega_1 - \omega_2 = \omega_p\left(1 + \frac{\omega_p}{2Q_p}\right) - \omega_p\left(1 - \frac{\omega_p}{2Q_p}\right)$$

so that

$$\boxed{\mathrm{BW} = \frac{\omega_p}{Q_p}} \qquad -3 \text{ db Bandwidth} \qquad (1.61)$$

1.6 LOWPASS, HIGHPASS, NOTCH, AND ALLPASS QUADRATIC FUNCTIONS

The normalized lowpass, normalized highpass, biquadratic notch (or band elimination), and normalized biquadratic allpass functions are as follows:

$$t_{\mathrm{LP}}(s) \equiv \frac{\omega_p^2}{s^2 + \dfrac{\omega_p s}{Q_p} + \omega_p^2} \qquad \text{Normalized Quadratic Lowpass Function} \qquad (1.62)$$

$$t_{\mathrm{HP}}(s) = \frac{s^2}{s^2 + \dfrac{\omega_p s}{Q_p} + \omega_p^2} \qquad \text{Normalized Quadratic Highpass Function} \qquad (1.63)$$

$$t_{\mathrm{N}}(s) = \frac{s^2 + \omega_z^2}{s^2 + \dfrac{\omega_p s}{Q_p} + \omega_p^2} \qquad \text{Normalized Biquadratic Notch Function} \qquad (1.64)$$

$$t_{\mathrm{AP}}(s) = \frac{s^2 - \dfrac{\omega_p s}{Q_p} + \omega_p^2}{s^2 + \dfrac{\omega_p s}{Q_p} + \omega_p^2} \qquad \text{Normalized Biquadratic Allpass Function} \qquad (1.65)$$

The corresponding gain and phase frequency response functions $M(\omega)$ and $\theta(\omega)$ may be obtained using the method described in section 1.5 for the

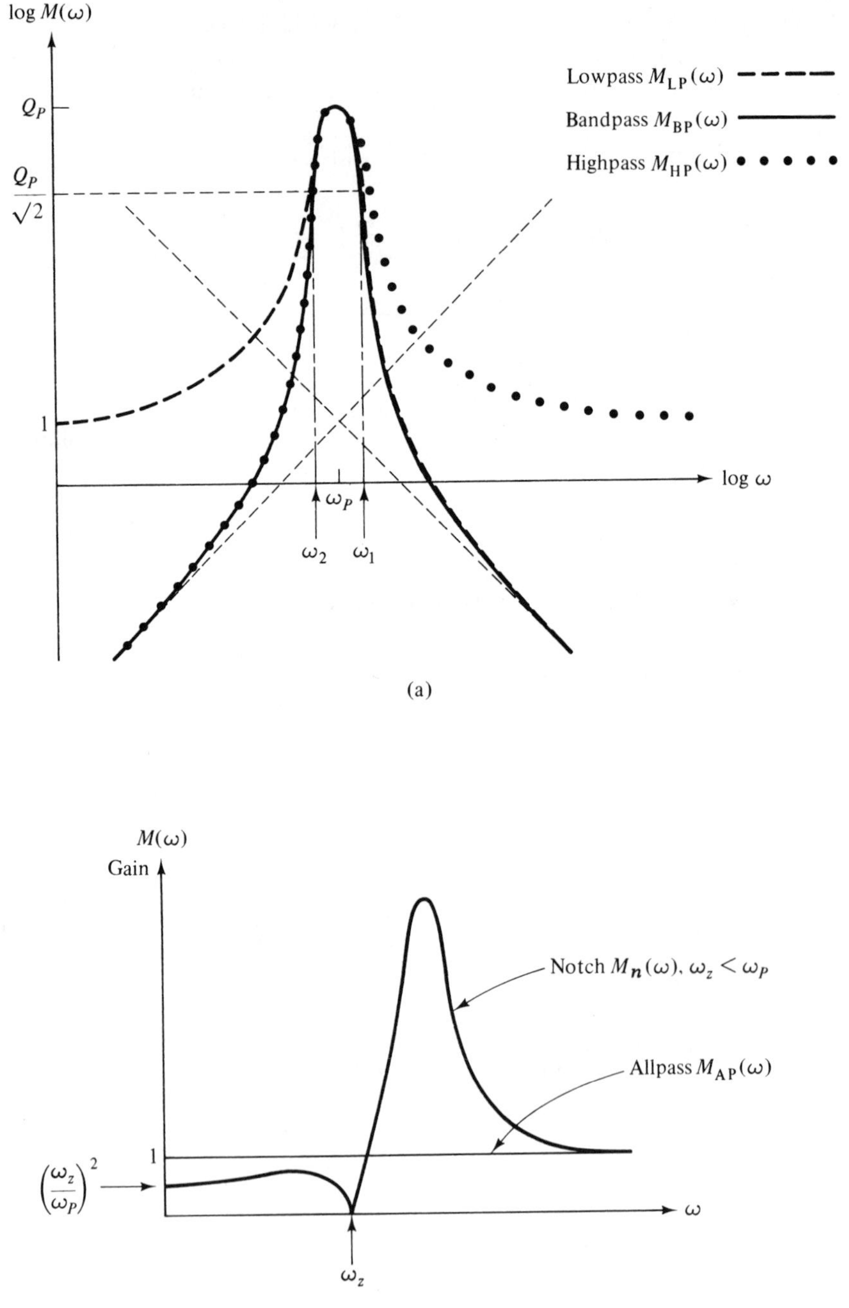

FIGURE 1.7 *(a) Normalized quadratic lowpass, bandpass, and highpass magnitude (gain) responses; (b) Allpass and notch magnitude responses.*

normalized bandpass function. The $M(\omega)$ functions are shown in Fig. 1.7(a) and 1.7(b); these sketches are given for $Q_p \gg 1$ from which it is observed that the lowpass function $M_{\text{LP}}(\omega)$ is asymptotic to unity as $\omega \to 0$, and the highpass function $M_{\text{HP}}(\omega)$ is asymptotic to unity as $\omega \to \infty$. Otherwise, $M_{\text{LP}}(\omega)$ and $M_{\text{HP}}(\omega)$ behave in a similar manner to the bandpass function $M_{\text{BP}}(\omega)$ for $Q_p \gg 1$. Cascaded connections of quadratic functions are used, as described in chapter 7, to implement transfer functions of orders greater than two.

The quadratic *notch* function $t_{\text{N}}(s)$ of equation (1.64) is used to give $t_{\text{N}}(j\omega) = 0$ at $\omega = \omega_z$; it can lead to a function with particularly high magnitude selectivity in the region between ω_z and ω_p. Cascade connections of this type are used to realize highly selective lowpass and highpass filters of order greater than two.

The quadratic *allpass* function $t_{\text{AP}}(s)$ is characterized by a constant (unity) magnitude function $M_{\text{AP}}(\omega)$. This function, and cascaded connections of such functions, are used to *control or shape the phase characteristics of a signal without affecting the magnitude characteristic.*

The reader is asked to investigate the properties of $t_{\text{LP}}(s)$, $t_{\text{BP}}(s)$, $t_{\text{HP}}(s)$, $t_{\text{N}}(s)$, and $t_{\text{AP}}(s)$ in the problems at the end of this chapter.

1-7 IMPEDANCE DENORMALIZATION AND FREQUENCY DENORMALIZATION

It is common practise to design and analyze *RC*-active circuits in their *normalized* form; that is, at an impedance level and over a frequency range such that the *RC* elements have *numerically convenient* ranges of values. For examples, resistance values in the range 0.1 to 10 ohms and capacitance values in the range 0.1 to 10 farads. Normalized *RC*-active filters are often designed such that the edge of the passband is at 1 radian per second; for example, the normalized network of Fig. 1.8(a) is a simple lowpass *RC* network with a voltage transfer function

$$H(s) \equiv \frac{V_2(s)}{V_1(s)} = \frac{1}{1 + sCR} = \frac{1}{1 + s} \tag{1.66}$$

so that

$$M(\omega) \equiv |H(j\omega)| = \frac{1}{(1 + \omega^2 C^2 R^2)^{1/2}} = \frac{1}{(1 + \omega^2)^{1/2}} \tag{1.67}$$

as sketched in Fig. 1.8(b). The edge of the passband is defined to be at $\omega = 1$ rad/sec, where $M(\omega) = 1/\sqrt{2}$.

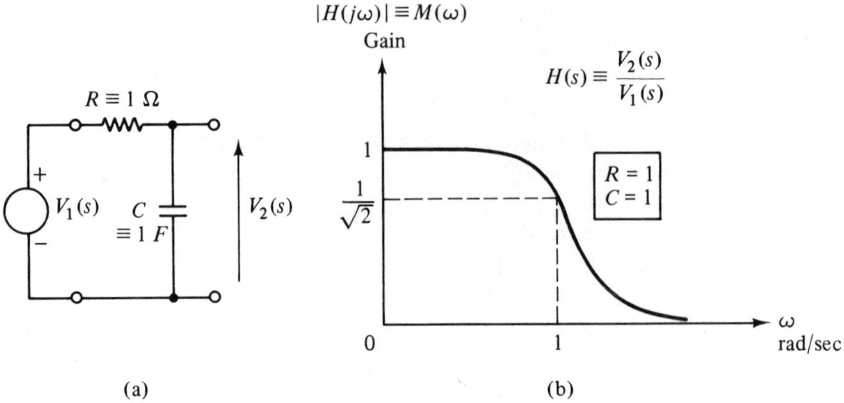

FIGURE 1.8 *(a) Normalized network; (b) Normalized gain response $M(\omega)$.*

Impedance denormalization by the factor K_1 is achieved by multiplying all resistance element values by K_1 and dividing all capacitance element values by K_1. *This operation leaves the voltage transfer function $H(s)$ unaltered.* Frequency denormalization by the factor K_2 is achieved by dividing all capacitance elements by K_2. This operation is equivalent to dividing s by K_2, resulting in the frequency-scaled transfer function $H(s/K_2)$; thus, the gain $M(\omega)$ is frequency-scaled (or denormalized) to $M(\omega/K_2)$. The combined effects of impedance and frequency denormalizations are summarized as follows:

$$\begin{aligned} &R \to K_1 R && \text{IMPEDANCE DENORMALIZATION BY } K_1 \\ &C \to \frac{C}{K_1 K_2} && \text{AND FREQUENCY DENORMALIZATION BY } K_2 \end{aligned} \tag{1.68}$$

The normalized network shown in Fig. 1.8(a) may be denormalized to a resistance value R of 10 kΩ and a cut-off frequency of 10^3 rad/sec by selecting $K_1 = 10^4$ and $K_2 = 10^3$ so that $R = 10^4\ \Omega$ and $C = 10^{-4} \times 10^{-3} = 0.1\ \mu$F.

Some practical examples of impedance and frequency denormalization are given at the end of this chapter.

1-8 SUMMARY

The concepts of linearity, time-invariance, the Laplace transform description of a linear time-invariant system, stability, s-plane behavior, poles, zeros, and steady-state sinusoidal frequency response are reviewed briefly in this chapter. If the reader is unfamiliar with these topics, it is particularly recommended that the problems at the end of this chapter be solved.

The commonly encountered second-order transfer functions are introduced, and it is found that the general shape and, particularly, the asymptotic behavior of the gain $M(\omega)$ and phase $\theta(\omega)$ functions may be estimated from the pole-zero pattern of the function $H(s)$.

Finally, it is stressed that the starting point for the practical application of the material in this book is a known transfer function $H(s)$. There is a wide range of textbooks available on filter approximation theory that provide tabulated data for the coefficients of $H(s)$ and for a variety of useful filter functions. Some typical functions $H(s)$ are introduced throughout the text as required to illustrate analysis, design, and synthesis problems.

PROBLEMS

1-1. Prove that the ideal inductance is a linear network element.

1-2. The output $y(t)$ is related to the input $x(t)$ via the relationship

$$y(t) = x^3(t)$$

Is the system linear? Is the system time-invariant? Explain your answers.

1-3. Prove that equation (1.17) describes a linear time-invariant network.

1-4. Given the following definition of the one-sided Laplace transform

$$\mathcal{L}[f(t)] \equiv F(s) = \int_0^\infty f(t)e^{-st}\, dt$$

Prove that equation (1.18) is valid if $y(t)$, $x(t)$, the first $(M-1)$ derivatives of $y(t)$ and the first $(N-1)$ derivatives of $x(t)$ are zero at time $t = 0^-$.

1-5. Find the poles and zeros of the following functions:

(a) $H(s) = \dfrac{(s+1)}{(s+2)(s+3)}$

(b) $H(s) = \dfrac{(s^2+1)}{\left(s^2 + \frac{1}{10}s + 2\right)}$

(c) $H(s) = \dfrac{1}{s}$

Comment on the BIBO stability and asymptotic stability of the systems in (a), (b), and (c).

1-6. A linear system having a transform transfer function $H(s)$ is excited by a steady-state complex exponential input waveform

$$x(t) = e^{j\omega t}$$

Show that the corresponding output complex waveform is given by

$$y(t) = H(j\omega)e^{j\omega t}$$

[*Hint:* Use equation (1.17) to show this result.]

1-7. Derive the gain frequency response $M(\omega)$ and phase frequency response $\theta(\omega)$ for the three functions of Problem 1.5. Plot these functions on a graph as functions of ω. Comment on the behavior of $M(\omega)$ and $\theta(\omega)$ in the regions of ω where $j\omega$ is close to a pole or a zero in the s-plane.

1-8. Given that

$$H(s) = \frac{s}{s^2 + 0.1s + 4}$$

calculate the Q factor Q_p and natural frequency ω_p. Construct a pole diagram on an s-plane and then *use geometrical construction in the s-plane* to evaluate equation (1.45) for $M(\omega)$ at $\omega = 2$, 2.05, and 1.95 by measuring M_{z1}, M_{p1}, M_{p2}. Comment on the significance of the three frequencies.

Use equation (1.45) for $\theta(\omega)$ to construct a plot of $\theta(\omega)$ versus ω in the region 1.9 to 2.1 rad/sec.

1-9. Repeat Problem 1.8 for the function

$$H(s) = \frac{16(s^2 + 15)}{s^2 + 0.1s + 4}$$

but plot $M(\omega)$ versus ω on logarithmic axes. Note that you may wish to employ an expanded scale in the region $1.9 < \omega < 2.1$. What are the high-frequency ($\omega \rightarrow \infty$) and low-frequency ($\omega \rightarrow 0$) asymptotes according to your graph of $M(\omega)$?

1-10. Assuming $Q_p = 10$, sketch $M(\omega)$ and $\theta(\omega)$ versus ω on logarithmic axes for the normalized functions $t_{LP}(s)$, $t_{BP}(s)$, $t_{HP}(s)$, $t_N(s)$ and $t_{AP}(s)$.

1-11. A normalized RC-active filter has the following RC-element values:

$$R_1 = 1.34\ \Omega, \qquad R_2 = 0.713\ \Omega$$
$$C_1 = 1.0\ \text{F}, \qquad C_2 = 0.1\ \text{F}$$

The filter has a lowpass transfer function $H(s)$ with a cut-off frequency of 1 rad/sec. Determine the values of R_1, C_1, and C_2 for a frequency-denormalized and impedance-denormalized version of this filter having a cut-off frequency of 1 kHz and an impedance level determined by $R_2 \equiv 10\ \text{k}\Omega$.

1-12. A normalized RC-active bandpass filter, having a center frequency ω_0 of 1 rad/sec and bandwidth BW of 0.1 rad/sec, has the following *RC*-element values:

$$R_1 = R_2 = R_3 = 1\ \Omega \qquad R_4 = 2\ \Omega$$
$$C_1 = 0.5\ \text{F}, \qquad C_2 = 1.0\ \text{F}$$

Denormalize this filter so that $\omega_0 = 10$ kHz and $R_4 = 20$ kΩ by calculating the denormalized values of R_1, R_2, R_3, C_1, and C_2. What is the bandwidth BW of the denormalized filter?

BIBLIOGRAPHY

VAN VALKENBURG, M. E., *Network Analysis* 3rd Edition, Prentice-Hall, 1974, chapters 1 and 2.

2

ELEMENTS AND SUBNETWORKS OF RC-ACTIVE NETWORKS

2-1 INTRODUCTION

The proliferation of published literature on *RC*-active networks has led to a variety of techniques for describing the network model of an *RC*-active circuit. The circuit complexity of typical modern high-quality *RC*-active filters has resulted in some generally acceptable classifications of one-port and two-port active subnetworks. These one-port and two-port subnetworks may be quite complicated (consisting of possibly two operational amplifiers and five or more *RC* elements), but they are sufficiently widely employed that it is useful to define the subnetwork as an element (or building-block).

In this chapter, the concepts of passivity, losslessness, and activity are defined and then employed to classify the more important one-port and two-port subnetworks or elements. The reader who is familiar with passive network theory will find that the *R*, *L*, and *C* elements are suitably classified along with the transformer and gyrator; the *active* subnetworks and elements are introduced as one-ports or two-ports in much the same way as in the passive case. In this manner it is hoped that the reader will not experience too much difficulty in understanding the terminal behavior of *negative-RLC* elements, FDNR elements, and the various generalized immittance converter (GIC) subnetworks.

Controlled sources and independent sources are important active subnetworks and are defined in this chapter. It is shown that finite-gain controlled sources are subnetworks that may be realized by means of nullors. *The importance of the nullor as a fundamental subnetwork in the classification of* RC-*active networks cannot be underestimated. It is shown in subsequent chapters of this book that the use of nullors will allow a simple and unified approach to the description of* RC-*active filter synthesis techniques*. The reader is, therefore, urged to attempt the nullor problems at the end of the chapter.

2-2 PASSIVITY, LOSSLESSNESS, AND ACTIVITY

Consider the general linear one-port network N, Fig. 2.1, with port voltage $v(t)$ and port current $i(t)$ as shown. The instantaneous power $p(t)$ delivered to the network N is defined by

$$p(t) \equiv v(t)i(t) \tag{2.1}$$

so that the net energy $E(t_0)$ delivered to the network N to time t_0 is

$$E(t_0) \equiv \int_{-\infty}^{t_0} v(t)i(t)\,dt \tag{2.2}$$

If $E(t_0)$ is positive, then at time t_0 the network N is said to have absorbed net positive energy. If $E(t_0)$ is negative, then at time t_0 the network N is said to have delivered net positive energy [equal to $-E(t_0)$] to the source. Passivity, losslessness, and activity are directly related to the ability of a network to absorb or deliver net positive energy. If a one-port network N does not, at any time, deliver net positive energy to the source ($E(t_0) \geqslant 0$) *over all possible waveforms* v(t) *and* i(t), then the network N is defined as a passive network. *An active network is defined as a network that is not passive.*

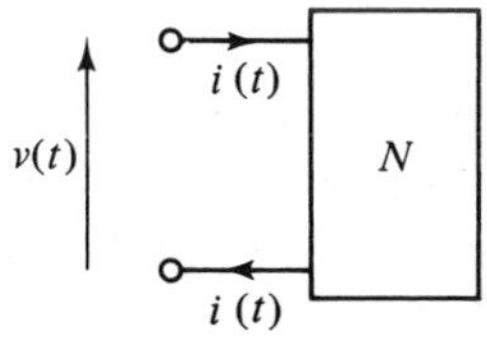

FIGURE 2.1

DEFINITION: *A one-port network N is a passive network if, for all possible excitation waveforms $v(t)$; $i(t)$ the net energy $E(t_0)$ delivered to the network is non-negative over all t_0; that is,*

$$E(t_0) \geqslant 0, \qquad \forall\, v(t);\, i(t) \tag{2.3}$$

PASSIVITY

Let $v(t)$ and $i(t)$ be constrained to be "square-integrable." That is,

$$\int_{-\infty}^{\infty} v^2(t)\,dt < \infty \quad \text{and} \quad \int_{-\infty}^{\infty} i^2(t)\,dt < \infty$$

SQUARE INTEGRABILITY CONDITIONS

Physically, this implies that

$$v(\infty) = v(-\infty) = i(\infty) = i(-\infty) = 0$$

so that the network is "relaxed" at both $t = \infty$ and $t = -\infty$. Under these conditions, we define *losslessness*. The one-port is lossless if and only if

$$E(\infty) = \int_{-\infty}^{\infty} v(t)i(t)\,dt = 0, \qquad \forall\, v(t);\, i(t) \tag{2.4}$$

LOSSLESSNESS

Thus, a lossless network *always* absorbs zero net energy $E(\infty)$ when the excitation waveforms are "square-integrable."

The definitions above of passivity, losslessness, and activity of a one-port network may be extended easily to the general case of the N-port network of Fig. 2.2, for which the net energy $E(t_0)$ delivered to (or absorbed by) all M-ports is, simply,

$$E(t_0) = \sum_{i=1}^{M} \int_{-\infty}^{t_0} v_i(t)i_i(t)\,dt \tag{2.5}$$

Then, passivity requires that, for all possible waveforms $v_i(t)$ and $i_i(t)$, the above net energy $E(t_0)$ is non-negative over all t_0. Similarly, losslessness requires that $E(\infty)$ is zero if $v_i(t)$ and $i_i(t)$ are square-integrable over all $i = 1, 2, 3, \ldots, M$. It is a simple matter to prove that an interconnection of passive (lossless) networks results in a network that is also passive (lossless).

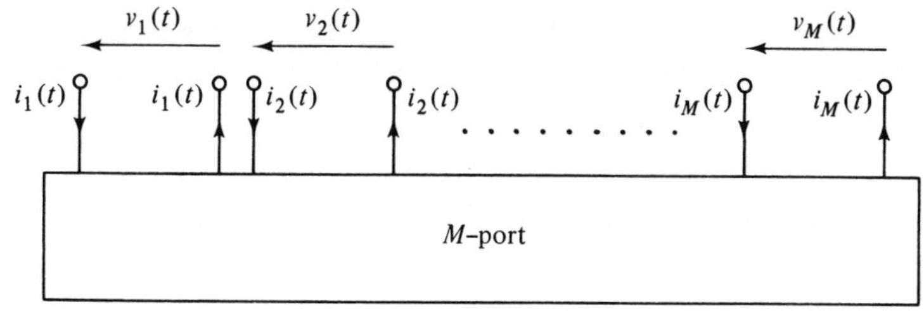

FIGURE 2.2

2-3 PASSIVE ONE-PORT ELEMENTS, *R*, *C*, AND *L*

The basic linear time-invariant *passive* one-port elements are the resistance element R, the capacitance element C, and the inductance element L. They are defined by the constraints that exist between $v(t)$ and $i(t)$ as shown in Table 2.1.

TABLE 2.1 *Passive one-port elements, R, C and L.*

Element	*Symbol*	*Definition*
Resistance, R Ohms, Ω	$i_R(t)$ R (resistor) $v_R(t)$ $R > 0$	$v_R(t) \equiv i_R(t)R$ Passive nonlossless
Capacitance, C Farads, F	$i_c(t)$ C (capacitor) $v_c(t)$ $C > 0$	$i_c(t) \equiv C \dfrac{dv_c(t)}{dt}$ or $v_c(t) = \dfrac{1}{C}\int_{-\infty}^{t} i_c(x)\,dx$ Passive lossless
Inductance, L Henries, H	$i_L(t)$ L (inductor) $v_L(t)$ $L > 0$	$v_L(t) \equiv L \dfrac{di_L(t)}{dt}$ or $i_L(t) = \dfrac{1}{L}\int_{-\infty}^{t} v_L(x)\,dx$ Passive lossless

The inherent passivity of the R, C, and L elements is proven from the defining equations. For example, in the case of the resistance element R,

$$v_R(t) \equiv i_R(t)R, \qquad R > 0 \tag{2.6}$$

so that the net energy delivered to the network is given by equation (2.2) as

$$E(t_0) \equiv \int_{-\infty}^{t_0} v_R(t)i_R(t)\,dt = \int_{-\infty}^{t_0} Ri_R^2(t)\,dt \tag{2.7}$$

and, since $Ri_R^2(t)$ is non-negative for all $i_R(t)$ and all t, it follows that $E(t_0)$ is non-negative for all $v_R(t)$ and $i_R(t)$ and, therefore, the resistance element is *passive*. Furthermore, for nonzero $v_R(t)$ and $i_R(t)$, $E(t_0)$ is clearly a *monotonically nondecreasing* positive function of t_0 so that it is not possible for $E(\infty)$ to equal zero; the resistance element is, therefore, *nonlossless*.

In the case of the capacitance element,

$$\begin{aligned} E(t_0) &\equiv \int_{-\infty}^{t_0} v_c(t)i_c(t)\,dt = C\int_{-\infty}^{t_0} v_c(t)\frac{dv_c(t)}{dt}\,dt \\ &= C\int_0^{v_c(t_0)} v_c(t)\,dv_c = \frac{1}{2}Cv_c^2(t_0) \end{aligned} \tag{2.8}$$

It follows that $E(t_0)$ is non-negative over all $v_c(t)$ and $i_c(t)$ and, therefore, the capacitance element is *passive*. It is also *lossless* because, for square-integrable $v_c(t)$, $v_c(\infty) \equiv 0$, and then equation (2.8) implies that

$$E(\infty) = \tfrac{1}{2}Cv_c^2(\infty) = 0, \qquad \forall\, v_c(t);\, i_c(t)$$

Similarly, the *inductance* element may be shown to be *passive* and *lossless*.

Q-Factor and dissipation factor of LC elements: In practice, it is not possible to make an inductance or a capacitance element that is *exactly* lossless. In physical terms, a small proportion of the energy absorbed by an L or C element is not recoverable to the source at $t = \infty$ but is irretrievably lost or *dissipated* (in the form of *heat* energy). We are not concerned here with the detailed mechanisms by which this dissipation occurs but present, without proof, the equivalent circuits of Fig. 2.3 for the nonideal (that is, lossy or nonlossless) L and C elements. The series resistance r_s and the parallel resistance r_p represent an accurate low-frequency model for the small loss that occurs in practical LC elements.

The Q-factor of an element is a measure of the energy stored compared with the energy that is dissipated under conditions of *steady-state*

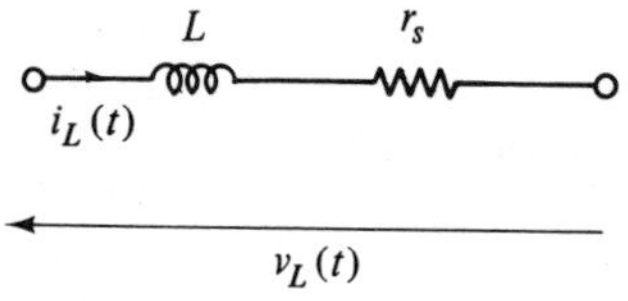

Lossy inductor

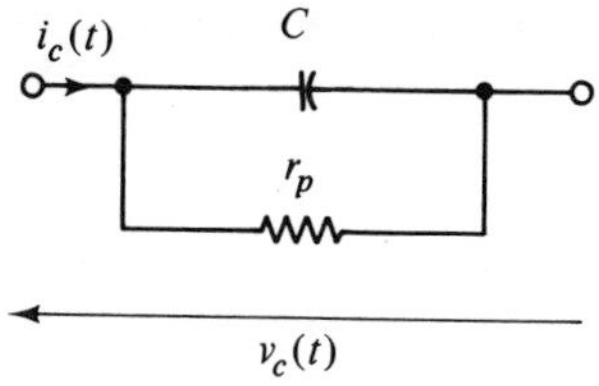

Lossy capacitor

FIGURE 2.3

sinusoidal excitation. Specifically,

$$Q \equiv 2\pi \times \frac{\text{Peak energy stored per cycle}}{\text{Total energy dissipated per cycle}} \tag{2.9}$$

The dissipation factor δ is defined as the reciprocal of the Q-factor,

$$\delta \equiv Q^{-1} \quad \text{DISSIPATION FACTOR} \tag{2.10}$$

The Q-factor of the capacitor model shown in Fig. 2.3 is derived as follows: Letting $v_c(t) \equiv V \sin \omega t$ leads, via equation (2.2), to a "peak energy stored per cycle" of $\frac{1}{2}CV^2$. The "energy dissipated (in r_p) per cycle" is derived from equation (2.2) as $V^2\pi/r_p\omega$, so that equation (2.9) results in a Q-factor given by

$$Q_c = 2\pi \times \frac{\frac{1}{2}CV^2}{V^2\pi/r_p\omega} = \omega C r_p \tag{2.11}$$

High-quality capacitors can be expected to possess a Q-factor that is much greater than unity over their usable range of ω so that, from equation (2.11), $r_p \gg 1/\omega C$.

A similar analysis of the inductor model shown in Fig. 2.3 leads to the inductance Q-factor Q_L given by

$$Q_L = \frac{\omega L}{r_s} \tag{2.12}$$

In this case, high-quality inductors have $r_s \ll \omega L$ over their usable range of ω.

2-4 PASSIVE TWO-PORT ELEMENTS

The transformer and the gyrator are the two most important two-port *passive* elements and they are shown in Table 2.2 with the corresponding definitions and symbols. The parameter n that characterizes the transformer is known as the *turns-ratio* because it corresponds physically to the ratio of the number of turns of the windings employed for the two coupled inductors that are used to make a transformer. The parameter g that characterizes the gyrator is known as the *transconductance* of the gyrator because it has the dimension of conductance and because it relates input quantities $(v_1(t), i_1(t))$ to output quantities $(v_2(t), i_2(t))$.

Both the transformer and the gyrator are shown to be passive and lossless. For the transformer, the net energy delivered is

$$E_T(t_0) = \int_{-\infty}^{t_0} [v_2(t)i_2(t) + v_1(t)i_1(t)]\,dt \tag{2.13}$$

TABLE 2.2 *Passive two-port elements.*

Element	*Symbol*	*Definition*
Transformer	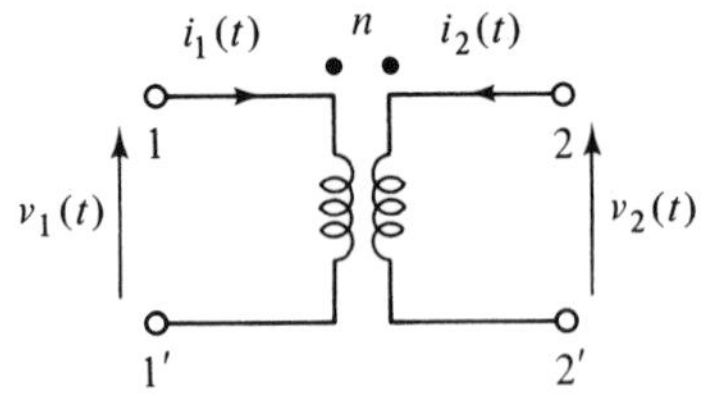	$v_1(t) = nv_2(t)$ $-i_2(t) = ni_1(t)$ n real $n > 0$
Gyrator	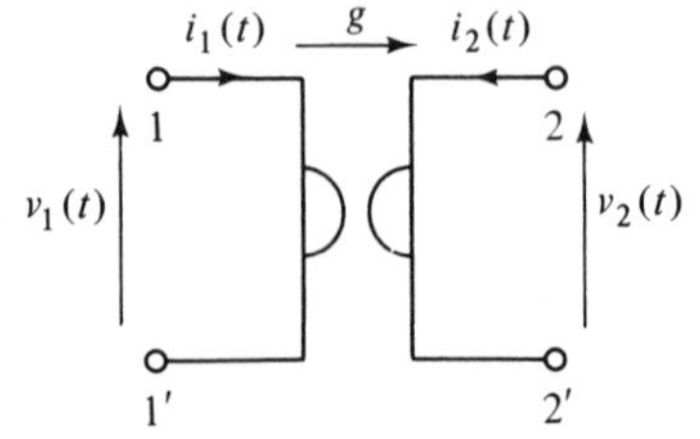	$i_1(t) = gv_2(t)$ $i_2(t) = -gv_1(t)$ g real $g > 0$

Note: (i) If $g < 0$, it is conventional to indicate this by reversing the direction of the arrow above the gyrator symbol.

(ii) If $n < 0$, it is conventional to indicate this by moving one of the dots above the transformer symbol to a similar position below the transformer.

and, by substitution into equation (2.13) of the defining relationships $v_2(t) = n^{-1}v_1(t)$ and $i_2(t) = -ni_1(t)$ it follows that $E_T(t_0) = 0$. The fact that the delivered energy is always zero implies that the ideal transformer is not only passive and lossless but also is *not able to store energy*.

The passivity and losslessness of the ideal gyrator are proven in a similar way to the transformer. Thus,

$$E_G(t_0) = \int_{-\infty}^{t_0} [v_2(t)i_2(t) + v_1(t)i_1(t)]\, dt = 0 \tag{2.14}$$

by substitution of the defining relationships for v_2, i_2, v_1, and i_1. The ideal gyrator is, therefore, similar to the ideal transformer in that it is passive and lossless and does not store energy.

Impedance transformation is an important property of both the transformer and the gyrator. In terms of the Laplace transform of the variables $v_1(t)$, $i_1(t)$, $v_2(t)$, $i_2(t)$ it is readily shown from the definitions in Table 2.2 that

$$\frac{V_1(s)}{I_1(s)} = -n^2\frac{V_2(s)}{I_2(s)} \qquad \text{for the transformer}$$

and

$$\frac{V_1(s)}{I_1(s)} = -g^{-2}\frac{I_2(s)}{V_2(s)} \qquad \text{for the gyrator}$$

Therefore, if these networks are terminated at 2-2′ in a transform impedance $Z_2(s) = V_2(s)/-I_2(s)$, it follows that the transform impedances seen at terminals 1-1′ are given by

$$Z_{1d}(s) \equiv \frac{V_1(s)}{I_1(s)} = n^2 Z_2(s) \qquad \text{for the transformer} \tag{2.15}$$

and

$$Z_{1d}(s) \equiv \frac{V_1(s)}{I_1(s)} = \frac{g^{-2}}{Z_2(s)} \qquad \text{for the gyrator} \tag{2.16}$$

Note that the transformer is an impedance *converting* device whereas the gyrator is an impedance *inverting* device.

EXAMPLE 2.1: Give a *time-domain* proof of the fact that terminating 2-2′ of the gyrator with a capacitance C_2 produces an effective inductance between terminals 1-1′.

SOLUTION: We derive the relationship between $v_1(t)$ and $i_1(t)$ by using the definitions of the gyrator and capacitance elements. Thus,

$$v_1(t) = -\frac{i_2(t)}{g} = -\frac{1}{g}\left[-C\frac{dv_2(t)}{dt}\right]$$
$$= \frac{C}{g}\left[\frac{d}{dt}\left(\frac{i_1(t)}{g}\right)\right] = \frac{C}{g^2}\left[\frac{di_1(t)}{dt}\right]$$

which is exactly the defining relationship for an inductance L of value C/g^2. This technique for realizing an effective (simulated) inductance is important in RC-active filter synthesis and design.

2-5 ACTIVE ONE-PORT ELEMENTS

In the study of RC-active networks it is often convenient to describe a particular active driving-point immittance in terms of an active network *element*. Five active one-port network elements are described in Tables 2.3 and 2.4; the $-R$, $-C$ and $-L$ elements are constrained by $v(t)$, $i(t)$ relationships that are similar to those that apply for conventional (passive) R, C, and L elements except that a negative sign is associated with the describing equations. Some properties of $-R$, $-C$, and $-L$ elements are pursued in Problems 2.10 to 2.16, where it may be shown that all three elements are active, the $-L$ and $-C$ elements are lossless and the $-R$ element delivers positive energy continuously to the driving source (it is not capable of dissipating positive energy over any time interval). Methods of designing $-R$, $-C$, and $-L$ elements are discussed in section 5.5.1.

TABLE 2.3 *Active one-port elements, -R, -C and -L.*

Element	*Symbol*	*Definition*
Negative resistance	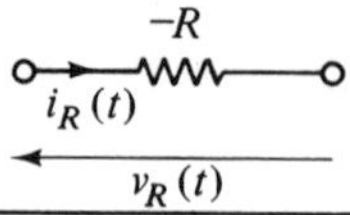	$v_R(t) = -Ri_R(t)$ $R > 0$
Negative capacitance	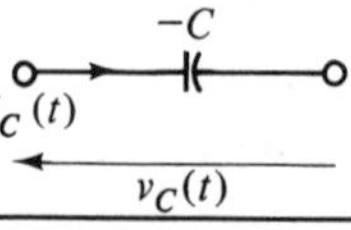	$v_C(t) = -\frac{1}{C}\int_{\infty}^{t} i_c(t)\,dt$ $C > 0$
Negative inductance	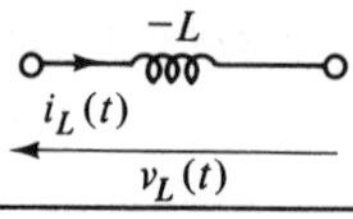	$v_L(t) = -L\frac{di_L(t)}{dt}$ $L > 0$

TABLE 2.4 *Active FDNR one-ports.*

Element	*Symbol*	*Definition*
FDNR D element	$i_D(t)$ D 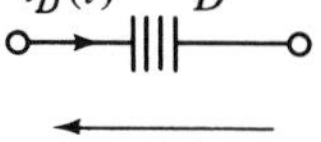 $v_D(t)$ $D>0$	$i_D(t) = D\dfrac{d^2v_D(t)}{dt^2}$ implying $Z_D(j\omega) = -\dfrac{1}{\omega^2 D}$
FDNR E element	$i_E(t)$ E 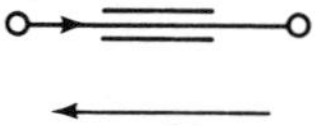 $v_E(t)$ $E>0$	$v_E(t) = E\dfrac{d^2i_E(t)}{dt^2}$ implying $Z_E(j\omega) = -\omega^2 E$

2-5.1 FDNR Elements

The frequency dependent negative resistance (FDNR) elements are of importance in the design of high-quality RC-active filters; they are driving-point immittances that are characterized by the terminal constraints

$$i_D(t) \equiv D\frac{d^2v_D(t)}{dt^2}, \qquad D > 0 \tag{2.17}$$

for the FDNR D element and

$$v_E(t) \equiv E\frac{d^2i_E(t)}{dt^2}, \qquad E > 0 \tag{2.18}$$

for the FDNR E element. The corresponding symbolic representation for these FDNR's is shown in Table 2.4. Taking the Laplace transforms of equations (2.17) and (2.18) and assuming zero initial conditions, we have

$$I_D(s) = s^2V_D(s)D \quad \text{and} \quad V_E(s) = s^2I_E(s)E$$

giving the transform impedances $Z(s)$ of the FDNR elements as

$$Z_D(s) \equiv \frac{V_D(s)}{I_D(s)} = \frac{1}{s^2D} \qquad D \text{ ELEMENT} \tag{2.19}$$

and

$$Z_E(s) = \frac{V_E(s)}{I_E(s)} = s^2E \qquad E \text{ ELEMENT} \tag{2.20}$$

Under conditions of steady-state sinusoidal excitation ($s = j\omega$) the impedances are

$$Z_D(j\omega) = -\frac{1}{\omega^2 D} \qquad \text{FDNR } D\text{-ELEMENT IMPEDANCE} \qquad (2.21)$$

and

$$Z_E(j\omega) = -\omega^2 E \qquad \text{FDNR } E\text{-ELEMENT IMPEDANCE} \qquad (2.22)$$

Note that these impedances are frequency-dependent, negative and resistive (that is, real); hence, the name FDNR.

It can be shown that the FDNR elements are essentially *active* by proving that there does exist at least one set of waveforms $v(t)$ and $i(t)$ at the terminals of the FDNR elements such that the net energy $E(t_0)$ is negative; this then establishes that the FDNR cannot be passive and, therefore, by definition, is active:

EXAMPLE 2.2: The voltage $v_D(t)$ across a D element is given by

$$v_D(t) \equiv \sin \omega t \qquad (2.23)$$

Calculate $E(t_0)$ and thereby prove that a D element corresponds to an *active* one-port network.

SOLUTION: Substituting equation (2.23) into equation (2.17), we obtain

$$i_D(t) = -\omega^2 D \sin \omega t$$

so that, by equation (2.2),

$$E(t_0) = \int_{-\infty}^{t_0} [\sin \omega t][-\omega^2 D \sin \omega t]\, dt$$

or

$$E(t_0) = -\omega^2 D \int_{-\infty}^{t_0} \sin^2 \omega t \, dt \qquad (2.24)$$

The integral term is positive for all ω and all t_0 so that $E(t_0)$ is negative over all ω, D, and t_0. Consequently, *for steady-state sinusoidal conditions*, the FDNR element delivers positive energy to the source for *all* time t_0. It is an *active* network because it is nonpassive; that is, there exists a $v_D(t)$, $i_D(t)$, and t_0 for which $E(t_0)$ is negative.

The preceding example proves activity for the D element by reference to sinusoidal steady-state conditions. However, a much stronger condition is apparent from the Example 2.2 than is implied in the definition of activity. It is recalled that activity (that is, nonpassivity) only requires that $E(t_0)$ be negative for at least *one* value of t_0. However, in the Example 2.2, $E(t_0)$ is negative over *all* values of t_0; in other words, positive energy is delivered to the source at all times.

EXAMPLE 2.3: Prove that $E(t_0)$ is negative for an FDNR D element whenever $\dfrac{v_D(t_0)\, dv_D(t_0)}{dt}$ is negative, and comment on the physical significance of the result.

SOLUTION: The net energy $E(t_0)$ is

$$E(t_0) = \int_{-\infty}^{t_0} D v_D(t) \frac{d^2 v_D(t)}{dt^2}\, dt$$

which, by the product rule for integration gives

$$E(t_0) = D v_D(t_0) \frac{dv_D(t_0)}{dt_0} - \int_{-\infty}^{t_0} D\left[\frac{dv_D(t)}{dt}\right]^2 dt \tag{2.25}$$

where $v_D(-\infty)$ is assumed to be zero. It follows by inspection of equation (2.25) that the integral term is always positive so that, wherever $\dfrac{v_D(t_0)\, dv_D(t_0)}{dt}$ is negative, $E(t_0)$ is negative. The physical significance of this result is that all *bounded* square-integrable waveforms possess some time t_0 at which $\dfrac{v_D(t_0)\, dv_D(t_0)}{dt}$ is negative; consequently, the FDNR delivers energy to the source at some time t_0 for all *bounded* square-integrable waveforms.

The *activity* of the FDNR E element may be proven by means of an example that is similar to Example 2.2.

Units and dimensions of FDNR elements D and E : The *units* and *dimensions* of the D and E elements may be derived from equations (2.17) and (2.18): The D element has dimensions of (amp) (sec)^2 (volt)$^{-1}$ and, since 1 farad is the capacitive unit for 1 amp per volt per second with dimensions (amp) (sec) (volt)$^{-1}$, we shall refer to the D element in units of *farad-seconds*, Fs. [Thus, wherever the second derivative of voltage $v_D(t)$ is 1 volt per second per second, the current $i_D(t)$ in a unit D element is 1 ampere.] In a similar way, the E element has units of *henry-seconds*, Hs.

EXAMPLE 2.4: Calculate the impedance $Z_D(j\omega)$ of a D element where

$$D \equiv \frac{1}{2\pi} \times 10^{-6}\ \text{Fs}$$

at a frequency f of 100 Hz.

SOLUTION: Since $f = 100$ Hz, we have

$$\omega = 2\pi \times 100 \text{ rad/sec}$$

and

$$D = \frac{1}{2\pi} \times 10^{-6} \text{ Fs}$$

giving, by equation (2.21),

$$Z_D(j\omega) = -\frac{1}{\omega^2 D} = -\frac{1}{(2\pi \times 100)^2\left(\frac{1}{2\pi} \times 10^{-6}\right)} \text{ ohm}$$

or

$$\boxed{Z_D(j\omega) = \frac{-100}{2\pi} \text{ ohm}} \tag{2.26}$$

Remarks: In the preceding discussion of the passive and active one-port elements, the topics of passivity, losslessness and activity are introduced and the $\pm R$, $\pm C$, $\pm L$, D, and E elements are defined. *Time-domain* analysis is used, although the reader who is familiar with the relationship between positive real functions and the properties of passive and lossless driving-point immittances could just as easily derive equivalent results in the complex frequency s-domain. (A *prerequisite* background in passive synthesis and the theory of positive real functions is intentionally avoided throughout this text.)

2-6 ACTIVE TWO-PORT ELEMENTS

The transmission matrix equations of a general linear lumped time-invariant two-port network may be written in the form

$$\begin{bmatrix} V_1(s) \\ I_1(s) \end{bmatrix} = \begin{bmatrix} A(s) & B(s) \\ C(s) & D(s) \end{bmatrix} \begin{bmatrix} V_2(s) \\ -I_2(s) \end{bmatrix} \tag{2.27}$$

where the functions $A(s)$, $B(s)$, $C(s)$, and $D(s)$ are real rational functions. Most of the important active two-port elements are obtained by applying suitable zero-constraints on these functions. The generalized immittance converters (GICs) and the nullor are the particular two-port elements that are defined in the following sections:

2-6.1 Generalized Immittance Converters (GICs)

DEFINITION: *A two-port network is a GIC if, when terminated at port 2 with an impedance $Z_2(s)$, the driving-point immittance $Z_{1d}(s)$ or $Y_{1d}(s)$ at port 1 is $K_2^{\pm 1}(s)Z_2(s)$, where $K_2^{\pm 1}(s)$ is not a function of $Z_2(s)$.*

The function $K_2^{\pm 1}(s)$ is the *conversion function* of the GIC operating on the termination $Z_2(s)$. The definition does not specify the property of the driving-point immittance at port 2 when an impedance $Z_1(s)$ is connected at port 1. Prior to consideration of this point, consider the general expression obtainable from equation (2.27) for the driving-point impedance Z_{1d} at port 1, which is

$$Z_{1d}(s) \equiv \frac{V_1(s)}{I_1(s)} = \frac{A(s)Z_2(s) + B(s)}{C(s)Z_2(s) + D(s)} \tag{2.28}$$

where $Z_2(s) = V_2(s)/-I_2(s)$.

It follows that the definition of the GIC may be satisfied in two possible ways. Either $A(s)$ and $D(s)$ are both zero or $B(s)$ and $C(s)$ are both zero; these two cases are considered separately.

Impedance-converting GIC

DEFINITION: *A GIC is an impedance-converting GIC if the functions $B(s)$ and $C(s)$ of its transmission matrix are zero over all s.*

It follows from equation (2.28) and the definition above that

$$Z_{1d} = \left[\frac{A(s)}{D(s)}\right] Z_2(s) \quad \text{IMPEDANCE-CONVERTING GIC} \tag{2.29}$$

so that the conversion function $K_2(s) = A(s)D^{-1}(s)$. Consider the impedance-converting GIC to be terminated at port 1 with an impedance $Z_1(s)$. Then the driving-point impedance Z_{2d} at port 2 may be derived as follows: From equation (2.27)

$$\begin{bmatrix} V_2(s) \\ -I_2(s) \end{bmatrix} = \frac{1}{A(s)D(s) - B(s)C(s)} \begin{bmatrix} D(s) & -B(s) \\ -C(s) & A(s) \end{bmatrix} \begin{bmatrix} V_1(s) \\ I_1(s) \end{bmatrix} \tag{2.30}$$

and, using $Z_1(s) \equiv V_1(s)/-I_1(s)$ and $Z_{2d}(s) = V_2(s)/I_2(s)$, we have

$$Z_{2d}(s) = \frac{B(s) + D(s)Z_1(s)}{A(s) + C(s)Z_1(s)} \tag{2.31}$$

For the *impedance-converting* GIC we substitute $B(s) = C(s) = 0$, so that

$$Z_{2d} = \left[\frac{D(s)}{A(s)}\right]Z_1(s) \equiv K_1(s)Z_1(s) \tag{2.32}$$

where $K_1(s)$ is the conversion function. Note that

$$\boxed{K_1(s)K_2(s) = 1} \quad \text{CONVERTING-TYPE GIC} \tag{2.33}$$

for impedance-converting GICs. The symbol used in this text for the converting-type GIC is shown in Table 2.5.

Impedance-Inverting GIC

DEFINITION: *A GIC is an impedance-inverting GIC if the functions $A(s)$ and $D(s)$ of its transmission matrix are zero over all s.*

It follows from equation (2.28) and the definition above that

$$Z_{1d}(s) = \left[\frac{B(s)}{C(s)}\right]\frac{1}{Z_2(s)} \equiv \frac{K_2(s)}{Z_2(s)} \tag{2.34}$$

where $K_2(s) = B(s)C^{-1}(s)$ and has dimensions of $(\text{ohm})^2$. An *impedance-inverting* GIC therefore inverts the terminating impedance $Z_2(s)$ and multiplies it by $K_2(s)$. It follows from equation (2.31) and the definition above that

$$Z_{2d}(s) = \left[\frac{B(s)}{C(s)}\right]\frac{1}{Z_1(s)} \equiv \frac{K_1(s)}{Z_1(s)} \tag{2.35}$$

where $K_1(s) = B(s)C^{-1}(s)$ and

$$\boxed{K_1(s) = K_2(s)} \quad \text{INVERTING-TYPE GIC} \tag{2.36}$$

The symbol used throughout this text for an inverting-type GIC is shown

TABLE 2.5 *The immittance converters.*

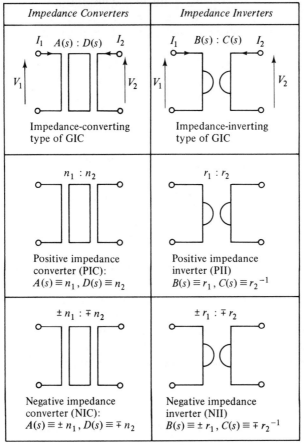

Impedance Converters	*Impedance Inverters*
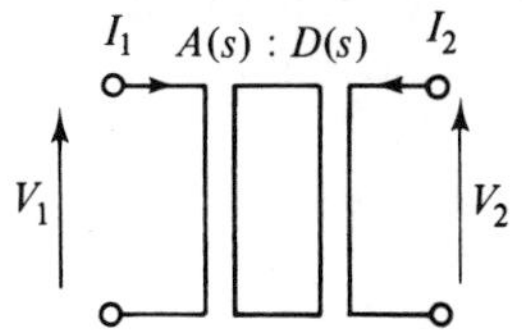 Impedance-converting type of GIC	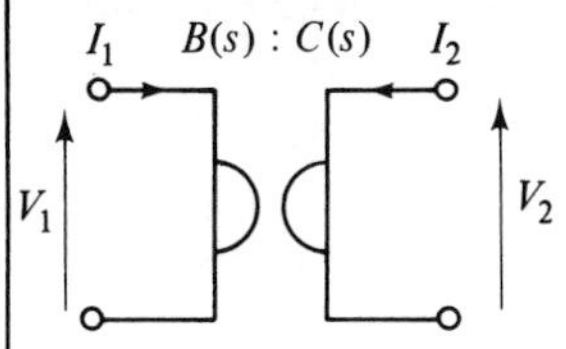 Impedance-inverting type of GIC
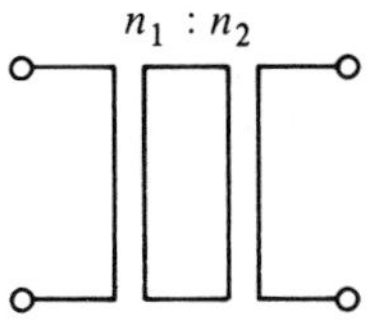 Positive impedance converter (PIC): $A(s) \equiv n_1, D(s) \equiv n_2$	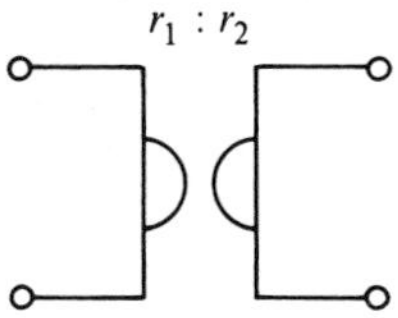 Positive impedance inverter (PII) $B(s) \equiv r_1, C(s) \equiv r_2^{-1}$
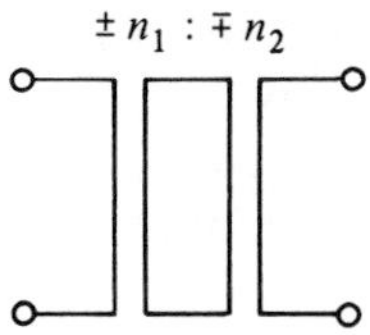 Negative impedance converter (NIC): $A(s) \equiv \pm n_1, D(s) \equiv \mp n_2$	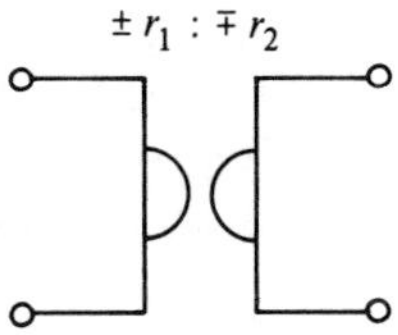 Negative impedance inverter (NII) $B(s) \equiv \pm r_1, C(s) \equiv \mp r_2^{-1}$

in Table 2.5. The symbol differs from that of the gyrator only by the designation of the functions $B(s)$ and $C(s)$—instead of the transconductance g of the gyrator—above the symbol.

2-6.2 GICs with Real Transmission Matrix Elements

In many applications, the transmission matrix elements A, B, C, and D are simple real constants that are written $\pm n_1$, $\pm r_1$, $\pm r_2^{-1}$, and $\pm n_2$, respectively. Then, the conversion function $K_2(s)$ of the converting-type GIC is simply $\pm n_1/n_2$. In such cases, the positive impedance converter (PIC) and the negative impedance converter (NIC) are defined as shown in

Table 2.5, so that the driving-point impedance at port 1, for example, is simply $n_1 Z_2(s)/n_2$ for the PIC and $-n_1 Z_2(s)/n_2$ for the NIC. In a similar way, the positive impedance inverter (PII) and the negative impedance inverter (NII) exhibit driving-point impedances at port 1 of $r_1 r_2/Z_2(s)$ and $-r_1/r_2 Z_2(s)$, respectively. The PII and the NII are also shown symbolically in Table 2.5, using a similar representation to that of the gyrator.

It is not difficult to prove that if $n_1 \neq n_2^{-1}$ and if $r_1 \neq r_2$, then the PIC and the PII are *active* two-port networks. For example, the energy delivered to the PIC is

$$E_{\mathrm{PIC}}(t_0) = \int_{-\infty}^{t_0} [v_1(t)i_1(t) + v_2(t)i_2(t)]\, dt$$

$$= \int_{-\infty}^{t_0} [1 - n_1 n_2] v_1(t) i_1(t)\, dt \qquad (2.37)$$

Except in the special case of the (passive) transformer, where $n_1 \equiv n_2^{-1}$, there exists the possibility that $E_{\mathrm{PIC}}(t_0) < 0$, which immediately requires that the PIC must be capable of delivering energy; it is therefore active. The reader may wish to verify that the PII is active if $r_1 \neq r_2$. Alternative descriptions of the PIC and PII are the *active-transformer* and the *active-gyrator*, respectively. Repetitions of the derivation above for the NIC and the NII lead immediately to the conclusion that they are both *active* two-ports.

2-6.3 Impedance Transforming Properties of GICs

A few simple examples of the use of GICs are now explained. Other applications are given as problems at the end of the chapter. The use of GICs in the realization of high-quality *RC*-active filters is described in chapters 8 and 9.

The converting-type GIC of Fig. 2.4 has a conversion function $K_2(s) = s^2/\omega_0^2$. The term ω_0 is a constant with dimensions of radians per second. Therefore, by equation (2.33), $K_1(s) = \omega_0^2/s^2$. Consequently, the driving-point impedance Z_{1d} of Fig. 2.4(a) is that of the FDNR E element $s^2 R/\omega_0^2$, and the driving-point impedance Z_{2d} of Fig. 2.4(b) is that of the FDNR D element $\omega_0^2/s^2 D$. In a similar way, the converting-type GIC may be used to realize active network driving-point impedances Z_{1d} that simulate the impedances sL/ω_0 and ω_0/sC of the L and C elements, respectively.

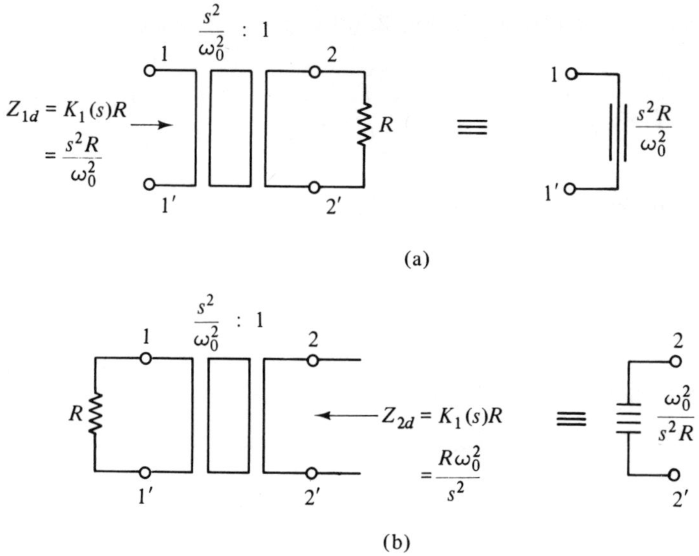

FIGURE 2.4

EXAMPLE 2.5: Given the network shown in Fig. 2.5(a), derive the functions $D_A(s)$ and $D_B(s)$ such that the overall transmission matrix T is

$$T = \begin{bmatrix} 1 & \frac{sR}{\omega_0} \\ 0 & 1 \end{bmatrix} \tag{2.38}$$

and thereby show that the network is equivalent to the simple single *floating inductance network* in Fig. 2.5(b).

SOLUTION: By definition, the transmission matrices of GIC A and GIC B are

$$T_A = \begin{bmatrix} 1 & 0 \\ 0 & D_A(s) \end{bmatrix}, \qquad T_B = \begin{bmatrix} 1 & 0 \\ 0 & D_B(s) \end{bmatrix} \tag{2.39}$$

and the transmission matrix T_R of the two-port, shown as a dashed line in Fig. 2.5(a) is

$$T_R = \begin{bmatrix} 1 & R \\ 0 & 1 \end{bmatrix} \tag{2.40}$$

from which it follows that the overall cascade of these three two-ports gives the transmission matrix T of Fig. 2.5(a) as $T = T_A T_R T_B$, or

$$T = \begin{bmatrix} 1 & 0 \\ 0 & D_A(s) \end{bmatrix} \begin{bmatrix} 1 & R \\ 0 & 1 \end{bmatrix} \begin{bmatrix} 1 & 0 \\ 0 & D_B(s) \end{bmatrix} = \begin{bmatrix} 1 & RD_B(s) \\ 0 & D_A(s)D_B(s) \end{bmatrix} \tag{2.41}$$

Now, the transmission matrix T_L of Fig. 2.5(b) is

$$T_L = \begin{bmatrix} 1 & \dfrac{sR}{\omega_0} \\ 0 & 1 \end{bmatrix} \tag{2.42}$$

so that, equating T and T_L, we get

$$D_B = \frac{s}{\omega_0}, \qquad D_A = \frac{\omega_0}{s} \tag{2.43}$$

Observe that in this example above the impedance R that is embedded between the GICs is impedance-scaled by the factor s/ω_0 to arrive at the equivalent network in Fig. 2.5(b). Furthermore, for GICA $K_{2A}(s) = s/\omega_0$ and for GICB $K_{1B} = s/\omega_0$. That is, the network is symmetrical and, in particular, the GICs are identical but simply reversed in the direction of the port connections. A generalized version of this application of GICs is described in chapter 9, section 9.7, for use in the synthesis of RC-active filters.

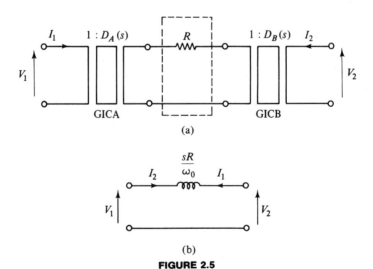

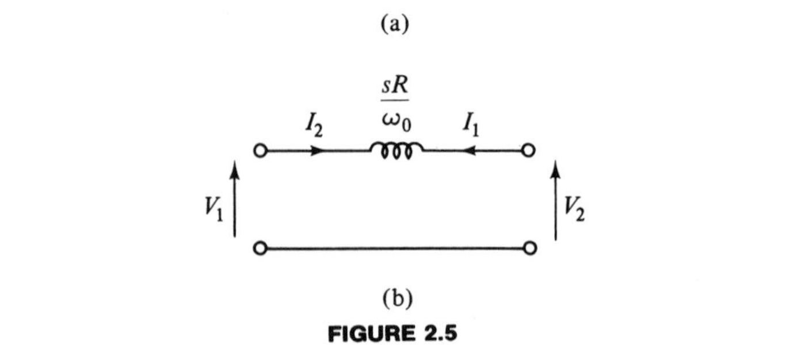

FIGURE 2.5

2-7 THE NULLOR AND THE INDEPENDENT SOURCES

The nullor is an important two-port; an understanding of its properties is of considerable use in the synthesis and analysis of *RC*-active networks. The nullor is used in many parts of this text as an aid to the development of insight into the relationships between the various types of *RC*-active networks.

DEFINITION: *A nullor is the two-port network that has the null transmission matrix; that is, it is characterized by the terminal equations*

$$\begin{bmatrix} V_1(s) \\ I_1(s) \end{bmatrix} = \begin{bmatrix} 0 & 0 \\ 0 & 0 \end{bmatrix} \begin{bmatrix} V_2(s) \\ -I_2(s) \end{bmatrix} \tag{2.44}$$

The symbol for the nullor is shown in Table 2.6. The concept of a null transmission matrix may at first thought seem to be a rather meaningless abstraction. Consider the first row of equation (2.44); that is, $V_1 = 0.V_2 + 0.(-I_2)$. In physical terms this implies that the input voltage V_1 is zero, regardless of the output variables V_2, I_2. Similarly, the second equation $I_1 = 0.V_2 + 0.(-I_2)$ implies that the input current is zero, regardless of the output variables. Therefore, equation (2.44) simply constrains $V_1(s)$ *and* $I_1(s)$ *to be zero* and does not in any way constrain $V_2(s)$ and $I_2(s)$; insofar as the definition is concerned, $V_2(s)$ *and* $I_2(s)$ *are both arbitrary*.

Let us consider the behavior at the *input port* 1-1′ of a nullor in more detail. To assist in further use of nullor theory, it is convenient to separate the input terminals 1-1′ and to think of the resultant one-port as an element with the property that it maintains zero voltage between its terminals and maintains zero current flow between its terminals; this element is called a *nullator*:

DEFINITION: *A nullator is a one-port network defined by the terminal constraints*

$$V_1(s) = I_1(s) = 0 \quad \text{or} \quad v_1(t) = i_1(t) = 0 \tag{2.45}$$

The symbol and definition for the nullator are shown in Table 2.6. Note that a nullator is certainly *not* a short circuit [because $i_1(t) = 0$], and it is certainly not an open circuit [because $v_1(t) = 0$].

TABLE 2.6 *The nullor, nullator, norator and independent sources.*

Element	*Symbol*	*Definition*
Nullor	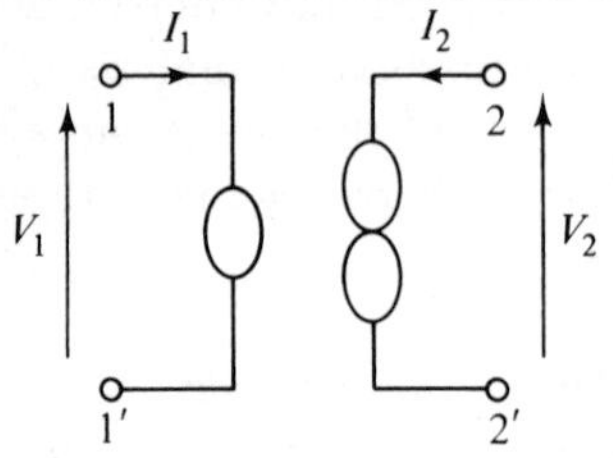	$V_1 \equiv I_1 \equiv 0$ $V_2 \equiv$ arbitrary $I_2 \equiv$ arbitrary
Nullator	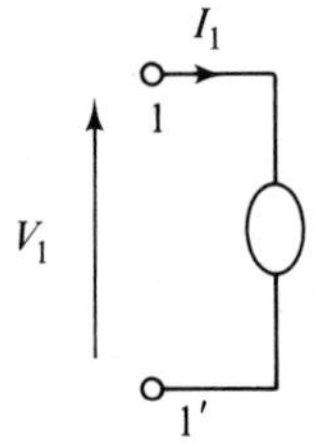	$V_1 \equiv I_1 \equiv 0$
Norator	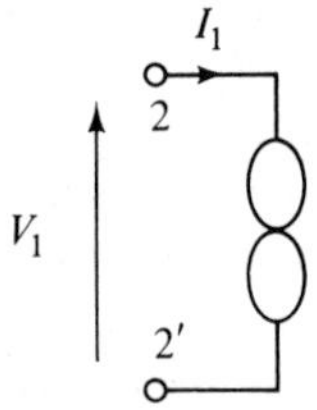	$V_1 \equiv$ arbitrary $I_1 \equiv$ arbitrary
Voltage source	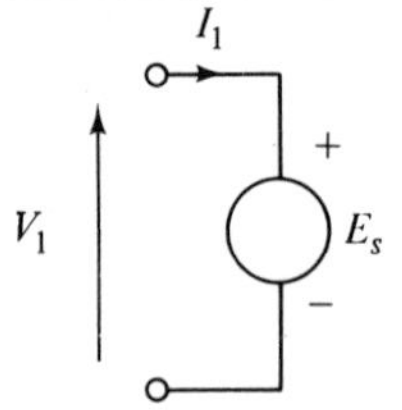	$V_1 \equiv E_s$ $I_1 \equiv$ arbitrary
Current source	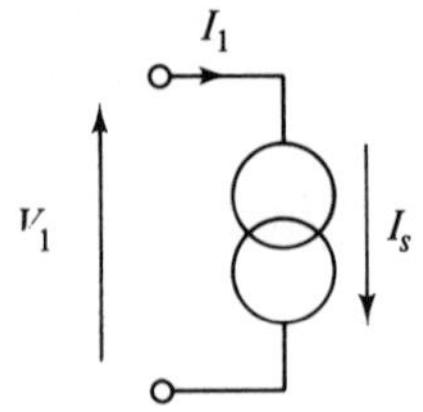	$V_1 \equiv$ arbitrary $I_1 \equiv I_s$

It is also convenient to separate the output terminals 2-2′ from the nullor and define a separate one-port, known as a *norator*:

DEFINITION: *A norator is a one-port network for which the terminal voltage $v_1(t)$ and the terminal current $i_1(t)$ are unconstrained; that is,*

$$\boxed{\begin{array}{ll} V_1(s) \equiv \text{arbitrary}, & I_1(s) \equiv \text{arbitrary} \\ \text{or} & \\ v_1(t) \equiv \text{arbitrary}, & i_1(t) \equiv \text{arbitrary} \end{array}} \tag{2.46}$$

This may, at first thought, seem very much like a nondefinition. However, the existence of such a network element implies a degree of freedom that does not exist for any other one-port. The symbol and definition of the norator are provided in Table 2.6.

The independent sources, which are the *excitation* sources of a network, are the voltage source $E_s(s)$ or $e_s(t)$, and the current source $I_s(s)$ or $i_s(t)$. The symbols and definitions of these independent sources are given in Table 2.6, from which it is noted that the current I_1 through the voltage source is arbitrary (that is, it depends on E_1 *and* the arbitrary network into which the voltage source is embedded, which is, of course, not defined). Similarly, the voltage V_1 across the terminals of the current source is arbitrary.

It is shown in chapter 5 that the previously defined FDNR elements, the inductance element and all of the GICs, including the passive transformer and gyrator, may be realized by means of a suitable interconnection of nullors, resistance elements, and capacitance elements. For this reason, *RC*-nullor networks are of fundamental importance, and it is therefore useful to study the means by which interconnections of *RC* elements and nullors may be analyzed. Prior to this, a digression to the subject of controlled sources, sometimes called dependent sources, is essential.

2-8 THE FOUR FINITE-GAIN CONTROLLED SOURCES

The previously discussed independent sources are characterized by the fact that the value of the voltage $e_s(t)$ or $E_s(s)$ across the terminals of the voltage source and the value of the current $i_s(t)$ or $I_s(s)$ flowing through the current source are defined independently of the network into which these sources are embedded. In the case of controlled (or dependent sources), this is inherently not the situation; the *controlled* voltage sources, of which there are two types, exhibit voltages across their output terminals that are specifically functions of some controlling voltage or some controlling current.

The *voltage-controlled voltage source* (VCVS) is a two-port network given by the transmission matrix T_{VCVS}, where

$$\begin{bmatrix} V_1 \\ I_1 \end{bmatrix} = \begin{bmatrix} \dfrac{1}{\mu} & 0 \\ 0 & 0 \end{bmatrix} \begin{bmatrix} V_2 \\ -I_2 \end{bmatrix} \quad \text{or} \quad T_{VCVS} \equiv \begin{bmatrix} \dfrac{1}{\mu} & 0 \\ 0 & 0 \end{bmatrix} \tag{2.47}$$

The network representation of a VCVS is shown in Table 2.7. The input and output voltages are related via the proportionality constant μ, and the input current I_1 is zero; the output current I_2 is arbitrary.

The *current-controlled voltage source* (CCVS) is defined by the transmission matrix

$$T_{CCVS} \equiv \begin{bmatrix} 0 & 0 \\ \dfrac{1}{r} & 0 \end{bmatrix} \tag{2.48}$$

where the network representation is shown in Table 2.7. In this case, the input voltage V_1 is zero and the output current I_2 is arbitrary; the input current I_1 and the output voltage V_2 are related by the constant *transresistance r* (ohm).

The *current-controlled current source* (CCCS) is defined by the transmission matrix

$$T_{CCCS} = \begin{bmatrix} 0 & 0 \\ 0 & -\dfrac{1}{\beta} \end{bmatrix} \tag{2.49}$$

The network representation is shown in Table 2.7. The input and output currents are related by the proportionality constant β, the input voltage V_1 is zero, and the output voltage V_2 is arbitrary.

Finally, the *voltage-controlled current source* (VCCS) is defined by the transmission matrix

$$T_{VCCS} = \begin{bmatrix} 0 & -\dfrac{1}{g} \\ 0 & 0 \end{bmatrix} \tag{2.50}$$

The network representation is shown in Table 2.7. The input current I_1 is zero, the output voltage V_2 is arbitrary, and the input voltage and output current are related via the *constant transconductance* g (ohm)$^{-1}$.

TABLE 2.7 *The four controlled sources.*

Symbol	*Definition*
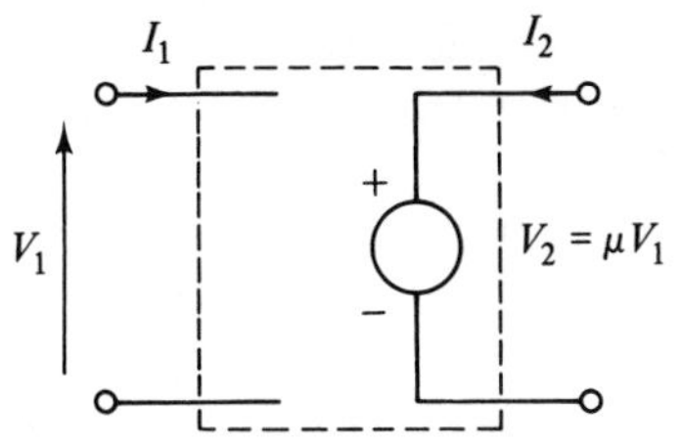 Voltage-controlled voltage source	$V_1 \equiv \frac{1}{\mu} V_2$ $I_1 \equiv 0$ $I_2 \equiv$ arbitrary
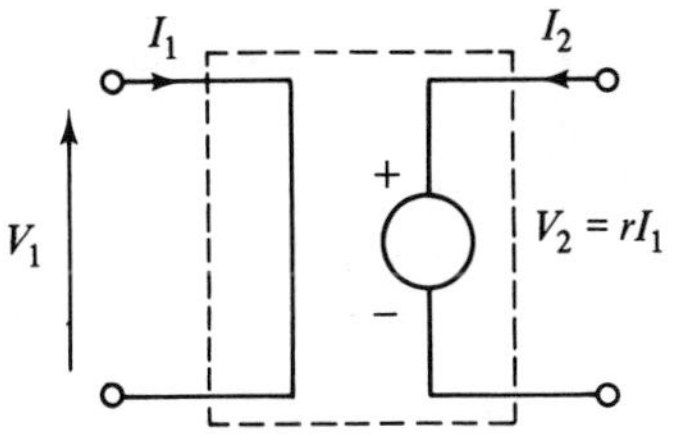 Current-controlled voltage source	$I_1 \equiv \frac{1}{r} V_2$ $V_1 \equiv 0$ $I_2 \equiv$ arbitrary
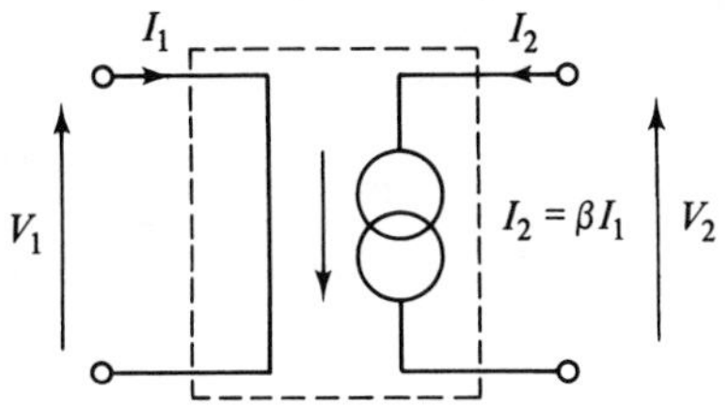 Current-controlled current source	$I_2 \equiv \beta I_1$ $V_1 \equiv 0$ $V_2 \equiv$ arbitrary
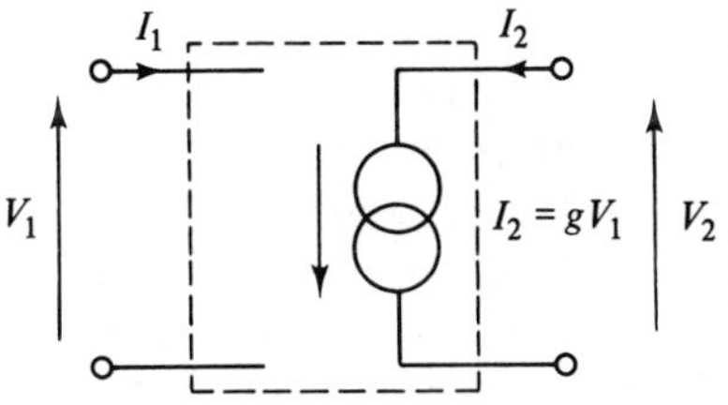 Voltage-controlled current source	$I_2 \equiv gV_1$ $I_1 \equiv 0$ $V_2 \equiv$ arbitrary

2-9 NULLOR MODELS OF CONTROLLED SOURCES

The four types of controlled source are used extensively in the analysis and synthesis of RC-active networks. For example, the VCCS is usually employed in the hybrid-π model of the bipolar transistor (see section 4.2 of chapter 4) and the VCVS is often used to represent the operational amplifier. It is now shown that *all four controlled sources may be realized by using nullor-resistor networks*. The principal advantages to be gained from replacing controlled sources by their corresponding R-nullor realizations are that circuit analysis is often much easier and that it is possible to explain the relationships between apparently unrelated RC-active networks by simply replacing the controlled sources by their nullor equivalents and then comparing the resultant RC-nullor networks.

The resistance-nullor equivalent networks for the four controlled sources are shown in Table 2.8. The validity of each of these networks may be checked by simply applying Kirchoff's voltage law (KVL) and Kirchoff's current law (KCL) in conjunction with the definitions of the nullator and the norator. This exercise provides the reader with an excellent opportunity to investigate the properties of nullors when embedded into some very simple networks. For example, it is now shown that the first network in Table 2.8, repeated in Fig. 2.6, is a valid representation of a VCCS. According to the VCCS terminal constraints in Table 2.8, it is necessary that the network in Fig. 2.6 has the following properties:

$$\left.\begin{array}{lll}\text{(i)} & I_1 = 0 & \\ \text{(ii)} & I_2 = gV_1, & g \text{ constant} \\ \text{(iii)} & V_2 = \text{arbitrary (unconstrained)} & \end{array}\right\} \tag{2.51}$$

It is obvious immediately that, in Fig. 2.6, $I_1 = 0$ because this is the current flowing in the nullators which, by definition of a nullator, is zero. Also, by definition of the nullators, it is known that the voltage drop across each nullator is zero, so that by application of KVL around loop I in Fig. 2.6 it follows that $V_g = V_1$; then, by Ohm's law, the current I_g in the conductance g is given by $I_g = gV_g$. Applying KCL at nodes X and Y, and recalling that the nullator currents are zero, we find that $I_2 = I_g = gV_g = gV_1$. The required input-output relationship $I_2 = gV_1$ is therefore proven; it remains to be shown that V_2 is arbitrary. This is in fact ensured by the norators because, by definition of the norator element, the norator voltages V_{N1} and V_{N2} are arbitrary (despite the fact that the norator currents are constrained by V_1 to equal gV_1). Thus, applying KVL to loop

TABLE 2.8 *Type 1 nullor controlled sources.*

Constraints	*Equivalent Nullor Network*	*Type 1*
$I_1 = 0$ $V_2 = \text{arb}^y$ $I_2 = gV_1$	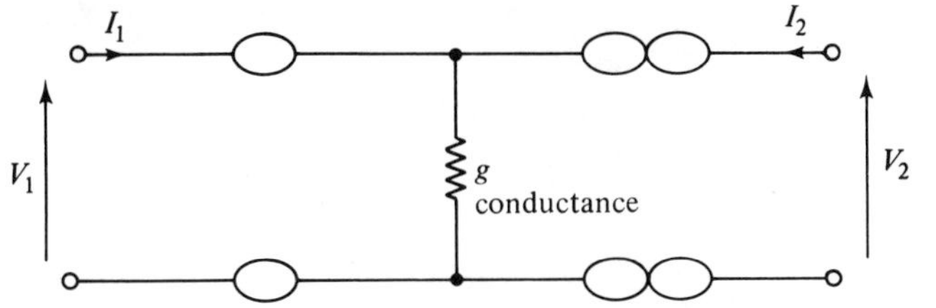	VCCS 1
$V_1 = 0$ $I_2 = \text{arb}^y$ $V_2 = rI_1$	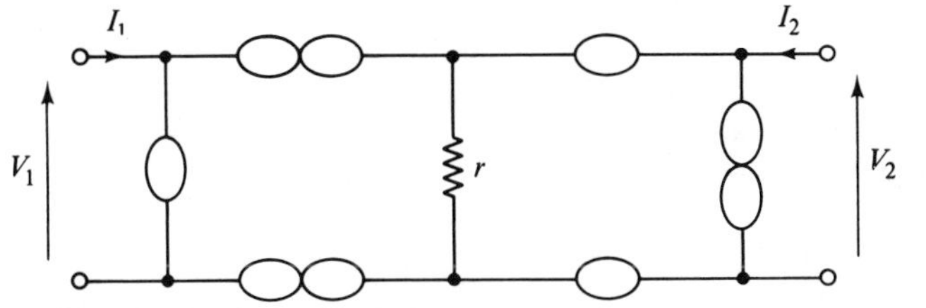	CCVS 1
$I_1 = 0$ $I_2 = \text{arb}^y$ $V_2 = -\frac{r_2}{r_1} V_1$	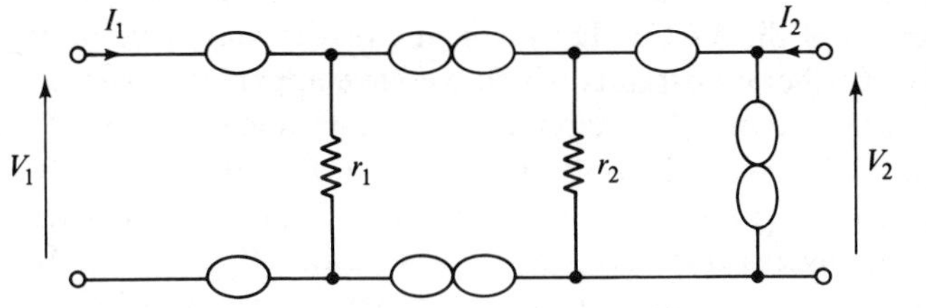	* VCVS 1
$V_1 = 0$ $V_2 = \text{arb}^y$ $I_2 = \frac{r_1}{r_2} I_1$	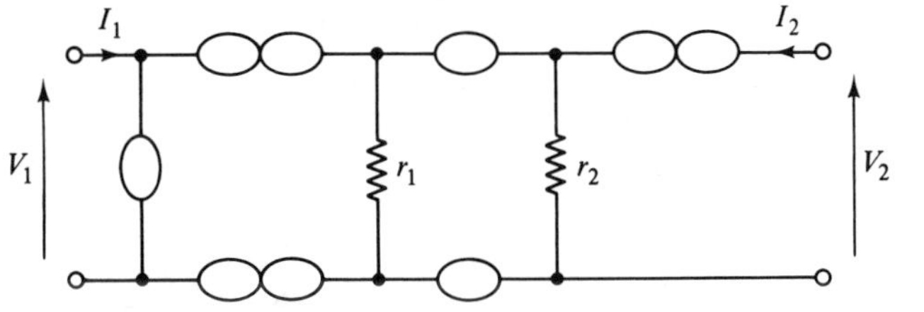	CCCS 1

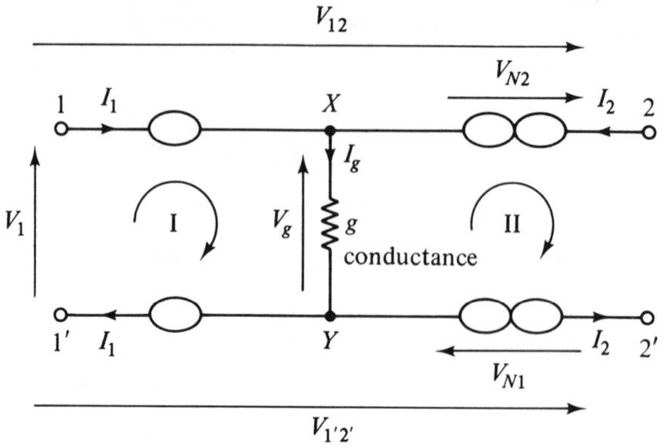

FIGURE 2.6

II, we get

$$
\begin{aligned}
V_2 &= V_g + V_{N1} + V_{N2} \\
&= V_1 + \text{(arbitrary voltages)} = \text{arbitrary voltage} \qquad (2.52)
\end{aligned}
$$

It has been shown that the network in Fig. 2.6 exhibits the three defining properties of the VCCS. However, the reader may have noticed that there are two further constraints which we have apparently overlooked? That is, there must be an open circuit between terminals 1 and 2 and an open circuit between terminals 1′ and 2′. Since $I_1 = 0$, this simply requires that the voltages V_{12} and $V_{1'2'}$ in Fig. 2.6 be arbitrary. The existence of the *two* norators ensures this because $V_{12} = V_{N2}$ and $V_{1'2'} = V_{N1}$. It is interesting to note that a nullator and a norator may be removed from the network if the VCVS has a short-circuit between terminals 1′ and 2′; the resultant *three-terminal VCCS* is shown as VCCS2 in Fig. 2.7 and also in Table 2.9.

At this point, the reader should be able to analyze *all* of the entries in Tables 2.8 and 2.9 to check that they are correctly classified as controlled sources. The type 2 networks in Table 2.9 are characterized by the fact that the elements r, g, r_1, and r_2 have a common input-output terminal 1′,

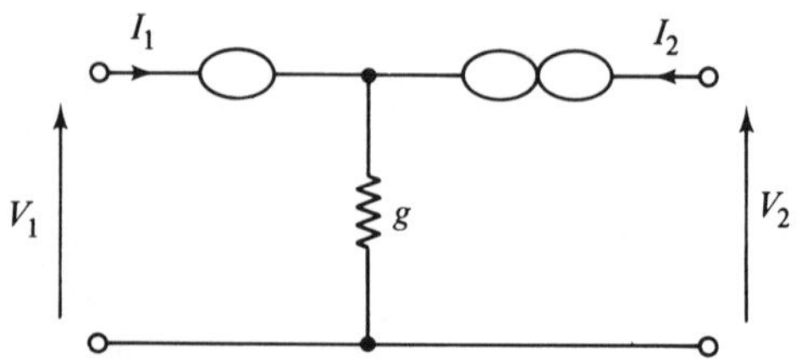

FIGURE 2.7

TABLE 2.9 *Type 2 and 3 nullor controlled sources.*

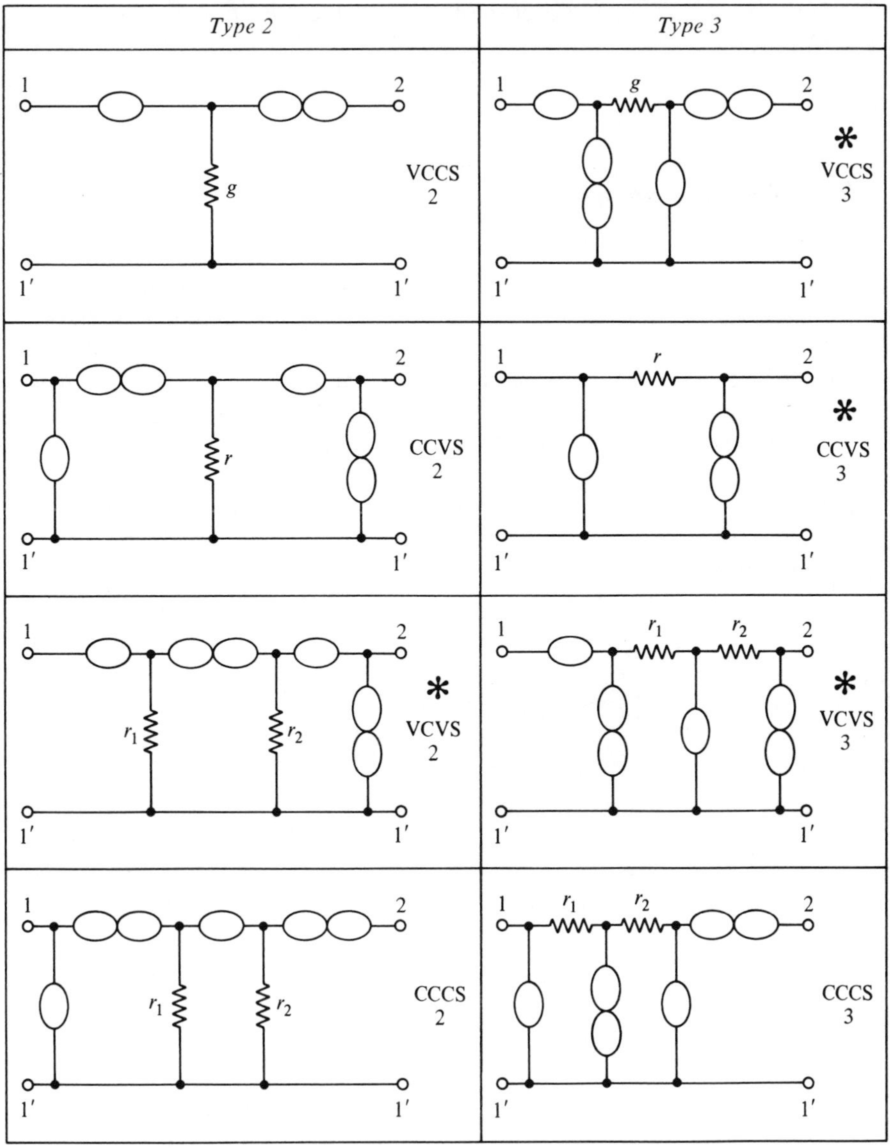

whereas the type 3 networks are characterized by the fact that none of the elements g, r, r_1, r_2 is connected to the common input-output terminal 1′. The three types of controlled sources are referred to in this book by the mnemonics VCCS1, VCCS2, VCCS3, etc., to correspond with Tables 2.8 and 2.9. The asterisks (*) associated with VCVS1, VCVS2, VCCS3, and

CCVS3 in Tables 2.8 and 2.9 are to indicate that the controlling factors are in fact the negative of those prescribed in Table 2.7. For example, $\mu = -r_2/r_1$ for VCVS1.

2-10 SIMPLE NULLOR TWO-PORTS WITHOUT *RC* ELEMENTS

The three-terminal *unity-gain* VCVS ($\mu = 1$) and the three-terminal *negative-unity-gain* CCCS ($\beta = -1$) may be realized with one nullor as shown in Fig. 2.8(a) and Fig. 2.8(b), respectively. These three-terminal networks may in fact be found as subnetworks of some of the networks shown in Table 2.9.

Although we have not yet discussed the synthesis of NICs, the reader may readily verify that the simple two-nullor networks shown in Fig. 2.9(a) and Fig. 2.9(b) are NICs for which $K_2(s) = -1$. Terminating either network with an impedance Z results in a driving-point impedance $-Z$ at the other port.

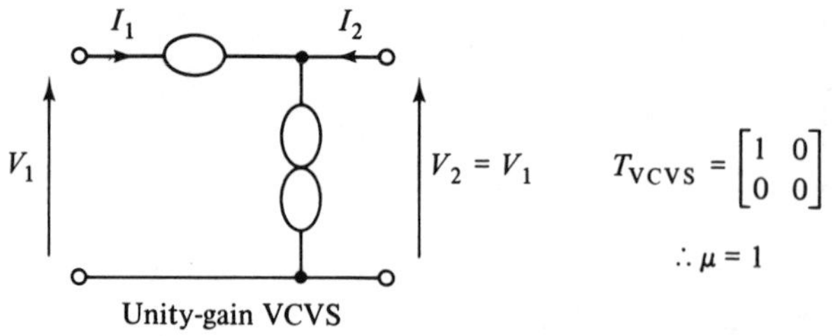

(a)

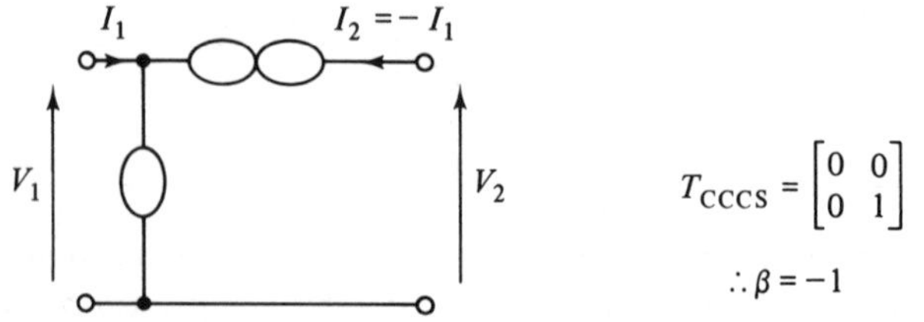

(b)

FIGURE 2.8

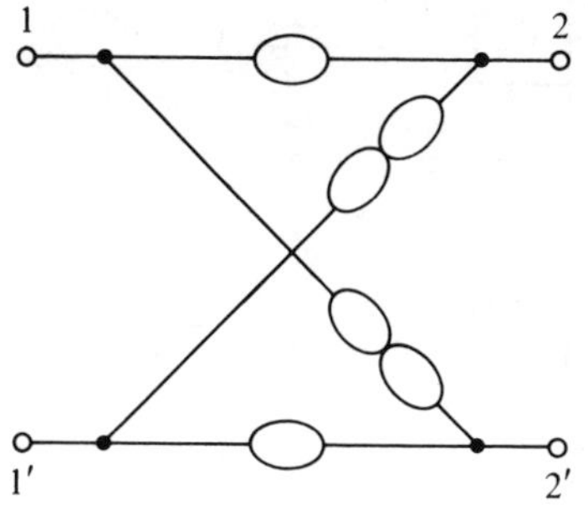

(a)

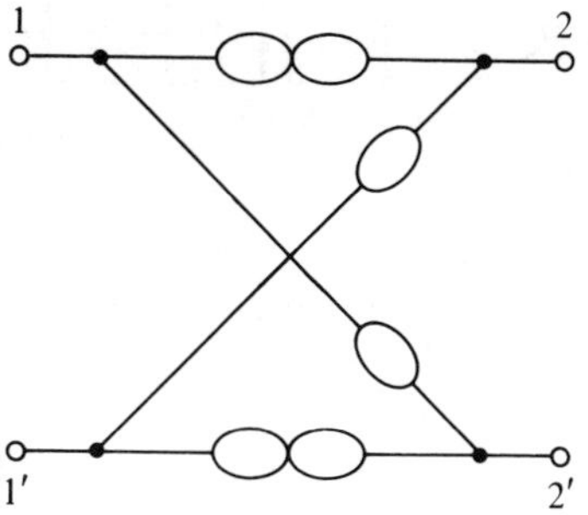

(b)

FIGURE 2.9

2-11 NULLOR EQUIVALENCES AND NETWORK SIMPLIFICATION

It is recalled that the *V-I* constraints applying to the nullator, norator, short-circuit and open-circuit are as follows:

TABLE 2-10 *Nullor equivalences.*

Element	*V*	*I*
Nullator	0	0
Norator	arb^y	arb^y
Short-circuit	0	arb^y
Open-circuit	arb^y	0

The constraints above are used to prove the following equivalences:

EQUIVALENCE 1: A series connection of $\pm R$, $\pm L$, $\pm C$ elements and at least one nullator is equivalent to a nullator, as shown in Fig. 2.10

The proof follows from the fact that, for zero initial conditions, $V = 0 + IZ_1 + IZ_2 + \ldots + IZ_N$, which is zero because I is zero.

EQUIVALENCE 2: A series connection of $\pm R$, $\pm L$, $\pm C$ elements and at least one norator is equivalent to a norator, as shown in Fig. 2.10.

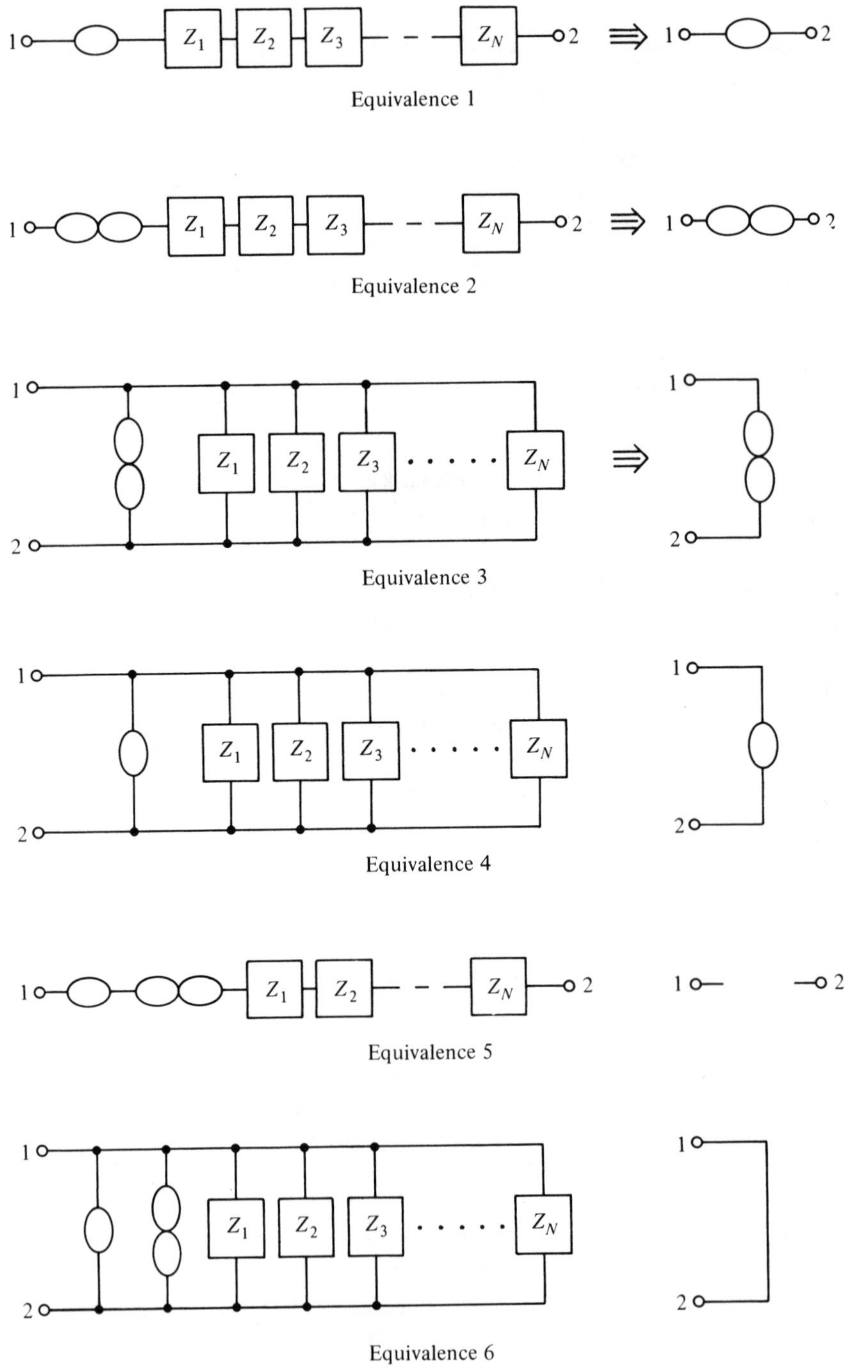
1
Z_1
Z_2
Z_3
Z_N
2
Equivalence 1
Equivalence 2
Equivalence 3
Equivalence 4
Equivalence 5
Equivalence 6

FIGURE 2.10

The proof follows from the fact that, for zero initial conditions, $V = IZ_1 + IZ_2 + \ldots + V_{N1} + \ldots + IZ_N$, where V_{N1} and, therefore, V is arbitrary for any I.

EQUIVALENCE 3: A parallel connection of $\pm R$, $\pm L$, $\pm C$ elements and at least one norator is equivalent to a norator, as shown in Fig. 2.10.

EQUIVALENCE 4: A parallel connection of $\pm R$, $\pm L$, $\pm C$ elements and at least one nullator is equivalent to a nullator, as shown in Fig. 2.10.

EQUIVALENCE 5: A series connection of $\pm R$, $\pm L$, $\pm C$ elements and at least one nullator and at least one norator is equivalent to an open-circuit, as shown in Fig. 2.10.

EQUIVALENCE 6: A parallel connection of $\pm R$, $\pm L$, $\pm C$ elements and at least one nullator and at least one norator is equivalent to a short-circuit, as shown in Fig. 2.10.

EQUIVALENCE 7: The star connection of two three-terminal nullors in Fig. 2.11 is equivalent to one four-terminal nullor.

The last five equivalences are not proven but left as an exercise for the reader. It is possible to use the above equivalences to simplify active networks and to demonstrate the equivalences between different RC-active circuit configurations. This subject is pursued in chapter 5.

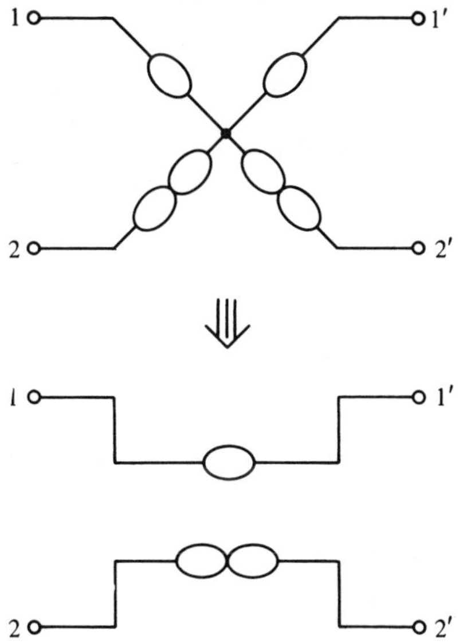

Equivalence 7

FIGURE 2.11

EXAMPLE 2.6: Simplify the network shown in Fig. 2.12 by removing as many redundant elements as possible.

SOLUTION: Applying Equivalence 1 to branches B3 and B4, Equivalence 2 to branch B7, and Equivalence 5 to branch B5 results in the network shown in Fig. 2.13. Then, applying Equivalence 6 to B4 and B6 (or Equivalence 7 to the star connection) in Fig. 2.13 leads to the network shown in Fig. 2.14 for which no further simplification is possible.

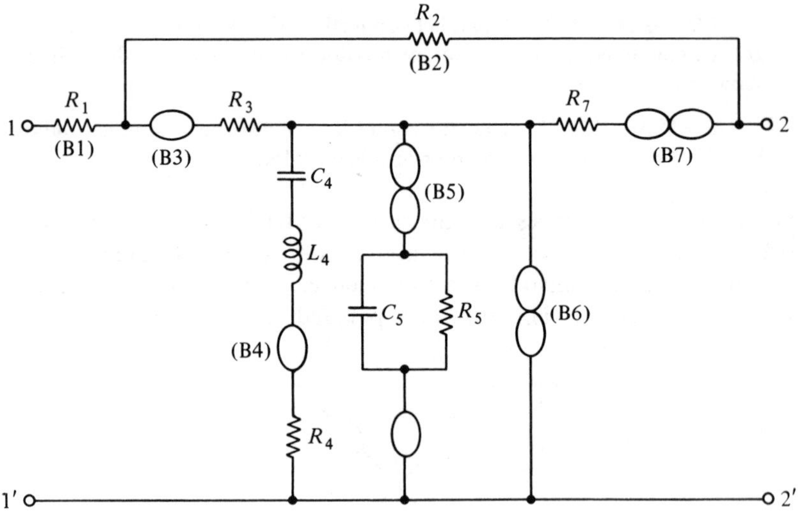

FIGURE 2.12

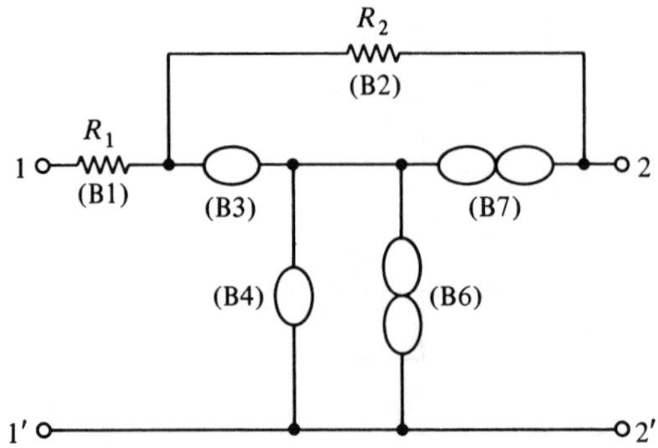

FIGURE 2.13

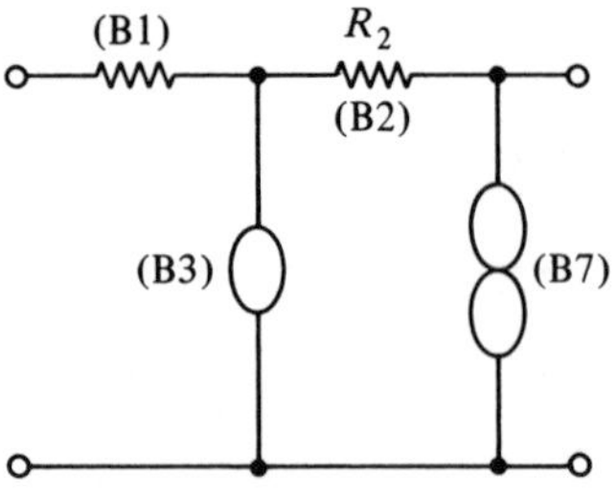

FIGURE 2.14

2-12 INFINITE-GAIN CONTROLLED SOURCES

It is a particularly useful result that

AN INFINITE-GAIN CONTROLLED SOURCE OF ANY OF THE FOUR TYPES IS EXACTLY EQUIVALENT TO A NULLOR.	(2.53)

The reader may check this result by setting $r_2 = \infty$, $r_1 = 0$, $g = \infty$, and $r = \infty$ in the networks shown in Tables 2.8 and 2.9. Then, by means of the above nullor equivalences it can be shown that each network is a nullor if its controlling-factor (or gain) is infinite. For example, the network VCVS1 is shown at the top of Fig. 2.15, with $r_2 = \infty$ and $r_1 = 0$; then by Equivalence 5 this is reduced to a nullor as shown in the diagram.

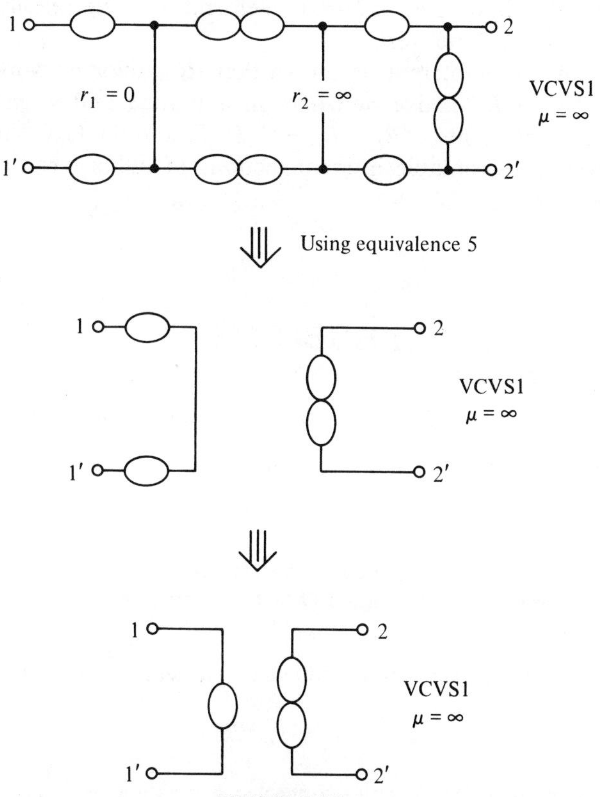

FIGURE 2.15

2-13 SUMMARY

The building-blocks of networks have been classified according to their passivity or activity. The one-port and two-port elements have been considered separately. The conventional *RLC* elements are the passive one-ports; the transformer and the gyrator are the passive lossless two-ports. The $-R$, $-L$, $-C$, and the FDNR D and E elements are classified as the active one-ports, where $-L$ and $-C$ are also lossless. Many of the active two-ports are described by using the concept of a generalized immittance converter (GIC) from which the active-transformer (or PIC) and the active-gyrator (or PII) may be derived by suitably constraining the transmission matrix elements of a general two-port network. The controlled sources are classified as active two-ports and the independent sources as active one-ports; the infinite-gain controlled sources of any of the four types are equivalent to a nullor, and it is shown that all finite-gain controlled sources may be realized using R-nullor networks. Finally, the concepts of nullor equivalence are used to demonstrate that redundant nullors and redundant $\pm R$, $\pm L$, $\pm C$ elements may be readily removed from a $\pm R$, $\pm C$, $\pm L$ network.

In subsequent chapters, it is shown that RC-nullor networks may be used to model all RC-active networks in a unified and straightforward way. The active elements $-R$, $-L$, $-C$, D, E, and the GIC subnetworks are all realizable by suitable interconnections of nullors, resistances and capacitances.

PROBLEMS

2-1. Prove that the ideal inductance element is both passive and lossless.

2-2. Prove that the Q-factor of a capacitance element C is given by

$$Q_C = \omega C r_p$$

where r_p is the effective parallel-connected dissipation resistance of Fig. 2.3. A capacitance of 1000 pF has a Q-factor Q_C of 2000 at 1 kHz. Calculate r_p at 1 kHz.

2-3. Prove that the Q-factor of an inductance element L is given by

$$Q_L = \frac{\omega L}{r_s}$$

where r_s is the effective series-connected dissipation resistance of Fig. 2.3.

An inductance of 1 mH has a Q-factor Q_L of 500 at 10 kHz. Calculate r_s at 10 kHz.

2-4. Given that a nonideal inductor L is exactly equivalent to the model in Fig. 2.3, we wish to derive an *equivalent* model consisting of a *parallel connection* of an inductance L' and resistance r_p'. Derive expressions for L' and r_p' in terms of L, r_s and ω. Comment on these expressions under the condition that $Q_L \gg 1$.

2-5. An ideal gyrator is terminated in a voltage source $e(t)$ at port 1 and a resistance R at port 2. The voltage source delivers power $p_e(t)$ to the network as follows:

$$p_e(t) = e^{-t}\, u(t) \text{ watts}$$

Derive an expression for the current $i(t)$ in the resistance R.

2-6. An ideal gyrator with transconductance g mho is terminated at port 2 in a series connection of an inductance L and resistance r_s. Prove that the driving-point impedance at port 1 is equivalent to a parallel connection of a capacitance C and resistance r_p. Derive expressions for C and r_p in terms of g, r_s, and L.

2-7. An ideal gyrator having transconductance g is terminated at port 2 with a nonideal capacitance having Q-factor Q_C. Derive the Q-factor Q_L of the effective nonideal inductance at port 1, and obtain an expression for the nonideal series resistance r_s of this inductance in terms of g, C, and Q_C.

2-8. A second-order tuned circuit is realized by terminating the ports of a gyrator as shown in Fig. P2.8. Derive an expression for the Q-factor Q_P of this circuit. How does the resonant frequency ω_P depend on the transconductance g?

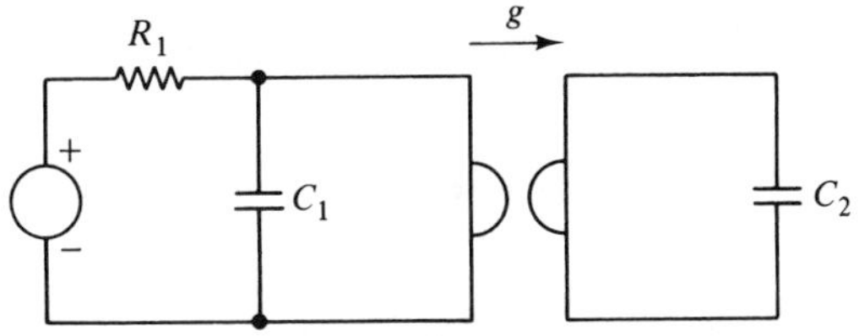

PROBLEM 2.8

2-9. A one-port network consists of a parallel connection of a 1000 pF capacitor and a 50 kΩ resistor. This network is excited by a steady-state sinusoidal voltage generator $v_1(t) = V \sin \omega t$, where $V = 1$ volt and $\omega = 10^5$ rad/sec. Assuming that the response of the network is in the sinusoidal steady-state, derive an expression for the energy delivered to the network since the time $t = 0$; thus, calculate $E(t) - E(0)$, $t > 0$. Sketch the resultant expression. Relate the sketch to the definition of the Q-factor Q_C of a capacitor.

2-10. Prove that both of the negative elements $-L$ and $-C$ are active and lossless.

2-11. Prove that the negative resistance element $-R$ is active and, furthermore, that it cannot absorb energy under any circumstances.

2-12. A network consists of a parallel connection of a capacitance C and a negative resistance $-R$. The voltage $v(t)$ across these two elements at time $t=0$ is written $v(0)$. Derive an expression for $v(t)$ in terms of $v(0)$, C and $-R$. Comment on the stability of the network.

2-13. A network consists of a parallel connection of an ideal *negative* resistance $-R$ and a nonideal capacitor C having Q-factor Q_P. Show that the effective Q-factor of this parallel connection is given by

$$Q_P' = \frac{Q_A Q_P}{Q_A + Q_P}$$

where $Q_A \equiv -\omega CR$. The term Q_A may be thought of as introducing an "active negative Q-factor Q_A." Comment on the stability of the network under three conditions:

$$-Q_A > Q_P, \qquad -Q_A < Q_P, \qquad -Q_A = Q_C$$

2-14. An FDNR element is given by

$$Z(j\omega) = -\left(\frac{\omega_0}{\omega}\right)^2 R_D, \qquad \omega_0 \text{ constant}$$

Sketch the functions $|Z(j\omega)|$ and $\arg Z(j\omega)$.

2-15. The FDNR element of Problem 2.14 is terminated in a resistance R. Derive the impedance of this parallel connection of $Z(j\omega)$ and R. By sketching the magnitude and phase of this impedance, investigate the behavior of the network as a resonant circuit. What are the Q-factor and the resonant frequency?

2-16. Let the voltage across the parallel connection in Problem 2.15 be $v(t)$ and let the total current flowing into this parallel connection be $i(t)$. Write the differential equation relating $v(t)$ and $i(t)$. With $v(t) \equiv V \sin \omega t$ and $\omega \equiv \omega_0 \sqrt{R_D/R}$, derive $i(t)$ from the differential equation above.

2-17. Prove that an inductance element L is passive and lossless.

2-18. Derive equation (2.30) from equation (2.27).

2-19. Prove that the NIC and the NII are both *active* two-ports.

2-20. A GIC has the transmission matrix

$$T = \begin{bmatrix} 1 & 0 \\ 0 & s+a \end{bmatrix}, \qquad a \quad \text{constant}$$

and is terminated in a capacitance C at port 2. Derive an expression for the input impedance at port 1.

2-21. A GIC has the transmission matrix

$$T=\begin{bmatrix} 1 & 0 \\ 0 & s^{-1} \end{bmatrix}$$

and is terminated at port 2 in a resistance R. Derive the input impedance at port 1.

2-22. The resistance R in Problem 2.21 is removed from port 2 and is instead used to terminate port 1. Derive the input impedance at port 2.

2-23. Derive the transmission matrices for each of the networks shown in Table 2.8 and, by comparison with equations (2.49), (2.50), (2.51), and (2.52), prove that these networks are correctly classified in Table 2.8 as controlled sources.

2-24. Repeat Problem 2.23 for Table 2.9.

2-25. Prove Equivalences 3, 4, 5, 6 and 7 of section 2.11.

2-26. Use the concepts of nullor equivalence to eliminate all redundant elements from the networks of Fig. P2.26:

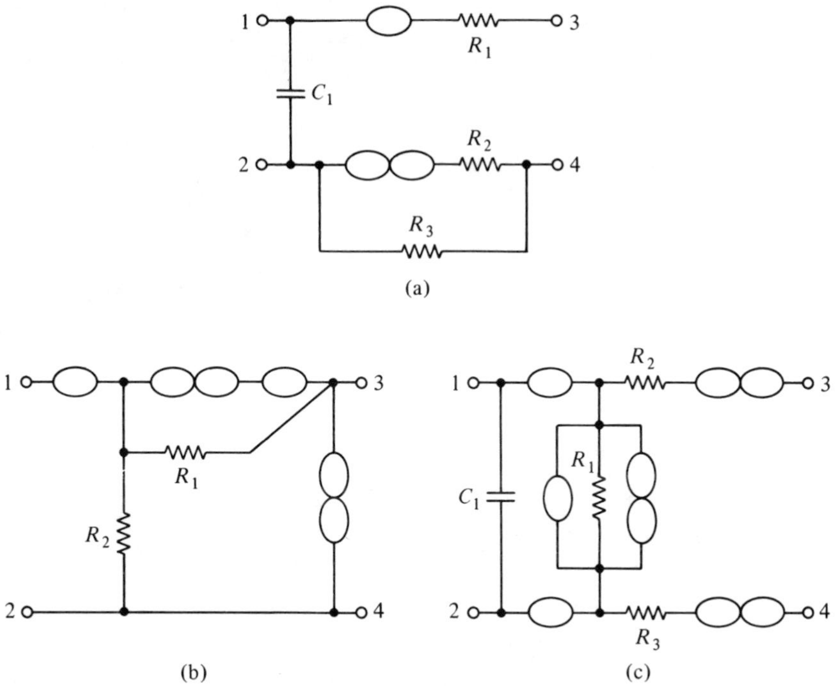

PROBLEM 2.26

BIBLIOGRAPHY

BENDIK, J., "Equivalent Gyrator Networks with Nullators and Norators," *IEEE Trans. Circuit Theory*, vol. CT-14, p. 908. 1967.

BRUTON, L. T., "Network Transfer Functions Using the Concept of Frequency Dependent Negative Resistance" *IEEE Trans. Circuit Theory*, vol. CT-16, p. 406–408, Aug. 1969.

DAVIES, A. C., "The Significance of Nullators, Norators and Nullors in Active Network Theory," *Radio Electron. Eng*., vol. 34, p. 259–267, 1967.

VAN VALKENBURG, M. E., *Network Analysis* 3rd Edition, Prentice-Hall, 1974, chapters 1 and 2.

3

ANALYSIS OF ACTIVE NETWORKS USING NULLORS

3-1 INTRODUCTION

The primary purpose of this chapter is to describe two *systematic* techniques that may be used to obtain and solve the network equations of *RLC*-nullor networks. These techniques are known as *nodal analysis* and *loop analysis*, and they are used to find the node voltages and the loop currents, respectively, of an arbitrary interconnection of inductance, resistance, capacitance and *nullor* elements.

The concept of the nullor as a network element is useful in the synthesis and analysis of active networks because the controlled sources and all of the GICs may be modelled by using *RC*-nullor networks, thereby implying that *RC-nullor networks may be used to model any interconnection of the basic elements or subnetworks that have been described in the previous chapter.*

The material in this chapter is *not* a prerequisite to any in subsequent chapters of this book. Consequently, if the reader is interested primarily in the modeling, synthesis, and design of *practical RC*-active filter networks it is possible to omit this chapter without loss of continuity. However, knowledge of the following methods of nodal and loop analysis is essential for the reader who wishes to analyze complicated *arbitrary* topological interconnections of nullors and L, C, and R elements.

It will be recalled that the GIC is defined in section 2.6.1 but, so far, the nullor modeling of GICs has not been described because it is preferable that the reader be able to analyze the nullor models of GICs in some detail. Therefore, this subject is introduced in section 3.5 of this chapter *after* the formal techniques of loop and nodal analysis are explained.

Finally, the problems at the end of the chapter are pertinent to an understanding of nullor networks. Many of the nullor networks in the problems form the basis of widely used transistor circuits and operational amplifier circuits.

3-2 NODAL ANALYSIS OF *RLC*-NULLOR NETWORKS

The nodal analysis method for *RLC*-nullor networks that is described in this section was proposed by A.C. Davies. It provides a systematic method for solving for the N independent nodal voltages $V_1, V_2, \ldots, V_j, \ldots, V_N$ of an $(N + 1)$ node network, where the network consists of an arbitrary interconnection of *RLC* elements, independent current sources and nullors. The method will be understood most readily by the reader who is familiar with the technique of nodal analysis that is used in the case of *passive RLC* networks because the Davies method is simply an extension of passive nodal analysis. In the following section, the usual technique of nodal analysis of *passive RLC* networks is reviewed briefly.

3-2.1 Review of the Nodal Analysis of Passive *RLC* Networks

The steps involved in the nodal analysis of an arbitrary interconnection of *RLC* elements and independent current sources in an $(N + 1)$ node network are well known. They are as follows:

STEP [1]: A datum reference node is selected and labeled node 0. Then, the procedure is to solve for the remaining node voltage with respect to the datum reference node.

STEP [2]: All other nodes are labeled from 1 to N. Then, the problem is to solve for the corresponding node voltages $V_1, V_2, V_3, \ldots, V_j, \ldots, V_N$, where these voltages are with respect to node 0.

STEP [3]: The nodal equations contain the current column vector $\mathbf{I} = \{I_1, I_2, \ldots, I_i, \ldots, I_N\}'$, where the ith component I_i is defined as the sum of the currents flowing into the ith node from the independent current sources. Therefore, each of the I_i terms is usually written by inspection and the current vector $\mathbf{I}$ is thereby obtained.

STEP [4]: The nodal admittance matrix $\mathbf{Y}_{N \times N} \equiv \{y_{ij}\}$ has dimensions $(N \times N)$ and may usually be written by inspection, using

$$y_{ii} \equiv \text{sum of transform admittances connected to node } i \tag{3.1}$$

$$-y_{ij} \equiv \text{sum of transform admittances connected between nodes } i \text{ and } j \tag{3.2}$$

It follows, therefore, that $\mathbf{Y}_{N \times N}$ is a *symmetric* matrix with negative off-diagonal terms of the form

$$y_{ij} = -\left[g_{ij} + sC_{ij} + \frac{1}{sL_{ij}} \right] = y_{ji}, \qquad j \neq i$$

Furthermore,

$$y_{ii} = \left[g_{ii} + sC_{ii} + \frac{1}{sL_{ii}} \right]$$

STEP [5]: The nodal equations of the network are then written in the matrix form

$$\mathbf{I} = \mathbf{Y}_{N \times N}\mathbf{V} \tag{3.3}$$

where $\mathbf{V}$ is the unknown column vector $\{V_1, V_2, V_3, \ldots, V_j, \ldots, V_N\}'$ of node voltages. Equation (3.3) is a set of N linear equations in N independent unknowns $V_1, \ldots, V_N$ for which a unique solution for $V_1, V_2, V_3, \ldots, V_N$ is obtainable by several well-known systematic procedures; for example, by using Cramer's rule or by Gauss elimination.

EXAMPLE 3.1: Derive the nodal equations for the passive network shown in Fig. 3.1(a).

SOLUTION: The first and second steps, which are the numbering of the nodes, are performed as in Fig. 3.1(a), with the datum reference node selected as shown. The method achieves a solution for the node voltages V_1, V_2, V_3 at nodes 1, 2, 3 with respect to node 0.

The third step results in the (3×1) $\mathbf{I}$ vector,

$$\mathbf{I} = \begin{bmatrix} 0 \\ +I_B \\ I_A - I_B \end{bmatrix} \tag{3.4}$$

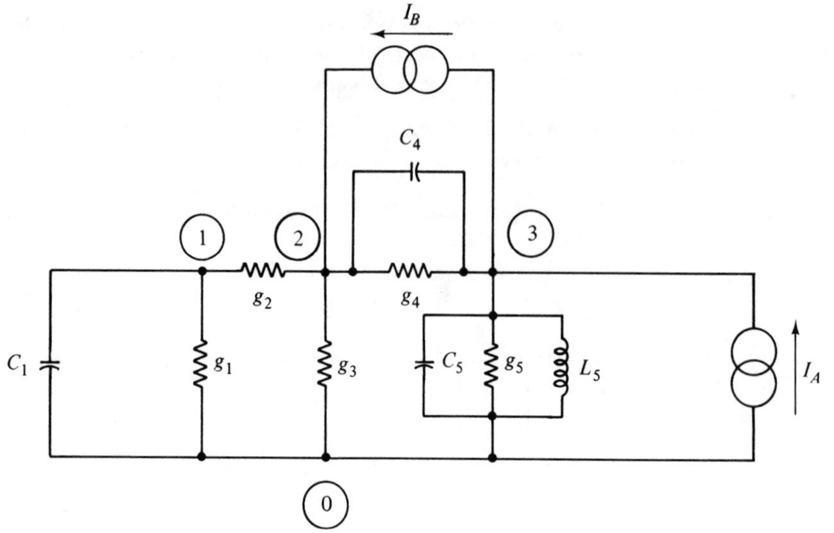

I_B
C_4
1
2
3
g_2
g_4
C_1
g_1
g_3
C_5
g_5
L_5
I_A
0

(a)

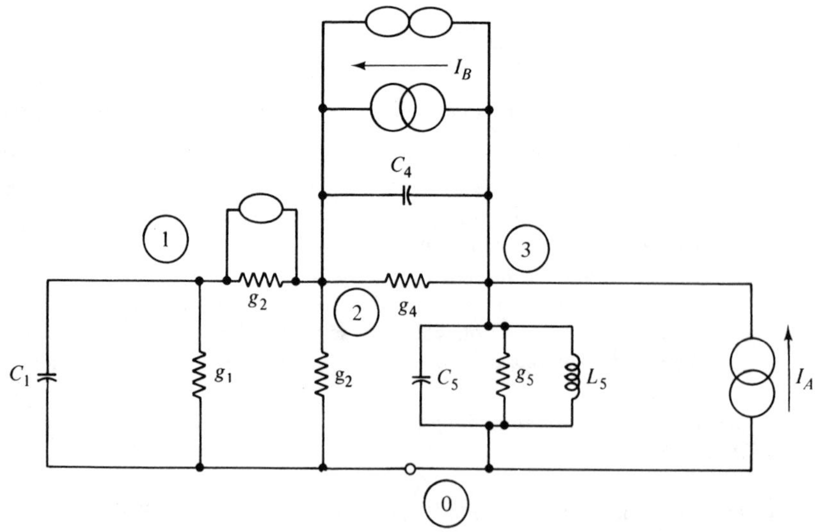

I_B
C_4
1
3
g_2
2
g_4
C_1
g_1
g_2
C_5
g_5
L_5
I_A
0

(b)

FIGURE 3.1

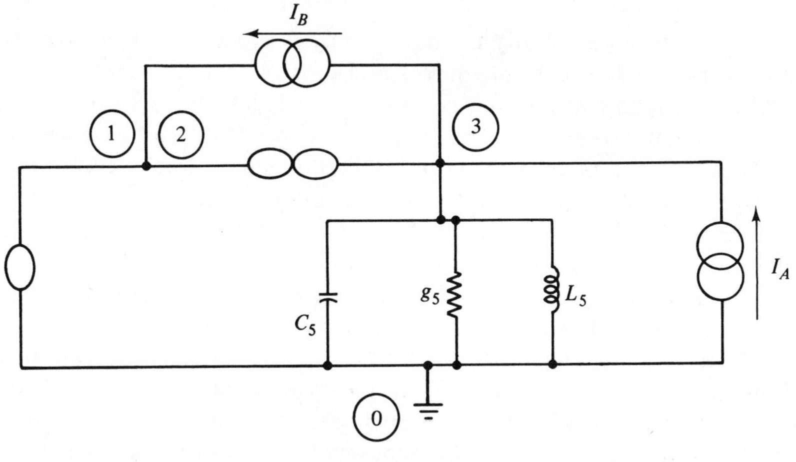

(c)

FIGURE 3.1 *(Continued).*

and the fourth step results in the $\mathbf{Y}_{3\times 3}$ nodal admittance matrix,

$$\mathbf{Y}_{3\times 3} = \begin{bmatrix} (g_1 + g_2 + sC_1) & -g_2 & 0 \\ -g_2 & (g_2 + g_3 + g_4 + sC_4) & -(g_4 + sC_4) \\ 0 & -(g_4 + sC_4) & \begin{pmatrix} g_5 + sC_5 + 1/sL_5 \\ + g_4 + sC_4 \end{pmatrix} \end{bmatrix} \quad (3.5)$$

leading to the final, fifth, step which is to write the nodal equations

$$\begin{bmatrix} 0 \\ +I_B \\ I_A - I_B \end{bmatrix} = \begin{bmatrix} (g_1 + g_2 + sC_1) & -g_2 & 0 \\ -g_2 & (g_2 + g_3 + g_4 + sC_4) & -(g_4 + sC_4) \\ 0 & -(g_4 + sC_4) & \begin{pmatrix} g_5 + sC_5 + \dfrac{1}{sL_5} \\ + g_4 + sC_4 \end{pmatrix} \end{bmatrix} \begin{bmatrix} V_1 \\ V_2 \\ V_3 \end{bmatrix}$$

(3.6)

which may be solved algebraically for the unknown transform voltages V_1, V_2, V_3.

3-2.2 RLC Elements and K Nullors

Having reviewed briefly the nodal analysis procedure for interconnections of independent current sources and passive *LCR* elements, it is now possible for us to consider the procedure that is involved in obtaining the nodal equations when there are *K* nullors connected arbitrarily between the nodes of the passive network. With the nullors removed from the network, the remaining passive nodal admittance matrix for an $(N + 1)$ node network has dimensions of $N \times N$ as described in step [4] above. The situation is illustrated diagrammatically in Fig. 3.2(a); a number of the $(N + 1)$ nodes is shown external to the network, and it is to these nodes that nullators and norators are connected. Recall that prior to the introduction of nullators and norators at these external nodes, the nodal admittance matrix $\mathbf{Y}_{N \times N} \equiv \{y_{ij}\}$ is the symmetric matrix described in step [4] above. A *single nullator* is now connected between nodes ⓟ and ⓠ as shown in Fig. 3.2(b). The consequence of this is that the voltage between nodes ⓟ and ⓠ is zero; that is, $V_p = V_q$ with respect to the datum reference node. Since V_p and V_q are now identical, we replace them by the variable $V_{pq} \equiv V_p = V_q$. The nodal equations now become

$$\begin{bmatrix} I_1 \\ I_2 \\ I_3 \\ \vdots \\ I_N \end{bmatrix} = \begin{bmatrix} y_{11} & y_{12} & \cdots & y_{1N} \\ y_{21} & & & \vdots \\ \vdots & & & \\ \vdots & & & \vdots \\ y_{N1} & \cdots & \cdots & y_{NN} \end{bmatrix} \begin{bmatrix} V_1 \\ V_2 \\ \vdots \\ V_{pq} \\ \vdots \\ V_{pq} \\ \vdots \\ V_N \end{bmatrix} \begin{matrix} \\ \\ \\ \leftarrow \text{replaces } V_p \\ \\ \leftarrow \text{replaces } V_q \\ \\ \\ \end{matrix} \tag{3.7}$$

where the two V_{pq} entries in the $\mathbf{V}$ column vector replace the entries V_p and V_q that existed prior to embedding the nullator. Now, equation (3.7) may be rewritten in terms of a reduced matrix of dimensions $N \times (N - 1)$ by adding the qth column of $\mathbf{Y}_{N \times N}$ to the pth column of $\mathbf{Y}_{N \times N}$ and then deleting the qth column. The resultant admittance matrix, denoted

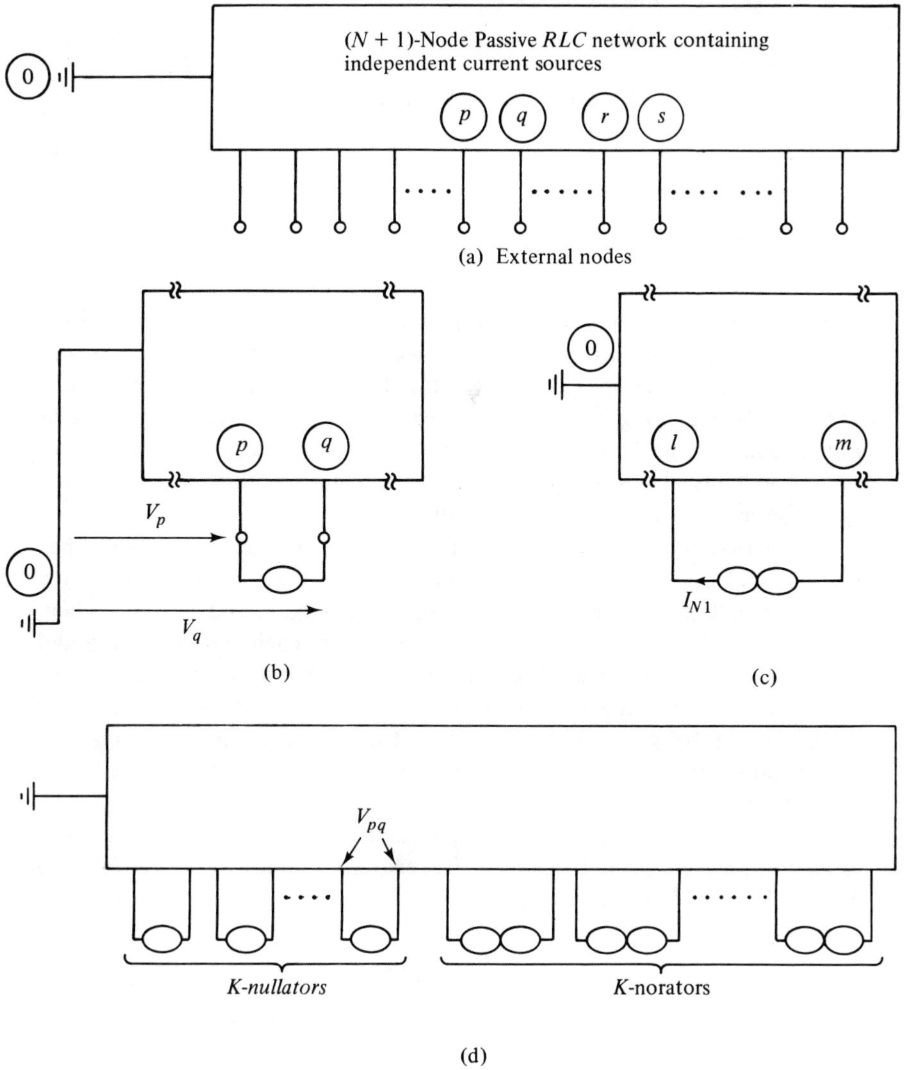
(N + 1)-Node Passive RLC network containing independent current sources
0
p
q
r
s
(a) External nodes
p
q
Vp
Vq
0
(b)
0
l
m
IN1
(c)
Vpq
K-nullators
K-norators
(d)

FIGURE 3.2

$\mathbf{Y}_{N,\,N-1}$, describes this system of equations:

$$\begin{bmatrix} I_1 \\ I_2 \\ I_3 \\ \vdots \\ I_N \end{bmatrix} = \begin{bmatrix} y_{11} & y_{12} & \cdots & (y_{1p} + y_{1q}) & y_{1,p+1} & \cdots & y_{1N} \\ y_{21} & y_{22} & & (y_{2p} + y_{2q}) & y_{2,p+1} & \cdots & y_{2N} \\ \vdots & & & & & & \vdots \\ y_{N1} & y_{N2} & \cdots & (y_{Np} + y_{Nq}) & y_{N,p+1} & \cdots & y_{NN} \end{bmatrix} \begin{bmatrix} V_1 \\ V_2 \\ \vdots \\ V_{pq} \\ \vdots \\ V_N \end{bmatrix} \tag{3.8}$$

where the pth column is now the sum of the previous pth and qth columns and where the previous qth column has been deleted. Note that the number of equations in (3.8) now exceeds by one the number of unknowns. *The preceding process is now continued until all K nullators have been introduced* into the network and the resulting nodal admittance matrix $\mathbf{Y}_{N,\,N-K}$ will have been formed by adding all pairs of columns corresponding to the nodal connections of the nullators.

Note that the possibility exists that at least one of the nullators has one terminal connected to the datum reference node 0. Suppose that a particular nullator is connected between the kth node and node 0. Then, clearly, $V_k = 0$ and therefore the term $y_{ik}V_k$ in each row of the nodal equations may be deleted. That is, the kth column of the admittance matrix is simply deleted if a nullator is introduced between the kth node and the datum reference node. Also note tthat the consequence is again to simply reduce the dimension of $\mathbf{Y}$ by one column for each nullator in the network.

The K nullators have now been added to the network. The first norator is now connected between, for example, nodes l and m. *Prior to the introduction* of the norator at nodes l and m, as shown in Fig. 3.2(c), the corresponding row equations have the form

$$I_l = \sum_r y_{lr} V_r \qquad \text{Nodal Equation At } l \text{ Before Introduction of a Norator} \tag{3.9}$$

and

$$I_m = \sum_r y_{mr} V_r \qquad \text{Nodal Equation At } m \text{ Before Introduction of a Norator} \tag{3.10}$$

where r indicates summation over all elements in that row and where V_r are the unknown components of the reduced $(N - K) \times 1$ node voltage vector. *After the introduction* of the first norator, it follows that the nodal equations at nodes l and m are modified by the unknown current I_{N1} that flows in this norator. This situation is shown in Fig. 3.2(c). Therefore,

$$I_l + I_{N1} = \sum_r y_{lr} V_r \quad \text{at node } l \tag{3.11}$$

and

$$I_m - I_{N1} = \sum_r y_{mr} V_r \quad \text{at node } m \tag{3.12}$$

Recall that we are attempting to solve for the node voltages; therefore, we do not particularly wish to retain I_{N1} as an unknown variable; it is eliminated by addition of equations (3.11) and (3.12) to give

$$I_l + I_m = \sum_r (y_{lr} + y_{mr}) V_r \tag{3.12A}$$

which is entirely equivalent to adding the lth row of the admittance matrix to the mth row and then deleting the lth row. The admittance matrix has therefore been reduced in dimension by one row as a result of introducing this first norator, so that it now has the form $\mathbf{Y}_{N-1,\,N-K}$. Continuing this process of adding rows of the admittance matrix that correspond to nodes to which norators are connected eventually leads, after K norators have been added, to an $(N - K) \times (N - K)$ nodal admittance matrix corresponding to $(N - K)$ equations in $(N - K)$ independent unknown node voltages describing the general $(N + 1)$ node K-nullor network shown in Fig. 3.2(d).

The reader may have noticed that this procedure of reducing rows in correspondence with the norator connections does not apply in the case of a norator that is connected between a node, say the ith node, and the datum reference node because it is not possible to identify a second row. The equation at this ith node is given by

$$I_i + I_{Ni} = \sum_r y_{ir} V_r \tag{3.13}$$

where I_i is the current entering node i from the independent current sources and I_{Ni} is the current entering node i from the norator. It is now argued that equations of the form shown in equation (3.13) are redundant and therefore the corresponding row may be deleted from the set of nodal equations. Assuming P norators that have one terminal connected to the datum node there are therefore $(N - K + P)$ equations in the $(N - K)$ unknown independent node voltages and in the P unknown independent norator currents I_{Ni}. However, the P equations containing the I_{Ni} terms are *not* required because we do not require a solution for the unknown currents I_{Ni}. Deleting all P of these equations leaves the *solvable* set of $(N - K)$ equations in $(N - K)$ unknown independent node voltages $V_1, V_2, \ldots, V_{pq}, \ldots, V_N$.

The above procedure for analyzing RC-nullor networks is now summarized:

STEP [1]: Remove all K nullors from the network, leaving a passive $(N + 1)$ node network, one node of which is the datum reference node.

STEP [2]: Write the $(N \times N)$ nodal admittance matrix equations for the remaining $(N + 1)$ node *passive* network. $\mathbf{Y}_{N\times N}$ is the passive nodal admittance matrix.

STEP [3]: For a nullator that was connected between the nodes p and q, for example, delete column q of $\mathbf{Y}$ and add the elements of column q to the elements of column p. The number of columns of the $\mathbf{Y}$ matrix is thereby reduced to $(N - 1)$. Repeat this process for every nullator not connected to the datum reference node.

STEP [4]: For a norator that was connected between the nodes l and m, for example, delete row l of the nodal equations and add the equation in row l to the equation in row m. This involves adding I_m to I_l and the mth row of the admittance matrix to the lth row of the admittance matrix. Repeat this process for every norator not connected to the datum reference node.

STEP [5]: For a nullator that is connected between, say, node k and the datum reference node, delete the kth column of the admittance matrix. Repeat this process for every nullator connected to the datum reference node.

STEP [6]: For a norator that is connected between, say, node i and the datum reference node, delete the equation in the ith row. This involves deleting I_i and deleting the ith row of the admittance matrix equations. Repeat this process for every norator.

STEP [7]: The preceding six steps result in the reduction of the $(N \times N)$ nodal admittance matrix of the passive network to the $(N - K) \times (N - K)$ nodal admittance matrix of the K-nullor $(N + 1)$ node RLC-nullor network. The corresponding $(N - K)$ equations may be solved for the $(N - K)$ independent node voltages. (The norator currents I_{Ni} and the branch currents are not obtained directly by this method.)

The method that has been described is now applied to several examples.

EXAMPLE 3.2: A single nullor network is created from Fig. 3.1(a) by connecting a nullator between nodes ① and ② and a norator between nodes ② and ③, as shown in Fig. 3.1(b). Derive the nodal equations for the resultant single-nullor network of Fig. 3.1(b) and thereby derive an expression for the node voltage V_3.

SOLUTION: The passive 3 × 3 nodal admittance matrix has already been calculated in equation (3.6). The nullator between nodes ① and ② requires addition of the first and second columns, as follows:

$$\begin{bmatrix} 0 \\ I_B \\ I_A - I_B \end{bmatrix} = \begin{bmatrix} (g_1 + sC_1) & 0 \\ (g_3 + g_4 + sC_4) & -(g_4 + sC_4) \\ -(g_4 + sC_4) & (g_4 + sC_4 + g_5 + sC_5 + 1/sL_5) \end{bmatrix} \begin{bmatrix} V_{1,2} \\ V_3 \end{bmatrix} \tag{3.14}$$

and the norator between nodes 2 and 3 requires the third row of the equation above to be added to the second row and subsequent deletion of the redundant third row, giving

$$\begin{bmatrix} 0 \\ I_A \end{bmatrix} = \begin{bmatrix} (g_1 + sC_1) & 0 \\ g_3 & (g_5 + sC_5 + 1/sL_5) \end{bmatrix} \begin{bmatrix} V_{1,2} \\ V_3 \end{bmatrix} \tag{3.15}$$

as the required nodal equations of the single-nullor network. Solving for V_3, we obtain

$$V_3 = \frac{I_A}{g_5 + sC_5 + \dfrac{1}{sL_5}} \tag{3.16}$$

EXAMPLE 3.3: Repeat the problem above, but this time obtain V_3 by employing the nullor equivalences of section 2.11 to simplify the nullor network prior to writing the nodal equations.

SOLUTION: By Equivalence 4 the parallel connection of a nullator and g_2 allows g_2 to be removed; by Equivalence 1 the parallel combination of C_1 and g_1 is replaced by a short-circuit; by Equivalence 3 the parallel combination of C_4 and g_4 is replaced by an open-circuit. The resultant simplified network is shown in Fig. 3.1(c) and has nodal equation

$$I_A = V_3\left(sC_5 + g_5 + \frac{1}{sL_5}\right) \tag{3.17}$$

or

$$V_3 = \frac{I_A}{\left(sC_5 + g_5 + \dfrac{1}{sL_5}\right)} \tag{3.18}$$

Equation (3.18) agrees with the result in equation (3.16).

EXAMPLE 3.4: Derive the nodal equations for the network in Fig. 3.3 and, thereby, prove that the driving-point admittance Y_{1d} is as follows:

$$Y_{1d} = -\frac{Y_2 Y_4}{Y_3} \tag{3.19}$$

Use this result to explain the possible application of this network as a GIC.

SOLUTION: Removing the nullor shown in Fig. 3.3 and writing the nodal equations for the passive network, we obtain

$$\begin{bmatrix} I_A \\ 0 \\ 0 \end{bmatrix} = \begin{bmatrix} Y_2 & -Y_2 & 0 \\ -Y_2 & (Y_2 + Y_3) & -Y_3 \\ 0 & -Y_3 & (Y_3 + Y_4) \end{bmatrix} \begin{bmatrix} V_1 \\ V_2 \\ V_3 \end{bmatrix} \tag{3.20}$$

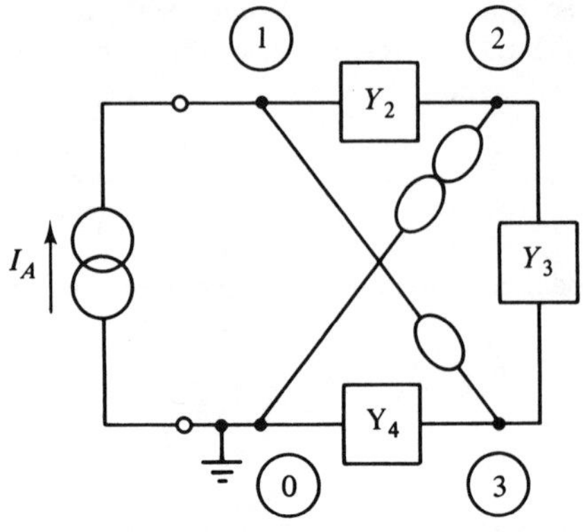

FIGURE 3.3

Adding the first and third column (due to the nullator) and deleting the second equation (due to the norator between ② and ⓪), we have

$$\begin{bmatrix} I_A \\ 0 \end{bmatrix} = \begin{bmatrix} Y_2 & -Y_2 \\ (Y_3 + Y_4) & -Y_3 \end{bmatrix} \begin{bmatrix} V_{1,3} \\ V_2 \end{bmatrix} \tag{3.21}$$

which may be solved for $V_{1,2}$ to obtain

$$V_{1,3} = -\frac{I_A Y_3}{Y_2 Y_4}$$

so that the driving-point admittance $Y_{1d} \equiv I_A / V_{1,3}$ is

$$Y_{1d} = -\frac{Y_2 Y_4}{Y_3} \tag{3.22}$$

The network is a GIC by virtue of the definition in section 2.6.1. It is an impedance-converting type if Y_2 or Y_4 is regarded as the termination and an impedance inverting type if Y_3 is regarded as the termination with conversion functions $-Y_{2,4}/Y_3$ and $-Y_2Y_4$, respectively.

3-3 LOOP ANALYSIS OF *RLC*-NULLOR NETWORKS

The technique of loop analysis of nullor networks was proposed by A.C. Davis. The method is quite similar to the loop analysis method for *passive* *RLC* networks. In this section, the loop analysis of a *planar* network is summarized without proof. The network is assumed to consist only of *RLC* elements, independent voltage sources and nullors. The network has

L loops (or meshes) and K nullors, and the objective is to determine the N unknown loop currents $I_1, I_2, \ldots, I_N$:

STEP [1]: Label the N loops of the nullor network $I, II, III, \ldots,$ etc., clockwise with unknown loop currents $I_1, I_2, I_3, \ldots, I_N$. The objective is to calculate the loop currents.

STEP [2]: *Replace each norator and each nullator by a short-circuit,* thereby resulting in a general N-loop *passive* network. *Write loop equations of this passive network* in the form

$$\mathbf{E} = \mathbf{ZI} \tag{3.23}$$

where $\mathbf{E} = \{E_1, E_2, \ldots, E_N\}'$ is the column vector of known independent loop voltages and $\mathbf{I} = \{I_1, I_2, \ldots, I_N\}'$ *is the column vector of unknown loop currents and* $\mathbf{Z}$ *is the* $(N \times N)$ *loop impedance matrix.*

STEP [3]: *Add pairs of columns of* $\mathbf{Z}_{N \times N}$ *that correspond to loop pairs that share nullators in the original nullor network.* Thus, if a nullator appears between loops j and k in the nullor network, then it follows that $I_j = I_k \equiv I_{jk}$, so that column j may be added to column k and then column j may be deleted.

STEP [4]: *Add pairs of rows of* $\mathbf{E} = \mathbf{ZI}$ *that correspond to loop pairs that share norators in the original nullor network.* The validity of this step is explained as follows: if a norator appears between loops p and q and if the voltage across the norator is V_N, then the KVL equations for loops p and q are

$$E_p + V_N = \sum_r Z_{rp} I_r \tag{3.24}$$

and

$$E_q - V_N = \sum_r Z_{rq} I_r \tag{3.25}$$

so that, by addition of the pth and qth rows,

$$E_p + E_q = \sum_r (Z_{rq} + Z_{rp}) I_r \tag{3.26}$$

STEP [5]: *Delete columns of* $\mathbf{Z}$ *corresponding to nullators that are* ***not*** *shared between loop pairs.* The validity of this step is proven readily, because a nullator in loop p that is *not* shared with some other loop immediately implies that $I_p = 0$ and, therefore, allows the deletion of the pth column.

STEP [6]: *Delete rows of* $\mathbf{E} = \mathbf{ZI}$ *corresponding to norators that are* ***not*** *shared between loop pairs.* The validity of this step is justified because it always follows that a norator in loop k that is *not* shared with some other loop results in a *redundant* loop equation of the form

$$E_k + V_N = \sum_r Z_{rk} I_r \tag{3.27}$$

Elimination of equations of this type always results in $(N - K)$ remaining equations in $(N - K)$ independent loop currents, where K is the number of nullors. The resultant equations are the required loop equations of the K-nullor N-loop network.

The preceding procedure is now demonstrated by means of several examples.

EXAMPLE 3.5: Derive the loop equations for the single-nullor network shown in Fig. 3.4(a).

SOLUTION: Loops, *I*, *II*, and *III* with loop currents I_1, I_2, and I_3 are identified in Fig. 3.4(a). Employing step [2], obtain the passive network shown in Fig. 3.4(b) and write the loop equations for this network; it has three loops ($N = 3$), and the loop equations are

$$\begin{bmatrix} E_1 \\ 0 \\ 0 \end{bmatrix} = \begin{bmatrix} (R_1 + R_2) & -R_2 & -R_1 \\ -R_2 & (R_2 + R_3 + R_4) & -R_3 \\ -R_1 & -R_3 & (R_1 + R_3) \end{bmatrix} \begin{bmatrix} I_1 \\ I_2 \\ I_3 \end{bmatrix} \tag{3.28}$$

By step [3], the nullator shown in Fig. 3.4(a) is shared between loops *II* and *III*, so the third column of the matrix above may be added to the second column, thereby creating the variables $I_{2,3} = I_2 = I_3$ in the reduced loop equations,

$$\begin{bmatrix} E_1 \\ 0 \\ 0 \end{bmatrix} = \begin{bmatrix} (R_1 + R_2) & -(R_1 + R_2) \\ -R_2 & (R_2 + R_4) \\ -R_1 & R_1 \end{bmatrix} \begin{bmatrix} I_1 \\ I_{2,3} \end{bmatrix} \tag{3.29}$$

Step [4] is omitted because the only norator is *not* shared between loop pairs. Step [5] is omitted because the only nullator *is* shared between loop pairs. Step [6] is applied to the norator which is only associated with one loop current, I_3, therefore requiring deletion of the third row of the equation above, giving the required two equations in two unknowns $(N - K = 3 - 1 = 2)$

$$\begin{bmatrix} E_1 \\ 0 \end{bmatrix} = \begin{bmatrix} (R_1 + R_2) & -(R_1 + R_2) \\ -R_2 & (R_2 + R_4) \end{bmatrix} \begin{bmatrix} I_1 \\ I_{2,3} \end{bmatrix} \tag{3.30}$$

as the required loop equations of the nullor network shown in Fig. 3.4(a).

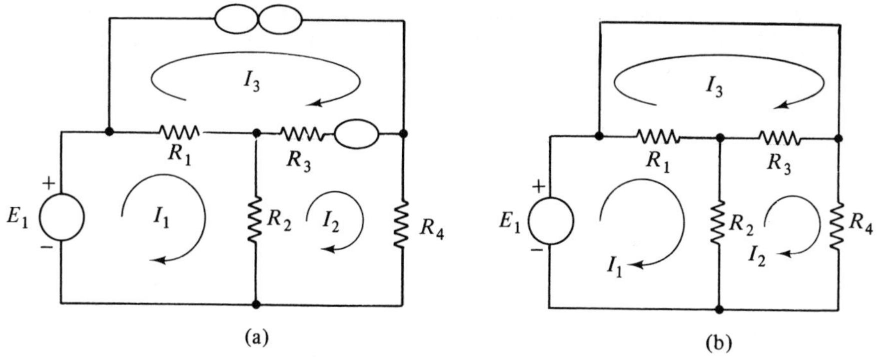

FIGURE 3.4

EXAMPLE 3.6: Use loop analysis to prove that, for the two-nullor GIC network shown in Fig. 3.5(a),

$$V_6 = -\frac{E_1 Z_3}{Z_2} \tag{3.31}$$

and

$$Z_{1d} = \frac{Z_2 Z_4 Z_6}{Z_3 Z_5} \tag{3.32}$$

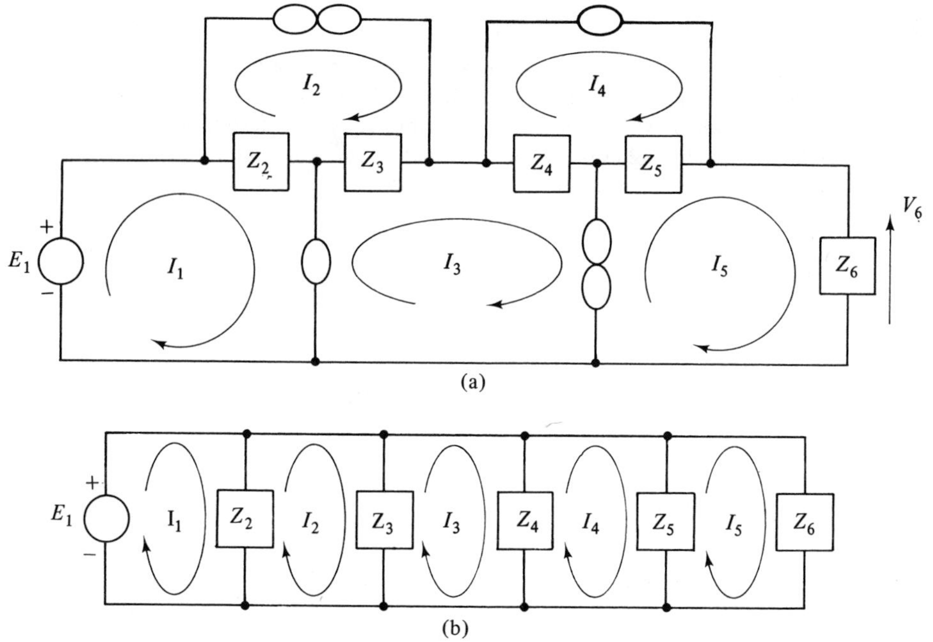

FIGURE 3.5

where Z_{1d} is the driving-point impedance at the port to which E_1 is connected.

SOLUTION: Short-circuiting all nullators and norators leads to Fig. 3.5(b), for which the loop equations are

$$\begin{bmatrix} E_1 \\ 0 \\ 0 \\ 0 \\ 0 \end{bmatrix} = \begin{bmatrix} Z_2 & -Z_2 & 0 & 0 & 0 \\ -Z_2 & (Z_2 + Z_3) & -Z_3 & 0 & 0 \\ 0 & -Z_3 & (Z_3 + Z_4) & -Z_4 & 0 \\ 0 & 0 & -Z_4 & (Z_4 + Z_5) & -Z_5 \\ 0 & 0 & 0 & -Z_5 & (Z_5 + Z_6) \end{bmatrix} \begin{bmatrix} I_1 \\ I_2 \\ I_3 \\ I_4 \\ I_5 \end{bmatrix} \tag{3.33}$$

Steps [3] and [4] of the loop analysis procedure refer to the nullator between loops *I* and *III* and the norator between loops *III* and *V*. In the equation above, column 3 is added to column 1 and then column 3 is deleted; row 5 is added to row 3 and then row 5 is deleted. Thus, we have

$$\begin{bmatrix} E_1 \\ 0 \\ 0 \\ 0 \end{bmatrix} = \begin{bmatrix} Z_2 & -Z_2 & 0 & 0 \\ -(Z_2 + Z_3) & (Z_2 + Z_3) & 0 & 0 \\ (Z_3 + Z_4) & -Z_3 & -(Z_4 + Z_5) & (Z_5 + Z_6) \\ -Z_4 & 0 & (Z_4 + Z_5) & -Z_5 \end{bmatrix} \begin{bmatrix} I_{1,3} \\ I_2 \\ I_4 \\ I_5 \end{bmatrix} \tag{3.34}$$

Steps [5] and [6] of the loop analysis procedure refer to the remaining nullator and norator, respectively. Deleting column 4 *of the original passive matrix equation* (3.33) and row 2 of this equation, corresponding to the locations of the nullator and norator that do *not* share loop pairs, results in the final reduction of the loop equations to

$$\begin{bmatrix} E_1 \\ 0 \\ 0 \end{bmatrix} = \begin{bmatrix} Z_2 & -Z_2 & 0 \\ (Z_3 + Z_4) & -Z_3 & (Z_5 + Z_6) \\ -Z_4 & 0 & -Z_5 \end{bmatrix} \begin{bmatrix} I_{1,3} \\ I_2 \\ I_5 \end{bmatrix} \tag{3.35}$$

It is now possible to solve for the loop currents $I_{1,3}$, I_2, I_5 and thereby obtain $V_6 = I_5 Z_6$ and $Z_{1d} = E_1/I_{1,3}$. The determinant Δ_Z of the 3×3 impedance matrix above is,

$$\Delta_z = Z_2 Z_4 Z_6 \tag{3.36}$$

so that, by Cramer's rule,

$$I_5 = \frac{\begin{vmatrix} Z_2 & -Z_2 & E_1 \\ (Z_3 + Z_4) & -Z_3 & 0 \\ -Z_4 & 0 & 0 \end{vmatrix}}{\Delta_z} \quad \text{and} \quad I_{1,3} = \frac{\begin{vmatrix} E_1 & -Z_2 & 0 \\ 0 & -Z_3 & (Z_5 + Z_6) \\ 0 & 0 & -Z_5 \end{vmatrix}}{\Delta_z} \tag{3.37}$$

giving,

$$I_5 = -\frac{E_1 Z_3}{Z_2 Z_6} \quad \text{and} \quad I_{1,3} = \frac{E_1 Z_3 Z_5}{Z_2 Z_4 Z_6} \tag{3.38}$$

Finally,

$$V_6 = I_5 Z_6 = -\frac{E_1 Z_3}{Z_2} \quad \text{and} \quad Z_{1d} \equiv \frac{E_1}{I_{1,3}} = \frac{Z_2 Z_4 Z_6}{Z_3 Z_5} \tag{3.39}$$

3-4 UNEQUAL NUMBERS OF NULLATORS AND NORATORS

The reader may have noticed that it has been assumed thus far that the *number* of nullators, say K_Z, is equal to the number of norators, say K_A, so that $K_Z \equiv K_A \equiv K$. If $K_Z \neq K_A$, then the nodal analysis procedure, for example, reduces to a set of equations of the form $\mathbf{I} = \mathbf{YV}$, where the dimensions of **I**, **Y**, and **V** are

$$\left.\begin{array}{ll} \mathbf{I}: & (N - K_A) \times 1 \\ \mathbf{Y}: & (N - K_A) \times (N - K_Z) \\ \mathbf{V}: & (N - K_Z) \times 1 \end{array}\right\} \tag{3.40}$$

so that the number of equations in $\mathbf{I} = \mathbf{YV}$ is simply $(N - K_A)$, and the number of independent unknowns in $\mathbf{I} = \mathbf{YV}$ is simply $(N - K_Z)$. Consequently, if $K_Z > K_A$, the number of equations exceeds the number of unknowns, so that there does not exist a nontrivial (that is, nonzero) solution vector **V**. Conversely, if $K_Z < K_A$, the number of unknowns exceeds the number of equations. The network therefore possesses an infinite number of solution vectors **V**. Therefore, *a necessary and sufficient condition for an interconnection of independent sources, nullators, norators, and RC elements to possess a unique solution of the network equations is that the number of norators K_A is equal to the number of nullators K_Z.*

EXAMPLE 3.7: Prove that the network shown in Fig. 3.6(a) does not possess a nontrivial solution for the node voltages V_1 and V_2.

SOLUTION: Removing the nullator and writing the nodal equations of the resultant passive network, we have

$$\underset{(2\times 1)}{\begin{bmatrix} I_A \\ 0 \end{bmatrix}} = \underset{(2\times 2)}{\begin{bmatrix} G_1 & -G_1 \\ -G_1 & (G_1 + G_2) \end{bmatrix}} \underset{(2\times 1)}{\begin{bmatrix} V_1 \\ V_2 \end{bmatrix}} \tag{3.41}$$

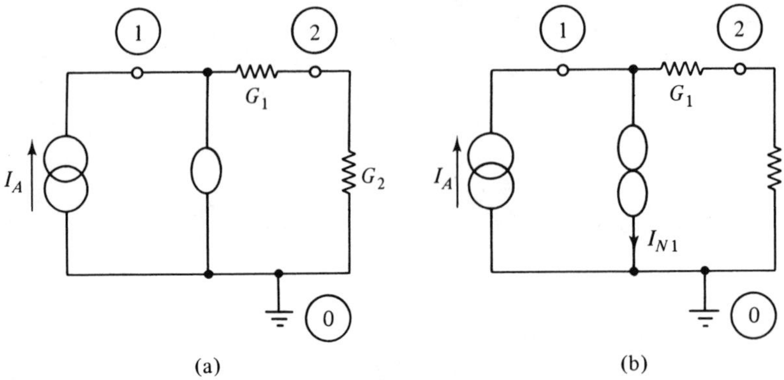

FIGURE 3.6

and, on replacing the nullator, the first column may be removed; thus,

$$\underset{(2\times1)}{\begin{bmatrix} I_A \\ 0 \end{bmatrix}} = \underset{(2\times1)}{\begin{bmatrix} -G_1 \\ (G_1 + G_2) \end{bmatrix}} \underset{(1\times1)}{[V_2]} \tag{3.42}$$

The only solution of these two equations in the single unknown V_2 is the trivial solution $V_2 = 0$, occurring if and only if I_A and/or G_1 equals zero. Then, of course, $V_1 = 0$ and is a trivial solution.

EXAMPLE 3.8: Prove that the network of Fig. 3.6(b) possesses an *infinite* number of solutions for the node voltages V_1 and V_2.

SOLUTION: It follows from equation (3.41) and the location of the norator shown in Fig. 3.6(b) that removal of the first row of the equation is required, leading to

$$\underset{(1\times1)}{[0]} = \underset{(1\times2)}{[G_1(G_1 + G_2)]} \underset{(2\times1)}{\begin{bmatrix} V_1 \\ V_2 \end{bmatrix}} \tag{3.43}$$

which *does not possess a unique solution for* V_1 *or* V_2. Rearranging, we obtain

$$-V_1 = \left[\frac{G_1 + G_2}{G_1}\right] V_2 \tag{3.44}$$

which clearly has an *infinity* of possible solutions for V_1 and V_2.

Having established that networks with unequal numbers of nullators and norators do not possess unique nonzero solutions, it is hereafter assumed that $K_A = K_Z$.

3-5 ON THE NULLOR MODELS OF GICs

The GIC is an important subnetwork from which *RC*-active filters are synthesized. The nullor modeling of GICs leads directly to corresponding transistor and operational amplifier GIC circuits and is, therefore, of significance to the circuit designer. In this section, the nullor models of GICs are used to *classify* the major types of GIC networks (in much the same way as the nullor models of controlled sources are classified in section 2.9). The reader will find that the nodal analysis technique previously described in section 3.2 is a useful means of analyzing the nullor model of a GIC.

3-5.1 Calculation of the Driving-Point Admittance Y_{id} from the Nodal Admittance Matrix

To understand the behavior of GIC networks it is often necessary to derive the driving-point admittance $Y_{id}(s)$. Usually it is possible to obtain $Y_{id}(s)$ indirectly from the nodal admittance matrix **Y** as follows: the transform nodal equations **I** = **YV** may be written

$$\mathbf{V} = \mathbf{Y}^{-1}\mathbf{I} = \left(\frac{\text{adj } \mathbf{Y}}{\det \mathbf{Y}}\right)\mathbf{I} \tag{3.45}$$

so that a *finite* current excitation vector $\mathbf{I} = \{I_1, I_2, \ldots, I_N\}'$ results in an *infinite* response vector $\mathbf{V} = \{V_1, V_2, \ldots, V_N\}'$ if and only if

$$\det \mathbf{Y} = 0 \tag{3.46}$$

In terms of Fig. 3.7, where $I_i(s)$ is assumed to be the *only* independent source,

$$V_i = \frac{I_i}{Y_i + Y_{id}}$$

from which it follows that a finite excitation transform $I_i(s)$ leads to an

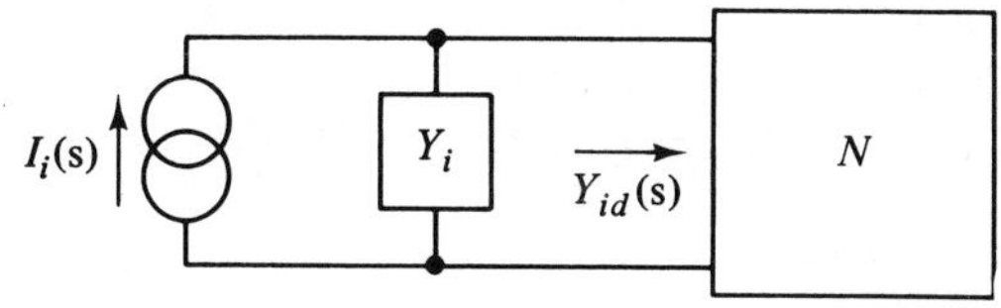

FIGURE 3.7

infinite response $V_i(s)$ if and only if

$$Y_i + Y_{id} = 0 \tag{3.47}$$

The simultaneous solution of equations (3.46) and (3.47), which are both valid under the *same* conditions of infinite V and finite I, leads to an expression for Y_{id}. An example is given in the following section.

3-5.2 Type 1 GICs

The type 1 GICs are characterized by the fact that they possess a network determinant $\Delta_y \equiv \det \mathbf{Y}$ of the form

$$\boxed{\Delta_y \equiv \det \mathbf{Y} = -Y_1Y_3 + Y_2Y_4} \quad \text{TYPE 1 GIC} \tag{3.48}$$

where Y_1 is the admittance that terminates port 1 and where *one* of Y_2, Y_3, or Y_4 terminates port 2. All the four admittances $Y_1, \ldots, Y_4$ are branches of a terminated type 1 GIC.

Three type 1 GICs are shown in Table 3.1 and are classified there as types 1A, 1B, and 1C. Note that the type 1A GIC is characterized by the fact that it employs only one nullor, whereas types 1B and 1C require two three-terminal nullors. The reader may wish to use the nodal analysis technique of section 3.2 to verify that all the type 1 GICs shown in Table 3.1 have a network determinant Δ_y provided by equation (3.48). Simultaneous solution of equation (3.46), equation (3.47) with $i = 1$ and equation (3.48) leads directly to

$$\boxed{Y_{1d} = -\frac{Y_2Y_4}{Y_3}} \quad \text{TYPE I GIC} \tag{3.49}$$

so that the conversion function $K(s)$ depends on which of the branches Y_2, Y_3, or Y_4 is defined as the terminating branch of port 2. For example, the type 1A GIC of Table 3.1 may be used as an immittance-*inverting* GIC with $K(s)$ equal to $-Y_2Y_4$ by defining port 2 as shown in Fig. 3.8(a). Alternatively, it may be used as an immittance-*converting* GIC with $K(s) = -Y_2/Y_3$ as shown in Fig. 3.8(b). The reader may check equation (3.49) by the nodal or loop analysis techniques that are described earlier in this chapter.

TABLE 3.1 *Type 1 GICs*

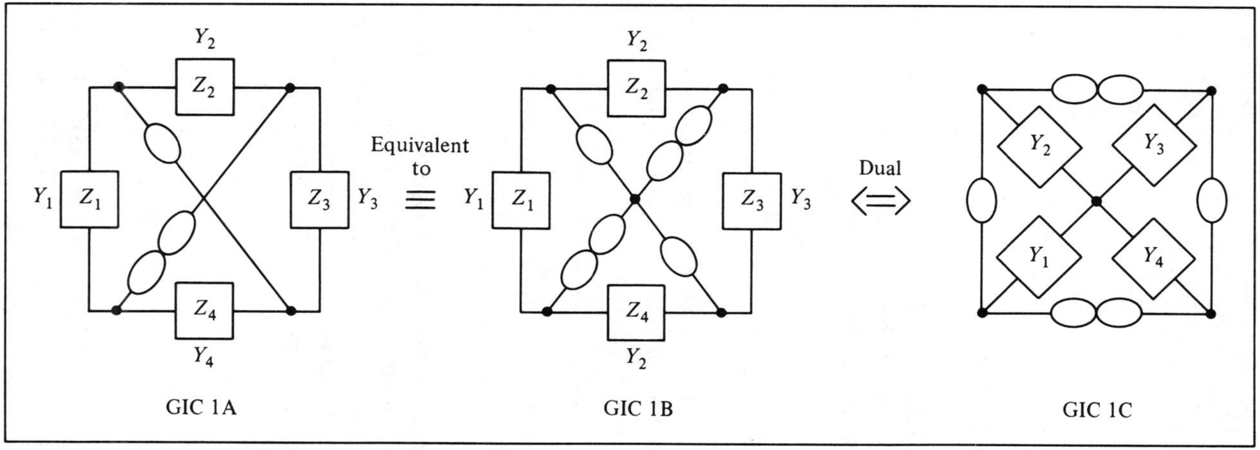

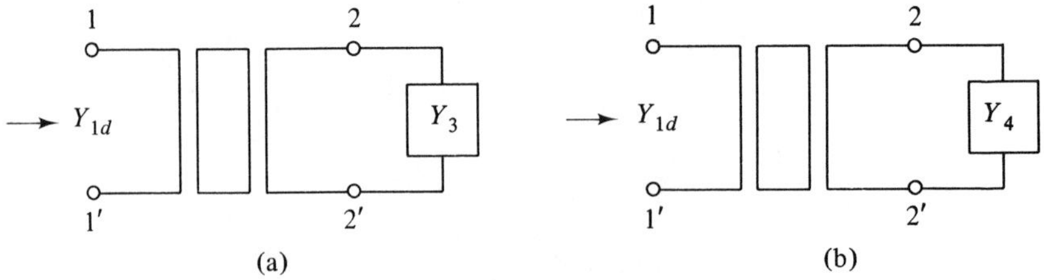

FIGURE 3.8

3-5.3 Type 2 GICs

The type 2 GICs form the basis for much of the modern work in the synthesis and design of high-quality *RC*-active filters. They are characterized by the network determinant

$$\boxed{\Delta_y = Y_1 Y_3 Y_5 + Y_2 Y_4 Y_6} \quad \text{TYPE 2 GIC} \tag{3.50}$$

where $Y_1, Y_2, \ldots, Y_6$ are branch admittances of the terminated GIC. Simultaneous solution of equation (3.46), equation (3.47) with $i = 1$ and equation (3.50) leads directly to

$$\boxed{Y_{1d} = +\frac{Y_2 Y_4 Y_6}{Y_3 Y_5}} \quad \text{TYPE 2 GIC} \tag{3.51}$$

The nullor implementations of type 2 GICs are readily derived from the previously considered type 1 GICs. Four useful type 2 configurations are shown in Table 3.2 as type 2A, 2B, 2C, and 2D; the derivation for GIC 2A, for example, follows by simply replacing Y_3 of the type 1A GIC of Table 3.2 by another similar type 1A GIC containing admittances Y_3, Y_4, and Y_5 as shown in Fig. 3.9. It follows directly from Fig. 3.9 and an equation of the form of equation (3.49) that

$$Y_{1d} = -\frac{Y_2 Y_6}{Y_{3d}} = -Y_2 Y_6\left(\frac{1}{-Y_3 Y_5 / Y_4}\right) = \frac{Y_2 Y_4 Y_6}{Y_3 Y_5} \tag{3.52}$$

as required by equation (3.51) for a type 2 GIC. The reader may check the validity of equation (3.51) for the type 2B, 2C, and 2D GICs of Table 3.2 by the method explained above.

Note that equation (3.47) is valid at all branches so that we could equally as well derive the general *i*th driving-point admittance Y_{id} for $i = 2, 3, \ldots, 6$. Thus, letting $i = 2$ we have, from equations (3.46), (3.47)

TABLE 3.2 *Type 2 GICs.*

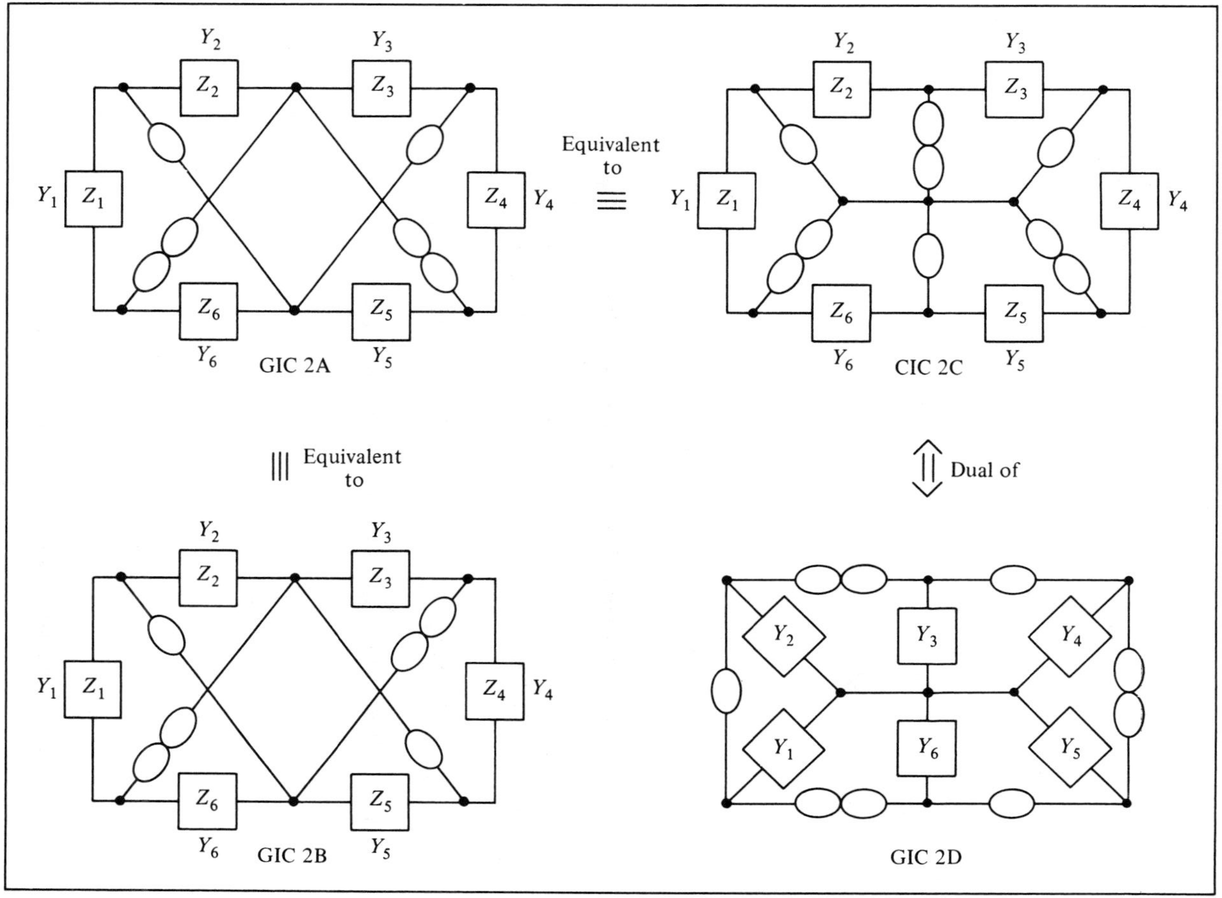

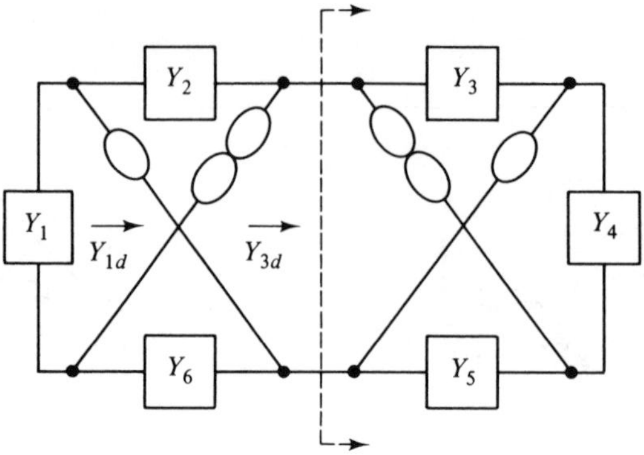

FIGURE 3.9

and (3.51)

$$Y_{2d} = \frac{Y_1 Y_3 Y_5}{Y_4 Y_6} \tag{3.53}$$

with similar expressions for $i = 3, 4, 5,$ and 6.

3-5.4 GICs with More Than Six Branches $Y_1 \ldots Y_{2N}$

Clearly, it is possible to cascade an indefinite number of GIC1As, GIC1Bs, GIC1Cs to obtain the general determinant

$$\Delta_y = (Y_1 Y_3 Y_5, \ldots, Y_{2N-1}) + (Y_2 Y_4 Y_6, \ldots, Y_{2N})(-1)^{N+1}$$

with a general driving-point admittance Y_{1d} given by

$$Y_{1d} = \frac{Y_2 Y_4 Y_6, \ldots, Y_{2N}}{Y_3 Y_5 Y_7, \ldots, Y_{2N-1}}(-1)^{N+1} \tag{3.54}$$

where a converting-type GIC is obtained by associating a numerator admittance with port 2 and an inverting-type GIC is obtained by associating a denominator admittance with port 2.

3-5.5 PIIs and Gyrators Using Nullors

The positive immittance inverter (PII) and the gyrator are defined in sections 2.6.2 and 2.4, respectively.

TABLE 3.3 *Type 2 positive impedance inverters.*

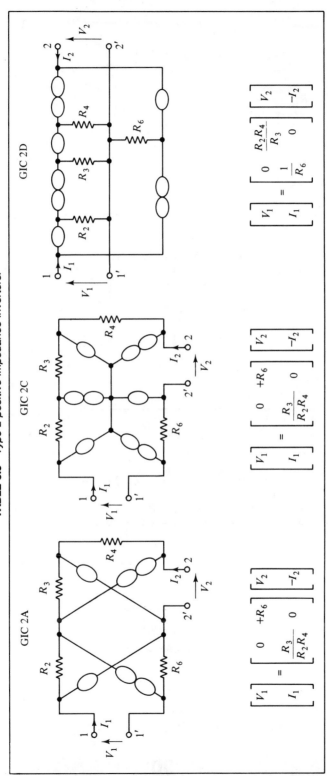

GIC 2A	GIC 2C	GIC 2D
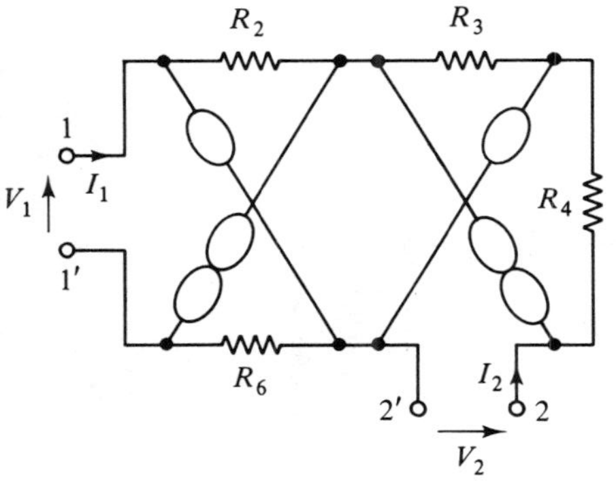	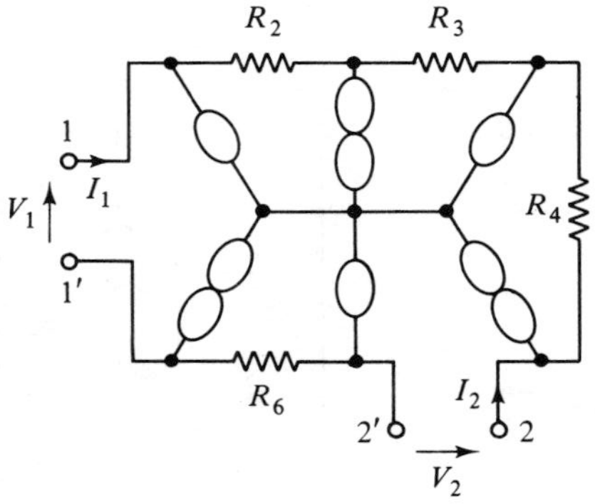	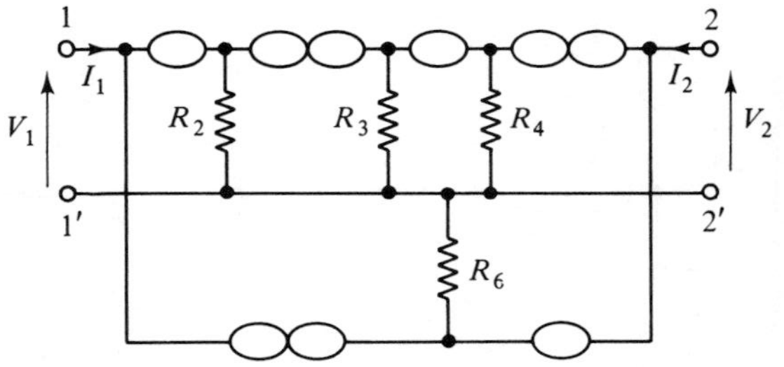
$\begin{bmatrix} V_1 \\ I_1 \end{bmatrix} = \begin{bmatrix} 0 & +R_6 \\ \dfrac{R_3}{R_2 R_4} & 0 \end{bmatrix} \begin{bmatrix} V_2 \\ -I_2 \end{bmatrix}$	$\begin{bmatrix} V_1 \\ I_1 \end{bmatrix} = \begin{bmatrix} 0 & +R_6 \\ \dfrac{R_3}{R_2 R_4} & 0 \end{bmatrix} \begin{bmatrix} V_2 \\ -I_2 \end{bmatrix}$	$\begin{bmatrix} V_1 \\ I_1 \end{bmatrix} = \begin{bmatrix} 0 & \dfrac{R_2 R_4}{R_3} \\ \dfrac{1}{R_6} & 0 \end{bmatrix} \begin{bmatrix} V_2 \\ -I_2 \end{bmatrix}$

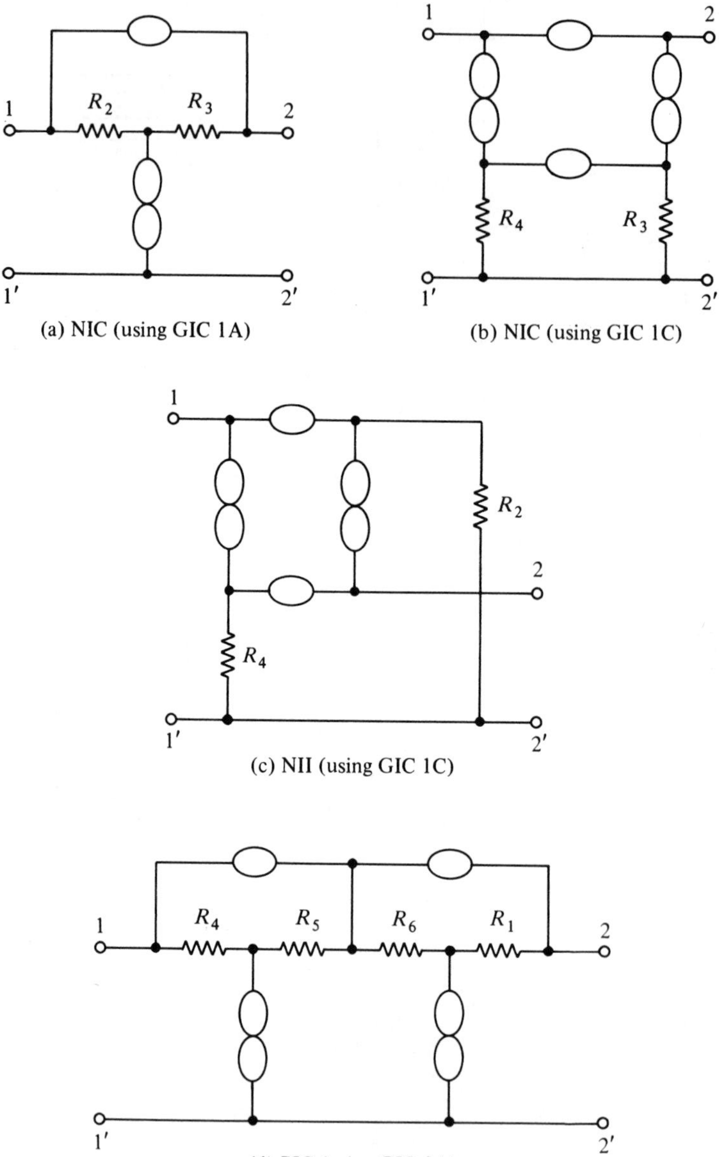

(a) NIC (using GIC 1A)

(b) NIC (using GIC 1C)

(c) NII (using GIC 1C)

(d) PIC (using GIC 2A)

FIGURE 3.10

The networks shown in Table 3.2 offer the possibility of realizing a variety of nullor realizations of PIIs and gyrators. For example, using Y_1 and Y_5 as the terminations and replacing Y_2^{-1}, Y_3^{-1}, Y_4^{-1}, and Y_6^{-1} by resistances R_2, R_3, R_4, and R_6 leads directly to the GIC realizations in Table 3.3, with the transmission matrices shown. These PII networks become gyrators on choosing $R_2 = R_3 = R_4 = R_6$. The three networks shown in Table 3.3 are important in the synthesis of high-quality *RC*-active filters.

It is relevant to note that these networks possess the transmission matrices of PIIs and gyrators, but they suffer from the extra constraint that the *voltages between terminal 1-2 and 1′-2′ are not generally arbitrary*. Consequently, considerable care must be taken to allow for these additional constraints when interconnecting networks of this type.

PICs, NICs, and NIIs using nullors: It is a straightforward matter to use the networks shown in Tables 3-1 and 3-2 to realize a variety of PICs, NICs, and NIIs. In Fig. 3.10 a number of such networks are shown where the corresponding GIC type is indicated. The reader is asked to analyze these networks and pursue some elementary applications in Problems 3-2 to 3-8 at the end of this chapter.

3-6 SUMMARY

Nodal and loop analysis techniques are described for *RLC*-nullor networks; the methods involve writing the network equations of a corresponding passive network using either the $(N \times N)$ nodal admittance matrix or the $(N \times N)$ loop impedance matrix. Introducing K nullors into the networks reduces the $(N \times N)$ passive network matrix to an $[(N - K) \times (N - K)]$ matrix description of the K-nullor active network.

The analysis techniques that are described here may be used to analyze the various nullor models of GICs that are shown in Tables 3.1, 3.2, and 3.3.

PROBLEMS

3-1. Write the nodal equations for the single nullor network in Fig. P3.1 in terms of the node voltages V_2 and $V_{1,3}$. Solve these node equations for $V_{1,3}$ in terms of C_1, R_2, R_3, and R_4. Thereby, determine the driving-point impedance Z_{1d}.

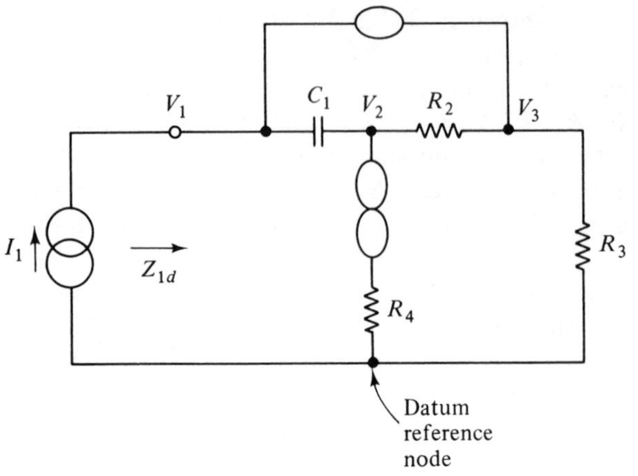

FIGURE P3.1

3-2. Prove that each of the GIC networks shown in Table 3.1 has a network determinant Δ_y given by

$$\Delta_y = -Y_1Y_3 + Y_2Y_4$$

3-3. Prove that each of the GIC networks shown in Table 3.2 has a network determinant Δ_y given by

$$\Delta_y = Y_1Y_3Y_5 + Y_2Y_4Y_6$$

3-4. Verify that the three GIC networks shown in Table 3.3 have the corresponding transmission matrices shown in that table. If each of these GIC networks is terminated at port 2 in a capacitance element C_5, derive the corresponding expressions for the driving-point impedance $Z_{1d} \equiv V_1/I_1$ at port 1. What is the effective inductance at port 1 for each of these GIC circuits?

3-5. Derive the transmission matrices of the four impedance converting networks shown in Fig. 3.10 on page 91.

3-6. The NIC Fig. 3.10(a) is terminated at port 2 in a resistance R_4. Prove that the driving-point impedance at port 1 is $-R_2R_4/R_3$.

3-7. The NIC in Fig. 3.10(b) is terminated at port 2 in a resistance R_2. Prove that the driving-point impedance at port 1 is $-R_4R_2/R_3$.

3-8. The NII in Fig. 3.10(c) is terminated in a capacitance C_3 at port 2. Prove that the driving-point impedance at port 1 is equivalent to an ideal *negative inductance*.

3-9. Write the loop equations of the network shown in Fig. P3.9 and thereby derive an expression for the *loop* current I_2.

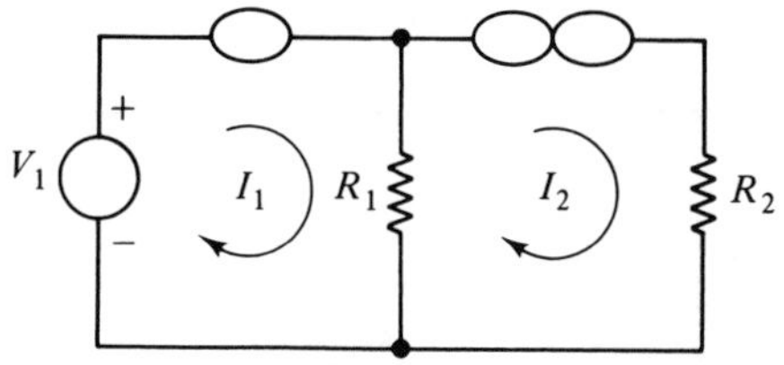

FIGURE P3.9.

3-10. Repeat Problem 3.9 for the networks shown in Fig. P3.10.

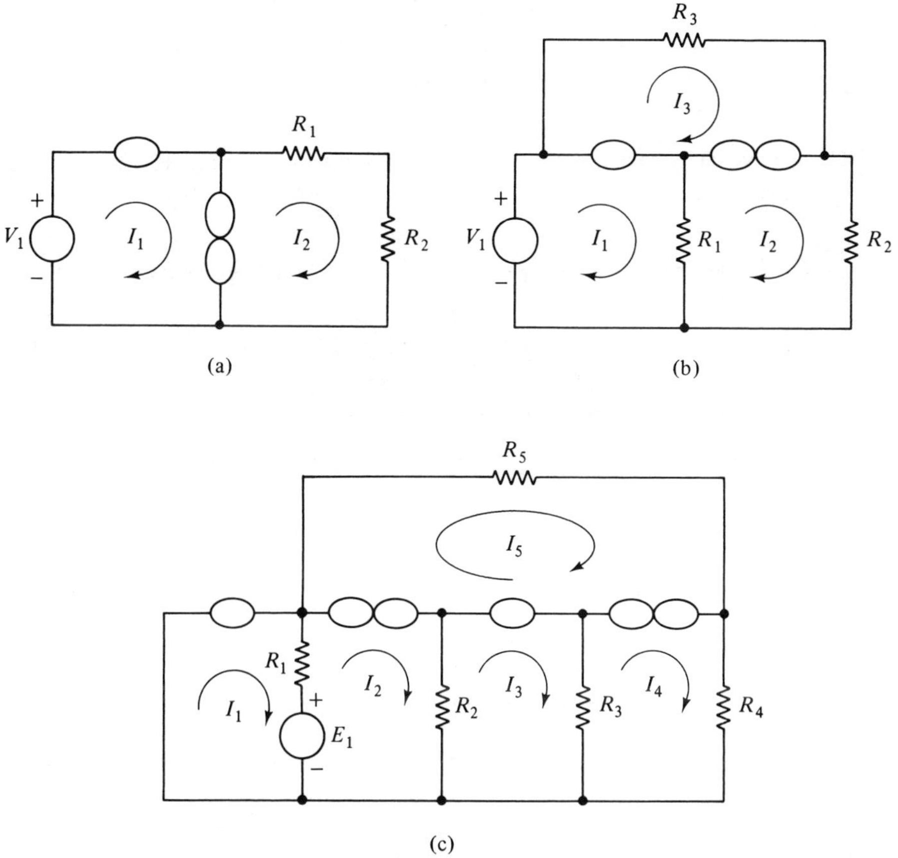

FIGURE P3.10

3-11. Write the loop equations of the network shown in Fig. P3.11 and then solve these equations for the loop current I_4 and thereby derive the voltage gain V_2/V_1 of the network in terms of R_1, R_2, R_3, R_4 and R_5.

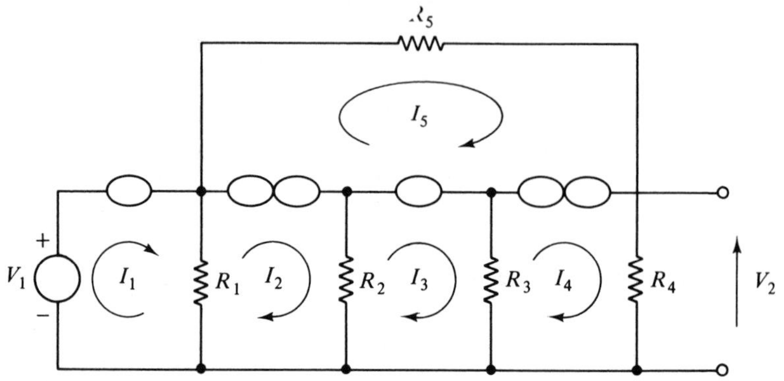

FIGURE P3.11

3-12. Derive the voltage gains V_2/V_1 of the networks shown in Fig. P3.12.

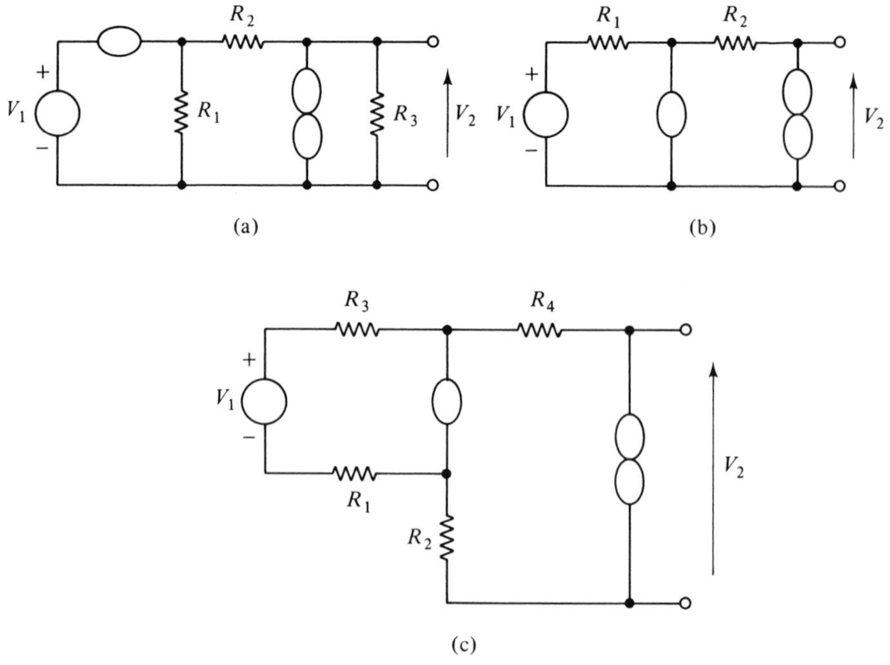

FIGURE P3.12

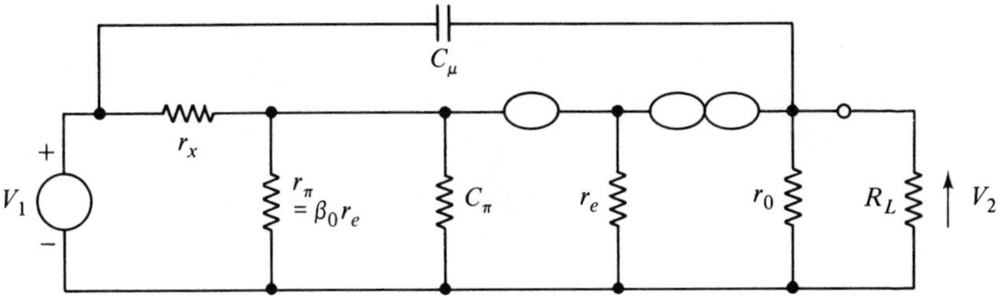

(d)

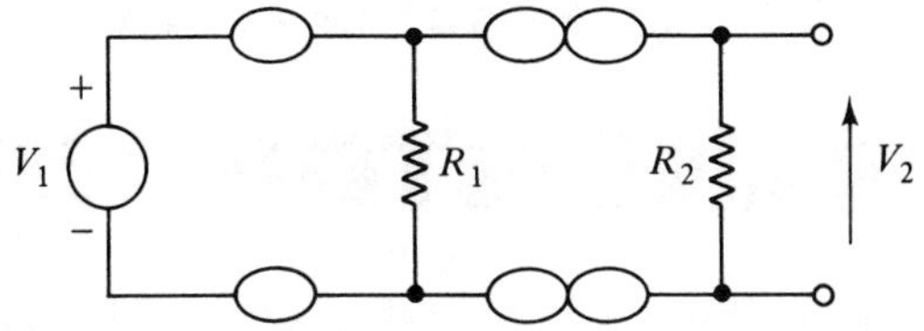

(e)

FIGURE P3.12 *(Continued).*

BIBLIOGRAPHY

Van Valkenburg, M. E., *Network Analysis* 3rd Edition, Prentice-Hall, 1974, chapter 3.

Davies, A. C., "The Significance of Nullators, Norators and Nullors in Active Network Theory," *Radio Electron. Eng.*, vol. 34, p. 259–267, 1967.

Antoniou, A., "Gyrators Using Operational Amplifiers," *Electronics Letters*, vol. 3, p. 350–352, 1967.

4

THE BIPOLAR TRANSISTOR AND LINEAR APPLICATIONS

4-1 INTRODUCTION

The modern electronic age was effectively launched by the invention of the bipolar transistor device. In this chapter, the electrical characteristics of the transistor are described briefly, without reference to the semiconductor physics that explain the nature of these characteristics. The transistor possesses a region of its voltage-current characteristics, known as the *linear region*, where the relationships between voltage and current variables are approximately linear. The transistor is useful in the linear region for the purpose of *linear signal amplification* and it is this particular application that is pursued in this chapter. The nullor is used as an element to describe the approximate direct-current (static) model of transistor circuits and also to describe an approximate dynamic (small-signal) model. The concept of an *ideal* transistor is found to be identical to that of a *three-terminal nullor* and this is used to develop techniques for rapid "first-approximation" analysis of transistor amplifier circuits.

The controlled-source nullor models that are described in chapter 2 are used directly in this chapter to obtain a total of 12 fundamental transistor implementations of the controlled sources. These transistor circuit configurations form the basis of transistor amplifier circuit design. The *differential* voltage transistor amplifier is an important circuit in active network design. It is a basic subcircuit of the modern operational amplifier and for this reason it is analyzed in some detail.

Transistor circuit implementations of generalized immittance converters (GICs) have applications in *RC*-active filter design and are, therefore, classified and described in sections 4-5 and 4-6. Further applications of transistor circuit analysis in the implementation of active subnetworks are pursued in the problems at the end of the chapter.

4-2 THE BIPOLAR TRANSISTOR DEVICE

A *detailed* study of the bipolar transistor as an amplification device requires a knowledge of the physics of semiconductors. The primary purpose of this explanation of the transistor is to delineate its electrical characteristics in a way that will enable the reader to design and analyze most linear transistor amplifier circuits with a minimal knowledge of the physics of the device.

There are two types of bipolar transistor: npn and pnp. They are both three-terminal devices and are described using the symbols shown in Fig. 4.1. The letters *b*, *e*, and *c* denote the three terminals which are referred to as the *base*, *emitter*, and *collector*. (The reader may consult a text on transistor physics in order to learn the physical significance of the base, emitter, and collector insofar as they describe the internal operation and structure of the device). The arrow associated with the emitters in Fig. 4.1 implies that under correct "linear" operating conditions, the average positive emitter current $\overline{i_e(t)}$ flows outward in the case of the npn transistor and inward in the case of the pnp transistor. The following discussion refers to the npn transistor.

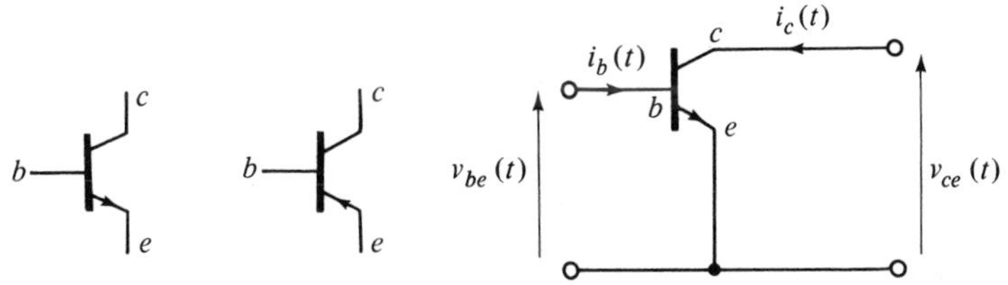

(a) npn Transistor (b) pnp Transistor (c) Voltage-current variables

FIGURE 4.1

4-2.1 The Nonlinear Static Characteristic

The electrical characteristics of the transistor may be described completely by establishing the relationship between the variables $v_{be}(t)$, $i_b(t)$, $v_{ce}(t)$ and $i_c(t)$ shown in the two-port representation of the transistor in Fig. 4.1(c). The *static characteristic* is the relationship between these variables when they are constants (that is, nontime-varying). Denoting these static quantities as V_{BE}, I_B, V_{CE}, and I_C, it is possible to express the relationship between these four independent variables in the form

$$I_C = f_1(V_{CE}, I_B), \qquad V_{BE} \text{ constant} \tag{4.1}$$

and

$$V_{BE} = f_2(I_B, V_{CE}), \qquad I_C \text{ constant} \tag{4.2}$$

Equation (4.1) is defined as the *static input characteristic*, and equation (4.2) is defined as the *static output characteristic*. If three of the four variables are known and the nonlinear functions f_1 and f_2 are known, it is always possible to find the fourth variable.

The static output characteristic and the static input characteristic are shown in Figs. 4.2 and 4.3 for a typical modern silicon bipolar npn

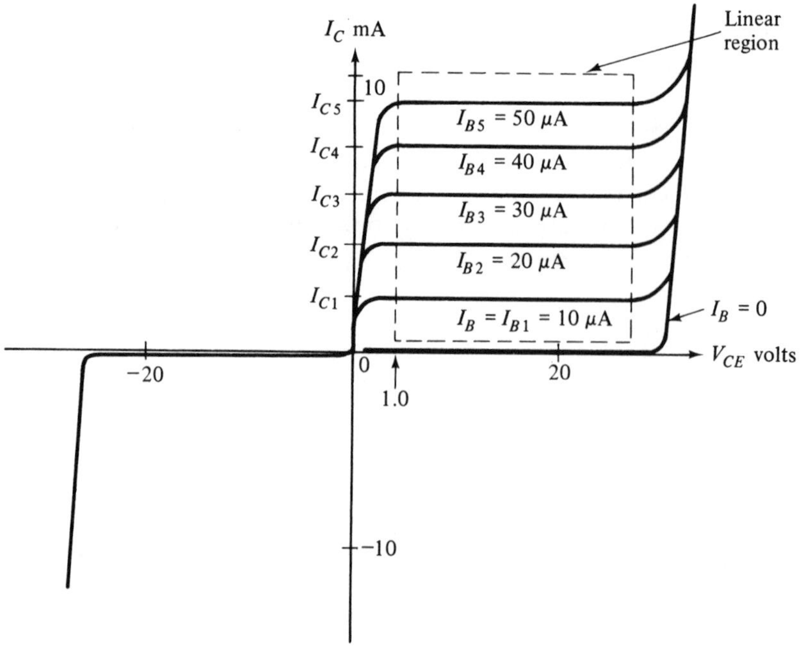

FIGURE 4.2 *Static output characteristic of an npn transistor.*

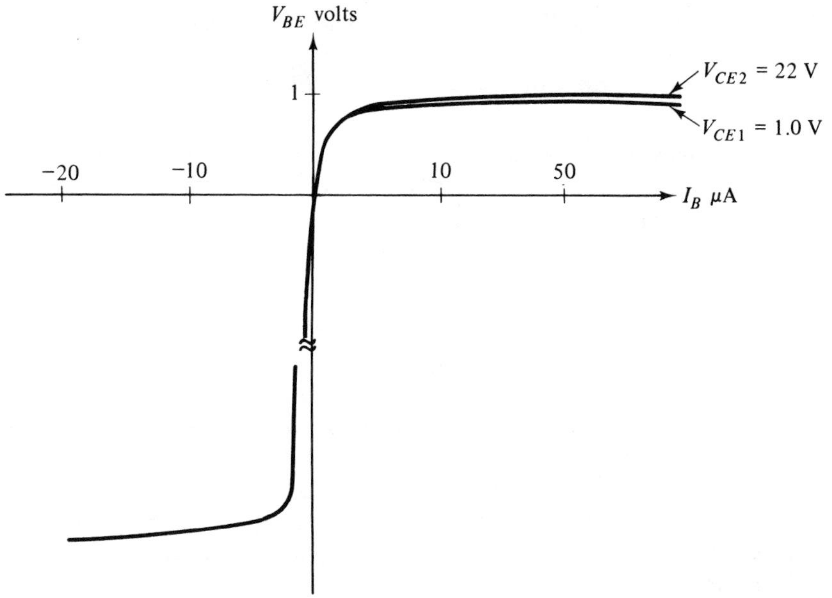

FIGURE 4.3 *Static input characteristic of a (silicon) npn transistor.*

transistor. The output characteristic possesses several features that are emphasized. First, *within the region enclosed by the dashed lines*, the collector current I_C has the following properties:

(i) it is virtually independent of V_{CE};

(ii) it is approximately proportional to the base current I_B; the dc (static) current gain $\beta_0 \equiv I_C/I_B$.

(iii) it is much larger than I_B; that is, $\beta_0 \gg 1$.

These three properties indicate that, within the dashed-line region, the *static output characteristic is that of a current-controlled current source* (CCCS) with controlling-factor (gain) β_0.

The static input characteristic has the important feature that it is almost independent of V_{CE}; the two curves for $V_{CE1} = 1.0$ V and $V_{CE} =$ 22.0 V are almost identical. Furthermore, the input voltage V_{BE} is constrained to a small range for all values of I_B within the linear region in Fig. 4.2. Thus, in the linear region

$$1 \leqslant I_B \leqslant 50\ \mu\text{A} \tag{4.3}$$

$$1 \leqslant V_{CE} \leqslant 22\ \text{V} \tag{4.4}$$

and, for this range of I_B and V_{CE}, it is clear from Fig. 4.3 that

$$0.5 \leqslant V_{BE} \leqslant 0.8 \text{ V}, \qquad \textit{silicon} \tag{4.5}$$

The *idealized* static characteristics, which are only *approximately valid in the linear region* are shown in Fig. 4.4, for which V_{BE} is a constant (V_{BEO}) and β_0 is a constant that is much greater than unity. The relevant equations for these idealized conditions are

$$v_{be}(t) \equiv V_{BE} = V_{BEO} \tag{4.6}$$

$$i_c(t) \equiv I_C = \beta_0 I_B \tag{4.7}$$

which correspond to the equivalent circuits shown in Figs. 4.5(a) and (b). It is usually valid to assume that for transistors fabricated with germanium, $V_{BEO} \simeq 0.25$ V; for most modern transistors, $50 < \beta_0 < 500$. Both V_{BEO} and β_0 are highly temperature dependent. For this reason, it is good engineering practice to desensitize the circuit performance to variations of V_{BEO} and β_0.

Modern bipolar transistors are often employed in circuits such that the dc current gain β_0 may be assumed to be of infinite magnitude without significantly affecting the node voltages or loop currents within the network. Note that if $\beta_0 = \infty$, then the CCCS in Fig. 4.5(b) becomes a nullor (as shown in section 2.12 of chapter 2) so that the infinite-β_0 static equivalent circuit contains only a constant voltage generator V_{BEO} and a nullor as shown in Fig. 4.5(c).

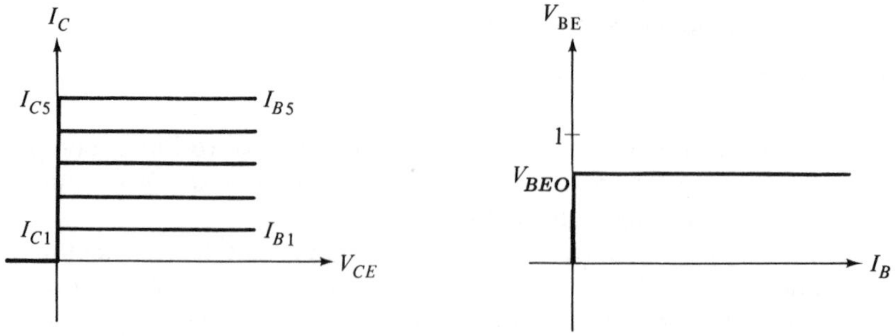

(a) Ideal nonlinear static output characteristic

(b) Ideal nonlinear static input characteristic

FIGURE 4.4

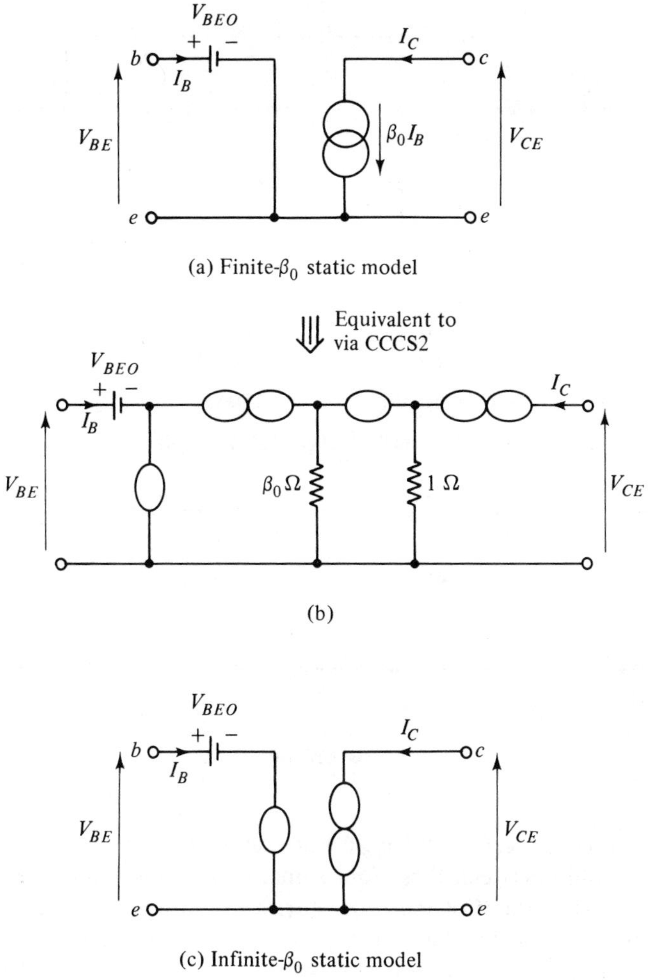

FIGURE 4.5

4-2.2 The Dynamic Characteristics

In most practical applications of transistor amplifiers, the *signal waveform* is superimposed on the aforementioned static (dc) waveforms. Then

$$\left.\begin{aligned} i_b(t) &\equiv I_B + i_B(t), & i_c(t) &\equiv I_C + i_C(t) \\ v_{be}(t) &\equiv V_{BE} + v_{BE}(t), & v_{ce}(t) &= V_{CE} + v_{CE}(t) \end{aligned}\right\} \tag{4.8}$$

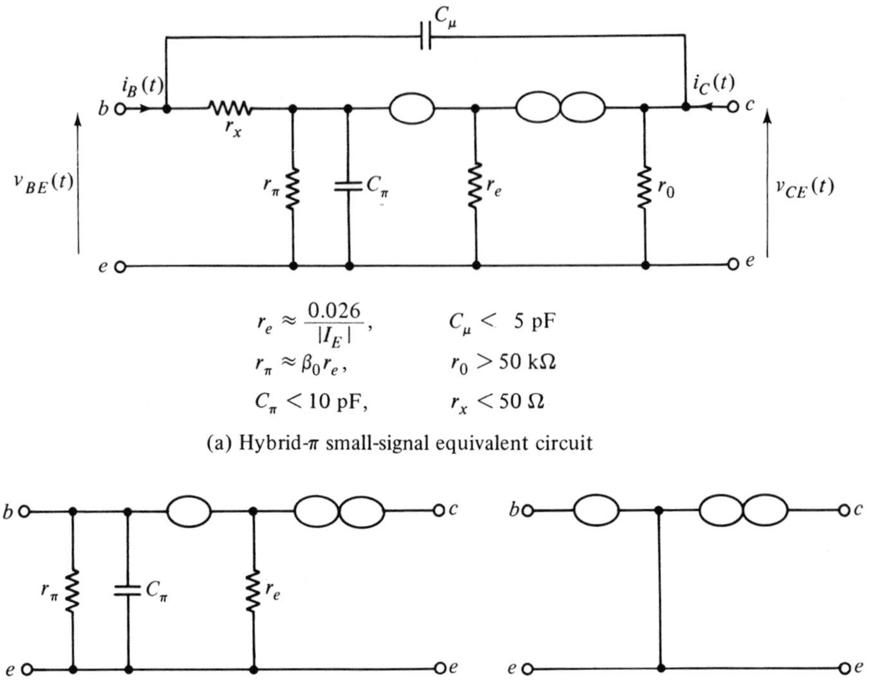

$r_e \approx \dfrac{0.026}{|I_E|}$, $C_\mu < 5$ pF

$r_\pi \approx \beta_0 r_e$, $r_0 > 50$ kΩ

$C_\pi < 10$ pF, $r_x < 50$ Ω

(a) Hybrid-π small-signal equivalent circuit

(b) Simplifed hybrid-π

(c) Ideal transistor $r_e = 0$

FIGURE 4.6

where $i_B(t)$, $i_C(t)$, $v_{BE}(t)$, and $v_{CE}(t)$ are defined as the signal waveforms. The relationship between these four signal waveforms can be derived from the physical electronics of the transistor; a useful equivalent circuit that is sufficiently accurate for the majority of practical applications is the small-signal hybrid-π equivalent circuit shown in Fig. 4.6(a). Typical numerical values for the parameters of the hybrid-π model are also shown in Fig. 4.6(a). The parameters r_x and C_μ may often be assumed to be zero and the output resistance r_0 may often be assumed to be infinitely large; if such assumptions are justified then the *simplified hybrid-π* shown in Fig. 4.6(b) may be used.

The ideal transistor: It is observed from Fig. 4.6(b) that the small-signal behavior of the bipolar transistor is approximately that of a VCCS2 controlled source (see Table 2-9) with a finite impedance $r_\pi \| C_\pi$ connected at the input terminals. Note that the resistance r_e of the VCCS is given by

$$r_e \approx \frac{0.026}{|I_E|} \tag{4.9}$$

At $|I_E| = 1$ mA, for example, r_e is 26 Ω. *In the ideal transistor model it is assumed that $r_e = 0$.* Then, the hybrid-π model shown in Fig. 4.6(b) may be further simplified; replacing r_e by a short-circuit in Fig. 4.6(b) results in a *redundant* impedance $r_\pi \| C_\pi$ across the nullator. Consequently, the *ideal* transistor possesses a small-signal equivalent circuit that is simply the three-terminal nullor shown in Fig. 4.6(c).

EXAMPLE 4.1: Derive the static collector current I_C and the static collector-emitter voltage V_{CE} for transistor T1 in Fig. 4.7(a). Assume $V_{BEO} = 0.7$ V, and use the *infinite-β_0 static model* shown in Fig. 4.5(c).

SOLUTION: The infinite-β_0 static model is substituted for T1, resulting in Fig. 4.7(b). This network is analyzed by using loop analysis or simply by

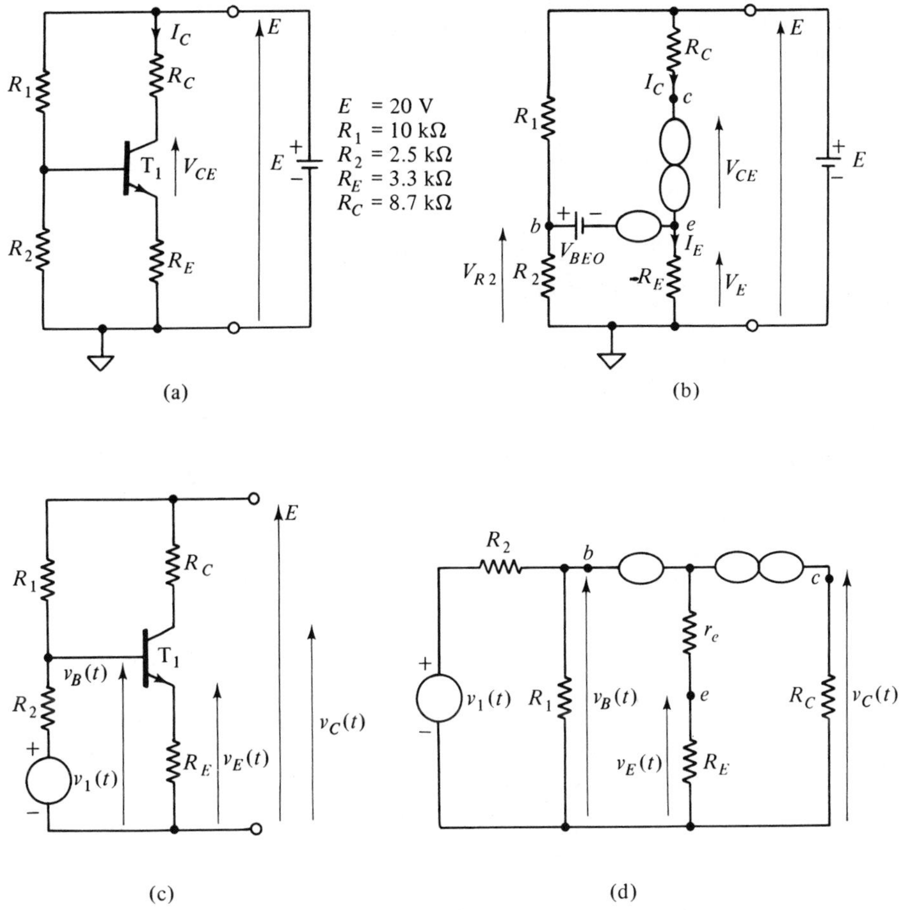

FIGURE 4.7

inspection of the network: since the nullator current is zero, the voltage V_{R2} is given by

$$V_{R2} = \left(\frac{R_2}{R_1 + R_2}\right)E \tag{4.10}$$

so, by KVL,

$$V_E = \left(\frac{ER_2}{R_1 + R_2}\right) - V_{BEO} \tag{4.11}$$

giving

$$I_E = \frac{1}{R_E}\left(\frac{ER_2}{R_1 + R_2} - V_{BEO}\right) \tag{4.12}$$

But, by KCL, $I_C = I_E$, so that I_C is also given by equation (4.12). Furthermore, by KVL

$$E = V_{CE} + I_C R_C + I_E R_E \tag{4.13}$$

and, therefore, rearranging and substituting the previously derived values of $I_C(= I_E)$, we have

$$V_{CE} = E - \left\{\left[\frac{ER_2}{R_E(R_1 + R_2)} - \frac{V_{BEO}}{R_E}\right]\left[R_E + R_C\right]\right\} \tag{4.14}$$

Substituting the numerical values shown in Fig. 4.7(a) into equations (4.12) and (4.14), we find that $V_{CE} = 8.0$ V and $I_C = 1$ mA.

EXAMPLE 4.2: A small-signal independent voltage source $v_1(t)$ is introduced into the circuit of Fig. 4.7(a) as shown in Fig. 4.7(c). Assuming that r_e is the only significant nonideal parameter of the transistor T1, derive expressions for the small-signal voltages $v_B(t)$, $v_C(t)$, and $v_E(t)$ at the terminals of the transistor in terms of R_1, R_2, R_E, R_C, r_e, and $v_1(t)$.

SOLUTION: The small-signal equivalent circuit is readily derived from Fig. 4.7(c) by replacing the independent source E by a short-circuit (why?) and transistor T1 by the simplified hybrid-π equivalent circuit, where $C_\pi = 0$ and $r_\pi = \infty$. This is equivalent to assuming that the transistor performance is independent of frequency ($C_\pi = 0$) and that $\beta_0 = \infty$ ($r_\pi = \beta_0 r_e$).

Analysis of Fig. 4.7(d) is straightforward. Since the nullator current is zero,

$$v_B(t) = v_1(t)\left(\frac{R_1}{R_1 + R_2}\right) \tag{4.15}$$

and, clearly, the nullator also ensures that the node that is shared by the nullator and the norator is also at potential $v_B(t)$. Consequently,

$$i_E(t) = \frac{v_1(t)}{(R_E + r_e)}\left(\frac{R_1}{R_1 + R_2}\right) \tag{4.16}$$

and

$$v_E(t) = v_1(t)\left(\frac{R_1}{R_1 + R_2}\right)\left(\frac{R_E}{R_E + r_e}\right) \tag{4.17}$$

and $v_C(t) = -i_C(t)R_C = -i_E(t)R_C$, giving

$$\boxed{v_C(t) = \frac{-v_1(t)R_1R_C}{(R_E + r_e)(R_1 + R_2)}} \tag{4.18}$$

It is interesting to note that the transistor parameter r_e appears in series with resistance R_E and that in many practical applications r_e is small enough that it may be neglected. From the dc analysis, the numerical values that have been selected lead to $|I_E| = 1$ mA, so that $r_e = 26\ \Omega$. Substituting numerical values into equation (4.18), we have

$$\frac{v_C(t)}{v_1(t)} = -2.092 \tag{4.19}$$

compared with the value of -2.109 that is obtained by assuming $r_e = 0$. Clearly, in this particular example, the ideal transistor nullor model shown in Fig. 4.6(c) provides an adequate model for predicting the low-frequency small-signal gain of Fig. 4.7(c).

4-3 TRANSISTOR AMPLIFIER CONFIGURATIONS

A variety of amplifier configurations are found by simply replacing the three-terminal nullors in the 12 R-nullor circuits of Tables 2-8 and 2-9 by *ideal* transistors. This leads directly to the corresponding 12 *ideal* small-signal transistor circuits shown in Table 4-1. The reader should study these circuits carefully to verify that they *are* directly equivalent to Tables 2-8 and 2-9 via substitutions of the ideal transistor small-signal equivalent circuit of Fig. 4.6(c).

The circuits shown in Table 4-1 are important because they are the basic building-blocks from which transistor amplifiers are often designed.

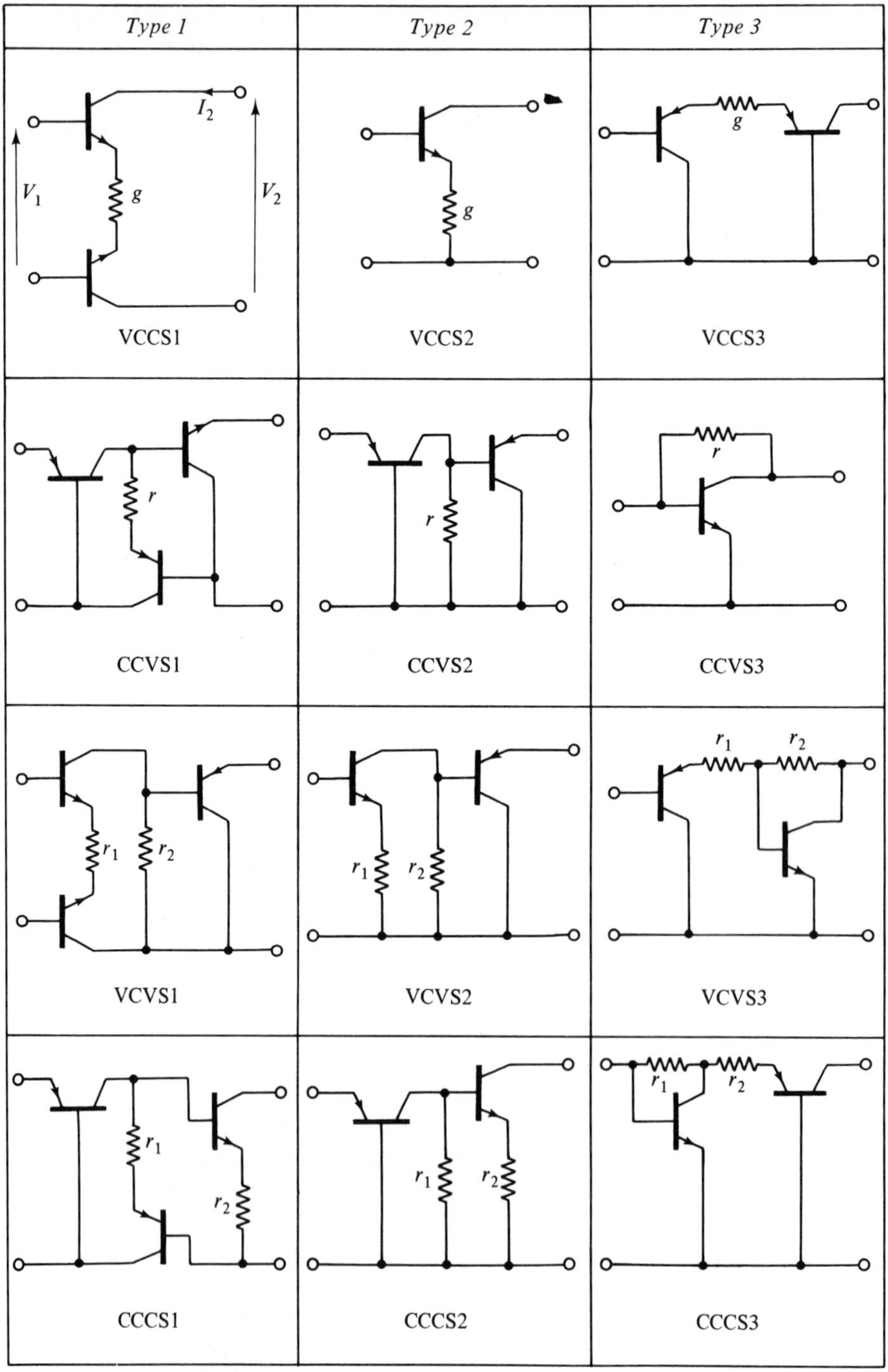

TABLE 4.1 *Basic transistor amplifier configurations.*

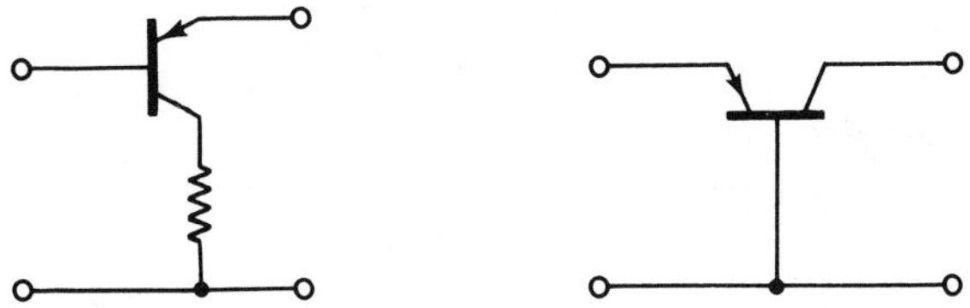

(a) Unity-gain VCVS (b) Unity-gain CCCS

FIGURE 4.8

The designer of a transistor amplifier is frequently concerned with the creative process of interconnecting the circuits of Table 4-1 in such a way that they may be biased in the linear region, with the minimum of additional circuitry, in order to meet the overall design objectives that may include the sinusoidal steady-state frequency domain behavior, the time-domain transient behavior, the input and output impedance levels, the dynamic range, the noise performance, the stability, the power dissipation and the allowable nonlinear distortion. Consideration of these matters is beyond the scope of this text. Particular circuit configurations that have been found to be useful are described briefly in the following sections and a number of related problems appear at the end of the chapter.

To complete the list of ideal transistor circuits, the *ideal* transistor versions of the *unity-gain* controlled sources of Fig. 2.8 are shown in Fig. 4.8.

4-3.1 Single-Transistor Amplifiers

The small-signal equivalent circuits of the three widely used single-transistor amplifiers are shown in Fig. 4.9: they are defined as the common-base, common-emitter, and common-collector configurations. The ideal voltage gains are determined by substituting nullors in place of the transistors, thus,

$$\frac{V_2(s)}{V_1(s)} = \frac{+R_C}{R_E} \qquad \text{COMMON-BASE CONFIGURATION} \tag{4.20}$$

and

$$\frac{V_2(s)}{V_1(s)} = -\frac{R_C}{R_E} \qquad \text{COMMON-EMITTER CONFIGURATION} \tag{4.21}$$

and

$$\frac{V_2(s)}{V_1(s)} = 1 \qquad \text{COMMON-COLLECTOR CONFIGURATION} \tag{4.22}$$

Problems 4-12 to 4-15 relate to the analysis of these configurations.

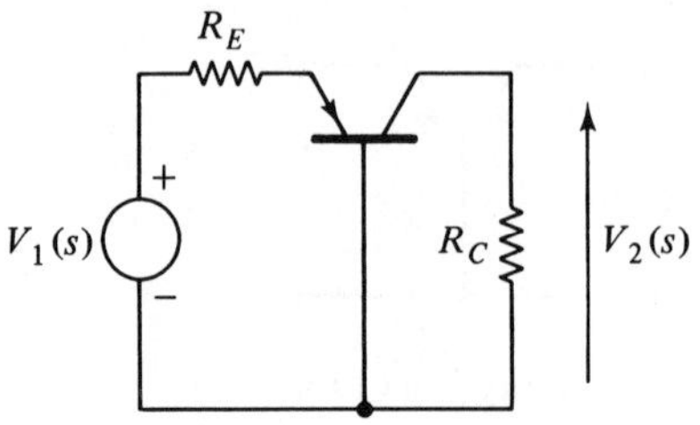

Common-base configuration
(CCVS2, part of)

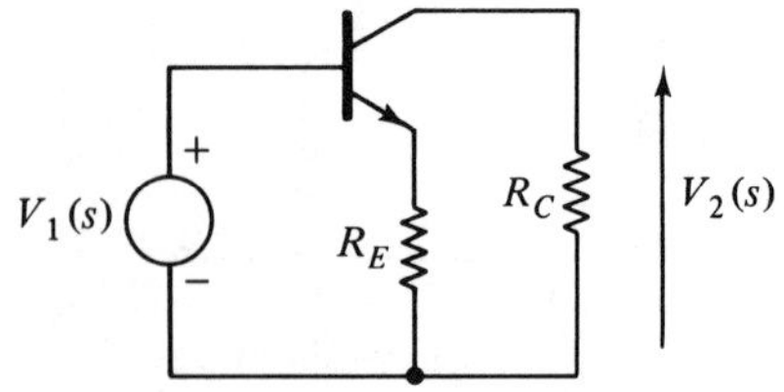

Common-emitter configuration
(VCCS 2)

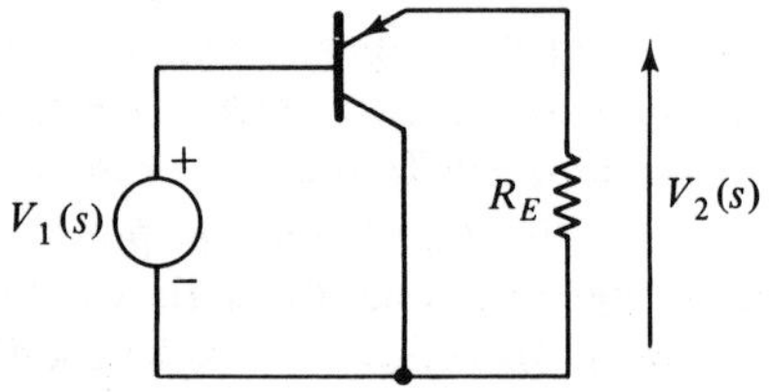

Common collector configuration

FIGURE 4.9 *Single transistor amplifier configurations.*

4-3.2 The Differential Amplifier

The basic feature of the voltage *differential* amplifier is that it amplifies the *difference* between two input signals $v_{B1}(t)$ and $v_{B2}(t)$. There are many different bipolar transistor circuits that are used as differential amplifiers. In this section, a typical circuit is analyzed in detail.

The two-transistor VCCS1 circuit shown in Table 4-1 provides the essential part of a voltage differential amplifier. For example, the transistors T1 and T2 in the differential amplifier circuit shown in Fig. 4.10(a) correspond to the two transistors of VCCS1 in Table 4-1. The static supply voltages E_1 and E_2, the resistors R_1, R_2, and R_3 and the transistor T3 are used solely for the purpose of biasing transistors T1 and T2. The circuit is designed to amplify the small-signal difference voltage $[v_{B1}(t) - v_{B2}(t)]$ with the output voltages appearing as the small-signal voltages $v_0(t)$ and $v_{C1}(t)$, $v_{C2}(t)$ at the collectors of transistor T1, T2.

The behavior of this differential amplifier circuit is best understood by deriving the static equivalent circuit and *then* the small-signal equivalent circuit. As the derivation proceeds, it is occasionally useful to refer to the following set of typical numerical values for the circuit parameters:

$$\left.\begin{array}{llll} E_1 = +20\text{ V}, & E_2 = -20\text{ V}, & R_1 = 30\text{ k}\Omega, & R_2 = 10\text{ k}\Omega \\ R_3 = 10\text{ k}\Omega, & R_E = 250\ \Omega, & R_C = 20\text{ k}\Omega, & V_{BEO} = 0.7\text{ V} \end{array}\right\} \quad (4.23)$$

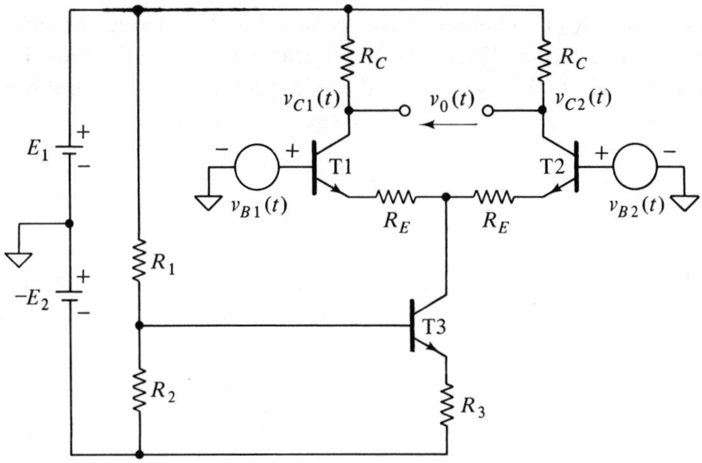

(a)

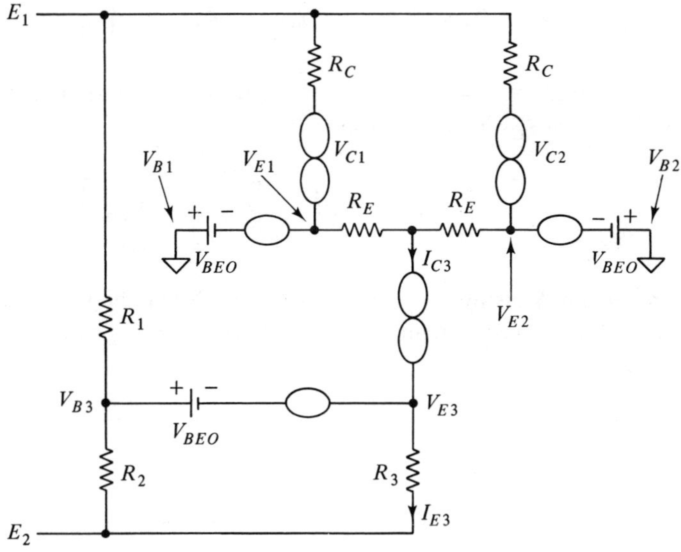

(b)

FIGURE 4.10 *The voltage differential amplifier.*

The static (dc) equivalent circuit is calculated by using the infinite-β_0 static equivalent circuit [Fig. 4.5(c)] for transistors T1, T2, and T3. This leads immediately to Fig. 4.10(b) and, by inspection of this figure it is clear that the static voltages V_{B1}, V_{B2}, V_{E1}, and V_{E2} are given by

$$V_{B1} = V_{B2} = 0 \tag{4.24}$$

$$V_{E1} = V_{E2} = -V_{BEO} \tag{4.25}$$

To establish the bias conditions $V_{CE1,2}$ and $I_{C1,2}$ for each transistor it is necessary that the collector current I_{C3} be determined. The voltage V_{B3} is given by

$$V_{B3} = E_2 + (E_1 - E_2)\left(\frac{R_2}{R_1 + R_2}\right) \tag{4.26}$$

which is -10.0 V, using the numerical values above. Then,

$$V_{E3} = V_{B3} - V_{BEO} = E_2 - V_{BEO} + (E_1 - E_2)\left(\frac{R_2}{R_1 + R_2}\right) \tag{4.27}$$

which is -10.7 V, using the numerical values. The current $I_{E3} = (V_{E3} - E_2)\,R_3 = I_{C3}$ leads, therefore, to

$$I_{C3} = \frac{1}{R_3}\left[(E_1 - E_2)\left(\frac{R_2}{R_1 + R_2}\right) - V_{BEO}\right] \tag{4.28}$$

which is 0.930 mA, using the numerical values. Since the resistors R_E have identical voltages across them, it follows that $I_{E1} = I_{E2} = I_{C3}/2 = I_{C1} = I_{C2}$, so that

$$\boxed{I_{C1} = I_{C2} = \frac{1}{2R_3}\left[(E_1 - E_2)\left(\frac{R_2}{R_1 + R_2}\right) - V_{BEO}\right]} \tag{4.29}$$

which is 0.465 mA, using the numerical values. Using $V_{C1,2} = E_1 - I_{C1,2}R_C$, we obtain

$$V_{C1} = V_{C2} = E_1 - \frac{R_C}{2R_3}\left[(E_1 - E_2)\left(\frac{R_2}{R_1 + R_2}\right) - V_{BEO}\right] \tag{4.30}$$

which is $+10.7$ V, using numerical values. Knowing $V_{C1,2}$ and $V_{E1,2}$ it follows that

$$
\boxed{\begin{aligned} V_{CE1} &= V_{CE2} = V_{C1,2} - V_{E1,2} \\ &= E_1 + V_{BEO} - \frac{R_C}{2R_3}\left[(E_1 - E_2)\frac{R_2}{R_1 + R_2} - V_{BEO}\right] \end{aligned}} \tag{4.31}
$$

which is 11.40 V, using the numerical values. The static (dc) analysis of the circuit is now complete; all node voltages have been determined.

The *small-signal equivalent circuit* may now be determined. This circuit is shown in Fig. 4.11(a), where the independent sources E_1 and E_2 are replaced by short-circuits. Note that the transistor T3 and associated resistors R_1, R_2, and R_3 in Fig. 4.11(a) are redundant and may be removed because T3 and R_3 are equivalent to the VCCS2 of Table 2-9 with a zero small-signal input voltage $v_{B3}(t)$; this implies that the circuit containing T3 and R_3 is simply a current source $i_{C3}(t)$ of zero magnitude which is, of course, an open-circuit. Removing the redundant open-circuit results in the reduced equivalent circuit shown in Fig. 4.11(b), where T1 and T2 are replaced by their nullor equivalents and *with the resistances* $r_{e1,2}$ *retained* as the only significant nonideal transistor parameters. By direct analysis of

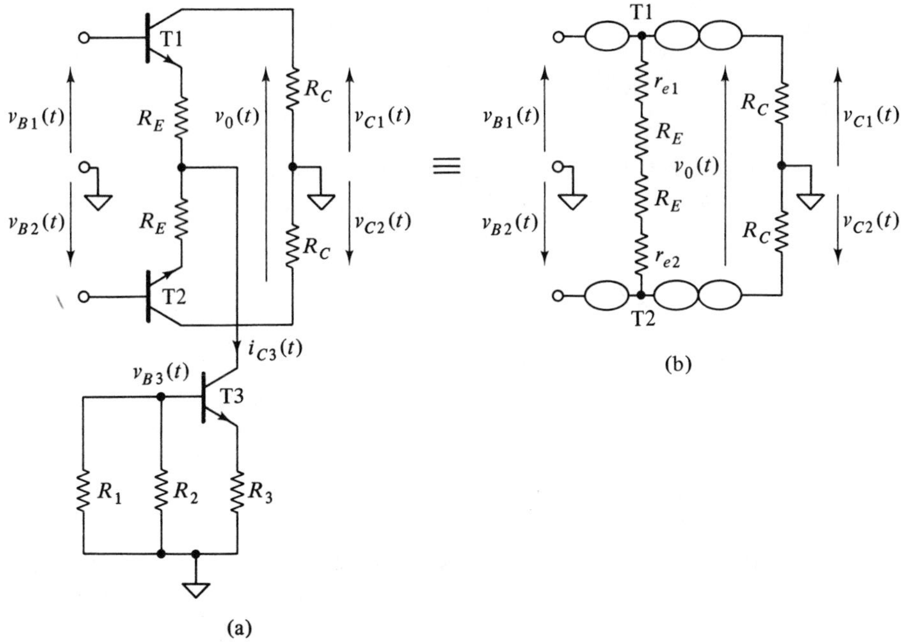

FIGURE 4.11 *Small-single circuits for the voltage differential amplifier.*

the equivalent circuit of Fig. 4.11(b), we find that

$$\boxed{\frac{v_{C1}(t)}{\Delta v_B(t)} = \frac{R_C}{2R_E + r_{e1} + r_{e2}} \equiv A_1} \tag{4.32}$$

where $\Delta v_B(t) \equiv v_{B2}(t) - v_{B1}(t)$ and is the *differential* input voltage. Similarly,

$$\boxed{\frac{v_{C2}(t)}{\Delta v_B(t)} = -\frac{R_C}{2R_E + r_{e1} + r_{e2}} \equiv A_2} \tag{4.33}$$

The voltages $v_{C1,2}(t)$ are the collector voltages of T1 and T2 (with respect to ground) and, therefore, A_1 and A_2 are defined as the *single-ended voltage gains*. Note that by neglecting C_π, r_π, and C_μ it has been shown that $A_1 = -A_2$. In general, it may be shown that it is necessary that T1 and T2 possess *identical* hybrid-π parameters in order to ensure that $A_1 = -A_2$ for all $\Delta v_B(t)$.

Knowing from equation (4.29) that $|I_{E1,2}| = 0.465$ mA leads, via equation (4.9), to $r_{e1,2} = 55.95\ \Omega$. Substituting numerical values for R_C, R_E, and $r_{e1,2}$ into equations (4.32) and (4.33), we get

$$\boxed{A_1 = -A_2 = 32.69} \tag{4.34}$$

4-4 TRANSISTOR REALIZATIONS OF GENERALIZED IMMITTANCE CONVERTERS (GICs) AND GYRATORS

The GIC is an important building-block in the design of *RC*-active networks, so it is worthwhile to consider the means by which the *ideal* transistor equivalent circuit of Fig. 4.6(c) may be used to derive *small-signal equivalent circuits* for the various GICs that are classified in Tables 3-1 and 3-2. In this way, it is found that most of the common gyrators, NIC, NII, PII and PIC transistor circuits may be readily understood.

4-4.1 Type 1 Transistor GICs

The three-terminal nullors of GIC1B and GIC1C in Table 3-1 may be replaced by the *ideal* transistor model of Fig. 4.6(c) in a variety of ways. Three possible transistor realizations are shown in Table 4-2.

The transistor implementation of GIC1A is not obvious because the nullator and norator do not share a common terminal. However, by

GIC 1A	GIC 1B	GIC 1C
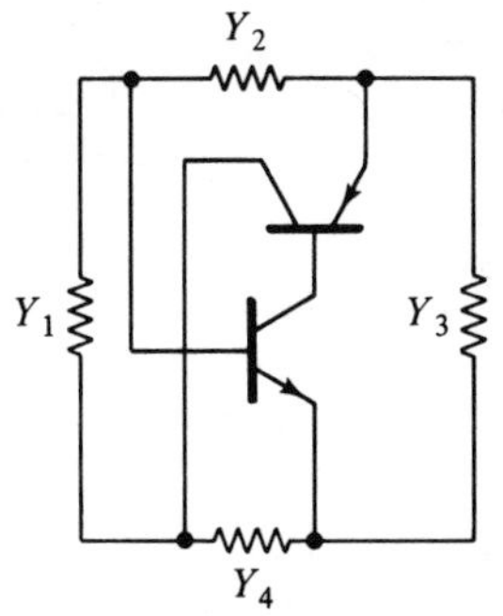	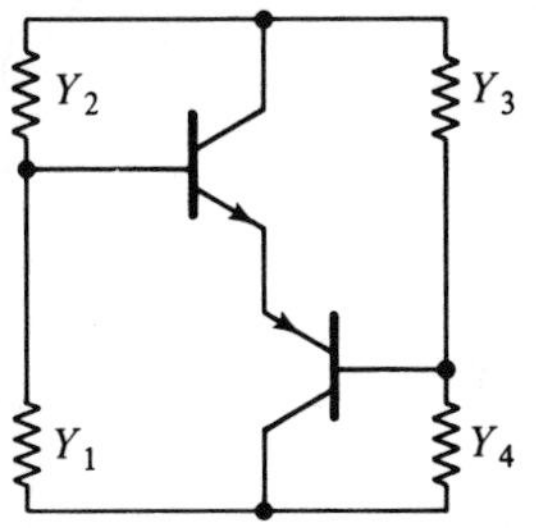	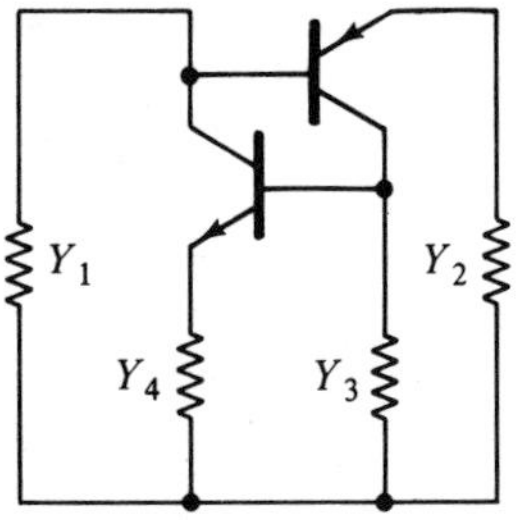

TABLE 4.2 *Transistor realizations of Type 1 GICs.*

adding a redundant open-circuit as shown in Fig. 4.12 it follows that an equivalent network containing two three-terminal nullors has been created. Direct replacement of these two nullors by ideal transistors leads to the GIC1A transistor circuit shown in Table 4-2. Practical versions of the small-signal equivalent circuits of Table 4-2 require careful design, especially with regard to the dc biasing circuitry. An example is now given of the use of GIC1C in Table 4-2 to implement a negative resistance.

Design of a negative resistance using GIC1C: The circuit shown in Fig. 4.13(a) is a biased version of GIC1C shown in Table 4-2. Note that I_{DC} is a dc current source that may be designed by using one transistor and three resistors as in the differential amplifier example (transistor T3) of section 4-3.2. The *ideal* driving-point impedance Z_d is given by equation (3.49) and is, simply,

$$Z_d = -\frac{R_1 R_3}{R_2} \tag{4.35}$$

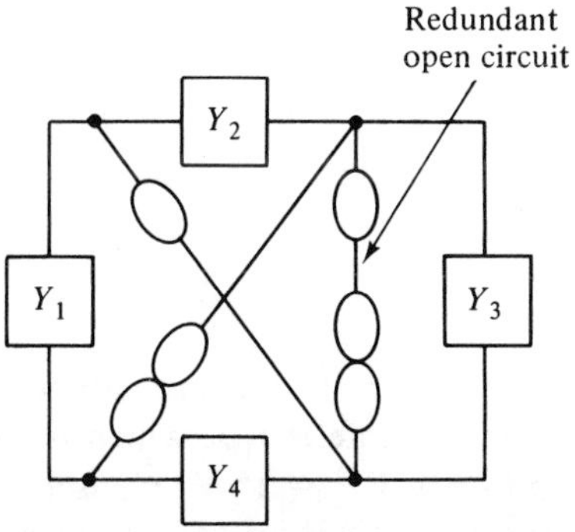

FIGURE 4.12

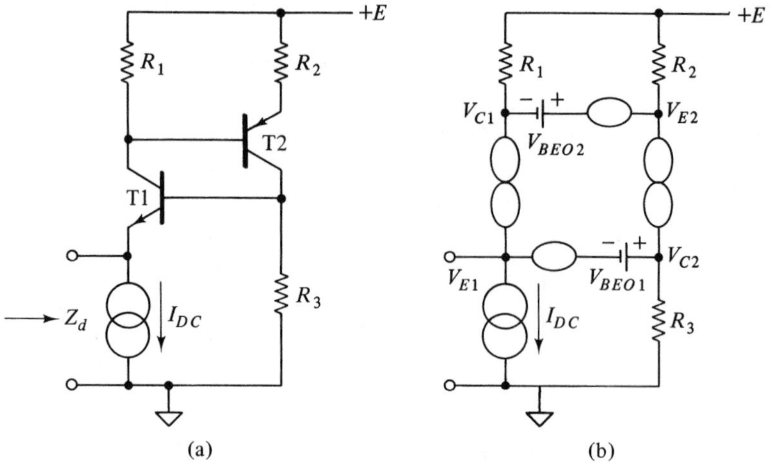

FIGURE 4.13 *Negative resistance circuit.*

The dc bias conditions may be calculated from Fig. 4.13(b) by using conventional nodal analysis for the unknown node voltages $V_{C1,2}$ and $V_{E1,2}$. It is more direct in this case to write by inspection the following four independent equations describing Fig. 4.13(b):

$$\left.\begin{aligned} V_{C1} &= E - I_{DC}R_1 \\ V_{E2} &= V_{C1} + V_{BEO2} \\ V_{C2} &= \frac{R_3}{R_2}\left[E - V_{E2}\right] \\ V_{E1} &= V_{C2} - V_{BEO1} \end{aligned}\right\} \tag{4.36}$$

which allow the designer to select $V_{C1,2}$ and $V_{E1,2}$ so as to provide satisfactory bias conditions while simultaneously constraining the resistors R_1, R_2, and R_3 to provide the desired value for Z_d in equation (4.35). For example, let

$$E \equiv 20.0\text{ V}, \qquad V_{BEO1} \equiv 0.7\text{ V}, \qquad V_{BEO2} \equiv 0.2\text{ V} \tag{4.37}$$

and assume that we require

$$Z_d = -10\text{ k}\Omega \tag{4.38}$$

Then, *selecting* $R_1 = R_2 = R_3 = 10$ kΩ immediately satisfies equation (4.35). The only remaining variable is I_{DC}, which is selected as $I_{DC} = 0.5$ mA so that, by equation (4.36), $V_{C1} = 15$ V. Consequently, $V_{E2} = 15.7$ V. $V_{C2} = 4.3$ V, and $V_{E1} = 4.1$ V. The bias conditions at each transistor are, therefore,

$$\left.\begin{aligned} V_{CE1} &\equiv V_{C1} - V_{E1} = 11.9\text{ V} \\ I_{C1} &= 0.5\text{ mA} \\ V_{CE2} &\equiv V_{C2} - V_{E2} = -11.4\text{ V} \\ I_{C2} &= \frac{V_{C2}}{R_3} = 0.43\text{ mA} \end{aligned}\right\} \tag{4.39}$$

4-4.2 Type 2 Transistor GICs

The type 2 GICs are useful because they may be employed to realize PIIs and, therefore, to simulate driving-point inductances and FDNR elements as explained in chapter 5. It follows by inspection of Table 3-2 that GIC2C and GIC2D are suitable for realization by direct substitution of ideal transistors for each of the three terminal nullors.

GIC2C transistor circuits: Inspection of GIC2C in Table 3-2 reveals that there are several ways in which the nullators and norators may be paired as three-terminal nullors. Therefore, it is relatively straightforward to develop *ideal* transistor versions of GIC2C. For example, the small-signal circuits in Fig. 4.14(a) and Fig. 4.14(b) are obtained by replacing the three-terminal nullors of GIC2C by *ideal* transistors. Both circuits possess the transmission matrix

$$T = \begin{bmatrix} 1 & 0 \\ 0 & \dfrac{Y_2 Y_4}{Y_3 Y_5} \end{bmatrix} \tag{4.40}$$

and are, therefore, small-signal circuit versions of an impedance-converting GIC (see section 2-6.1). The GIC circuits obtained by substituting *ideal* transistors in this way are often quite impractical because they are not easily biased. However, the particular realization in Fig. 4.14(a) *has* been implemented successfully and used in *RC*-active filter design.

GIC2D transistor circuits: Direct substitution of *ideal* transistors into GIC2D of Table 3-2 results in the three-transistor PII shown in Fig. 4.15(a). This circuit was one of the first successful electronic gyrators and is a cascade of three common-emitter circuits, having the transmission matrix

$$T = \begin{bmatrix} 0 & \dfrac{R_2 R_4}{R_3} \\ \dfrac{1}{R_6} & 0 \end{bmatrix}, \quad \text{PII, Fig. 4.15(a)} \tag{4.41}$$

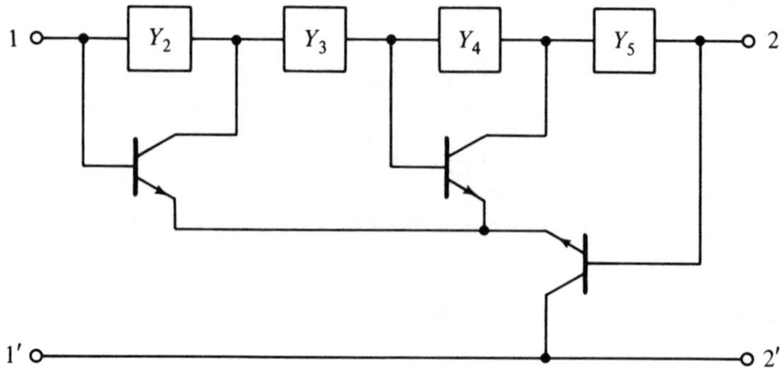

(a)

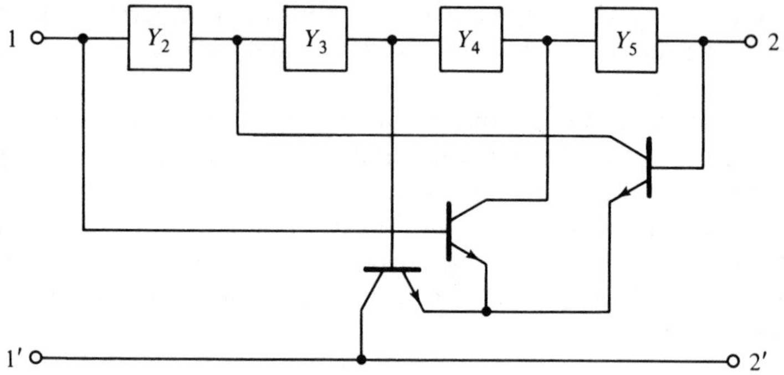

(b)

FIGURE 4.14

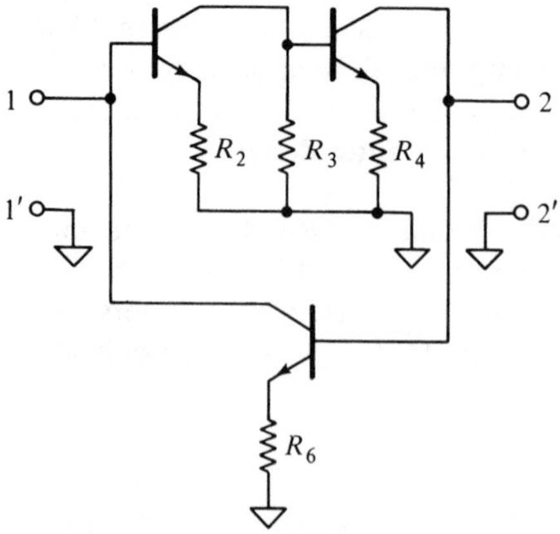

(a)

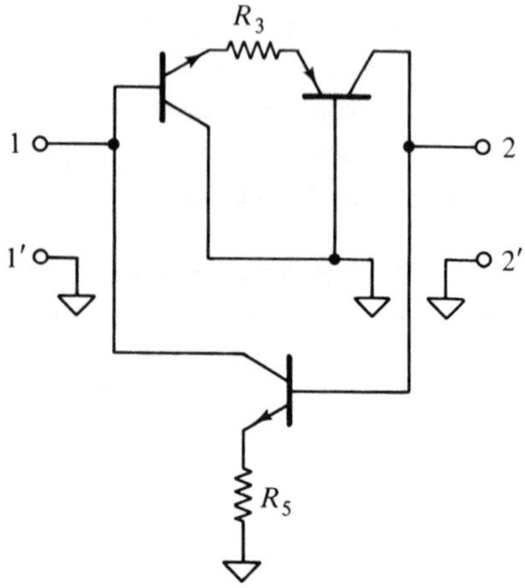

(b)

FIGURE 4.15 *Two transistor PII circuits.*

of a PII and was first used to simulate an inductance at port 1 by terminating port 2 in a capacitance (see Example 2.1 in chapter 2). A circuit similar to Fig. 4.15(a) employs a common-base stage and a common-collector stage in place of the upper two common-emitter stages and is shown in Fig. 4.15(b). The transmission matrix of this modified version of a GIC2D transistor circuit is as follows:

$$T = \begin{bmatrix} 0 & R_3 \\ \dfrac{1}{R_5} & 0 \end{bmatrix}, \quad \text{PII, Fig. 4.15(b)} \tag{4.42}$$

4-5 A PRACTICAL HIGH-QUALITY PII-GYRATOR USING TRANSISTORS

A PII that is highly successful in practical applications is the *differential amplifier* version of GIC2D. The circuit uses a cascade of the controlled sources VCCS1 and VCCS2 (that are shown in Tables 2-8 and 2-9) to realize two high-quality VCCSs, as shown in Fig 4.16. Transistors T1, T2, T3, and T4 are associated with the VCCS1 circuits and transistors T5, T6, and T7 with the VCCS2 circuits. The four dc current sources are required for biasing; typical numerical values for the resistors are

$$\left.\begin{aligned} R_{C1} &= R_{C3} = R_{C4} = 10\ \text{k}\Omega \\ R_{E1} &= R_{E2} = R_{E3} = R_{E4} = 5\ \text{k}\Omega \\ R_{E5} &= R_{E6} = R_{E7} = 1\ \text{k}\Omega \end{aligned}\right\} \tag{4.43}$$

This circuit is a versatile PII that is useful for inductance simulation because a grounded capacitance at port 2 appears as a *floating* (that is, ungrounded) *inductance* at port 1. The static dc analysis and the dynamic small-signal analysis are performed in a similar way to that employed to analyze the voltage differential amplifier in section 4-3.2. This PII is, in fact, two voltage differential controlled-current sources connected in parallel.

A brief explanation of the small-signal operation of this PII follows; a more detailed study of the circuit is requested in problems 4.20 to 4.23. Transistors T1, T2, and T5 form a small-signal differential input single-ended output VCCS given by

$$I_2(s) = +\frac{R_{C1}V_1(s)}{(R_{E1} + r_{e1} + R_{E2} + r_{e2})(R_{E5} + r_{e5})} \tag{4.44}$$

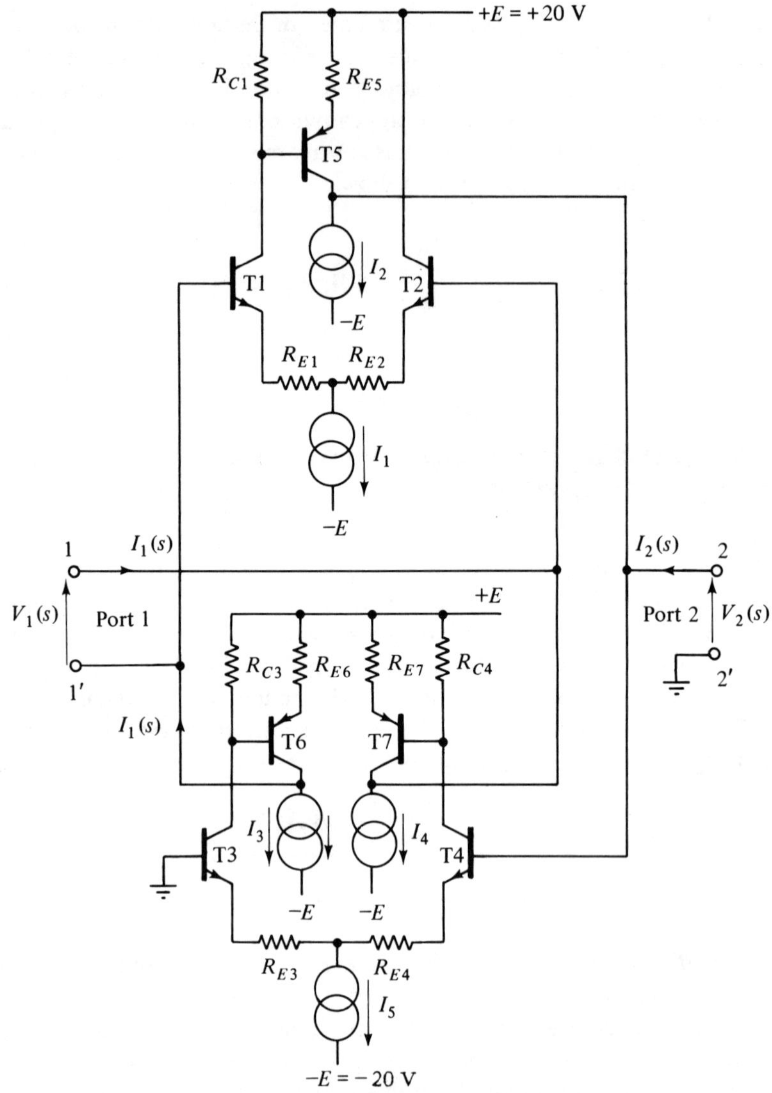

FIGURE 4.16 *A high quality transistor PII circuit.*

and transistors T3, T4, T6, and T7 from a small-signal differential input-differential output VCCS given by

$$I_1(s) = -\frac{R_{C3,4}}{(R_{E3} + r_{e3} + R_{E4} + r_{e4})(R_{E6} + r_{e6})} \tag{4.45}$$

where it is assumed that r_{e1}, r_{e2}, r_{e3}, r_{e4}, r_{e5}, and r_{e6} are the only nonideal parameters associated with the transistors and that $R_{C3} = R_{C4} \equiv R_{C3,4}$. It follows from equations (4-44) and (4-45) that the transmissions matrix T of the PII shown in Fig. 4.16 is

$$T = \begin{bmatrix} 0 & -\dfrac{(R_{E1} + r_{e1} + R_{E2} + r_{e2})(R_{E5} + r_{e5})}{R_{C1}} \\ \dfrac{-R_{C3,4}}{(R_{E3} + r_{e3} + R_{E4} + r_{e4})(R_{E6} + r_{e6})} & 0 \end{bmatrix} \tag{4.46}$$

If this circuit is employed to simulate a floating (ungrounded) inductance at port 1 by means of a capacitance C_2 at port 2, then it becomes a straightforward calculation to determine the value of that inductance L_{eq}. It follows directly from equations (4-46) and (2-28) that

$$Z_{1d}(s) = \frac{(R_{E5} + r_{e5})(R_{E6} + r_{e6})(R_{E1} + r_{e1} + R_{E2} + r_{e2})(R_{E3} + r_{e3} + R_{E4} + r_{e4})(sC_2)}{R_{C1}R_{C3,4}}$$

giving

$$\boxed{L_{eq} = \frac{(R_{E5} + r_{e5})(R_{E6} + r_{e6})(R_{E1} + r_{e1} + R_{E2} + r_{e2})(R_{E3} + r_{e3} + R_{E4} + r_{e4})C_2}{R_{C1}R_{C3,4}}} \tag{4.47}$$

Clearly, the simulated inductance L_{eq} is bias dependent via the relationships between $r_{e1,2,3,4,5,6}$ and I_E as described in equation (4.9). This dependence is minimized by ensuring that $r_{e1,2,3,4,5,6}$ are as small as possible, thus implying a large $|I_{E1,2,3,4,5,6}|$ and a correspondingly large power supply voltage E. The question of temperature dependence of L_{eq} is an important matter in RC-active filter inductance-simulation circuit design and is pursued in Problem 4.22.

Microelectronic versions of the transistor-GIC shown in Fig. 4.16 have been made successfully and high-quality filters have been constructed by replacing each of the inductors in a conventional passive *LC* filter by a capacitively terminated version of this circuit. The analysis and description of *inductance-simulation RC-active filter design* is contained in chapter 9.

4-6 SUMMARY

The bipolar transistor is a three-terminal device that may be biased in the linear region where its small-signal behavior may often be modeled by a three-terminal nullor. All of the controlled-source nullor circuits shown in Tables 2-8 and 2-9 have ideal transistor equivalent circuits, and many of those circuits that are shown in Table 4.1 are the basic transistor configurations from which transistor amplifiers are designed. In particular, VCCS1 is the basic building-block for the transistor voltage differential amplifier.

The type 1 and type 2 GIC networks are realizable by using transistors as shown in Table 4-2 and Figs. 4.14 and 4.15. A practical version of a PII-gyrator is described in section 4.5; microelectronic versions of this kind of gyrator circuit have been made and used to realize high-quality active inductance elements.

It has been found that bipolar transistors may be used to approximate three-terminal nullors. However, the practical problems associated with *biasing* the transistor realizations (see Table 4-1, for example) are usually quite difficult to solve, and the nonideal effects of the parameter r_e are significant impediments to the use of *simple* transistor realizations of *RC*-nullor networks. The *operational amplifier* is a device that employs a large number of transistors to realize a good approximation to a nullor. Furthermore, the operational amplifier is biased internally so that the user may virtually neglect the problems of biasing.

PROBLEMS

4-1. Prove that Fig. 4.5(b) is equivalent to Fig. 4.5(a).

4-2. Fig. P3.12(d) describes a hybrid-π small-signal representation of the bipolar transistor. Assume that

$$r_e = 26\ \Omega, \qquad R_L = 1\ \text{k}\Omega$$
$$C_\pi = 5\ \text{pF}, \qquad r_x = 25\Omega$$
$$C_\mu = 1\ \text{pF}, \qquad \beta_0 = 100$$
$$r_0 = 100\ \text{k}\Omega,$$

to calculate the steady sinusoidal voltage gain frequency response $V_2(j\omega)/V_1(j\omega)$. Plot the gain on logarithmic axes over the frequency range 100 Hz to 100 MHz. [*Note*: You may use the result of Problem 3.12(d) for this purpose.]

4-3. Repeat Problem 4.2, but by using the simplified hybrid-π model of Fig. 4.6(b). Comment on the range of frequencies over which Fig. 4.6(b), the *simplified* hybrid-π model, is a valid approximation to Fig. 4.6(a).

4-4. Determine the bias voltages V_{CE1} and V_{CE2} and the bias currents I_{C1} and I_{C2} for transistors T1 and T2 in each of the bias circuits shown in Fig. P4.4. Use the infinite-β_0 static model of Fig. 4.5(c) for each transistor. Assume $V_{BEO} = 0.7$ V for npn transistors and 0·2V for pnp transistors.

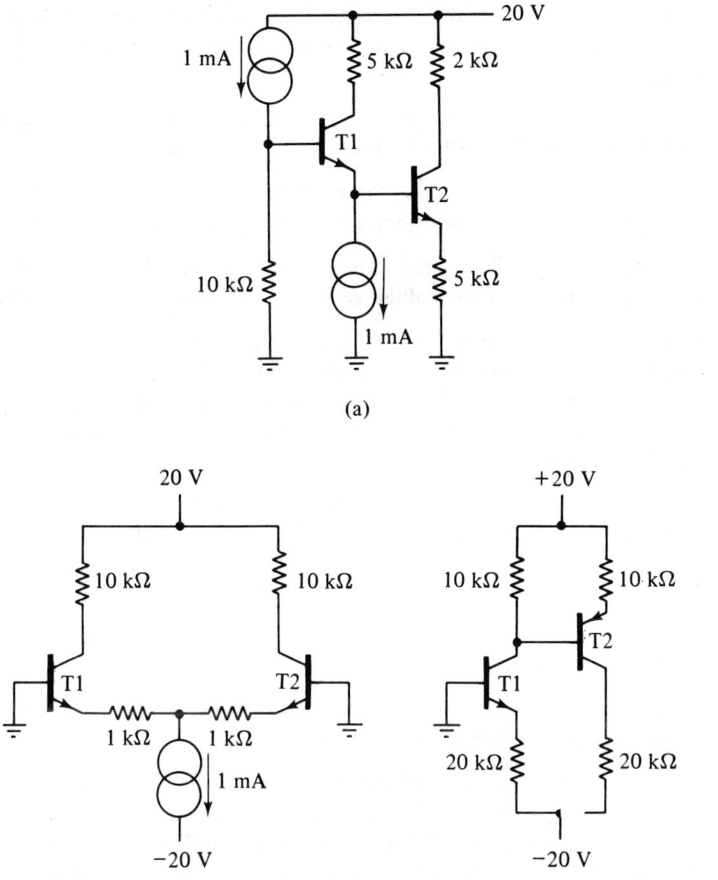

(b)

(c) *Note*: For pnp transistor T2 the voltage generator V_{BEO} in Fig. 4.5(a) is −0.2 V.

FIGURE P4.4

4-5. Write the *small-signal equivalent circuits* for the three transistor configurations of Problem 4-4. Assume that each transistor may be modeled by means of the ideal nullor network of Fig. 4.6(c).

4-6. Assuming ideal transistors [Fig. 4.6(c)], obtain the nullor versions of the small-signal equivalent circuits of Table 4-1 and then identify the resultant nullor two-port networks with the corresponding networks shown in Tables 2-8 and 2-9.

4-7. The nonzero values of r_e limit the accuracy of the transistorized controlled sources of Table 4-1. On the assumption that all the transistors shown in Table 4-1:

(a) have the same value for r_e, and

(b) that r_e is the only nonideal transistor parameter,

derive the nonideal controlling gain factors g' for the three VCCS circuits shown in Table 4-1. If $|I_E|$ is 1 mA for all transistors, how small must the conductance g be to ensure that g' is within approximately 1% of the ideal value?

4-8. Repeat Problem 4.7 for the three CCVS circuits shown in Table 4-1, deriving the nonideal controlling gain factor r', in terms of r and r_e.

4-9. Repeat Problem 4.7 for the three VCVS circuits shown in Table 4-1, deriving the nonideal controlling gain factor μ' in terms of r_2, r_1, and r_e.

4-10. Repeat Problem 4.7 for the three CCCS circuits shown in Table 4-1, deriving the nonideal controlling gain factor β' in terms of r_2, r_1, and r_e.

4-11. Comment on the implications, insofar as dc emitter bias currents are concerned, of the results obtained in Problems 4.7, 4.8, 4.9, and 4.10. For example, if $g^{-1}=r=r_2=r_1=10\ \text{k}\Omega$ and we require the controlling gain factors to be within 1% of their ideal values, what approximate values for $|I_E|$ are implied?

4-12. Use the small-signal equivalent circuit of Fig. 4.6(b) to derive the *nonideal* small-signal voltage gain $V_2(s)/V_1(s)$ of the *common-base* transistor amplifier configuration of Fig. 4.9 in terms of R_C, R_E, r_e, r_π, and C_π. Derive the pole locations of the transfer function $V_2(s)/V_1(s)$.

4-13. Repeat Problem 4.12 for the *common-emitter* configuration of Fig. 4.9. Comment and compare with Problem 4.12.

4-14. Repeat Problem 4.12 for the *common-collector* configuration of Fig. 4.9. Comment and compare with Problem 4.12.

4-15. Assuming $R_E=1\ \text{k}\Omega$, $R_C=10\ \text{k}\Omega$, $r_e=26\ \Omega$, $r_\pi=2600\ \Omega$, and $C_\pi=10$ pF, compare the pole locations and the sketched gain frequency response functions $|V_2(j\omega)/V_1(j\omega)|$ for the *common-base* and *common-emitter* configurations of Fig. 4.9. Which configuration has the larger bandwidth and which configuration has the higher input impedance?

4-16. Assuming r_e is the only nonideal transistor parameter for both transistors, prove that the driving-point admittance "seen" by the branch Y_1 of GIC 1C shown in Table 4-2 is given by

$$Y_{1d}=\frac{-\left(Y_4+\frac{1}{r_e}\right)\left(Y_2+\frac{1}{r_e}\right)}{Y_3}$$

A nonideal negative capacitance is realized by choosing $Y_4=sC_4$, $Y_2=G_2$, and $Y_3=G_3$. Derive the Q-factor of the nonideal negative capacitance at Y_{1d} in terms of C_4, G_2, G_3, and r_e, and calculate this Q-factor for $\omega=10^4$ rad/sec., $C_4=0.1\ \mu F$, $G_2=G_3=10^{-3}\ \Omega$, and $r_e=26\ \Omega$. Repeat the calculation of Q for $\omega=10^5$ rad/sec. Comment on the practical implications of these calculations.

4-17. Prove that both the small-signal equivalent GIC circuits shown in Fig. 4.14 have the ideal transmission matrix of equation (4-40).

4-18. Prove that Fig. 4.15(a) is a gyrator if $R_2=R_3=R_4=R_6\equiv g^{-1}$ and that the transconductance of the gyrator is g.

4-19. Prove that Fig. 4.15(b) is a gyrator if $R_3=R_5\equiv g^{-1}$ and that the transconductance of the gyrator is g.

4-20. The complete PII circuit of Fig. 4.16 may be used to realize an equivalent active inductance at port 1 by terminating it in a capacitance C_2 at port 2. Consider the dc analysis of this circuit under the following conditions:

$$I_1=I_5=2\ \text{mA}$$
$$I_2=I_3=I_4=1\ \text{mA}$$
$$+E=20\ \text{V},\quad -E=-20\ \text{V}$$
$$R_{C1}=R_{C3}=R_{C4}=10\ \text{k}\Omega$$
$$R_{E1}=R_{E2}=R_{E3}=R_{E4}=5\ \text{k}\Omega$$
$$R_{E5}=R_{E6}=R_{E7}=1\ \text{k}\Omega$$

Assume the infinite-β_0 dc models of Fig. P4.20 for the transistors.

Substitute the resistor values above and the dc transistor models below into Fig. 4.16 and, thereby, derive the base voltage V_B, collector voltage V_C,

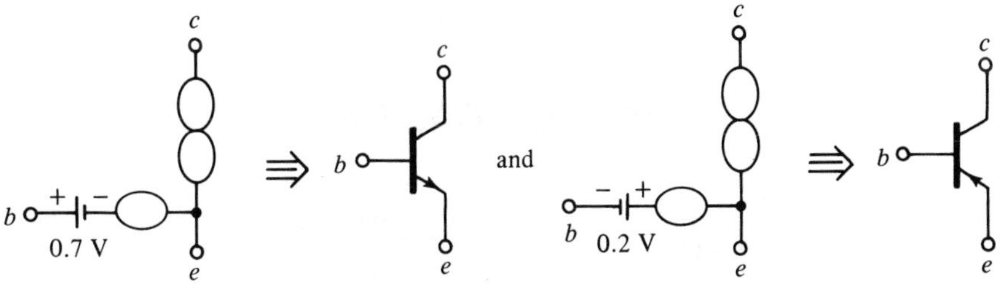

FIGURE P4.20

and emitter voltage V_E at each of the seven transistors T1, T2, T3, T4, T5, T6, T7. [*Hints*: Assume that terminals 1 and 1′ are at dc ground potential and that $I_{C1}=I_{C2}=I_1/2$, $I_{C3}=I_{C4}=I_5/2$. You may find it useful to calculate quantities in the following sequence: V_{E1}, V_{E2}, V_{C1}, V_{E5}, V_{E3}, V_{E4}, V_{C3}, V_{C4}, V_{E6}, V_{E7}.]

4-21. Prove that the small-signal transmission matrix of the PII shown in Fig. 4.16 is provided by equation (4.46).

4-22. Use equations (4.46) and (2.28) to prove that equation (4.47) is valid. Assuming a terminating capacitance C_2 of 10,000 pF in Problem 4.20, derive L_{eq} assuming $r_e=0\cdot 0/I_e|26|$ for each transistor at 300°K. Given that r_e is proportional temperature in °K., determine L_{eq} at 400°K.

4-23. Repeat Problem 4.22, but for an ideal value of zero for each r_e. Comment on the significance of r_e and T in contributing to the nonideal behavior of the circuit.

4-24. Assuming ideal transistors [Fig. 4.6(c)], show that the voltage gain e_2/e_1 of the small-signal equivalent circuit shown in Fig. P.4.24 is

$$\frac{e_2}{e_1}=\frac{R_3}{R_2}\left[1+\frac{R_5}{R_4}\right].$$

[*Hint*: R_1 and R_6 do not appear in the expression for e_2/e_1; you should expect that they may be eliminated using nullator-norator equivalences.]

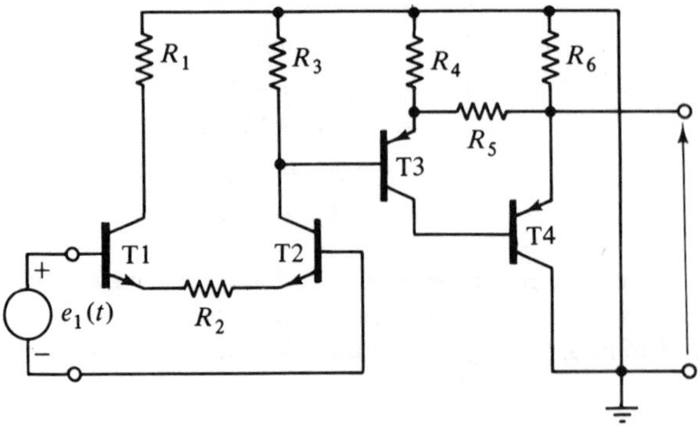

FIGURE P4.24

BIBLIOGRAPHY

MILLMAN, J., "*Microelectronics: Digital and Analog Circuits and Systems*" McGraw-Hill, 1979, chapters 3, 8, 10, 11, 12, 13, 14.

HEINLEIN, W., and H. HOLMES, "*Active Filters for Integrated Circuits,*" Prentice-Hall, chapter 5.

5

THE OPERATIONAL AMPLIFIER AND LINEAR CIRCUIT APPLICATIONS

5-1 INTRODUCTION

The operational amplifier is an important practical active building-block with which *RC*-active circuits are implemented. Interconnections of resistors, capacitors and operational amplifiers are used widely to make microelectronic *RC*-active filters. In this chapter, the operational amplifier is described and the corresponding nullor model is used to determine the ideal and the nonideal performance of the more widely used linear *RC*-active subcircuits.

5-2 THE OPERATIONAL AMPLIFIER DEVICE

The operational amplifier (OP AMP) is a high-gain electronic amplifier that can be fabricated as a low-cost microelectronic device. Progress in microelectronic technologies and high demand for the device have resulted in rapid decreases in cost and increases in performance. A typical modern OP AMP dissipates less than 50 mW with dissipations as low as 1 mW quite readily obtainable. The OP AMP operates over a wide range of power supply voltages with devices available that will operate on a supply voltage as low as 1.5 V and other special-purpose OP AMPs that operate with supply voltages as large as 100 V. As many as four separate OP AMPs may be fabricated on a single wafer of silicon and encapsulated in a single package. At present, a high-performance OP AMP costs less than a high-precision capacitor. For this reason, the designers of amplifiers and *RC*-active filters are motivated to use extra OP AMPs if this results in a saving of expensive precision capacitors or expensive large-valued capacitors. Techniques of amplifier design that require large coupling capacitors or large bypass capacitors are often disregarded in favor of multistage direct-coupled (capacitorless) circuits. The early designers of *RC*-active filters were constrained by cost to use only *one* amplifier; the availability of low-cost OP AMPs has, however, caused a drastic change in design techniques so that *multiple-OP AMP RC-active circuits* are often used if a saving in precision capacitors and/or precision resistors is achieved.

The OP AMP is also an important component in the design of *nonlinear* circuits such as rectifiers, modulators, analog multipliers, oscillators, trigger circuits, waveform generation circuits and timing circuits. It is, in fact, a fundamental building-block in electronic circuit design. In this chapter, the most important elementary properties of the OP AMP are explained, and linear applications are described with emphasis on the parameters that are of concern to the designer of *RC*-active filters.

5-2.1 The Ideal Operational Amplifier

The ideal OP AMP is equivalent to a nullor. It is, therefore, equivalent to an *infinite-gain* controlled source (see section 2-12) of any one of the four types: VCCS, CCCS, VCVS, or CCVS.

In practice, the OP AMP is usually designed with a differential input but with a single-ended output so that the corresponding nullor equivalent circuit possesses a norator that has one terminal at ground potential, as shown in Fig. 5.1(a). The symbol for the *ideal* OP AMP is also shown in Fig. 5.1(a): it is important to note that the grounded terminal 4 is *not* shown explicitly in the symbol.

The type 1 transistor amplifiers shown in Table 4-1 are clearly candidates for use as the input stage of an OP AMP because they possess *differential* inputs. At present, the voltage-controlled configurations VCCS1 and VCCS2 are used almost exclusively in the input stages of OP AMP circuits and lead to the type of differential amplifier circuit that is analyzed in section 4-3.2.

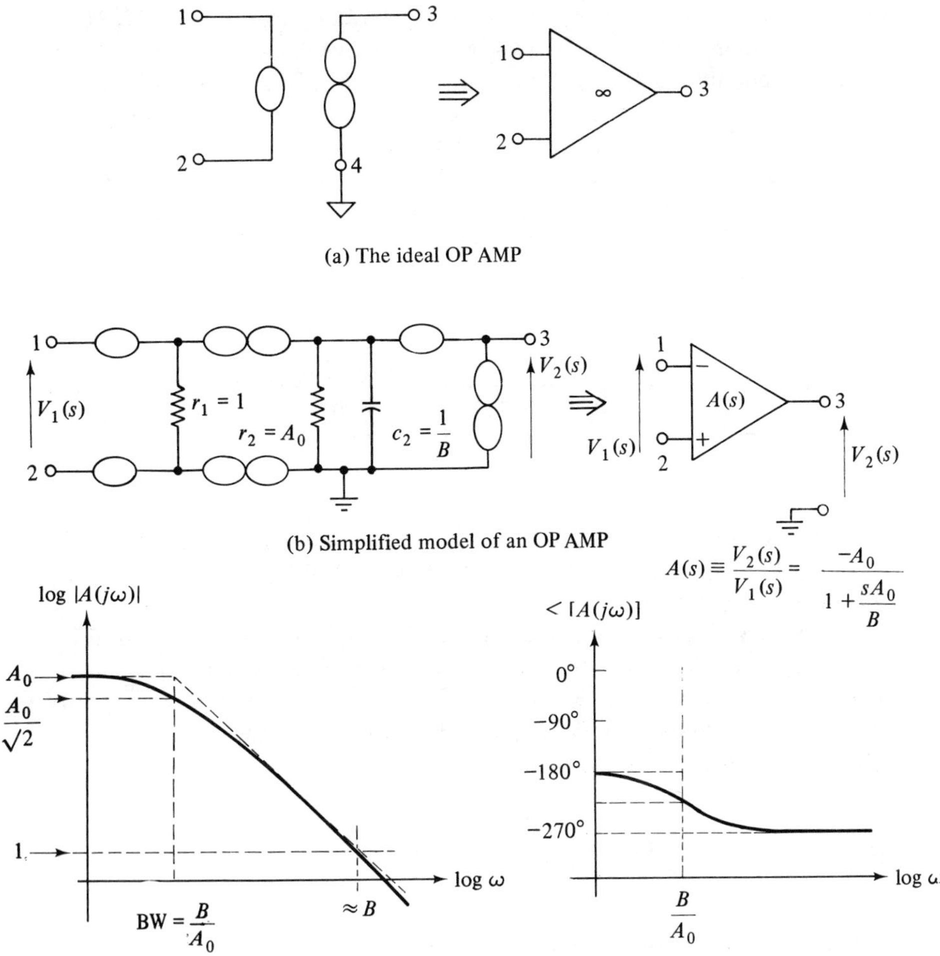

(a) The ideal OP AMP

(b) Simplified model of an OP AMP

(c) Magnitude frequency response $|A(j\omega)|$

(d) Phase frequency response $< [A(j\omega)]$

FIGURE 5.1

5-2.2 The Nonideal Operational Amplifier

An *ideal* OP AMP is not realizable in practice because the transistors themselves are nonideal. For example, the gain $-r_2/r_1$ of VCVS1 (Table 4-1) is limited by the fact that the emitter connections of the input transistors T1 and T2 possess resistances $r_{e1,2}$ in the corresponding small-signal transistor equivalent circuits [Fig. 4.6(b)]. Consequently, at low frequencies, the magnitude of the available voltage gain from VCVS1 is not greater than $r_2/(r_{e1} + r_{e2})$. Designers have overcome this limitation by *cascading* circuits such as VCVS1 and VCCS1 to realize the necessary high-gain. Several practical circuit implementations of the OP AMP are included in the problems at the end of this chapter.

The most widely used OP AMPs are designed to approximate a high-gain VCVS over a specific frequency range. A typical transform voltage transfer function of an OP AMP VCVS is the so-called *dominant-pole response*; that is,

$$A(s) = \frac{-A_0}{1 + s\left(\frac{A_0}{B}\right)} \qquad \text{DOMINANT-POLE RESPONSE} \tag{5.1}$$

The gain $A(j\omega)$ is closely approximated by equation (5.1) for all frequencies ω in the range where $|A(j\omega)| > 1$. As $\omega \to 0$, the gain $A(j\omega) \to -A_0$ where, of course, $A_0 \gg 1$.

The VCVS1 shown in Table 2-8 may be used to *model* equation (5.1) by simply providing an additional capacitance c_2 in parallel with r_2 to introduce the required pole at $s = -1/c_2r_2$. The resultant equivalent circuit and the symbol for the nonideal OP AMP are shown in Fig. 5.1(b), where the transfer function is given by

$$A(s) \equiv \frac{V_2(s)}{V_1(s)} = -\frac{r_2}{r_1}\left(\frac{1}{1 + sc_2r_2}\right) \tag{5.2}$$

so that, comparing equations (5.1) and (5.2), we may choose r_1, r_2, and c_2 in our model as follows:

$$r_1 \equiv 1\ \Omega, \qquad r_2 \equiv A_0\ \Omega, \qquad c_2 \equiv B^{-1}\ \text{F} \tag{5.3}$$

The steady-state sinusoidal frequency response $A(j\omega)$ is

$$A(j\omega) = \frac{-A_0}{1 + j\omega\left(\frac{A_0}{B}\right)} \tag{5.4}$$

from which it follows that the magnitude frequency response $|A(j\omega)|$ and the phase frequency response $< [A(j\omega)]$ are as follows:

$$|A(j\omega)| = \frac{A_0}{\sqrt{1 + \left(\frac{\omega A_0}{B}\right)^2}}$$

and

$$<\left[A(j\omega)\right] = -\pi - \arctan\left(\frac{\omega A_0}{B}\right) \tag{5.5}$$

These frequency response functions are shown in Figs. 5.1(c) and 5.1(d). The *bandwidth* BW of the response $|A(j\omega)|$ is usually defined as the frequency at which $|A(j\omega)| = (1/\sqrt{2})A_0$. Therefore, from equation 5.4 and the preceding definition of BW,

$$\boxed{\mathrm{BW} = \frac{B}{A_0}} \qquad \text{BANDWIDTH} \tag{5.6}$$

or

$$\boxed{B = A_0 \cdot \mathrm{BW}} \qquad \text{GAIN-BANDWIDTH PRODUCT} \tag{5.7}$$

Consequently, the parameter B is the product "bandwidth × dc gain A_0," and is, therefore, often known as the *gain-bandwidth product* (GBP) of the OP AMP. Clearly, the GBP is a measure of the ability of the amplifier to provide gain over a wide bandwidth. In the case of the dominant-pole response, GBP has a further physical interpretation: evaluating $A(j\omega)$ at $\omega = B$ we have

$$A(jB) = \frac{-A_0}{1 + jA_0} \tag{5.8}$$

so that, since $A_0 \gg 1$,

$$|A(jB)| \approx 1 \tag{5.9}$$

implying that B is also the radian frequency at which the magnitude response $|A(j\omega)|$ is approximately unity. For this reason, the GBP B is sometimes described as the *unity-gain frequency* of the OP AMP. The bandwidth BW and the GBP B are shown in Fig. 5.1(c). Note that, *for*

$\omega \gg BW$.

$$|A(j\omega)| \approx +\frac{B}{\omega} \quad \text{and} \quad <[A(j\omega)] \approx -\frac{3\pi}{2} \tag{5.10}$$

A typical modern low-cost OP AMP has the following parameters for the dc gain A_0 and the GBP B:

$$A_0 = 10^5, \qquad B = 10^7 \text{ rad/sec} \tag{5.10A}$$

This leads to a bandwidth BW of

$$\text{BW} = \frac{10^7}{10^5} = 10^2 \text{ rad/sec} \tag{5.10B}$$

so that, in this case, equations (5.10) are valid approximations in the frequency range well above 10^2 rad/sec or 15.9 Hz. It is important to be aware of the fact that the use of such an OP AMP at audio frequencies, say 50 Hz to 20 kHz, implies that the magnitude response $|A(j\omega)|$ is, approximately, inversely proportional to frequency throughout the frequency range with extremum values of 30,331 at 50 Hz and 79.5 at 20 kHz. The OP AMP is clearly not a particularly good approximation to a nullor at frequencies in excess of the low audio range. For this reason, it is usually justifiable to use the ideal nullor model of an OP AMP only in a first-order (ideal) analysis of a circuit. In consideration of nonideal behavior, we note that $A(s) \approx -B/s$, and the OP AMP behaves very much like a high-gain inverting integrator.

The parameters A_0 and B can vary significantly from one sample OP AMP to another. Manufacturers of OP AMPs usually specify minimum values and typical values for A_0 and B at a particular power supply voltage and temperature. In practice, this *spread* of A_0 and B between samples is typically about 3 : 1. The *temperature dependence* of A_0 and B can also be of practical significance. The typical dependence of A_0 and B on temperature is shown in Fig. 5.2. A common problem is to design an amplifier or an RC-active filter that has a tightly specified transfer function $H(j\omega)$. For example, $|H(j\omega)|$ may be required to an accuracy of 1% at all audio frequencies, for all OP AMP sample devices and over the entire temperature range $-55°$C to 125°C. The transfer function is then clearly required to be highly *insensitive* to perturbations of A_0 and B. This important problem of *desensitizing a network function* to perturbations of network parameters is significant in the design of *RC*-active filters and is the subject of chapters 6 and 7.

A more accurate model than Fig. 5.1(b) is shown in Fig. 5.3(a), where the nonideal input and output impedances Z_0, Z_I, Z_{CM1}, and Z_{CM2} are

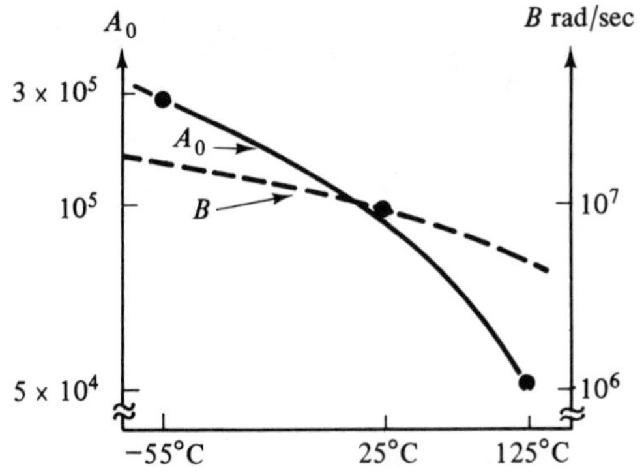

FIGURE 5.2 *Typical dependence of A_0 and B on temperature.*

included. The impedance Z_0 is the *output impedance*, Z_I the *differential-input impedance*, and $Z_{CM1,2}$ the *common-mode input impedances*. The effects of these impedances on circuit behavior are usually negligible by comparison with the effect of the finite frequency response $A(s)$. Typical numerical values, corresponding to the low-cost OP AMP described by equation (5.10A), are

$$Z_0 \equiv R_0 = 1\ \text{k}\Omega$$
$$Z_I \equiv R_I = 100\ \text{k}\Omega$$
$$Z_{CM1,2} \equiv R_{CM1,2} = 1\ \text{M}\Omega$$

Manufacturers of OP AMPs sometimes specify the input capacitances associated with Z_I and $Z_{CM1,2}$ so that they may be expressed as the parallel combinations

$$R_I \| \frac{1}{sC_I} \quad \text{and} \quad R_{CM1,2} \| \frac{1}{sC_{CM1,2}}$$

The corresponding VCVS model of the OP AMP is shown in Fig. 5.3(b), where the controlling factor of the VCVS is the frequency response function $A(s)$.

5-3 INVERTING OP AMP CIRCUITS

The OP AMP circuit shown in Fig. 5.4(a) is the inverting-type configuration. It is used extensively in the design of RC-active networks and may be used as an amplifier, integrator or differentiator. The *ideal* voltage transfer

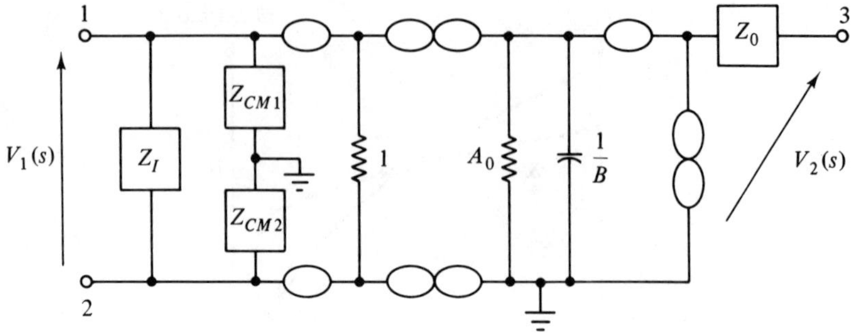

(a) Nullor model of the OP AMP

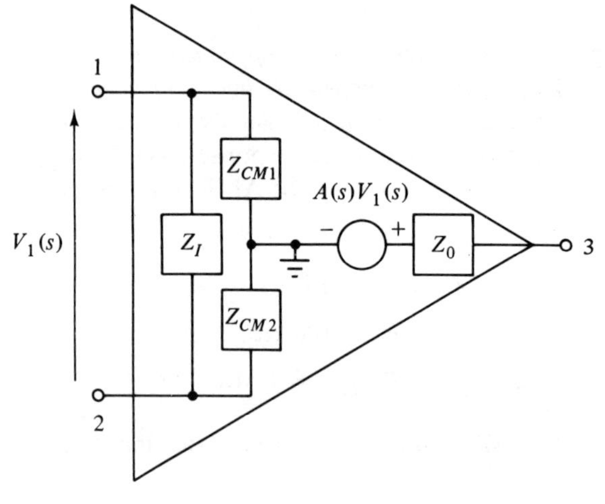

(b) VCVS model of the OP AMP

FIGURE 5.3

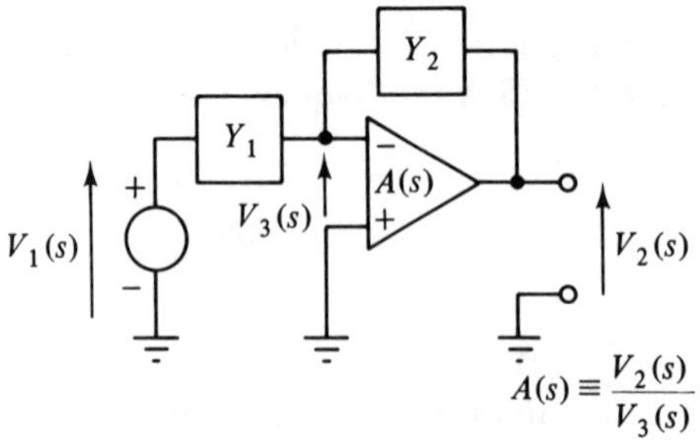

(a) Inverting-gain OP AMP circuit

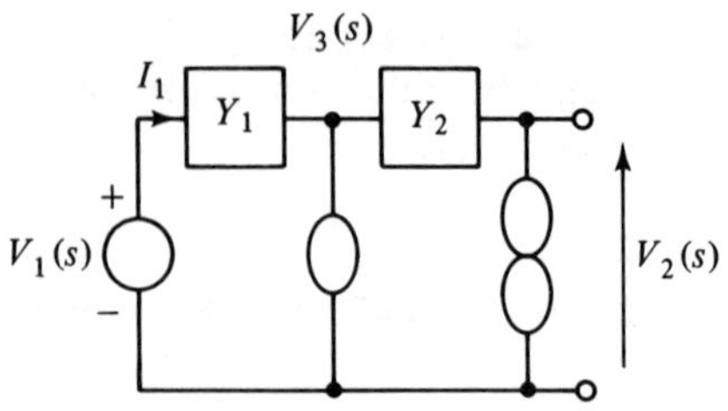

(b) Nullor equivalent of the ideal inverting-gain circuit, $A(s) = \infty$

FIGURE 5.4

function $H(s) = V_2(s)/V_1(s)$ is derived from the equivalent nullor model of the OP AMP that is shown in Fig. 5.4(b): the nullor and Y_2 form an inverting current-controlled voltage source (CCVS3 in Table 4-1), where the input current $I_1 = V_1Y_1$ so that

$$V_2 = -\frac{1}{Y_2}I_1 = -\frac{Y_1}{Y_2}V_1 \tag{5.11}$$

or

$$\boxed{H(s) \equiv \frac{V_2}{V_1} = -\frac{Y_1}{Y_2}} \quad \text{IDEAL TRANSFER FUNCTION} \tag{5.12}$$

The reader is requested, in Problem 5.2, to derive equation (5.12) by the nodal analysis technique of section 3.2.

The *ideal inverting-amplifier* is obtained if

$$Y_2 \equiv G_2, \qquad Y_1 \equiv KG_2, \qquad K \text{ real, positive} \tag{5.13}$$

so that

$$\boxed{H(s) = -K} \quad \text{IDEAL INVERTING-AMPLIFIER} \tag{5.13A}$$

corresponding to the circuit realization in Fig. 5.5(a).

The *ideal inverting-integrator* is defined by the input-output equation

$$v_2(t) = -\frac{1}{T}\int_{-\infty}^{t} v_1(t)\, dt \tag{5.14}$$

or

$$\boxed{H(s) \equiv \frac{V_2(s)}{V_1(s)} = -\frac{1}{sT}} \quad \text{IDEAL INVERTING INTEGRATOR} \tag{5.15}$$

which is easily achieved by selecting

$$Y_2 \equiv sC_2, \qquad Y_1 \equiv G_1 \ \text{ and } \ T \equiv \frac{C_2}{G_1} \tag{5.16}$$

in equation (5.12). The corresponding circuit is shown in Fig. 5.5(b).

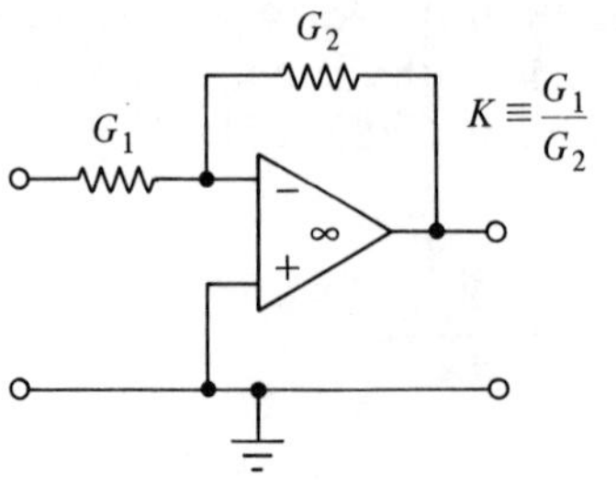

(a) Inverting amplifier

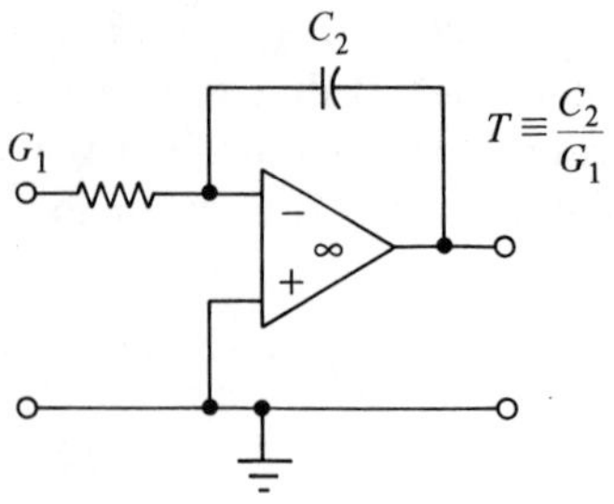

(b) Inverting integrator

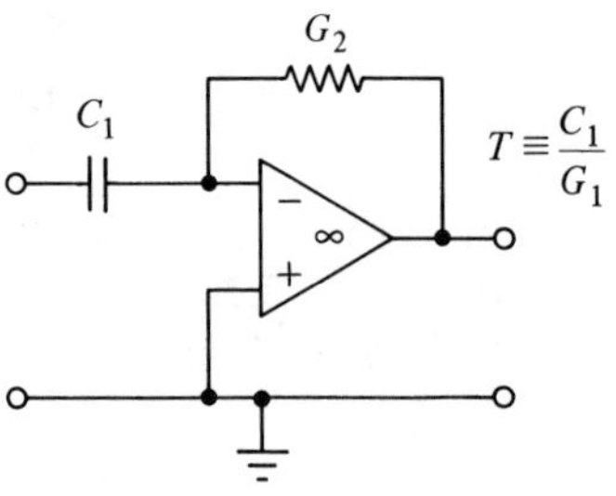

(c) Inverting differentiator

FIGURE 5.5

The *ideal inverting-differentiator* is defined by the input-output equation

$$v_2(t) = -T\frac{dv_1(t)}{dt} \tag{5.17}$$

or

$$\boxed{H(s) \equiv \frac{V_2(s)}{V_1(s)} = -sT} \qquad \text{IDEAL INVERTING DIFFERENTIATOR} \tag{5.18}$$

where the corresponding circuit is shown in Fig. 5.5(c).

Nonideal performance: The *finite* voltage gain $A(s)$ of the OP AMP is usually the most significant of the nonideal parameters in the equivalent circuit of the OP AMP, Fig. 5.3(b). Therefore, in this section, the nonideal impedances Z_I, $Z_{CM1,2}$, and Z_0 of the OP AMP are neglected and the OP AMP is assumed to be a VCVS with gain $A(s)$ as given by the dominant-pole expression in equation (5.1). Straightforward analysis of Fig. 5.4(a) is performed by writing Kirchoff's current law at the inverting input terminal of the OP AMP, hence:

$$[V_1(s) - V_3(s)]Y_1(s) = [V_3(s) - V_2(s)]Y_2(s) \tag{5.19}$$

Substituting $V_3(s) = V_2(s)/A(s)$ into equation (5.19) and rearranging, we have

$$H(s) \equiv \frac{V_2(s)}{V_1(s)} = -\frac{Y_1(s)}{Y_2(s)}\left[\frac{1}{1 - A(s)^{-1}\left[1 + \frac{Y_1(s)}{Y_2(s)}\right]}\right] \tag{5.20}$$

Comparing equations (5.20) and (5.12) reveals that the nonideal transfer function $H(s)$ differs from the ideal function by the *multiplicative* function $M(s)$, where

$$M(s)^{-1} \equiv 1 - A(s)^{-1}\left[1 + \frac{Y_1(s)}{Y_2(s)}\right] \tag{5.21}$$

and, of course, $M(s)$ is ideally unity if $A(s) = \infty$.

The *nonideal steady-state sinusoidal frequency response* $H(j\omega)$ is derived from equation (5.20) and is pursued here for the case of the *inverting amplifier*. Substituting equations (5.4) and (5.13) into equation (5.20), we have

$$H(j\omega) = -K\left[\frac{1}{1 + \left(A_0^{-1} + j\frac{\omega}{B}\right)(1 + K)}\right] \qquad \text{NONIDEAL GAIN OF INVERTING AMPLIFIER} \tag{5.22}$$

The corresponding magnitude gain function $|H(j\omega)|$ is shown in Fig. 5.6 for the typical OP AMP parameters of equation (5.10A). The curve for $K = \infty$ is simply the gain $|A(j\omega)|$ of the OP AMP. It follows from equation (5.22) that the high-frequency asymptote is

$$H(j\omega) \approx -\left(\frac{K}{1 + K}\right)\frac{B}{j\omega} \tag{5.23}$$

where $\omega \gg B/(1 + K)$ and $\omega \gg B/A_0$. Observe that, for $K \gg 1$, this asymptote is simply the function $-B/j\omega$.

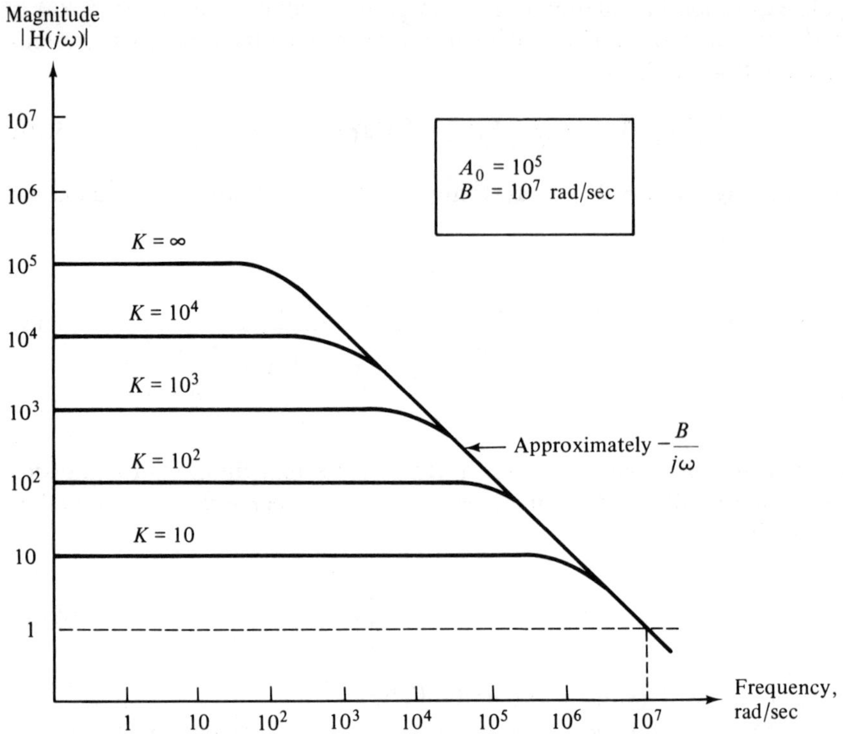

FIGURE 5.6 *Magnitude frequency response of the inverting-gain amplifier for different values of K.*

The bandwidths BW of the curves $|H(j\omega)|$ in Fig. 5.6 clearly decrease as K increases. The definition of BW implies that $< H(j\omega)$ is $-3\pi/2$ at ω equal to BW so that, from equation (5.22),

$$\frac{\text{BW}}{B}(1 + K) = 1 + A_0^{-1}(1 + K) \tag{5.24}$$

and, for $A_0 \gg K + 1$, equation (5.24) gives

$$\boxed{\text{BW} \approx \frac{B}{(1 + K)}} \qquad \text{BANDWIDTH OF THE INVERTING-GAIN AMPLIFIER CIRCUIT} \tag{5.25}$$

Note that the bandwidth BW is, approximately, inversely proportional to the ideal gain K if $K \gg 1$ and $A_0 \gg K + 1$.

In the design of high-quality *RC*-active filters, it is almost always necessary to employ this circuit such that $H(j\omega)$ is very close to the ideal value $-K$ over the range of ω for which the performance of the filter is of interest. That is, where

$$\omega \ll \text{BW} \tag{5.26}$$

and, therefore, where

$$M(j\omega) \approx 1 \tag{5.27}$$

Under these conditions, it is nevertheless often important to *know* the small deviation of $H(j\omega)$ from the ideal value $-K$. Consequently, substituting equations (5.13) and (5.1) into equation (5.21) we obtain

$$M(j\omega)^{-1} = 1 + \left(A_0^{-1} + j\frac{\omega}{B}\right)(1 + K) \tag{5.28}$$

and, since $A_0 \gg 1$ and $A_0 \gg K + 1$,

$$|M(j\omega)| \approx \left[1 + \frac{\omega^2(1+K)^2}{B^2}\right]^{-1/2} \quad \text{and} \quad < M(j\omega) \approx -\arctan\frac{\omega(1+K)}{B} \tag{5.29}$$

The functions $|M(j\omega)|$ and $< M(j\omega)$ are useful in *RC*-active filter design because they provide measures of the departure of $H(j\omega)$ from the ideal value $-K$. In high-gain applications, where $K \gg 1$ and $\omega K/B \ll 1$,

$$M(j\omega) \approx 1 - \frac{\omega^2 K^2}{2B^2} \quad \text{and} \quad < M(j\omega) \approx -\frac{\omega K}{B} \tag{5.30}$$

The equations above are especially useful in the determination of the dependence of multiple-OP AMP *RC*-active filter transfer functions on the GBP B of the OP AMPs.

This completes the discussion of the nonideal performance of the inverting-amplifier. A similar analysis may be performed for the inverting-integrator and the inverting-differentiator OP AMP circuits. Problems 5.6 and 5.7 request that the reader derive the corresponding $M(j\omega)$ functions shown in Table 5-1. Note that all the phase error functions $< M(j\omega)$ shown in Table 5-1 are *negative*. This has an important practical consequence in terms of the nonideal performance of *RC*-active filters that are realized by interconnecting inverting amplifiers and inverting-integrators. The subject is pursued in chapter 10.

TABLE 5.1

Circuit	*Ideal H(s)*	$\lvert M(j\omega)\rvert$	$< M(j\omega)$	*Assumptions*
Inverting-amplifier	$-K \equiv -\dfrac{R_2}{R_1}$	$1-\dfrac{\omega^2(K+1)^2}{2B^2}$	$-\arctan\left[\dfrac{\omega(K+1)}{B}\right]$	$A_0 \gg 1$ $A_0 \gg K+1$
Inverting-integrator	$-\dfrac{1}{sT} \equiv -\dfrac{1}{sC_2R_1}$ $T \equiv C_2R_1$	$1-\dfrac{1}{BT}-\left(\dfrac{\omega^2}{2B^2}\right)$	$-\arctan\left[\dfrac{\omega}{B}\right]$	$A_0 \gg BT$ $A_0 \gg \dfrac{B}{\omega^2T}$ $A_0 \gg 1$ $B \gg \omega^2T$
Inverting-differentiator	$-sT \equiv -sC_1R_2$ $T \equiv C_1R_2$	$1+\dfrac{\omega^2T}{B}-\left(\dfrac{\omega^2}{2B^2}\right)$	$-\arctan\left[\dfrac{\omega}{B}\right]$	$A_0 \gg \dfrac{B}{\omega^2T}$ $A_0 \gg BT$ $B \gg \omega^2T$ $B \gg \dfrac{1}{T}$

The summation amplifier circuit: The OP AMP summation circuit and the corresponding ideal nullor equivalent circuit are shown in Fig. 5.7. The ideal relationship between the output $V_2(s)$ and the multiple-inputs $V_{1A}(s)$, $V_{1B}(s)$, $V_{1C}(s)\ldots$ is derived from Fig. 5.7 by noting that the network is a CCVS (see CCVS3 in Table 2-9) with multiple-input currents $V_{1A}(s)Y_{1A}(s)$, $V_{1B}(s)Y_{1B}(s), \ldots,$ hence,

$$V_2(s) = -\frac{1}{Y_2(s)}\left[V_{1A}(s)Y_{1A}(s) + V_{1B}(s)Y_{1B}(s) + \ldots\right] \quad (5.31)$$

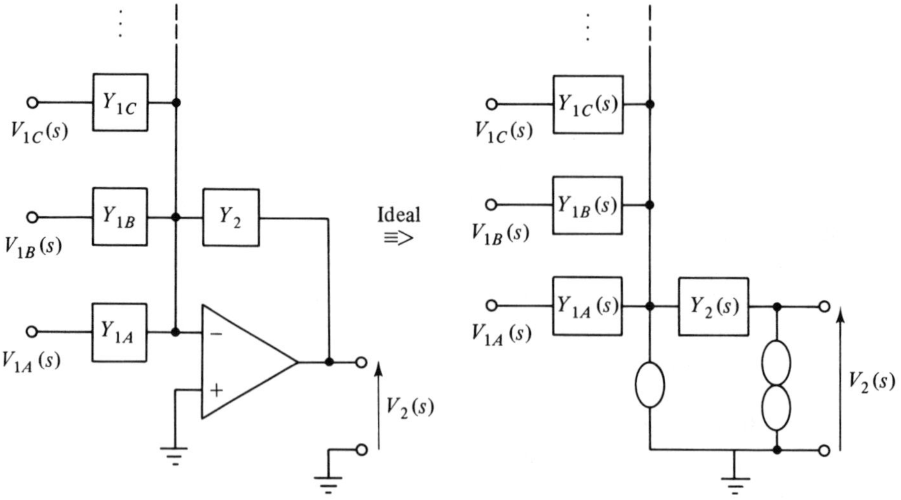

FIGURE 5.7 *Inverting Summer circuit.*

A broadly used application of this circuit is to provide inverting-summation of a set of voltages $V_{1A}(s)$, $V_{1B}(s) \ldots$ by selecting $Y_2(s) = Y_{1A}(s) = Y_{1B}(s) \ldots \equiv G$, so that

$$V_2(s) = -[V_{1A}(s) + V_{1B}(s) + V_{1C}(s) + \ldots] \tag{5.32}$$

The nonideal performance of this circuit is similar to that of the single-input inverting amplifier.

5-4 NONINVERTING OP AMP CIRCUITS

In this section, the noninverting amplifier and the noninverting integrator circuits are explained and the nonideal performance is derived in terms of the $M(j\omega)$ function for each circuit.

5-4.1 The Noninverting Amplifier

The *noninverting amplifier* circuit is shown in Fig. 5.8 with the corresponding ideal nullor equivalent circuit, for which it is straightforward to show that the ideal voltage gain $H(s)$ is

$$\boxed{H(s) \equiv \frac{V_2(s)}{V_1(s)} = \frac{R_1 + R_2}{R_1}} \tag{5.33}$$

so that the circuit behaves as a noninverting VCVS of gain $+K$ if R_2 is selected as $(K - 1)R_1$.

The *nonideal* gain of Fig. 5.8 may be derived by noting that the voltage across the differential-input terminals of the finite-gain OP AMP is

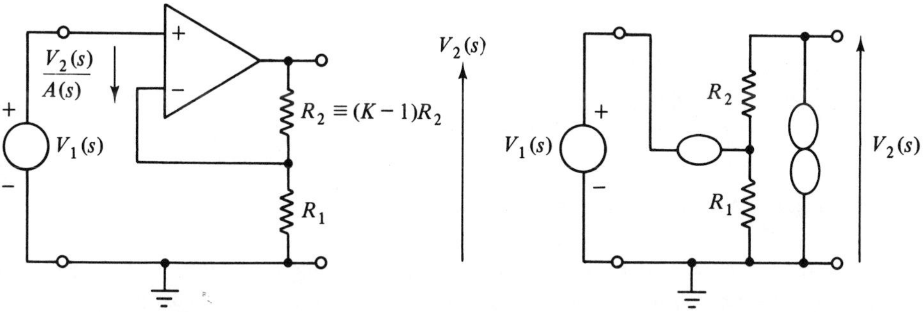

FIGURE 5.8 *Noninverting amplifier circuit.*

given by $V_2(s)/A(s)$, where $A(s)$ is the voltage gain of the OP AMP; then, writing KVL at the input side of the OP AMP, we have

$$V_1(s) = V_2(s)\left(\frac{R_1}{R_1 + R_2}\right) - \frac{V_2(s)}{A(s)} \tag{5.34}$$

Rearranging and using $R_2 = (K - 1)R_1$ leads to

$$H(s) \equiv \frac{V_2(s)}{V_1(s)} = K\left[\frac{1}{1 - A^{-1}(s)K}\right] \tag{5.35}$$

Substituting the dominant-pole frequency response of equation (5.1) leads directly to

$$\boxed{H(j\omega) \approx K\left[\frac{1}{1 + j\dfrac{\omega K}{B}}\right]} \quad \text{FREQUENCY RESPONSE OF NONINVERTING AMPLIFIER} \tag{5.36}$$

where it is assumed that $A_0 \gg K$. Consequently,

$$M(j\omega) \approx \left[1 + j\frac{\omega K}{B}\right]^{-1} \tag{5.37}$$

This function is useful in the analysis of the nonideal performance of RC-active filters containing this amplifier. The bandwidth BW of $H(j\omega)$ is obtained from equation (5.36) by equating the term $\omega K/B$ to unity so that

$$\boxed{\text{BW} \approx \frac{B}{K}} \tag{5.38}$$

Comparing equations (5.38) and (5.25) it is clear that for $K \gg 1$ the inverting and noninverting OP AMP circuits possess approximately the same bandwidth BW. In the case where $K = 1$, the noninverting circuit possesses twice the bandwidth BW of the inverting circuit.

5-4.2 Noninverting Integrator Circuits

There are at least two types of *noninverting integrator circuits* that are used in the realization of RC-active filters. The important distinguishing features of the two circuits are the nonideal functions $M(j\omega)$. These circuits, types 1A and 1B, employ two OP AMPs and are equivalent to the two-nullor network shown in Fig. 5.9. It is readily shown that the transfer function of this ideal noninverting integrator network is

$$H(s) = \frac{K}{sT} \tag{5.39}$$

where $T \equiv R_1C_2$ and $K \equiv R_4/R_3$. In fact, the network is simply the cascade of two of the previously considered *inverting-gain* OP AMP circuits.

The two-nullor network (Fig. 5.9) may be used to design *two distinct types* of noninverting integrators.

Type 1A phase-lag noninverting integrator: An ideal OP AMP is associated with nullator a and norator x in Fig. 5.9, and a second ideal OP AMP is associated with nullator b and norator y. This leads to the type 1A noninverting integrator circuit shown in Table 5-2. Of course, the circuit is nothing but the cascade connection of an inverting-amplifier $-K$ and an inverting-integrator $-1/sT$ with the ideal transfer function of equation (5.39). The *nonideal* transfer function need not be derived in detail, because as shown in Table 5-1, the $M(j\omega)$ functions have been derived for both the inverting-amplifier and the inverting-integrator. Therefore, multiplying the previously derived $M(j\omega)$ functions for these latter two circuits leads directly to the $|M(j\omega)|$ and $< M(j\omega)$ functions for the type 1A noninverting integrator:

$$|M(j\omega)| = \left[1 - \frac{\omega^2(K+1)^2}{2B_1^2}\right]\left[1 - \frac{1}{B_2T} - \frac{\omega^2}{2B_2^2}\right] \tag{5.40}$$

which may be simplified by using the assumptions

$$B \equiv B_1 = B_2, \quad \omega \ll B, \quad BT \gg 1, \quad K \equiv 1 \tag{5.40A}$$

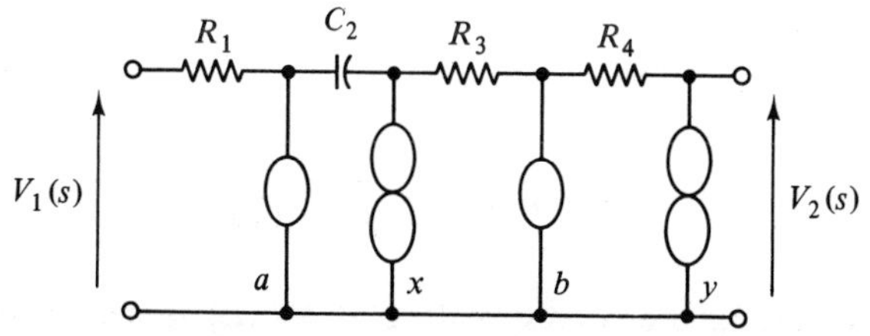

FIGURE 5.9 *Noninverting integrator nullor network.*

TABLE 5.2 *Type 1A and Type 1B noninverting integrator circuits.*

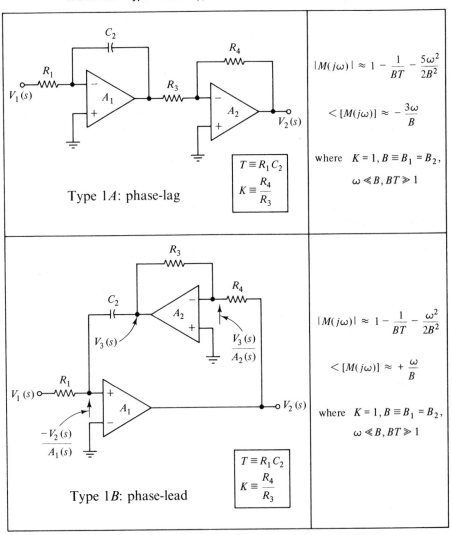

Circuit	Response
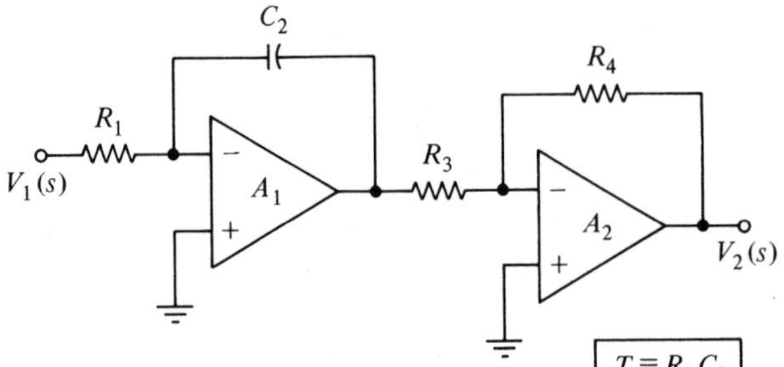 $T \equiv R_1 C_2$, $K \equiv \frac{R_4}{R_3}$ Type 1A: phase-lag	$\lvert M(j\omega) \rvert \approx 1 - \frac{1}{BT} - \frac{5\omega^2}{2B^2}$ $< [M(j\omega)] \approx -\frac{3\omega}{B}$ where $K = 1, B \equiv B_1 = B_2,$ $\omega \ll B, BT \gg 1$
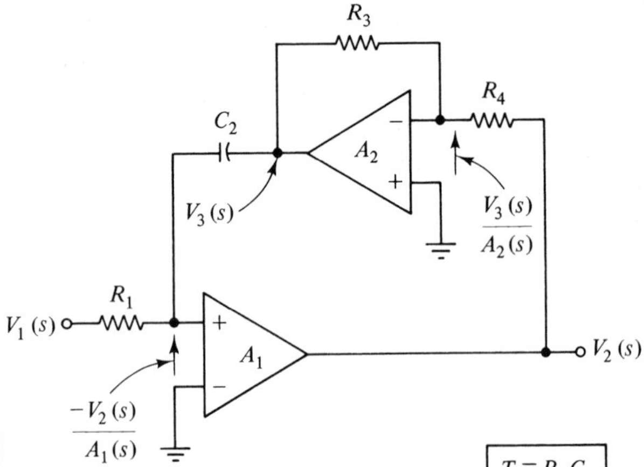 $T \equiv R_1 C_2$, $K \equiv \frac{R_4}{R_3}$ Type 1B: phase-lead	$\lvert M(j\omega) \rvert \approx 1 - \frac{1}{BT} - \frac{\omega^2}{2B^2}$ $< [M(j\omega)] \approx +\frac{\omega}{B}$ where $K = 1, B \equiv B_1 = B_2,$ $\omega \ll B, BT \gg 1$

to

$$|M(j\omega)| \approx 1 - \frac{1}{BT} - \frac{5\omega^2}{2B^2} \tag{5.41}$$

Similarly, by adding the corresponding $< M(j\omega)$ terms corresponding to the inverting-amplifier and inverting-integrator shown in Table 5-1 it follows that, for the type 1A noninverting integrator,

$$\boxed{\angle M(j\omega) \approx -\frac{3\omega}{B}} \qquad \text{PHASE-LAG ERROR TERM OF TYPE 1A CIRCUIT} \tag{5.42}$$

under the assumptions in equation (5.40A). The fact that $\angle M(j\omega)$ is negative leads to the description of the type 1A noninverting integrator as the PHASE-LAG NONINVERTING INTEGRATOR. Equation (5.42) has important consequences in the design of high-quality RC-active filters, as explained in chapters 8 and 9.

Type 1B phase-lead noninverting integrator: If ideal OP AMPs are realized by pairing nullator a with norator y and nullator b with norator x, then a different two-OP AMP circuit from the type 1A is obtained; the resultant circuit is shown as type 1B in Table 5-2. This circuit has the same *ideal* transfer function as the type 1A version, but the nonideal transfer function as represented by $M(j\omega)$ differs from the type 1A circuit. The derivation of $M(j\omega)$ is now given, using the assumptions given in equation (5.40A). The transfer function $V_3(s)/V_2(s)$ of the inverting-amplifier of the type 1B circuit is obtained directly from equation (5.20):

$$\frac{V_3(s)}{V_2(s)} = \frac{-1}{1 - 2A_2^{-1}(s)}, \qquad K \equiv 1 \tag{5.43}$$

where, ideally, $A_2(s)^{-1}$ is zero and the transfer function is simply -1 ($\equiv -K$). Then, by application of KCL at the noninverting input of OP AMP A_1, it follows that

$$\frac{1}{R_1}\left[V_1(s) + \frac{V_2(s)}{A_1(s)}\right] = \left[\frac{-V_2(s)}{1 - 2A_2^{-1}(s)} + \frac{V_2(s)}{A_1(s)}\right](-sC_2) \tag{5.44}$$

Rearranging, we have

$$H(s) \equiv \frac{V_2(s)}{V_1(s)} = \frac{1}{sT}\left[-\frac{1}{A_1(s)sT} + \frac{1}{1 - 2A_2^{-1}(s)} - \frac{1}{A_1(s)}\right]^{-1} \tag{5.45}$$

By substituting

$$A_{1,2}(s)^{-1} = -\left(A_{01,2}^{-1} + \frac{s}{B_{1,2}}\right)$$

into equation (5.45) we have

$$H(s) = \frac{1}{sT}\left[\frac{1}{sT}\left(\frac{1}{A_{01}} + \frac{s}{B_1}\right) + \frac{1}{A_{01}} + \frac{s}{B_1} + \frac{1}{\left(1 + \dfrac{2}{A_{02}} + \dfrac{2s}{B_2}\right)}\right]^{-1} \tag{5.46}$$

The term $[\ \ ]^{-1}$ in equation (5.46) is the nonideal multiplicative function $M(s)$ which is, of course, unity for $A_{01,2} = B_{1,2} = \infty$. Substituting $s = j\omega$ into equation (5.46), we have

$$M(j\omega) = \left[\frac{1}{j\omega T}\left(A_{01}^{-1} + \frac{j\omega}{B_1}\right) + A_{01}^{-1} + \frac{j\omega}{B_1} + \frac{1}{1 + 2A_{02}^{-1} + \dfrac{2j\omega}{B_2}}\right]^{-1}$$

where $H(j\omega) = M(j\omega)/j\omega T$. We now make the usual assumptions that

$$\frac{\omega}{B_{1,2}} \ll 1 \quad \text{and} \quad A_{01,2} \gg 1$$

which allow simplification to

$$M(j\omega) \approx \left[1 + \frac{1}{B_1 T} + j\omega\left(\frac{1}{B_1} - \frac{2}{B_2} + \frac{1}{\omega^2 T A_{01}}\right)\right]^{-1} \tag{5.47}$$

The A_{01} term is usually sufficiently large that

$$A_{01} \gg \frac{B_{1,2}}{\omega^2 T} \tag{5.47A}$$

and if both OP AMPs are from the same semiconductor chip (that is, if we use a dual- or quad-OP AMP package) then the assumption that

$$B_1 \approx B_2 \equiv B \tag{5.47B}$$

is accurate to within a few per cent and then equations (5.47), (5.47A), and (5.47B) indicate that

$$M(j\omega) \approx \left(1 + \frac{1}{BT} - j\frac{\omega}{B}\right)^{-1} \tag{5.48}$$

Using $BT \gg 1$ and $\omega/B \ll 1$, we have

$$\boxed{\angle[M(j\omega)] \approx +\frac{\omega}{B}} \qquad \text{PHASE LEAD ERROR TERM OF TYPE 1B CIRCUIT} \tag{5.49}$$

and

$$|M(j\omega)| \approx 1 - \frac{1}{BT} - \frac{\omega^2}{2B^2} \tag{5.50}$$

It is important to note that the phase term in equation (5.49) for the noninverting type 1B integrator is of equal magnitude and opposite sign to that of the inverting-integrator described in Table 5-1. This fact is used to advantage in the design of ultrahigh-quality RC-active ladder filters as described in chapter 10, where it is found that phase-cancellation techniques may be employed to drastically desensitize the dependence of the filter transfer function to the gain bandwidth products B of the OP AMPs.

5-5 OP AMP RELAIZATIONS OF GENERALIZED IMMITTANCE CONVERTERS (GICs) AND GYRATORS

The OP AMP realizations of GICs are useful building-blocks in the synthesis of high-quality RC-active filters. Such circuits are currently manufactured as single-chip microelectronic devices. A two-OP AMP GIC microcircuit has a variety of practical applications and is particularly useful for realizing RC-active simulated-inductance elements and FDNR elements.

5-5.1 Single-OP AMP GIC Circuits

By inspection of Tables 3-1 and 3-2 it is clear that *only* GIC1A may be realized with *one* nullor. (The remaining GICs from GIC1B through GIC2D require two or three nullors.) An OP AMP realization constrains one terminal of the norator of GIC1A to be at ground potential, because it is assumed that the OP AMP possesses a single-ended output according to the model of Fig. 5.1(a). Consequently, the direct substitution of an OP AMP for the nullor of GIC1A leads to the GIC1A OP AMP circuit shown in Fig. 5.10, where the ideal transmission matrix T of the GIC two-port

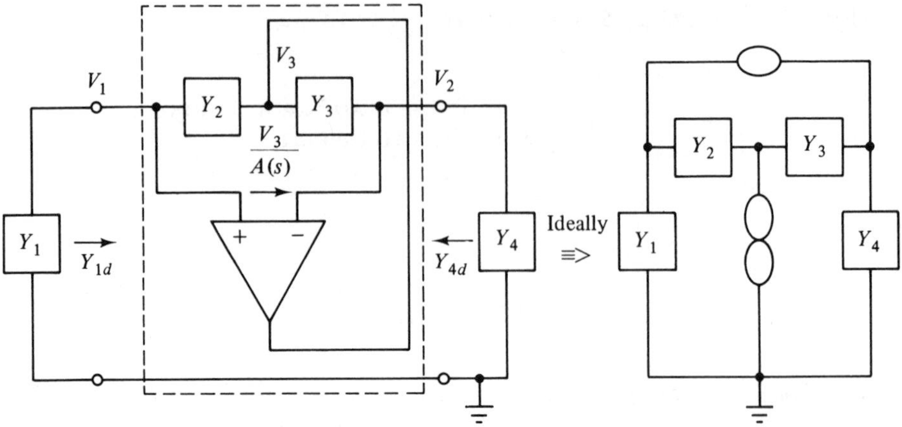

FIGURE 5.10 *Type GIC 1A OP AMP circuits.*

enclosed within the dashed lines is

$$T = \begin{bmatrix} 1 & 0 \\ 0 & -\dfrac{Y_2}{Y_3} \end{bmatrix} \tag{5.51}$$

so that, according to the analysis in section 2.6.1, $A(s) = 1$ and $D(s) = -Y_2/Y_3$ and the *ideal* conversion function $K_2(s) = -Y_3/Y_2$. The *ideal* driving-point admittances Y_{1d} and Y_{4d} are given by equation (3.49); that is,

$$Y_{1d} = -\frac{Y_2 Y_4}{Y_3} \quad \text{and} \quad Y_{4d} = -\frac{Y_1 Y_3}{Y_2} \tag{5.52}$$

The circuit shown in Fig. 5.10 has been used essentially in two ways. As a two-port network, with $Y_2 \equiv G_2$ and $Y_3 \equiv G_3$, it has been employed as a single-OP AMP negative impedance converter (NIC) in the synthesis of transimpedance functions and voltage transfer functions. A more important application of the GIC1A (Fig. 5.10) is for the realization of negative elements $-R$, $-C$, or $-L$ by appropriate choices of Y_2, Y_3, Y_4 (or Y_1, Y_3, Y_2) in equations (5.52). For example, $-R$ elements may be used to provide circuit cancellation of the unavoidable *positive* resistance associated with lossy inductors or lossy transmission lines, and recently $-R$ networks have been proposed for the realization of bidirectional amplifiers. The use of $-C$ elements has been proposed for the realization of high-quality bandpass filters where the circuit shown in Fig. 5.10 has been employed successfully. In the following section, the performance and limitations of the GIC1A circuit as a means of realizing negative resistances is considered in some detail.

The nonideal performance of Fig. 5.10 is predominantly due to the *finite* OP AMP gain $A(s)$. Analysis of Fig. 5.10, which is requested in Problem 5.9, gives the driving-point admittance $Y_{1d}(s)$ at port 1

$$Y_{1d}(s) = -\frac{Y_2Y_4}{Y_3}\left[\frac{1 + A^{-1}(s)\left(1 + \dfrac{Y_3}{Y_4}\right)}{1 - A^{-1}(s)\left(1 + \dfrac{Y_4}{Y_3}\right)}\right] \tag{5.53}$$

where the nonideal multiplicative function $M(s)$ is associated with the term in the square brackets.

5-5.2 The Negative-Resistance Single-OP AMP Circuit

A negative *ideal* driving-point resistance at port 1 is obtained as shown in Fig. 5.11 by selecting $Y_2 \equiv G_2$, $Y_3 \equiv G_3$, $Y_4 \equiv G_4$ so that, for *infinite* $A(s)$, $Y_{1d} = -G_2G_4/G_3$.

The nonideal driving-point admittance Y_{1d} is obtained from equation (5.53) as follows:

$$Y_{1d}(s) \approx -\frac{G_2G_4}{G_3}\left[\frac{1 - \dfrac{s}{B}\left(1 + \dfrac{G_3}{G_4}\right)}{1 + \dfrac{s}{B}\left(1 + \dfrac{G_4}{G_3}\right)}\right] \tag{5.54}$$

where it is assumed that $A(s)$ is given by the dominant-pole response of

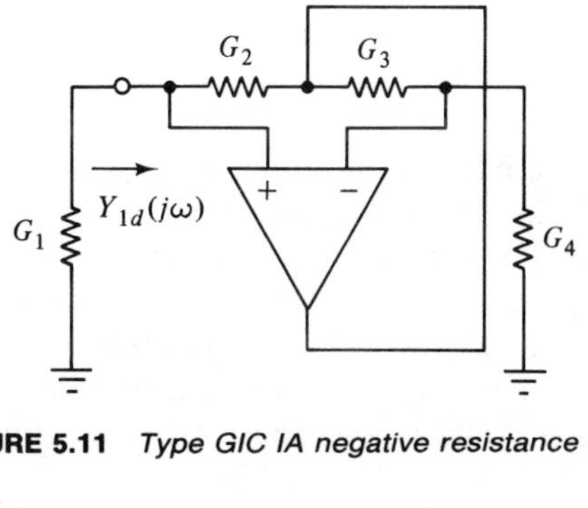

FIGURE 5.11 *Type GIC IA negative resistance circuit.*

equation (5.1) and that

$$A_0 \gg 1 + \frac{G_3}{G_4}, \qquad A_0 \gg 1 + \frac{G_4}{G_3} \tag{5.55}$$

Note that the accuracy with which equation (5.54) approaches the ideal value of $-G_2G_4/G_3$ may be increased by increasing the GBP B; however, $Y_{1d}(s)$ *inherently* possesses a left-half plane pole and a right-half plane zero and this has important practical implications insofar as the stability of the circuit is concerned. For example, let the terminating source admittance Y_1 be given by a positive conductance G_1 so that the pole of the resultant circuit is given by the solution of equation (3.47); that is, by the value(s) of s satisfying

$$Y_{1d}(s) + G_1 = 0 \tag{5.56}$$

Then, from equations (5.54) and (5.56), it is easily shown that the pole location is given by

$$s = B\left[\frac{G_2G_4 - G_1G_3}{G_1G_3 + G_1G_4 + G_2G_4 + G_2G_3}\right] \quad \text{POLE LOCATION} \tag{5.57}$$

This real pole is in the right-half plane, and the circuit is therefore unstable, if

$$\boxed{G_2G_4 > G_1G_3} \quad \text{INSTABILITY CONDITION} \tag{5.58}$$

and, conversely, the circuit is stable if

$$\boxed{G_2G_4 < G_1G_3} \quad \text{STABILITY CONDITION} \tag{5.59}$$

In terms of the *ideal* driving-point admittance $-G_2G_4/G_3$, it follows that the network is stable for $G_1 > |Y_{1d}|$ and unstable for $G_1 < |Y_{1d}|$.

The accuracy of the nonideal $-R$ element, as provided by equation (5.54), may be interpreted from the locus of $Y_{1d}(j\omega)$ that is shown in Fig. 5.12. The admittance $Y_{1d}(0)$ is approximately at the ideal point $-G_2G_4/G_3$ and, for increasing ω, rotates to a position on the positive real axis at $\omega = \infty$. The magnitude and phase errors are obtained from equa-

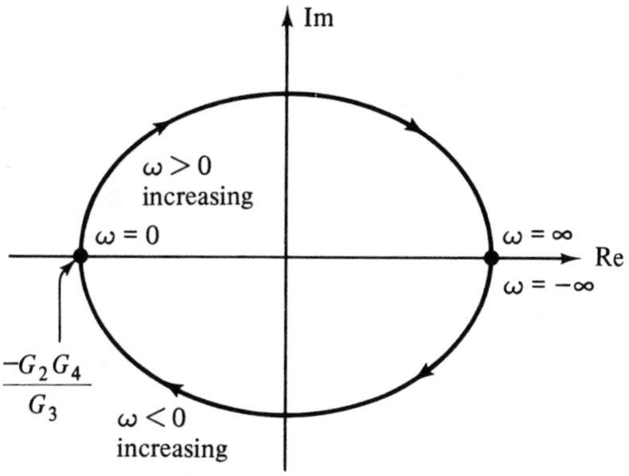

FIGURE 5.12 *Locus of $Y_{1d}(j\omega)$ for Figure 5.11.*

tion (5.54) in terms of the $M(j\omega)$ function

$$M(j\omega) = \frac{1 - \dfrac{j\omega}{B}\left(1 + \dfrac{G_3}{G_4}\right)}{1 + \dfrac{j\omega}{B}\left(1 + \dfrac{G_4}{G_3}\right)} \tag{5.60}$$

For example, at *low* frequencies

$$\frac{\omega G_4}{BG_3} \ll 1, \qquad \frac{\omega G_3}{BG_4} \ll 1, \qquad \frac{\omega}{B} \ll 1 \tag{5.61}$$

so that, from equation (5.60),

$$|M(j\omega)| \approx 1 - \frac{\omega^2}{2B^2}\left(\frac{G_4}{G_3} - \frac{G_3}{G_4}\right) \tag{5.62}$$

and

$$<[M(j\omega)] \approx -\frac{\omega}{B}\left(2 + \frac{G_4}{G_3} + \frac{G_3}{G_4}\right) \tag{5.63}$$

Note that, for $G_4 = G_3$, the magnitude function $|M(j\omega)| = 1$, implying zero error in $|Y_{1d}(j\omega)|$ and minimum error in $< [M(j\omega)]$.

A similar analysis to the above may be performed for $-C$ and $-L$ simulated elements. Problems 5.11, 5.12, and 5.13 are concerned with these circuits, where it is found that the stability problem is particularly severe if real terminating impedances are used.

This completes the discussion of single-OP AMP GIC circuits.

5-5.3 On the Number of Possible OP AMP Circuits

It may well have occurred to the reader that the two nonideal OP AMP circuits shown in Table 5-2 are not the only possible OP AMP implementations of the two-nullor network shown in Fig. 5.9. Although two possible ways of pairing two nullators with two norators are considered, the *choice of polarities* at the differential input terminals of A_1 and A_2 is assigned as $\mp$ and $\mp$ for the type 1A circuit and $\pm$ and $\mp$ for the type 1B circuit, respectively. By appropriately reversing the polarity at the input of A_1 and then at A_2 it is clearly possible to generate a total of eight different circuits that are ideally equivalent to the two-nullor network of Fig. 5.9. Of course, for *infinite* $A_{1,2}$, all eight circuits possess the *same* transfer function [equation (5.39)]; the *nonideal* transfer function is represented by the $M(s)$ function and is quite different for each circuit. The type 1A circuit and the type 1B circuit possess different $M(j\omega)$ functions and, in chapter 10, this fact is used to advantage in *RC*-active filter design.

The question arises as to the *number* of possible nonideal OP AMP circuits that may be found for the general case of a K-nullor network possessing K grounded norators. Clearly, K nullators and K norators may be paired to form K ideal OP AMPs in $K!$ ways; *each* one of the resultant $K!$ ideal OP AMP circuits may be used to generate 2^K further circuits by appropriately employing each of the two possible polarity connections at the inputs of each of the K OP AMPs. Therefore, the total number N_T of possible OP AMP implementations of a K-nullor network is given by

$$N_T = 2^K K! \tag{5.64}$$

The nullor representation of multiple-OP AMP circuits is therefore of fundamental importance because it becomes possible for the circuit designer to consider N_T possible OP AMP circuit implementations of a given nullor network. The *ideal* performance is identical for each of the N_T circuits but, in general, the nonideal performance that is derived by allowing $A_1, A_2, \ldots, A_K$ to be finite is different for each of the N_T circuits. It is by consideration of this profileration of *possible* OP AMP circuits that novel and extremely useful aspects of nonideal performance have been discovered and applied in the design of *RC*-active filters. The fact that $\angle [M(j\omega)]$ is positive for the type 1B noninverting integrator is an essential feature of the ladder filter synthesis techniques described in chapter 9. In the following section, it is particularly important that consideration be given to all N_T realizations.

5-5.4 Two OP AMP GIC Circuits

In this section, the two-OP AMP GIC circuits are analyzed and the nonideal driving-point admittance is used to derive some elementary conclusions with respect to the stability of these circuits.

The two-OP AMP GIC circuits are important in the design of high-quality *RC*-active filters. They are used to realize simulated-inductance elements and FDNR elements by appropriate selections of the branches $Y_2 \ldots Y_6$ in equation (3.51) so that, for example,

$$Y_{1d} = \frac{G_2 G_4 G_6}{s C_5 G_3} \qquad \text{SIMULATED INDUCTANCE ELEMENT} \tag{5.65}$$

or

$$Y_{1d} = \frac{s^2 C_2 C_4 G_6}{G_3 G_5} \qquad \text{FDNR } D \text{ ELEMENT} \tag{5.66}$$

The applications of GICs in filter design are considered in chapters 8 and 9 where a knowledge of the *nonideal performance* of GICs is important. In the following discussion, two types of OP AMP GIC are introduced and defined as the *uncoupled-GIC* and the *coupled-GIC*. The nonideal driving-point admittance Y_{1d} and the corresponding pole positions are discussed briefly, and it is explained that the coupled-GIC is the preferred circuit. In subsequent sections, the detailed analysis of stimulated-inductance and FDNR circuits is performed.

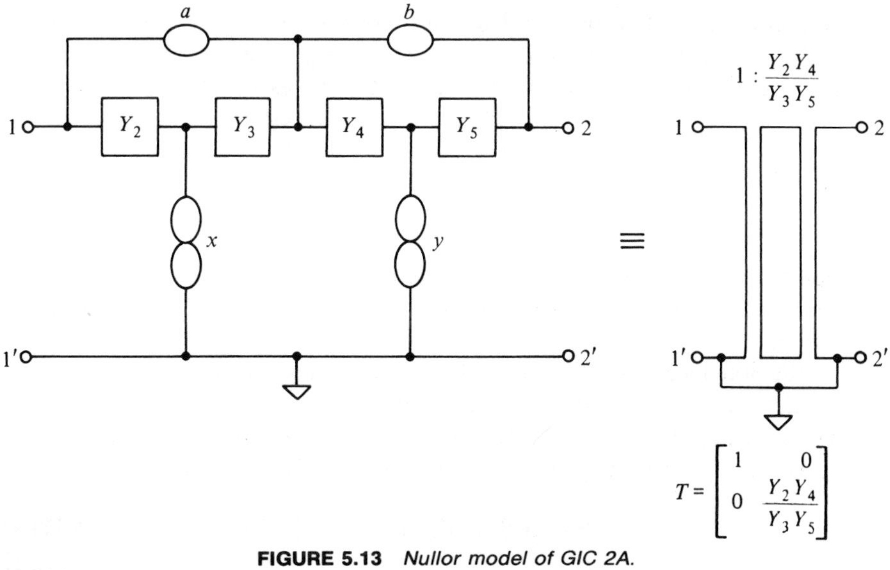

FIGURE 5.13 *Nullor model of GIC 2A.*

The starting point for the discussion of two-OP AMP GICs is a brief inspection of the GIC-nullor networks shown in Table 3-2. Only GIC2A is a direct candidate for implementation with two OP AMPs because it is the only GIC which contains two nullors and where *both* norators share a common (ground) terminal. GIC2A is shown in Fig. 5.13, with its ideal transmission matrix T and the corresponding symbolic notation. Now, it follows from the discussion in section 5-5.3 that there is a total (N_T) of eight OP AMP circuit implementations of Fig. 5.13 and, clearly, four of these result from pairing nullator a with norator x to form OP AMP A_1 and nullator b with norator y to form OP AMP A_2. One of these four possible circuits is shown in Table 5-3 as the *uncoupled-GIC* and is defined as an *uncoupled-GIC* because it is nothing more than an uncoupled cascade of two of the single-OP AMP GICs that are discussed in section 5-5.1; the three other uncoupled-GICs are obtained by exploiting the three remaining polarity combinations at the inputs of A_1 and A_2.

The four coupled-GIC circuits are obtained by pairing nullator a with norator y to form OP AMP A_2 and nullator b with norator x to form OP AMP A_1. One of the four possible coupled-GIC circuits is shown in Table 5-3 where, as before, the remaining three circuits are obtained by exploiting the three remaining combinations of input polarities of A_1 and A_2. The essential feature of the coupled-GIC circuits is the cross-coupling of the outputs of the OP AMPs as indicated by the circuit diagram.

The *nonideal* driving-point admittance Y_{1d} at port 1 of the uncoupled- and coupled-GICs may be derived in terms of the branch admittances Y_2, Y_3, Y_4, Y_5, Y_6 and the OP AMP gains $A_1(s)$ and $A_2(s)$. It is assumed that $A_{1,2}(s)$ are characterized by the dominant-pole frequency response of equation (5.1), so that the nonideal nature of the circuits is represented by the four parameters $A_{01,2}$ and $T_{1,2}$. Direct analysis of the two-OP AMP GIC circuits shown in Table 5-3 leads to the expressions for Y_{1d} in equations (5.67) and (5.69) of Tables 5-4 and 5-5 respectively. The details of the analysis are requested in Problems 5.15 and 5.16. It is observed that $Y_{1d} \to Y_2Y_4Y_6/Y_3Y_5$ as $A_{01,2} \to \infty$.

Stability: A *detailed* analysis of the stability of the two-OP AMP GIC circuits is available in the literature. The stability problem is now reviewed briefly.

The pole locations of the two GIC circuits shown in Table 5-3 may be derived by using the method that is employed in section 3-6.1. That is, if port 1 is terminated in an admittance $Y_1(s)$, then the poles of the resultant GIC circuit are given by the values of s that satisfy the equation

$$Y_1(s) + Y_{1d}(s) = 0 \tag{5.71}$$

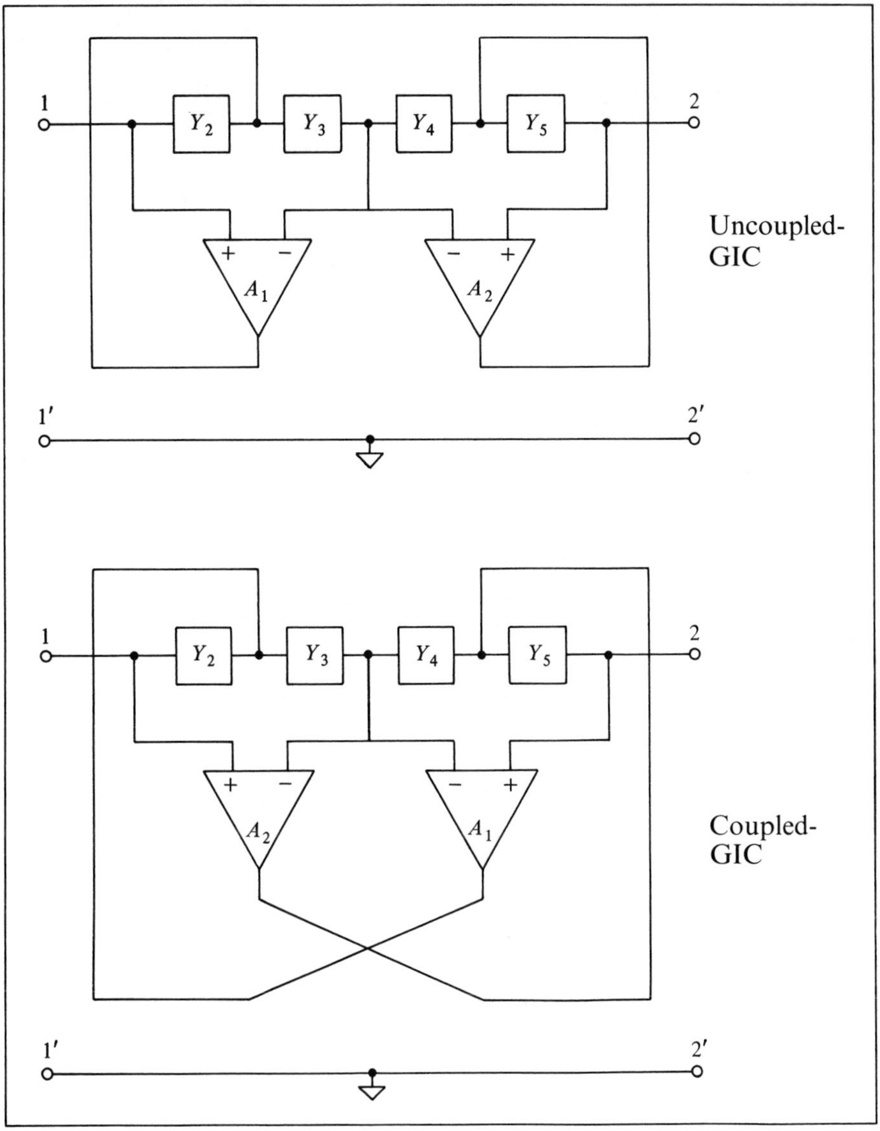

TABLE 5.3 *The uncoupled and coupled type 2 GIC circuits.*

Since we have expressions for $Y_{1d}(s)$, in equations (5.67) and (5.69), it is possible to write equation (5.71) as

$$Y_1(s) + \frac{N_{1d}(s)}{D_{1d}(s)} = 0 \tag{5.72}$$

where $Y_{1d}(s) \equiv N_{1d}(s)/D_{1d}(s)$. Then, the poles of the network are given

TABLE 5.4 *Coupled-GIC*

$$Y_{1d} = Y_2\left[\frac{A_{01}A_{02}Y_4Y_6 + A_{01}(1 + sT_2)Y_3(Y_5 + Y_6) + A_{02}(1 + sT_1)Y_4(Y_5 + Y_6) + (1 + sT_1)(1 + sT_2)(Y_3 + Y_4)(Y_5 + Y_6)}{A_{01}A_{02}Y_3Y_5 + A_{01}(1 + sT_2)Y_3(Y_5 + Y_6) + A_{02}(1 + sT_1)Y_4(Y_5 + Y_6) + (1 + sT_1)(1 + sT_2)(Y_3 + Y_4)(Y_5 + Y_6)}\right] \tag{5.67}$$

and

$$\begin{aligned}\Delta(s) = {} & A_{01}A_{02}[Y_1Y_3Y_5 + Y_2Y_4Y_6] + A_{01}(1 + sT_2)[Y_3(Y_1 + Y_2)(Y_5 + Y_6)] \\ & + A_{02}(1 + sT_1)[Y_4\ (Y_1 + Y_2)(Y_5 + Y_6)] + (1 + sT_1)(1 + sT_2)[(Y_1 + Y_2)(Y_3 + Y_4)(Y_5 + Y_6)]\end{aligned} \tag{5.68}$$

TABLE 5.5 *Uncoupled-GIC*

$$Y_{1d} = Y_2\left[\frac{A_{01}A_{02}Y_4Y_6 + A_{01}(1 + sT_2)Y_4(Y_5 + Y_6) + A_{02}(1 + sT_1)(Y_3Y_5 - Y_4Y_6) - (1 + sT_1)(1 + sT_2)(Y_0 + Y_4)(Y_5 + Y_6)}{A_{01}A_{02}Y_3Y_5 - A_{01}(1 + sT_2)Y_3(Y_5 + Y_6) + A_{02}(1 + sT_2)(Y_3Y_5 - Y_4Y_6) - (1 + sT_1)(1 + sT_2)(Y_3 + Y_4)(Y_5 + Y_6)}\right] \tag{5.69}$$

and

$$\begin{aligned}\Delta(s) = {} & A_{01}A_{02}[Y_1Y_3Y_5 + Y_2Y_4Y_6] - A_{01}(1 + sT_2)[Y_1Y_3(Y_5 + Y_6) - Y_2Y_4(Y_5 + Y_6)] \\ & + A_{02}(1 + sT_1)[Y_3Y_5(Y_1 + Y_2) - Y_4Y_6(Y_1 + Y_2)] - (1 + sT_1)(1 + sT_2)[(Y_1 + Y_2)(Y_3 + Y_4)(Y_5 + Y_6)]\end{aligned} \tag{5.70}$$

by the values of s satisfying the equation

$$Y_1(s)D_{1d}(s) + N_{1d}(s) \equiv \Delta(s) = 0 \tag{5.73}$$

Some preliminary conclusions concerning the stability of these GICs are now deduced. It is well known that the poles are in the left-half s-plane if and only if $\Delta(s)$ is a strictly Hurwitz polynomial. Furthermore, a *necessary* condition for $\Delta(s)$ to be strictly Hurwitz is that all the coefficients of s be present *and of the same sign*. The expressions for $\Delta(s)$, derived from equations (5.72), (5.69), and (5.67) are given in equations (5.68) and (5.70). Note that for the coupled-GIC [equation (5.68)], *all* of the terms are associated with positive signs. That is, if $Y_1 \ldots Y_6$ are strictly Hurwitz polynomials, then $\Delta(s)$ has coefficients in s that are all present and of the same sign; it therefore also follows that the particular coupled-GIC circuit shown in Table 5-3 possesses the property that, if the branches Y_i have the form

$$Y_i = \sum_{j=0}^{N} a_{ji}s^j, \qquad a_{ji} > 0 \tag{5.74}$$

then the characteristic equation $\Delta(s)$ *possesses coefficients in s that are all present and of the same sign for all positive real values of* a_{ji}, A_{01}, A_{02}, T_1, T_2. This is a *necessary* condition for the stability of the circuit over all possible positive real values of the parameters a_{ji}, A_{01}, A_{02}, T_1, T_2. None of the other seven possible two-OP AMP implementations of Fig. 5.13 satisfy this condition; that is, they allow right-half plane poles for specific combinations of a_{ji}, A_{01}, A_{02}, T_1, T_2. This has led to a preference for the coupled-GIC circuit shown in Table 5-3. It was found, during the early research into this matter, that problems of transient instability and latch-up are caused by the migration of the poles into the right-half plane during the transient switch-on of the power supply voltages when $A_{01,2}$ are migrating from zero to their steady-state values (of about 10^5). Such problems are avoided by employing the coupled-GIC shown in Table 5-3.

EXAMPLE 5.1: A simulated inductance is approximated at port 1 of the coupled-GIC of Table 5-3 by choosing the branches as G_2, G_3, G_4, C_5, and G_6. Port 1 is terminated in a capacitance C_1. Assume that the OP AMPs have *real* gains $A_{01,2}$ and, thereby, prove that there does not exist any combination of A_{01}, A_{02}, C_1, G_2, G_3, G_4, C_5, G_6 that causes the poles to enter the right-half s-plane.

SOLUTION: It follows by direct substitution into equation (5.68) that

$$\begin{aligned}\Delta(s) = {} & A_{01}A_{02}\left[s^2C_1G_3C_5 + G_2G_4G_6\right] \\ & + A_{01}[G_3(sC_1 + G_2)(sC_5 + G_6)] \\ & + A_{02}[G_4(sC_1 + G_2)(sC_5 + G_6)] \\ & + (sC_1 + G_2)(G_3 + G_4)(sC_5 + G_6) \end{aligned} \tag{5.75}$$

It follows that $\Delta(s)$ is *quadratic* in s of the form $\alpha s^2 + \beta s + \gamma$, where $\alpha, \beta, \gamma > 0$. The poles of the network are given by

$$\alpha s^2 + \beta s + \gamma = 0 \tag{5.76}$$

or

$$s_{1,2} = -\frac{\beta}{2\alpha}\left(1 \pm \sqrt{1 - \frac{4\alpha\gamma}{\beta^2}}\right) \tag{5.77}$$

from which it is easily shown that

$$\text{Re}[s_{1,2}] < 0 \tag{5.78}$$

over all positive α, β, and γ. Consequently, there cannot exist any combination of A_{01}, A_{02}, C_1, G_2, G_3, G_4, C_5, and G_6 that will cause $\text{Re}[s_{1,2}] > 0$ and thereby result in right-half plane poles. This is a most fortunate result because the poles are *ideally* ($A_{01} = A_{02} = \infty$) *on the imaginary axis* at

$$s_{1,2} = \pm j\left(\frac{G_2G_4G_6}{C_1G_3G_2}\right) \qquad \text{IDEAL POLE POSITIONS, } A_{01} = A_{02} = \infty \tag{5.79}$$

and yet we are assured that there exists no combination of A_{01} and A_{02} that can "nudge" them into the right-half plane and cause consequential instability.

To summarize this brief discussion of stability, it is found that the coupled-GIC shown in Table 5-3 is useful because it is stable over a wide range of operating conditions. It meets a *necessary* condition for stability over all positive real parameter values that is not met by the remaining seven configurations. The interested reader is referred to the recommended reading at the end of the chapter for a more detailed discussion of the stability of two-OP AMP GICs. In this text, the coupled-GIC of Table 5-3 is used exclusively as the preferred GIC configuration.

5-6 INDUCTANCE SIMULATION USING GICs

The coupled-GIC shown in Table 5-3 is especially suitable for the realization of a grounded simulated inductance L_1 at port 1 because of the preferred stability properties of the circuit. An *ideal* inductance is realized at port 1 by selecting either $Y_3 = sC_3$ or $Y_5 = sC_5$, and the remaining branches as conductances. The two possible circuits are defined as type 1 and type 2 and are shown in Tables 5-6 and 5-7.

Inductance simulation is one of the principal applications of GICs and, therefore, the analysis of the type 1 and type 2 circuits is considered here in some detail. The general idea of inductance simulation is simply to replace the inductance elements of a passive LC filter by GIC-simulated inductances. As one would expect, the nonideal performance of the simulated inductance becomes of critical importance in most active filters that are designed in this way. In this section, the nonidealness of the simulated inductance L_1 is characterized in terms of the Q factor $Q_l(\omega)$ and the fractional deviation $\epsilon_l(\omega)$ of the inductance from its ideal value. By way of introduction, consider the type 1 circuit of Table 5-6 for which, *ideally*,

$$Y_{1d}(s) = \frac{G_2 G_4 G_6}{sC_3 G_5}, \quad \text{IDEAL } A_{01,2} \to \infty \tag{5.80}$$

and, therefore,

$$L_1 = \frac{C_3 G_5}{G_2 G_4 G_6}, \quad \text{IDEAL } A_{01,2} \to \infty \tag{5.81}$$

The principal parameters that determine the nonideal function $Y_{1d}(s)$ are the OP AMP parameters $A_{01,2}$ and $B_{1,2}$. The analysis technique is to use equation (5.67) to express the nonideal $Y_{1d}(s)$ in the form

$$Y_{1d}(s) = \frac{G_2 G_4 G_6}{sC_3 G_5}[M(s)] \tag{5.82}$$

where the nonideal multiplicative function $M(s)$ is a function of $A_{01,2}$, $B_{1,2}$, and the branch parameters $G_{2,4,5,6}$ and C_3. The *Q-factor* $Q_l(\omega)$ of a nonideal inductance is defined as

$$Q_l(\omega) \equiv \frac{\text{Im}[Z_{1d}(j\omega)]}{\text{Re}[Z_{1d}(j\omega)]} = -\frac{\text{Im}[Y_{1d}(j\omega)]}{\text{Re}[Y_{1d}(j\omega)]} \tag{5.83}$$

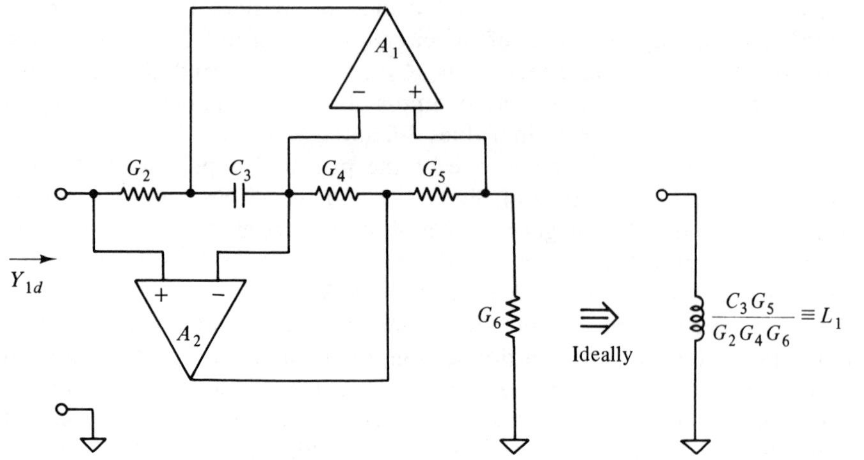

Dissipation factor; Q factor: $Q_l(\omega) \equiv d_l^{-1}(\omega)$

$$d_l(\omega) \approx \underbrace{(G_5 + G_6)\left[\frac{\omega C_3}{A_{02} G_4 G_6} + \frac{G_4}{A_{01} C_3 G_5}\right]}_{d_{Ll}(\omega)} + \underbrace{\omega(G_5 + G_6)\left[\frac{1}{B_1 G_6} - \frac{1}{B_2 G_5} - \frac{\omega^2 C_3}{B_1 B_2 G_4 G_6}\right]}_{d_{Hl}(\omega)}$$

Enhancement condition:

$$B_1 G_6 = B_2 G_5$$

Inductance fractional error:

$$\epsilon_{Hl}(\omega) \approx \epsilon_l(\omega) \approx (G_5 + G_6)\left[\frac{\omega^2 C_3}{B_2 G_4 G_6} + \frac{G_4}{B_1 C_3 G_5}\right]$$

TABLE 5.6 *Type 1 inductance-simulation circuit.*

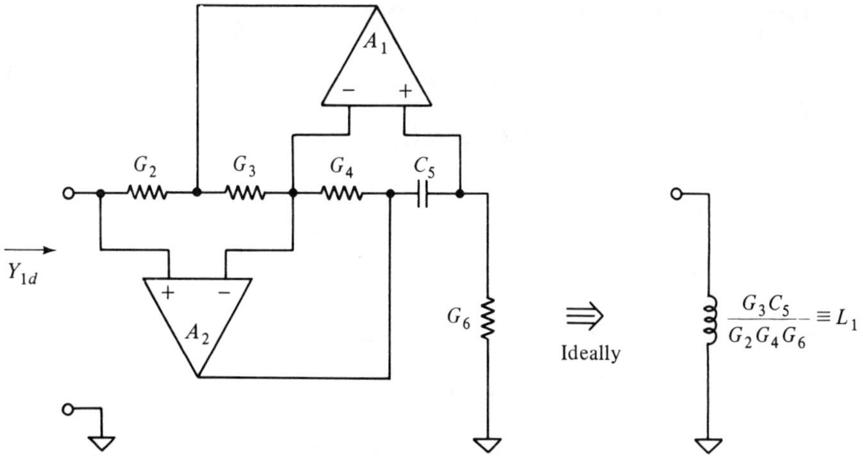

Dissipation factor; Q factor: $Q_l(\omega) \equiv d_l^{-1}(\omega)$

$$d_l(\omega) \approx \omega \frac{C_5}{G_6}\left[\frac{1}{A_{01}} + \frac{G_3}{A_{02}G_4}\right] + \frac{G_6}{\omega C_5}\left[\frac{G_4}{A_{01}G_3} + \frac{1}{A_{02}}\right]$$
$$+ \omega\left[\frac{1}{B_2}\left(\frac{G_3}{G_4} - 1\right) + \frac{1}{B_1}\left(1 - \frac{G_4}{G_3}\right)\right] - \omega^3 \frac{C_5}{B_1B_2G_6}$$

Enhancement condition:

$$G_3 = G_4$$

Inductance fractional error:

$$\epsilon_{Hl}(\omega) \approx \epsilon_l(\omega) \approx \frac{1}{B_1}\left[\frac{G_4G_6}{G_3C_5} + \omega^2 \frac{C_5}{G_6}\right] + \frac{1}{B_2}\left[\frac{G_6}{C_5} + \omega^2 \frac{G_3C_5}{G_4G_6}\right]$$

TABLE 5.7 *Type 2 inductance-simulation circuit.*

where it is sometimes more convenient to employ the reciprocal expression, known as the *dissipation factor* $d_l(\omega)$, where

$$d_l(\omega) \equiv Q_l^{-1}(\omega) \tag{5.84}$$

It follows from equations (5.82) and (5.83) that

$$d_l(\omega) \equiv Q_l^{-1}(\omega) = \tan[< M(j\omega)] \qquad \text{DISSIPATION OR } Q\text{-FACTOR} \tag{5.85}$$

and

$$Q_l^{-1}(\omega) \approx\ < M(j\omega), \qquad |Q_l(\omega)| \gg 1. \tag{5.85A}$$

where the identical result applies to *any* nonideal simulated inductance.

The *phase* error associated with $Y_{1d}(j\omega)$ is clearly incorporated into the definition above of dissipation factor $d_l(\omega)$. The *magnitude* error associated with $Y_{1d}(j\omega)$ is represented by calculating the *effective* nonideal inductance $L_{1\text{eff}}$, as opposed to the ideal inductance L_1 of equation (5.81). Thus,

$$\omega L_{1\text{eff}} \equiv \text{Im}[Z_{1d}] \tag{5.86}$$

Equations (5.86), (5.82), and (5.81) give

$$L_{1\text{eff}} = L_1 \frac{\text{Re}[M(j\omega)]}{|M(j\omega)|^2} \tag{5.87}$$

A more convenient representation of the deviation of $L_{1\text{eff}}$ from the ideal value L_1 is to define the *fractional error of the inductance* $\epsilon_l(\omega)$ as

$$\epsilon_l(\omega) \equiv \frac{L_{1\text{eff}} - L_1}{L_1} \tag{5.88}$$

so that, by equations (5.87) and (5.88),

$$\epsilon_1(\omega) = \frac{\text{Re}[M(j\omega)]}{|M(j\omega)|^2} - 1 \approx 1 - \text{Re}[\text{M}(j\omega)], \qquad \text{M}(j\omega) \approx 1 \tag{5.89}$$

FRACTIONAL INDUCTANCE ERROR

5-6.1 The Type 1 Inductance-Simulation GIC Circuit

Analysis of this circuit is given here in detail; a similar analytical procedure applies to *all* of the simulated-inductance and FDNR GIC circuits and, therefore, particular attention is given to the assumptions used in the analysis. Substituting G_2, sC_3, G_4, G_5, and G_6 for the branch admittances Y_2, Y_3, Y_4, Y_5, and Y_6, respectively, in equation (5.67) leads directly to

$$Y_{1d} = G_2\left[\frac{A_{01}A_{02}G_4G_6 + D(s)}{A_{01}A_{02}sC_3G_5 + D(s)}\right] = \frac{G_2G_4G_6}{sC_3G_5}\overbrace{\left[\frac{1 + \dfrac{D(s)}{A_{01}A_{02}G_4G_6}}{1 + \dfrac{D(s)}{A_{01}A_{02}sC_3G_5}}\right]}^{M(s)} \tag{5.90}$$

where the term in square brackets is $M(s)$ and where

$$\begin{aligned} D(s) = {} & s^3[(G_5 + G_6)T_1T_2C_3] \\ & + s^2[(G_5 + G_6)(A_{01}T_2C_3 + T_1T_2G_4 + (T_1 + T_2)C_3)] \\ & + s[(G_5 + G_6)(A_{01}C_3 + A_{02}T_1G_4 + (T_1 + T_2)G_4 + C_3)] \\ & + [(G_5 + G_6)(A_{02}G_4 + G_4)] \end{aligned} \tag{5.91}$$

This rather unwieldy expression for Y_{1d} may be much simplified by means of the following assumptions. First, the dc open-loop gains $A_{01,2}$ are typically in the region of 10^5, so that it may be assumed that

$$A_{01,2} \gg 1 \tag{5.92}$$

and

$$A_{01,2} \gg \frac{G_i}{G_j} \tag{5.93}$$

where G_i and G_j are *any* of the branch conductances of the GIC (in this case, G_2, G_4, G_5, and G_6). The latter assumption implies that the ratio between these conductances does not exceed about 10^4.

The second assumption concerns the time constants τ_{ij} formed by the GIC branch conductances G_j and branch capacitances C_i (in this case, simply C_3), defined by

$$\tau_{ij} \equiv \frac{C_i}{G_j} \text{ seconds} \tag{5.94}$$

It is almost always a good assumption that

$$B_{1,2} \equiv \frac{A_{01,2}}{T_{1,2}} \gg \tau_{ij}^{-1} \tag{5.95}$$

for all possible time constants τ_{ij} that may be formed (for example, C_3/G_2, C_3/G_4, C_3/G_5, C_3/G_6). Applying the assumptions in equations (5.92), (5.93), (5.94) and, (5.95) to equation (5.91) leads to the following simplified expression for $D(s)$:

$$D(s) \approx s^3[(G_5 + G_6)T_1T_2C_3] + s^2[A_{01}T_2C_3(G_5 + G_6)] + s[(G_5 + G_6)(A_{01}C_3 + A_{02}T_1G_4)] + [(G_5 + G_6)A_{02}G_4] \tag{5.96}$$

Using equations (5.90) and (5.96) we obtain

$$\boxed{M(s) \approx \frac{1 + (G_5 + G_6)[a_3s^3 + a_2s^2 + a_1s + a_0]}{1 + (G_5 + G_6)[b_2s^2 + b_1s + b_0 + b_{-1}s^{-1}]}} \tag{5.97}$$

where

$$\left.\begin{aligned}
a_3 &= \frac{C_3}{B_1B_2G_4G_6}, & a_2 &= \frac{C_3}{B_2G_4G_6} \\
a_1 &= \frac{C_3}{A_{02}G_4G_6} + \frac{1}{B_1G_6}, & a_0 &= \frac{1}{A_{01}G_6} \\
b_2 &= \frac{1}{B_1B_2G_5}, & b_1 &= \frac{1}{B_2G_5} \\
b_0 &= \frac{1}{A_{02}G_5} + \frac{G_4}{B_1C_3G_5}, & b_{-1} &= \frac{G_4}{A_{01}C_3G_5}
\end{aligned}\right\} \tag{5.98}$$

The expression in equation (5.97) for $M(s)$ may be used to derive the nonideal behavior of the simulated inductance at Y_{1d}.

Q-factor calculation: It follows from equation (5.97) that

$$M(j\omega) \approx \frac{1 + (G_5 + G_6)\left[(a_0 - a_2\omega^2) + j\omega(a_1 - a_3\omega^2)\right]}{1 + (G_5 + G_6)\left[(b_0 - b_2\omega^2) + j\omega(b_1 - b_{-1}\omega^{-2})\right]} \tag{5.99}$$

and, since $|Q_l(\omega)| \gg 1$ it follows from equation (5.85) that the imaginary terms of the numerator and denominator of equation (5.99) are normally much less than unity, so that

$$M(j\omega) \approx 1 + (G_5 + G_6)\left[(a_0 - b_0) - \omega^2(a_2 - b_2)\right.$$
$$\left. + j\omega(a_1 - b_1) - j\omega(a_3\omega^2 - b_{-1}\omega^{-2})\right] \tag{5.100}$$

and, consequently, from equation (5.85) the dissipation factor $d_l(\omega)$ and Q factor $Q_l(\omega)$ are given by

$$d_l(\omega) \equiv Q_l^{-1}(\omega) = \frac{\omega\left[(a_1 - b_1) - (a_3\omega^2 - b_{-1}\omega^{-2})\right](G_5 + G_6)}{1 + (G_5 + G_6)\left[(a_0 - b_0) - \omega^2(a_2 - b_2)\right]} \tag{5.101}$$

Substitution of the a_i, b_i coefficients from equation (5.98) into equation (5.101) gives

$$d_l(\omega) \approx \frac{\omega(G_5 + G_6)\left[\left(\dfrac{C_3}{A_{02}G_4G_6} + \dfrac{1}{B_1G_6} - \dfrac{1}{B_2G_5}\right)\right.}{1 + (G_5 + G_6)\left[\left(\dfrac{1}{A_{01}G_6} - \dfrac{1}{A_{02}G_5} - \dfrac{G_4}{B_1C_3G_5}\right)\right.} \cdots$$

$$\cdots \frac{\left. - \left(\dfrac{C_3\omega^2}{B_1B_2G_4G_6} - \dfrac{G_4\omega^{-2}}{A_{01}C_3G_5}\right)\right]}{\left. - \omega^2\left(\dfrac{C_3}{B_2G_4G_6} - \dfrac{1}{B_1B_2G_5}\right)\right]} \tag{5.102}$$

This expression for the dissipation factor may be much simplified by restricting the range of ω over which an accurate description is required.

That is, it is assumed that

$$\omega \ll B_{1,2} \tag{5.103}$$

and

$$\omega^2 \ll B_{1,2}\tau_{ij}^{-1} \tag{5.104}$$

(Rearranging the two conditions above gives $B_{1,2}/\omega \gg 1$ and $B_{1,2}/\omega \gg \omega\tau_{ij}$; since $|A_{1,2}(j\omega)| \approx B_{1,2}/\omega$, it follows that the constraints in equations (5.103) and (5.104) are effectively equivalent to assuming that the *upper frequency of interest remains in the frequency region well below* $B_{1,2}$, where the magnitude of the gains of the OP AMPs is significantly in excess of unity and of $\omega\tau_{ij}$.) The assumptions in equations (5.103) and (5.104) allow the denominator of equation (5.102) to be set to unity, so that

$$\boxed{\begin{aligned} d_l(\omega) \approx & \underbrace{(G_5 + G_6)\left[\frac{\omega C_3}{A_{02}G_4G_6} + \frac{G_4}{A_{01}C_3G_5\omega}\right]}_{d_{Ll}(\omega)} \\ & \underbrace{+\omega(G_5 + G_6)\left[\frac{1}{B_1G_6} - \frac{1}{B_2G_5} - \frac{\omega^2 C_3}{B_1B_2G_4G_6}\right]}_{d_{Hl}(\omega)} \end{aligned}} \tag{5.105}$$

It is useful to consider this expression in two parts. The terms containing $A_{01,2}$ form the *low-frequency dissipation factor* $d_{Ll}(\omega)$, and the terms containing $B_{1,2}$ form the *high-frequency dissipation factor* $d_{Hl}(\omega)$, so that

$$d_l(\omega) = d_{Ll}(\omega) + d_{Hl}(\omega) \tag{5.106}$$

In practice, it usually turns out that $|d_{Hl}(\omega)| \gg d_{Ll}(\omega)$, except at very low frequencies, so that the question of minimizing $|d_{Hl}(\omega)|$ is pertinent. By the assumption in equation (5.104), it follows that the term $\omega^2 C_3/B_1B_2G_4G_6$ in equation (5.105) is much less than both $1/B_1G_6$ and $1/B_2G_5$. Consequently, by selecting

$$\boxed{B_1G_6 = B_2G_5} \quad Q_{Hl}(\omega), d_{Hl}(\omega) \text{ ENHANCEMENT CONDITION} \tag{5.107}$$

the dissipation factor $d_{Hl}(\omega)$ may be suppressed significantly toward zero. That is, the Q-factor $Q_{Hl}(\omega)$ may be enhanced significantly toward infinity.

Of course, it is not difficult to match conductances to within a small fraction of a per cent, so that $G_5 \approx G_6$. The matching of GBPs B_1 and B_2 is best achieved by employing *dual* OP AMPs, because the semiconductor circuitry for OP AMPs A_1 and A_2 are adjacent on the same semiconductor substrate, thereby ensuring substantial similarity of performance. Measurements have shown that B_1 and B_2 are typically matched to better than 3% in the case of dual OP AMPs. Let the GBP matching ratio m be defined as

$$m \equiv \frac{B_1}{B_2} \quad \text{GBP MATCHING FACTOR} \tag{5.108}$$

Then from equations (5.108) and (5.105),

$$d_{Hl}(\omega) \approx \frac{2\omega}{B_1}\left(1 - m - \frac{\omega^2 C_3 m}{B_1 G_4}\right), \qquad G_5 = G_6 \tag{5.109}$$

where it is assumed that $G_5 = G_6$. It follows by inspection of equation (5.109) that, since $\omega^2 C_3 m / B_1 G_4 \ll 1$, the suppression of $d_{Hl}(\omega) \to 0$ is achieved over a wide frequency range by ensuring that $m \to 1$.

Interpretation of these results is readily achieved in the case of *uniform conductances*; that is, where $G_2 = G_4 = G_5 = G_6 \equiv G$. Then, writing $\tau = C/G$, equation (5.105) becomes

$$d_l(\omega) \approx \underbrace{\frac{2}{A_0}\left(\omega\tau + \frac{1}{\omega\tau}\right)}_{d_{Ll}(\omega)} + \underbrace{\frac{2\omega}{B_1}\left(1 - m - \frac{\omega^2 \tau m}{B_1}\right)}_{d_{Hl}(\omega)} \quad \text{UNIFORM-CONDUCTANCE CASE} \tag{5.110}$$

The dissipation factors $d_{Ll}(\omega)$ and $d_{Hl}(\omega)$ are shown graphically in Fig. 5.14, where the curves for $d_{Hl}(\omega)$ correspond to the three conditions: $m > 1$, $m < 1$, and $m = 1$. The essential conclusion is that substantial enhancement of $d_{Hl}(\omega)$ toward zero is possible by matching B_1 and B_2 ($m = 1$) *and* by selecting $G_5 = G_6$. A similar conclusion follows for the FDNR circuit that is analyzed in section 5-7.1.

Inductance fractional error $\epsilon_l(\omega)$ calculations: The nonideal effective inductance $L_{1\text{eff}}$ is derived from equations (5.89) and (5.100). From equation (5.100),

$$\text{Re}[M(j\omega)] \approx 1 + (G_5 + G_6)[a_0 - b_0 - \omega^2(a_2 - b_2)] \tag{5.111}$$

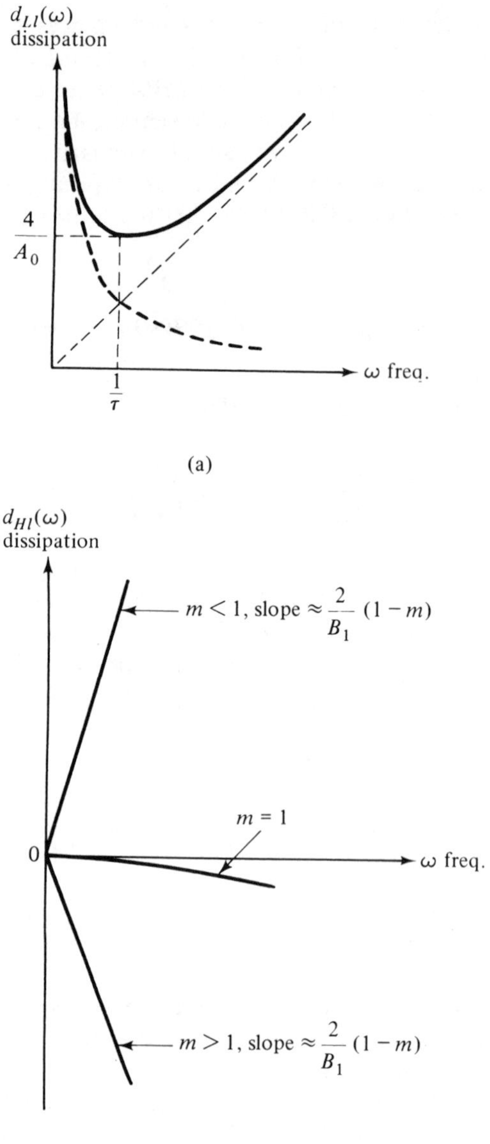

FIGURE 5.14 *(a) Low-frequency dissipation d_{Ll} uniform conductance case; (b) high-frequency dissipation d_{Hl} uniform conductance case.*

where it is assumed that the terms containing b_0, a_0, b_2, and a_2 are much less than unity, so Re $[M(j\omega)]$ is close to the ideal value of unity. Then, substituting b_0, a_0, b_2, and a_2 from equation (5.98) and subtracting from unity according to equation (5.89), yields

$$\epsilon_l(\omega) \approx (G_5 + G_6)\left[\frac{1}{A_{02}G_5} - \frac{1}{A_{01}G_6} + \frac{\omega^2 C_3}{B_2 G_4 G_6} + \frac{G_4}{B_1 C_3 G_5}\right] \quad (5.112)$$

which, due to the high values of $A_{01,2}$, may usually be approximated very accurately by neglecting the terms containing A_{01} and A_{02}. This gives the high-frequency inductance error term

$$\epsilon_{Hl}(\omega) \approx (G_5 + G_6)\left[\frac{\omega^2 C_3}{B_2 G_4 G_6} + \frac{G_4}{B_1 C_3 G_5}\right] \qquad \text{FRACTIONAL INDUCTANCE ERROR} \quad (5.113)$$

which, for the previously considered *uniform-conductance* design simplifies to

$$\epsilon_{Hl}(\omega) \approx 2\left[\frac{\omega^2 \tau}{B_2} + \frac{1}{B_1 \tau}\right] \qquad \text{UNIFORM-CONDUCTANCE INDUCTANCE ERROR} \quad (5.114)$$

This function is shown in Fig. 5.15 for various values of inductance L_1 where, from equation (5.81),

$$L_1 = \frac{\tau}{G} = \frac{C}{G^2} \quad (5.115)$$

Thus, for uniform conductances G, the various curves of Fig. 5.15 are achieved by appropriately selecting the capacitance C $(= C_3)$. As C and therefore L_1 are increased, the fractional error $\epsilon_l(\omega)$ decreases at low frequencies and increases at high frequencies. If, in a particular application, it is necessary that $L_{1\text{eff}}$ should remain essentially constant over a wide frequency range, then it may be important to ensure that τ is selected so that the $1/B_1\tau$ term dominates equation (5.114).

The *accuracy* of these derivations of $d_l(\omega)$ and $\epsilon_l(\omega)$ is not verified at this point. Numerical verification is obtained subsequent to the following analysis of the type 2 circuit.

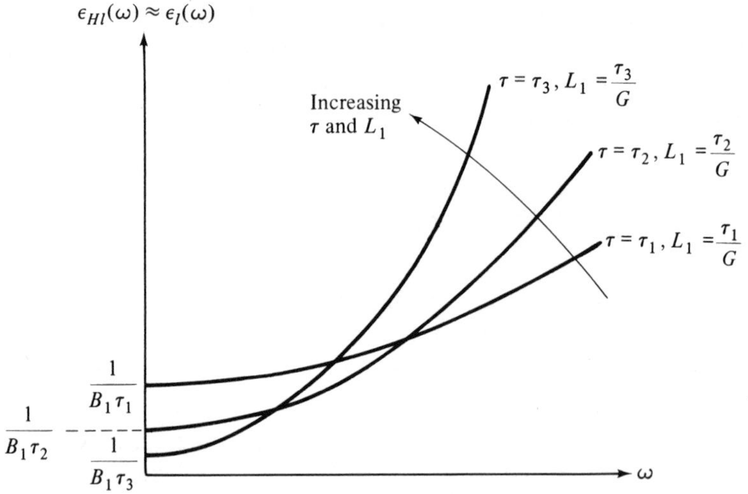

FIGURE 5.15 *High-frequency inductance fractional errors for various τ and L_1.*

5-6.2 The Type 2 Inductance-Simulation GIC Circuit

This circuit differs from the type 1 circuit in the location of the capacitance element. As shown in Table 5-7, the capacitance is at C_5. The analysis of this circuit follows exactly the procedure that is given above for the type 1 circuit and, therefore, is not presented here in detail. Problems 5.21 to 5.23 request that all the analytic steps be performed that lead to the corresponding expressions for $d_l(\omega)$ and $\epsilon_l(\omega)$. Table 5-7 contains the expressions for $d_l(\omega)$ and $\epsilon_{Hl}(\omega)$. An essential and important difference between the type 2 circuit and the type 1 circuit is in the behavior of the dissipation function $d_l(\omega)$. From Table 5-7, for the type 2 circuit:

$$d_l(\omega) \approx \omega\frac{C_5}{G_6}\left(\frac{1}{A_{01}} + \frac{G_3}{A_{02}G_4}\right) + \frac{G_6}{\omega C_5}\left(\frac{G_4}{A_{01}G_3} + \frac{1}{A_{02}}\right) + \omega\left[\frac{1}{B_2}\left(\frac{G_3}{G_4} - 1\right) + \frac{1}{B_1}\left(1 - \frac{G_4}{G_3}\right)\right] - \omega^3\frac{C_5}{B_1B_2G_6} \tag{5.116}$$

DISSIPATION FACTOR TYPE 2 CIRCUIT

The low-frequency dissipation $d_{Ll}(\omega)$ part of this expression is similar to that obtained for the type 1 circuit, but the remaining high-frequency dissipation term $d_{Hl}(\omega)$ is significantly different from the type 1 circuit from equation (5.116)

$$d_{Hl}(\omega) \approx \omega\left[\frac{1}{B_2}\left(\frac{G_3}{G_4}-1\right)+\frac{1}{B_1}\left(1-\frac{G_4}{G_3}\right)\right]-\omega^3\frac{C_5}{B_1B_2G_6} \quad (5.117)$$

and it is clearly possible to suppress the term in square brackets to zero simply by selecting

$$\boxed{G_3 = G_4} \quad \text{ENHANCEMENT CONDITION, TYPE 2 CIRCUIT} \quad (5.118)$$

It is not necessary to match the GBPs B_1 and B_2 as is required for the type 1 circuit. At high frequencies, it is expected that the term $\omega^3\, C_5/B_1B_2G_6$ will dominate the overall dissipation $d_l(\omega)$ and at low-frequencies it is expected that the term in equation (5.116) that contains ω in the denominator will dominate $d_l(\omega)$.

The *uniform conductance* case again allows straightforward interpretation of the behavior of the circuit. Thus, substituting $G_2 = G_3 = G_4 = G_6 = G$, $C_5 = C$, and $\tau \equiv C/G$ into equation (5.116) and assuming $A_{01,2} \approx A_0$ gives

$$d_l(\omega) \approx \underbrace{\frac{2}{A_0}\left(\omega\tau+\frac{1}{\omega\tau}\right)}_{d_{Ll}(\omega)} - \underbrace{\omega^3\frac{C_5}{B_1B_2G_6}}_{d_{Hl}(\omega)} \quad \text{UNIFORM CONDUCTANCE DISSIPATION FACTOR} \quad (5.119)$$

The dissipation $d_l(\omega)$ is evidently independent of the mismatch factor m that appears in the corresponding equation (5.110) for the type 1 circuit. The low-frequency term $d_{Ll}(\omega)$ is identical for the uniform conductance type 1 and type 2 circuits, and the graphical description of equation (5.119) is, therefore, described adequately by Fig. 5.14(a) and the $m = 1$ curve of Fig. 5.14(b).

The complete expression for the fractional inductance error $\epsilon_l(\omega)$ of the type 2 circuit, the derivation of which is requested in Problem 5.23, is given by

$$\epsilon_l(\omega) \approx \underbrace{\frac{1}{A_{01}}\left(\frac{G_4}{G_3}-1\right)+\frac{1}{A_{02}}\left(1-\frac{G_3}{G_4}\right)}_{\epsilon_{Ll}(\omega)} + \underbrace{\frac{1}{B_1}\left(\frac{G_4G_6}{G_3C_5}+\omega^2\frac{C_5}{G_6}\right)+\frac{1}{B_2}\left(\frac{G_6}{C_5}+\omega^2\frac{G_3C_5}{G_4G_6}\right)}_{\epsilon_{Hl}(\omega)} \quad (5.120)$$

where it is almost always valid to neglect $\epsilon_{Ll}(\omega)$. The expression for $\epsilon_{Hl}(\omega)$ is given in Table 5-7, with the corresponding expression for $d_l(\omega)$. Reconsidering the inductance error $\epsilon_l(\omega)$ of equation (5.120) for the uniform conductance, case we have

$$\epsilon_{lH}(\omega) \approx \left(\frac{1}{B_1} + \frac{1}{B_2}\right)\left(\frac{1}{\tau} + \omega^2\tau\right) \quad \text{INDUCTANCE ERROR UNIFORM CONDUCTANCE CASE} \tag{5.121}$$

which is similar in general form to the type 1 result of equation (5.114).

This completes the analysis of the type 1 and type 2 inductance-simulation circuits. A typical numerical example is now considered in order to compare the *exact* dissipation and inductance error with the approximate expressions that have been derived and tabulated in Tables 5-6 and 5-7.

EXAMPLE 5.2: A Type 2 Inductance-Simulation Circuit. The *exact* dissipation factor $d_l(\omega)$ and inductance error $\epsilon_l(\omega)$ of the type 2 inductance-simulation circuit may be obtained from equation (5.67). In this section, the *algebraic approximations* for d and Q shown in Table 5-7 are compared with the exact expressions for three separate sets of circuit parameters. The values of the circuit parameters are summarized in Table 5-8, from which it is observed that all three sets of conditions, A, B, and C, lead to an *ideal* inductance L_1 of 15.915 mH. Condition A corresponds to a uniform conductance design *with enhancement* ($G_3 = G_4$) and *perfect matching* of GBPs ($B_1 = B_2$, $m = 1$). Condition B corresponds to a situation similar to that of Condition A, except that the GBPs are mismatched, so that $B_2 = \frac{1}{3}B_1$. Condition C corresponds to a *nonenhanced implementation* ($G_4 = 4G_3$) under conditions of *mismatched GBPs*.

The *exact* dissipation functions $d_l(\omega)$, as derived from equation [(5.67)], are given as the solid curves A, B, and C in Fig. 5.16. The corresponding dissipation functions that are given aproximately by equation (5.116) or equation (5.119) are shown with the dashed lines. The agreement between the exact result and the simplified algebraic expressions is generally acceptable. The nonenhanced mismatched-GBP result of curve C is dominated by the high-frequency *negative* dissipation term in equation (5.116) at frequencies above 1 kHz. Below 1 kHz, there is a cancellation effect between the positive function $d_{Ll}(\omega)$ and the negative function $d_{Hl}(\omega)$ in equation (5.116). At *very* low frequencies (< 100 Hz), the function $d_{Ll}(\omega)$ dominates the dissipation due to the term $G_6G_4/\omega A_{01}C_5G_3$ that increases as $\omega \to 0$.

As expected, Conditions A and B result in a much improved dissipation $d_l(\omega)$, because enhancement is maintained; Q factors $d_l^{-1}(\omega)$ in excess of 500 are achieved over the frequency range 100 Hz to 10 kHz. (This compares with a Q-factor of -19.5 at 10 kHz for the nonenhanced Condition C.)

The inductance errors $\epsilon_l(\omega)$ are shown in Fig. 5.17 for Conditions A and B; the algebraic expression in equation (5.119) leads to the dashed curves

TABLE 5.8

Condition	$\frac{1}{G_2}$ kΩ	$\frac{1}{G_3}$ kΩ	$\frac{1}{G_4}$ kΩ	C_5 pF	$\frac{1}{G_6}$ kΩ	L_2 mH	B_1 rad/sec	B_2 rad/sec	*Comments*	*Pertinent Equations*
A	10	10	10	159.15	10	15.915	$2\pi \times 10^7$	$2\pi \times 10^7$	Enhanced; Matched GBP's; Uniform conductances	(5.110) (5.114)
B	10	10	10	159.15	10	15.915	$2\pi \times 10^7$	$\frac{2\pi}{3} \times 10^7$	Enhanced; Mismatched GBP's; Uniform conductances	(5.110) (5.114)
C	20	20	5	318.30	10	15.915	$2\pi \times 10^7$	$\frac{2\pi}{3} \times 10^7$	Nonenhanced; Mismatched GBP's; Nonuniform conductances	(5.109) (5.113)

N.B. $A_{01} = A_{02} = 10^5$.

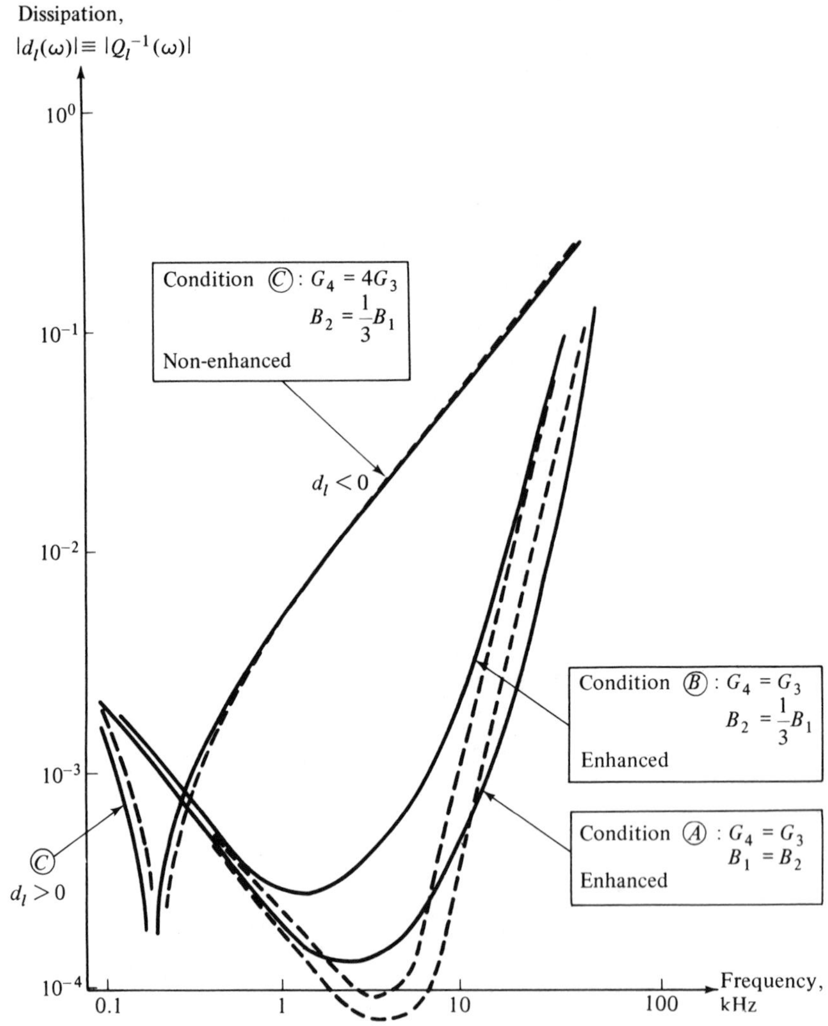

FIGURE 5.16 *Dissipation functions for conditions A, B and C of Example 5.2, type 2 inductance simulation GIC circuit.*

and these are in good agreement with the exact result. The error $\epsilon_l(\omega)$ is relatively constant and less than 0.1% over the frequency range 100 Hz to 10 kHz, increasing to 1% at about 30 kHz for the Condition B.

In summary, it is found that for a type 2 circuit with mismatched GBPs of 10 MHz and $3\frac{1}{3}$ MHz, it is possible to realize a high-quality inductance ($|Q| > 500$, $\epsilon_l < 1\%$) over the frequency range 100 Hz to 10 kHz, *provided that Q enhancement is maintained* by ensuring that $G_3 = G_4$.

Similar results are obtained for the type 1 circuit, except that it is necessary that $B_1 \approx B_2$, implying that the matching factor m must lie in the range 0.97 to 1.03 in order to attain the enhancement that is achievable with the type 2 circuit.

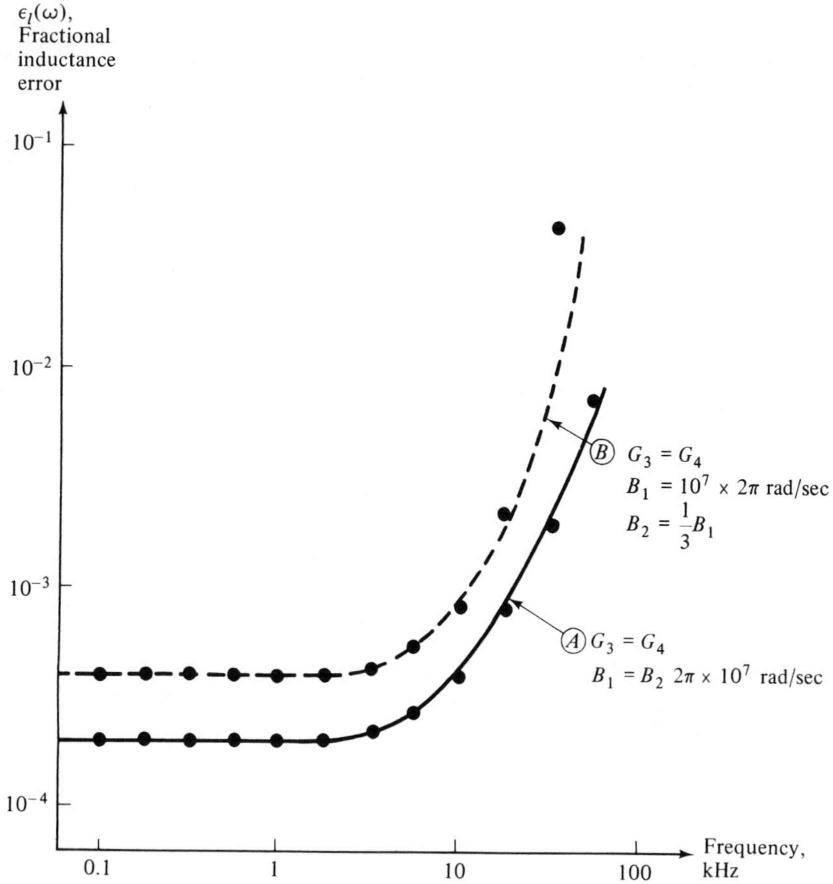

FIGURE 5.17 *Fractional inductance errors for conditions A and B of Example 5.2, type 2 inductance simulation circuit.*

5-7 FDNR CIRCUITS USING GICs

An ideal FDNR of the form

$$Y_{1d} = s^2 D \tag{5.122}$$

may be realized by using the coupled-GIC of Table 5-3 in three different ways: the branch admittances may be chosen to realize the three types of FDNR circuit as follows:

$$\boxed{Y_{1d} = \frac{s^2 C_2 C_4 G_6}{G_3 G_5}, \quad Y_{1d} = \frac{s^2 G_2 C_4 C_6}{G_3 G_5}, \quad Y_{1d} = \frac{s^2 C_2 G_4 C_6}{G_3 G_5}} \tag{5.123}$$

Type 1   Type 2   Type 3

The analysis of the corresponding nonideal circuits follows the same procedure that has been described in detail for the simulated-inductance circuits. That is, $Y_{1d}(s)$ is derived by substituting for $Y_2 \ldots Y_6$ in equation (5.67), and the resultant expression is simplified by means of the assumptions in equations (5.92), (5.93), (5.94), and (5.95). The corresponding nonideal multiplicative function $M(s)$ is used to obtain the dissipation factor $d_d(\omega)$ and Q factor $Q_d(\omega)[\equiv d_d^{-1}(\omega)]$ via the defining relationship

$$d_d(\omega) \equiv \tan < [M(j\omega)] = Q_d^{-1}(\omega) \qquad \text{FDNR DISSIPATION AND } Q\text{-FACTOR} \qquad (5.124)$$

and then, by a similar analysis to that leading to equation (5.89), it is possible to obtain the fractional error of D from the relationship

$$\epsilon_d(\omega) \approx \mathrm{Re}[M(j\omega)] - 1 \qquad \text{FRACTIONAL ERROR IN } D \qquad (5.125)$$

The type 1 circuit is shown in Table 5-9, and the following analysis outlines the main steps in the derivation of $d_d(\omega)$ and $\epsilon_d(\omega)$ for this widely used circuit. This particular circuit has been produced as a microelectronic device by manufacturers.

5-7.1 The Type 1 FDNR GIC Circuit (Table 5.9):

Substituting C_2, C_4, G_3, C_5, and G_6 for Y_2, Y_4, Y_3, Y_5, and Y_6 into equation (5.67) and use of the assumptions of equations (5.92) (5.93), (5.94), and (5.95) gives

$$Y_{1d} \approx \frac{s^2 C_2 C_4 G_6}{G_3 G_5} \underbrace{\left[\frac{1 + \dfrac{D(s)}{A_{01}A_{02}sC_4G_6}}{1 + \dfrac{D(s)}{A_{01}A_{02}G_3G_5}} \right]}_{M(s)} \qquad (5.126)$$

where

$$D(s) \approx s^3[T_1T_2C_4(G_5 + G_6)] + s^2[A_{02}T_1C_4(G_5 + G_6)] + s[A_{01}T_2G_3 + A_{02}C_4)(G_5 + G_6)] + [A_{01}G_3(G_5 + G_6)] \qquad (5.127)$$

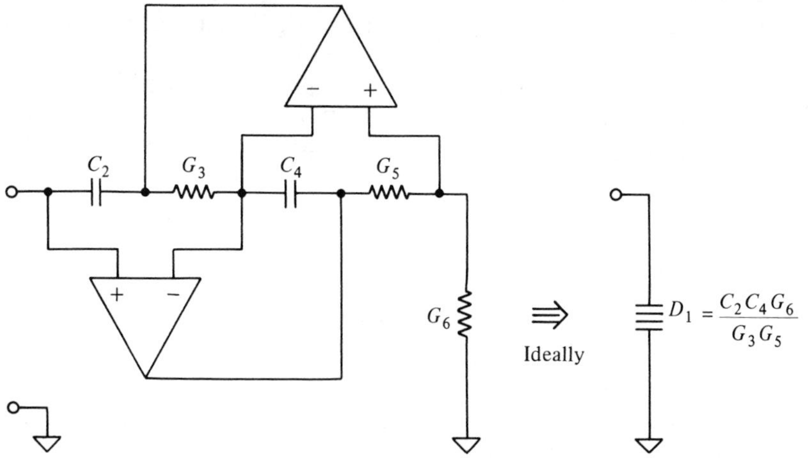

Dissipation factor; Q factor: $Q_d(\omega) \equiv d_d^{-1}(\omega)$

$$d_d(\omega) \approx \underbrace{(G_5 + G_6)\left[\frac{\omega C_4}{A_{01} G_3 G_5} + \frac{G_3}{\omega A_{02} C_4 G_6}\right]}_{d_{Ll}(\omega)} \underbrace{- \omega(G_5 + G_6)\left[\frac{1}{B_1 G_6} - \frac{1}{B_2 G_5} + \frac{\omega^2 C_4}{B_1 B_2 G_3 G_5}\right]}_{d_{Hl}(\omega)}$$

Enhancement condition:

$$B_1 G_6 = B_2 G_5$$

FDNR D-element fractional error:

$$\epsilon_{Hd}(\omega) \approx \epsilon_d(\omega) \approx (G_5 + G_6)\left[\frac{G_3}{B_2 C_4 G_6} + \frac{\omega^2 C_4}{B_1 G_3 G_5}\right]$$

TABLE 5.9 *Type 1 FDNR circuit.*

Then, employing equations (5.126), (5.127), (5.124), and (5.125) and using the previous assumptions in equations (5.103) and (5.104) leads directly to the expressions for $d_d(\omega)$ and $\epsilon_d(\omega)$ that are found in Table 5-9. The most significant result is that the high-frequency dissipation factor $d_{Hd}(\omega)$ is similar in form to that obtained for $d_{Hl}(\omega)$ in equation (5.105). The enhancement condition is given by equation (5.107) once more, and it is therefore generally necessary to *use dual OP AMPs* (so that $0.97 < m < 1.03$) *and select* $G_6 = G_5$ in order to achieve significant enhancement of $Q_d(\omega)$.

The *uniform conductance-uniform capacitance* case leads to simple expressions for $d_d(\omega)$: assuming $G_3 = G_5 = G_6 \equiv G$ and $C_2 = C_4 \equiv C$, and $\tau \equiv C/G$, the expression for $d_d(\omega)$ in Table 5.9 simplifies to

$$d_d(\omega) \approx \underbrace{+\frac{2}{A_0}\left(\omega\tau + \frac{1}{\omega\tau}\right)}_{d_{Ld}(\omega)} \underbrace{- \frac{2\omega}{B_1}\left(1 - m + \frac{\omega^2\tau m}{B_1}\right)}_{d_{Hd}(\omega)} \qquad \text{UNIFORM } G\text{-UNIFORM } C \tag{5.128}$$

which is similar to the type 1 inductance-simulation circuit [equation (5.110)] except for the negative sign associated with the high-frequency dissipation term. The dissipation factor $d_d(\omega)$, therefore, behaves very much like that of the type 1 inductance-simulation circuit.

The expression in Table 5-9 for the fractional error $\epsilon_d(\omega)$ in the value of D is also similar to that obtained for the simulated-inductance circuits. The complete expression for $\epsilon_d(\omega)$, where

$$\epsilon_d(\omega) \equiv \frac{D_{1eff} - D_1}{D_1} = \text{Re}[M(j\omega)] - 1 \tag{5.129}$$

is

$$\epsilon_d(\omega) \approx \underbrace{\left[\frac{1}{A_{01}G_6} - \frac{1}{A_{02}G_5}\right](G_5 + G_6)}_{\epsilon_{Ld}(\omega)} + \underbrace{\left[\frac{G_3}{B_2C_4G_6} + \omega^2\frac{C_4}{B_1G_3G_5}\right](G_5 + G_6)}_{\epsilon_{Hd}(\omega)} \tag{5.130}$$

Derivation of this result is requested in Problem 5-27. Only the dominant $\epsilon_{Hd}(\omega)$ term is given in Table 5-9. Note that for the uniform conductance-uniform capacitance case, the fractional error $\epsilon_{Hd}(\omega)$ simplifies to

$$\epsilon_{Hd}(\omega) \approx 2\left(\frac{\omega^2\tau}{B_1} + \frac{1}{B_2\tau}\right) \qquad \text{UNIFORM } G/\text{UNIFORM } C \tag{5.131}$$

which is similar to the behavior of $\epsilon_{Hl}(\omega)$ in equation (5.114) for the type 1 inductance-simulation circuit.

Q factor estimation: *Type 1 FDNR Circuit:* Suppose that the type 1 FDNR circuit of Table 5-9 is used to realize a D element, where $C = C_2 = C_4$ and $G = G_3 = G_5 = G_6$, so that

$$D_1 = \frac{C^2}{G} \tag{5.132}$$

Let $C = 1000$ pF and $G = 10^{-4}\ \text{k}\Omega^{-1}$ so that $D_1 = 10^{-14}$ Fs and $\tau \equiv C/G = 10^{-5}$ sec. Then, consider that economy OP AMPs are used for which

$$B_1 = 2\pi \times 10^6 \text{ rad/sec}, \qquad A_{01,2} = 10^5 \tag{5.133}$$

and $\frac{1}{2} \leqslant m \leqslant 1$. By direct substitution of these parameters into equations (5.92), (5.93), (5.94), (5.95), (5.103), and (5.104) it follows that all seven assumptions are valid if

$$\omega \ll 7.9 \times 10^5 \text{ rad/sec} \quad \text{or} \quad f \ll 126\ k\text{Hz} \tag{5.134}$$

With this restriction in mind, direct substitution of these numerical values for A_0, τ, B_1, and m into equation (5.128) gives the three Q-factors curves shown in Fig. 5.18 for $m = \frac{1}{2}$, 1, and 2. As expected, the enhancement condition ($m = 1$) results in large-valued Q-factor magnitudes $|Q_d(\omega)|$. Suppose that, in a given application, it is required that $|Q_d(\omega)| \geqslant 1000$. Then, if $m = 1$, it is possible to use the FDNR at frequencies in excess of 10^5 rad/sec. Whereas for $m = \frac{1}{2}$, the upper frequency limit is approximately 10^4 rad/sec, and for $m = 2$ it is not possible to achieve $|Q_d(\omega)| \geqslant 10^3$.

5-7.2 The Type 2 *FDNR* Circuit

This circuit is realized by choosing the GIC branches as indicated in equation (5.123). The reader is asked to prove in Problem 5.28 that the high-frequency dissipation term is given by

$$d_{Hd}(\omega) \approx \omega\left(\frac{C_4C_6}{B_1G_3G_5}\right) + \omega\left(\frac{1}{B_1} - \frac{1}{B_2}\right) - \frac{1}{\omega}\left(\frac{G_3G_5}{B_2C_4C_6}\right) \tag{5.135}$$

This circuit is not as useful as the type 1 and type 3 FDNR circuits because it is not possible to suppress $d_{Hd}(\omega)$ over a wide frequency range; the term $(\frac{1}{B_1} - \frac{1}{B_2})$ *is* suppressed by using matched Bs, but the remaining terms containing ω and $1/\omega$ do not lend themselves to cancellation over a range of ω.

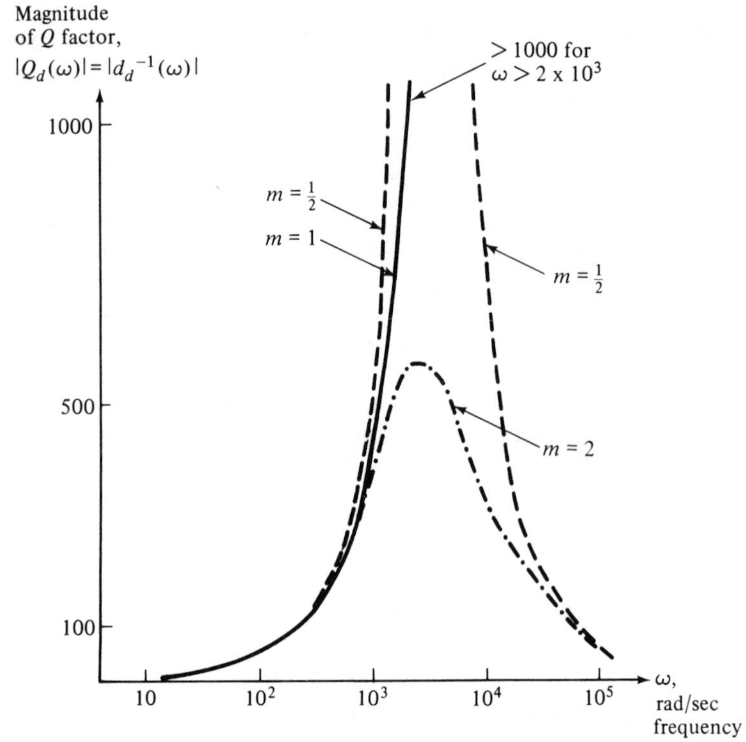

FIGURE 5.18 *Q factors for type 1 FDNR circuit, m* $=\frac{1}{2}$*, 1 and 2.*

5-7.3 The Type 3 *FDNR* Circuit

This circuit is especially useful. It is realized by selecting $Y_2 = sC_2$ and $Y_6 = sC_6$ and the remaining GIC branches as conductances. The reader is asked to analyze this circuit, using equation (5.67), to show that the expression for dissipation factor $d_d(\omega) \equiv Q_d^{-1}(\omega)$ and fractional magnitude error $\epsilon_d(\omega)$ shown in Table 5-10 are correct. (Note the similarity of these expressions for $d_d(\omega)$ and $\epsilon_d(\omega)$ with the corresponding expressions in Table 5-7 for the type 2 inductance-simulation circuit.) The enhancement condition $G_3 = G_4$ does *not* require matching of *B*s; for this reason, this FDNR circuit is highly recommended.

5-8 SUMMARY

The OP AMP is a three-terminal microelectronic device that is readily available at low cost. It is ideally equivalent to a nullor, but in practice it usually possesses a dominant-pole frequency response characteristic. The

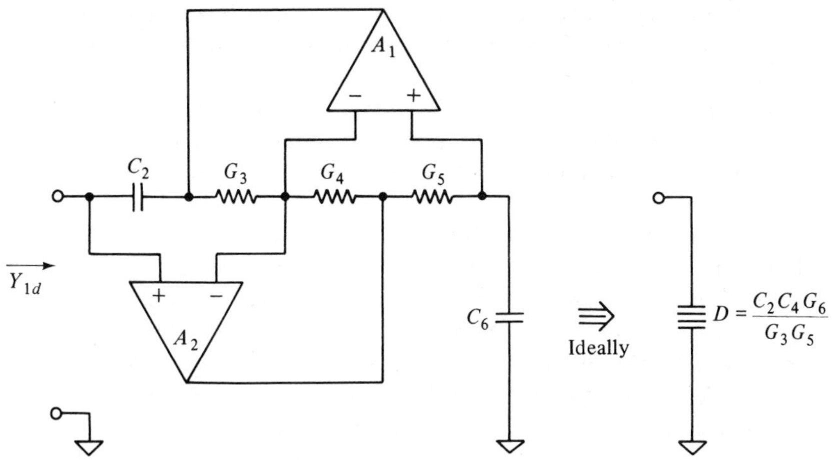

Dissipation factor; Q factor: $Q_d(\omega) \equiv d_d^{-1}(\omega)$

$$d_d(\omega) \approx \underbrace{-\left[\frac{\omega C_6 G_4}{A_{01} G_3 G_5} + \frac{\omega C_6}{A_{02} G_5}\right] - \left[\frac{G_5}{A_{01}\omega C_6} + \frac{G_3 G_5}{A_{02}\omega C_6 G_4}\right]}_{d_{Ld}(\omega)} + \underbrace{\frac{\omega}{B_2}\left[\frac{G_3}{G_4} - 1\right] + \frac{\omega}{B_1}\left[1 - \frac{G_4}{G_3}\right]}_{d_{Hd}(\omega)}$$

Enhancement condition:

$$G_3 = G_4$$

FDNR $\bar{D}$ element fractional error:

$$\epsilon_d(\omega) \approx \frac{1}{A_{02}}\left[\frac{G_3}{G_4} - 1\right] + \frac{1}{A_{01}}\left[1 - \frac{G_4}{G_3}\right] + \frac{1}{B_1}\left[\frac{G_5}{C_6} + \omega^2 \frac{G_4 C_6}{G_3 G_5}\right] + \frac{1}{B_2}\left[\frac{G_3 G_5}{G_4 C_6} + \omega^2 \frac{C_6}{G_5}\right]$$

TABLE 5.10 *Type 3 FDNR circuit.*

gain of the OP AMP is approximately equal to $-B/s$ over most of the usable frequency range, and it is this fact that often leads to major discrepancies between ideal and nonideal performance of linear OP AMP circuits and RC-active filters.

The inverting-type and the noninverting-type circuits may be used for amplification, integration, summation, and differentiation of signal waveforms. The nonideal performance of these circuits is summarized in Table 5-1 and equation (5.37).

The two-OP AMP coupled-GIC circuit shown in Table 5-3 is a useful subnetwork that may be used directly in transfer function synthesis or to realize FDNR and simulated-inductance elements. The nonideal performance of coupled-GIC FDNR and simulated-inductance circuits is summarized by the expressions for dissipation and magnitude errors that are given in Tables 5-6, 5-7, and 5-9. The concept of Q enhancement is important if it is necessary to maintain a high Q-factor over a wide frequency range and, therefore, the conditions for this enhancement are supplied in the tables. The results that are derived in this chapter are the basic modeling equations that are used to estimate the nonideal performance of RC-active filters that employ these circuits as subcircuits.

The reader has now reached the point in this book at which the classification, modeling, and analysis of the basic circuits and subnetworks is virtually complete. This essentially concludes the first part of the book. The next two chapters are rather general in nature: Chapter 6 is devoted entirely to the topic of network sensitivity. It turns out that the sensitivity performance of an RC-active filter circuit is an important criterion in the design process. The material in chapter 7 is a summary of the major signal-flow graph structures from which RC-active filter circuits are realized. Both chapters 6 and 7 may be omitted if the reader's objective is to apply immediately the circuits described in this chapter to the design of RC-active filters. A far deeper understanding of the design process is achieved by reading chapters 6 and 7 before proceeding to chapter 8.

PROBLEMS

5-1. An economy OP AMP with a dominant-pole frequency response has

$$A_0 = 200{,}000, \qquad \frac{B}{2\pi} = 1.5 \text{ MHz}$$

Derive the magnitude frequency response $|A(j\omega)|$ and the phase frequency response $<[A(j\omega)]$ at the frequencies 1 Hz, 10 Hz, 100 Hz, 100 kHz, and 1 MHz. What is the bandwidth BW?

Use equation (5.1) to calculate the frequency at which $|A(j\omega)|$ is unity. Show that this frequency is a good approximation to the term B in equation (5.1).

5-2. Replace the ideal OP AMP shown in Fig. 5.4(a) by a nullor and thereby verify equation (5.12) using, (i) the nullor loop analysis method and (ii) the nullor nodal analysis method.

5-3. The nonideal performance of the inverting-amplifier OP AMP circuit of Fig. 5.4(a) may be estimated by replacing the OP AMP with the dominant-pole nullor model of Fig. 5.1(b) as shown in Fig. P5.3.

Eliminate the redundant nullator and redundant norator and transform the series combination of E_1 and admittance Y_1 to the Norton equivalent circuit consisting of a current source $E_1 Y_1$ in parallel with an admittance Y_1. Then, use the nodal analysis method of section 3-2.2 to show that

$$\begin{bmatrix} E_1 Y_1 \\ 0 \end{bmatrix} = \begin{bmatrix} (Y_1 + Y_2) & -Y_2 \\ 1 & Y \end{bmatrix} \begin{bmatrix} V_{1,2} \\ V_{3,4} \end{bmatrix}, \qquad Y \equiv \frac{1}{A_0} + \frac{j\omega}{B}$$

and thereby confirm the validity of equation (5.20).

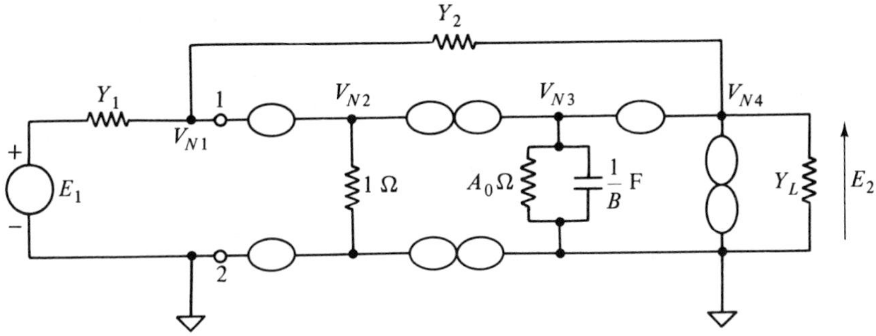

FIGURE P5.3

5-4. Use equation (5.22) to sketch on log-log paper, the closed-loop gain frequency responses $|H(j\omega)|$ for the nonideal inverting-amplifier with

$$A_0 \equiv 10^5, \qquad \frac{B}{2\pi} \equiv 1 \text{ MHz}$$

and $K \equiv G_1/G_2 = 10$, 10^3, and 10^5. What are the bandwidths BW for each of these three values of K?

Show that equations (5.23) and (5.25) are good approximations for the above three values of K.

5-5. Investigate the accuracy of equations (5.29) for the numerical values given in Problem 5.4. Thereby, verify that equations (5.30) are accurate for $K \gg 1$ and $\omega K/B \ll 1$.

5-6. Use equation (5.21), with $Y_1 \equiv G_1$, $Y_2 \equiv sC_2$, and $T \equiv C_2/G_1$ to prove that the nonideal inverting-integrator [Fig. 5.5(b)] has the approximate $M(j\omega)$ functions shown in Table 5-1.

5-7. Use equation (5.21), with $Y_1 \equiv sC_1$, $Y_2 \equiv G_2$, and $T \equiv C_1/G_2$ to prove that the nonideal inverting-differentiator [Fig. 5.5(c)] has the approximate $M(j\omega)$ functions shown given in Table 5-1.

5-8. The *ideal* gain of the noninverting amplifier shown in Fig. 5.8 is given in equation (5.33) and the *nonideal* gain in equation (5.35). The nonideal equivalent circuit, using Fig. 5.1(b) for the OP AMP, is shown in Fig. P5.8. Noting that $V_{N1} = E_1$, show that equation (5.35) is valid.

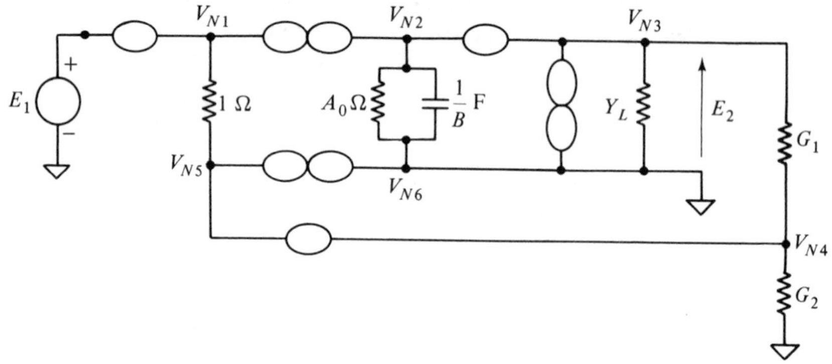

FIGURE P5.8

5-9. Replace Y_1 in Fig. 5.10 by an independent current source I_1 and then calculate the voltage V_1 across port 1 as a function of $A(s)$, Y_2, Y_3, and Y_4. Thereby, prove the validity of equation (5.53).

5-10. The nonideal transform driving-point admittance $Y_{1d}(s)$ of the circuit shown in Fig. 5.11 is given by equation (5.54). Assume

$$G_2 = G_3 = G_4 = 10^{-4}\ \text{ohm}^{-1}$$

$$B = 2\pi \times 10^6\ \text{rad/sec}$$

$$A_0 = 10^5$$

and sketch the input admittance $Y_{1d}(j\omega)$ locus as shown in Fig. 5.12, marking the frequencies 1 kHz, 10 kHz, and 100 kHz on the sketched locus.

5-11. A *negative capacitance* element may be simulated approximately by selecting

$$Y_2 = G_2, \qquad Y_3 = G_3, \qquad Y_4 = sC_4$$

in Fig. 5.10. Use equation (5.53) to derive an expression for the nonideal transform admittance $Y_{1d}(s)$.

5-12. For

$$G_2 = G_3 = 10^{-4}\ \text{ohm}^{-1}$$
$$C_4 = 10{,}000\ \text{pF}$$
$$A_0 = 10^5$$
$$B = 2\pi \times 10^6\ \text{rad/sec}$$

calculate $Y_{1d}(s)$ and the locations in the s-plane of the poles and zeros of $Y_{1d}(s)$.

Calculate the Q-factor of this nonideal negative capacitance $Y_{1d}(j\omega)$ at 10 kHz, using

$$Q_{-C}(\omega) = \omega(-C)r_p$$

where r_p is the effective parallel resistance r_p (see the definition of a Q-factor in section 2-3). How does $Q_{-C}(\omega)$ vary as a function of ω?

5-13. Let the source termination at port 1 in Problem 5.11 be a resistance R_s. Show that this effective parallel connection of a resistance R_s and a nonideal admittance $Y_{1d}(s)$ is unstable if and only if the roots of the equation

$$\frac{1}{R_s} + Y_{1d}(s) = 0$$

are in the right-half s-plane. Thereby, use the numerical values in Problem 5.12 to show that R_s must be impractically low-valued if this configuration is to be stable.

5-14. Let the polarity of the differential input terminals shown in Fig. 5.10 be reversed so that we effectively replace $A(s)$ by $-A(s)$ in all expressions for $Y_{1d}(s)$. Consider the consequence of this on the $Y_{1d}(j\omega)$ expression that was calculated in Problem 5.11 and on the range of values of source termination R_s over which the circuit is stable.

5-15. Prove that the driving-point transform admittance $Y_{1d}(s)$ of the terminated coupled-GIC shown in Table 5-3 is given by equation (5.67). (This is a lengthy proof.)

5-16. Prove that the driving-point transform admittance $Y_{1d}(s)$ of the terminated uncoupled-GIC shown in Table 5-3 is given by equation (5.69). (This is a lengthy proof.)

5-17. Prove that the expressions for $\Delta(s)$, derived from equation (5.73), are valid in equations (5.68) and (5.70).

5-18. Prove that the "definition" of Q-factor given in equation (5.83) is consistent with the earlier definition of equations (2.9) and (2.12).

5-19. Investigate the importance of matched-gain bandwidth products B_1 and B_2 in the type 1 inductance-simulation circuit of Table 5-6 by plotting the dissipation factor $d_l(\omega)$, as given in equation (5.110) for the uniform conductance case, as a function of frequency ω for the conditions $A_0 = 10^5$,

$B=2\pi\times10^6$ rad/sec, $\tau=10^{-5}$ sec, and the following three mismatch conditions: (i) $m=\frac{1}{2}$, (ii) $m=1$ (matched conditions), (iii) $m=2$.

Plot $d_l(\omega)$ on log-log paper over the frequency range from $\omega=10^{-2}$ rad/sec to $\omega=2\pi\times10^5$ rad/sec. Check that equation (5.110) and, therefore, equation (5.105) are valid for the above numerical values by checking that the inequalities (5.92), (5.93), (5.94), (5.95), (5.103), and (5.104) *are* satisfied.

Finally, plot the low-frequency and the high-frequency asymptotes of $d_l(\omega)$ for the three values of m given above.

5-20. For the numerical values in Problem 5.19, calculate $\epsilon_l(\omega)$, using equation (5.112). Assume $m=1$. Sketch $\epsilon_l(\omega)$.

5-21. A type 2 inductance-simulation circuit is shown in Table 5-7. Use equation (5.69) and $Y_2=G_2$, $Y_3=G_3$, $Y_4=G_4$, $Y_5=sC_5$, $Y_6=G_6$ to derive $M(s)$, where $M(s)$ is defined by the relationship

$$Y_{1d}(s)=s\frac{C_5G_2G_4G_6}{G_3}[M(s)]$$

5-22. Use the expression for $M(s)$ in Problem 5.21 to derive a corresponding approximate expression for $M(j\omega)$. [*Hint:* This exercise is performed in section 5-6.1 for the type 1 circuit; a similar analysis is required here.]

5-23. Use the approximate expression for $M(j\omega)$ in Problem 5.22 to prove that equations (5.116) and (5.120) are valid for the type 2 inductance-simulation circuit.

5-24. Transform the series connection of the voltage source E_1 and admittance Y_1 in Fig. P5.24 to a Norton equivalent circuit consisting of a current generator E_1Y_1 in parallel with an admittance Y_1. Then, use RC-nullor nodal analysis

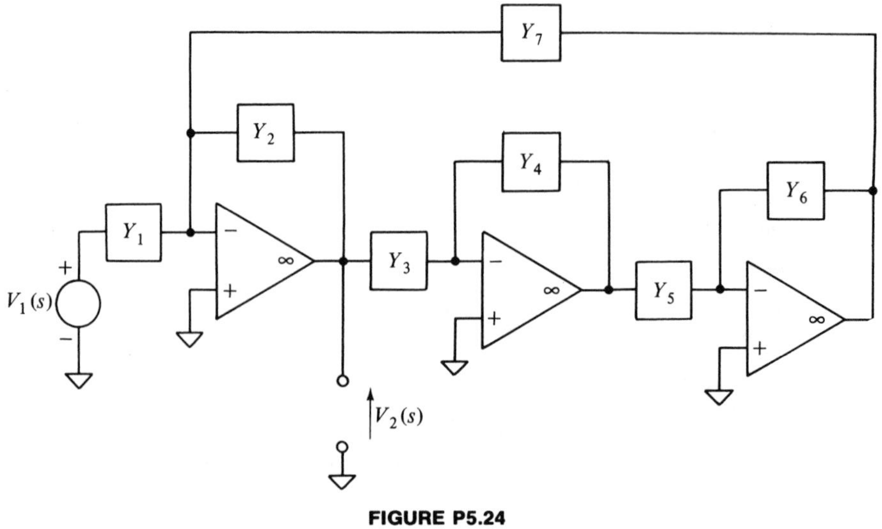

FIGURE P5.24

to prove that

$$\frac{V_2(s)}{V_1(s)} = -\frac{Y_1Y_4Y_6}{Y_2Y_4Y_6 + Y_3Y_5Y_7}$$

5-25. The amplifier shown in Fig. P5.25 has voltage $[V_1(s) + V_2(s)]$ applied to terminal 1 and $V_2(s)$ applied to terminal 2. Thus, $V_1(s)$ is the *differential* voltage between terminals 1 and 2 and $V_2(s)$ is the *common* voltage at terminals 1 and 2. Use RC-nullor loop analysis to derive $V_3(s)$ in terms of $V_1(s)$, $V_2(s)$, R_1, R_2, R_3, and R_4. Show that the condition $R_1 = R_2 = R_3 = R_4$ gives $V_3(s) = -V_1(s)$, implying that the common mode voltage $V_3(s)$ is suppressed by the circuit under these conditions.

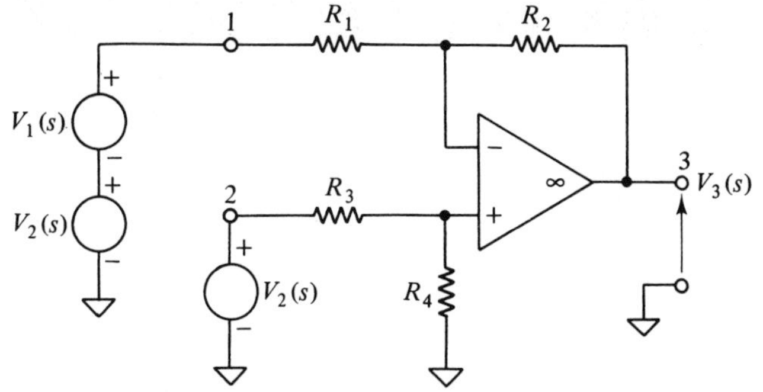

FIGURE P5.25

5-26. A type 1 FDNR circuit is shown in Table 5-9. Use equation (5.67) to verify that the function $M(s)$ is given by equations (5.126) and (5.127).

5-27. Use the result in Problem 5.26 to verify the expressions for $d_d(\omega)$ and $\epsilon_{Hd}(\omega)$ in Table 5-9.

5-28. Verify equation (5.135) for $d_{Hd}(\omega)$ of a type 2 FDNR circuit.

BIBLIOGRAPHY

MILLMAN, J., "*Microelectronics: Digital and Analog Circuits and Systems*" McGraw-Hill, 1979, chapters 15 and 16.

ANTONIOU, A., "Realizations of Gyrators Using Operational Amplifiers and Their Use in RC-Active Network Synthesis," *Proc. Inst. Elec. Eng.*, vol. 116, 1838–1850, Nov. 1969.

BRUTON, L. T., "Nonideal Performance of Two-Amplifier Positive Impedance Converters," *IEEE Trans. Circuit Theory*, vol. CT-17, p. 541–549, Nov. 1969.

AKERBERG, D., and K., MOSSBERG, "A Versatile Active RC Building Block with Inherent Compensation for the Finite Bandwidth of the Amplifier," *IEEE Trans. Circuits Syst.*, vol. CAS-23, p. 68–72, Feb. 1976.

6

NETWORK SENSITIVITY

6-1 INTRODUCTION

In previous chapters of this book, emphasis is placed on the description, design, and modeling of the *basic building-blocks* from which RC-active circuits are made. RC-active networks can always be modeled as interconnections of RC elements and nullors and a nullor is an infinite-gain controlled source. Finite-gain controlled sources and GICs, for example, may be realized in a variety of ways by interconnecting Rs, Cs, and nullors. Particular circuits have been described in chapters 4 and 5 because they are often used as building-blocks in the synthesis and design of more complicated circuits. It has been shown that the bipolar transistor and the OP AMP are devices that may be used to design inverting and noninverting amplifiers, inverting and noninverting integrators, simulated-inductance elements, FDNR elements, negative-RLC elements, generalized immittance converters (GICs), gyrators, etc. In most cases, the nonideal performance of these circuits may be characterized by a nonideal multiplicative function $M(s)$.

So far, we have not attempted to determine how these basic circuits might be interconnected and thereby used for some useful purpose. The *analysis* of any particular interconnection of these circuits is straightforward; the methods of loop and nodal analysis (chapter 3) may be used. In this and subsequent chapters, the synthesis, design and applications of *RC*-active networks are described. By use of the word *synthesis*, we imply the derivation of a topological interconnection of *ideal* *R*'s, *C*'s and nullors. In most practical cases, this may simply be equivalent to a topological interconnection of *R*s, *C*s, and OP AMPs. For any given network function $F(s)$, there usually exists a variety of synthesis techniques (and corresponding *RC*-nullor networks) that realize that function. By use of the word *design*, we imply the calculations of specific circuit parameter values that are necessary to ensure that, in practice, the network function $F(s)$ is within acceptable design constraints. For example, if $F(s)$ is a voltage transfer function, the *real* magnitude function $|F(j\omega)|$ may be specified within certain tolerance limits, and the design procedure therefore involves the selection of the resistance values, capacitance values and perhaps OP AMP parameters (A_0 and B) that ensure that the required function $|F(j\omega)|$ is achieved with sufficient accuracy. The values of the resistances and capacitance elements, as well as A_0 and B, are functions of time (aging) and ambient conditions (temperature, humidity). It is, therefore, often necessary to take into account the fact that any network parameter value x_i possesses an associated tolerance or worstcase fractional deviation ε_i such that

$$x_{iN}(1 - \epsilon_i) \leqslant x_i \leqslant x_{iN}(1 + \epsilon_i) \tag{6.1}$$

A major *practical* problem, associated with the manufacture of *RC*-active filter circuits, is that the transfer functions are often intolerably *sensitive* to perturbations of the circuit parameters. The *theoretical* synthesis problem was solved in 1961 by Linvill when it was proved that *any* real causal rational transfer function may be realized as the transimpedance $Z_{21}(s)$ of an electrical network consisting of just *one* negative immittance converter (NIC) and a suitable interconnection of *RC* elements. If it was not for the horrendously difficult *tolerance problems* associated with these early NIC networks, it is quite possible that the modern approach to *RC*-active network synthesis would not have been developed. Experimenters and designers soon found, sometimes to their confusion and ultimate dismay, that a circuit designed by using the Linvill-NIC method could always be found, but that if a capacitance of say 1732.14 pF was required, for example, then it was quite possibly *essential* that this capacitance be accurate in practice to within 0.1 pF! A tolerance problem of a totally discouraging magnitude was therefore associated with many of the early synthesis procedures. The problem was found to be particularly acute if high-quality networks (Q pole-pairs greater than 20) were required or if the devices (transistors or OP AMPs) departed in a significant way from a good approximation to a nullor.

With the development of a relatively low-cost microelectronic technology, the transistor and the OP AMP became less costly and they offered better performance in terms of gain, size and power dissipation. New structures and new synthesis techniques were simultaneously developed that offered better tolerance performance at the expense of employing more than one nullor (OP AMP). It was often found that the cost of using several OP AMPs in conjunction with 1% tolerance RC elements was less than the cost of employing an alternative circuit that used only one OP AMP but required 0.01% tolerance RC elements. *This trend toward using synthesis techniques that suppress the tolerance problem at the expense of increasing the number of OP AMPs has continued over the last two decades*; it is now often economically viable to use as many as $2N$ OP AMPs to realize a particular stringent Nth-order transfer function if a significant improvement in tolerance performance is achievable.

During recent years it has become increasingly evident that the tolerance problem, which may be thought of as the *sensitivity* of the required network function $F(s)$ to its physical parameters x_i, is intimately related to the *structure* of the network. This chapter commences with a definition of the classical sensitivity formulas and a derivation of the important properties of these formulas. In particular, the *invariance properties* are shown to be of fundamental importance in the determination of the lower-bound tolerance performance of electrical networks. Some of the more basic and important structures are described in chapter 7 in terms of their signal flow-graph representations and some relationships between flow-graph structure and sensitivity are described. By this approach, the reader should develop some insight and intuition for the probable tolerance behavior of the specific RC-active circuits that appear in the subsequent chapters and in the published literature.

6-2 SENSITIVITY

Let $F(s, x_i)$ be a network function, where s is the usual complex frequency operator and x_i is one of the set $x_1 \ldots x_i \ldots x_n$ of n *real* physical parameters of the network. For example, $F(s, x_i)$ might be the voltage transfer function of a network and x_i might represent the values of the RC elements. The tolerance or sensitivity problem is associated with determining the perturbed network function $F(s, x_i + \Delta x_i)$ that results from a finite perturbation of a particular x_i by the incremental quantity Δx_i. If $F(s, x_i)$ and $F(s, x_i + \Delta x_i)$ are analytic, then by the Taylor series,

$$F(s, x_i + \Delta x_i) - F(s, x_i) = \sum_{n=1}^{\infty} \frac{\partial^n F(s, x_i)(\Delta x_i)^n}{n! \partial x_i^n} \tag{6.2}$$

Recognizing the left-hand side as the *change* in the network function $\Delta F(s, x_i, \Delta x_i)$ and writing this change as ΔF, we have

$$\Delta F = \left[\frac{\partial F}{\partial x_i} \Delta x_i\right] + \left[\frac{\partial^2 F(\Delta x_i)^2}{2! \partial x_i^2}\right] + \left[\frac{\partial^3 F(\Delta x_i)^3}{3! \partial x_i^3}\right] + \ldots \tag{6.3}$$

where $F(s, x_i)$ is written for simplicity as F. Then,

$$\frac{\Delta F}{F} = \left[\frac{x_i \partial F}{F \partial x_i}\right]\left(\frac{\Delta x_i}{x_i}\right) + \frac{1}{2!}\left[\frac{x_i^2 \partial^2 F}{F \partial x_i^2}\right]\left(\frac{\Delta x_i}{x_i}\right)^2 + \ldots$$

$$\ldots \frac{1}{n!}\left[\frac{x_i^n \partial^n F}{F \partial x_i^n}\right]\left(\frac{\Delta x_i}{x_i}\right)^n + \ldots \tag{6.4}$$

If

$$\frac{\partial F}{\partial x_i} \neq 0 \tag{6.5}$$

then

$$\lim_{\Delta x_i \to 0}\left[\frac{\Delta F}{F}\right] = \left[\frac{x_i \partial F}{F \partial x_i}\right]\left(\frac{\Delta x_i}{x_i}\right) \tag{6.6}$$

The term in square brackets on the right-hand side of equation (6.6) is defined as the *first-order sensitivity function*. For simplicity, we use the notation

$$S_{x_i}^F \equiv \frac{x_i \partial F}{F \partial x_i} = \frac{\partial[\ln F]}{\partial[\ln x_i]} \tag{6.7}$$

FIRST-ORDER SENSITIVITY FUNCTION

so that the fractional deviation of the network function F, for *arbitrarily small* Δx_i is given [subject to the condition in equation (6.5)] by

$$\boxed{\frac{\Delta F}{F} \approx S_{x_i}^{F}\left(\frac{\Delta x_i}{x_i}\right)} \tag{6.8}$$

EXAMPLE 6.1: Given that

$$F(s, x_1) \equiv s^2 + x_1 s + x_2 \tag{6.9}$$

derive $S_{x_1}^F$ and thereby calculate the fractional change in F at $s = 2$, $x_1 = x_2 = 1$ due to a $+1\%$ change in x_1. Compare the answer obtained from equation (6.8) with the answer obtained directly from perturbing x_1 in equation (6.9).

SOLUTION: First, it is evident that we may use equation (6.8) because the expression for $F(s, x_i)$ meets both requirements: (they are, $\partial F/\partial x_i \neq 0$ at $s = 2$, $x_1 = x_2 = 1$, and $F(s, x_1)$ *is* analytic at $x_1 = 1$ and $x_1 = 1.01$.) Therefore, from equation (6.7),

$$S_{x_1}^F = \frac{s x_1}{s^2 + x_1 s + x_2} \tag{6.10}$$

and, since $\Delta x_1/x_1 \equiv \frac{1}{100}$ then equation (6.8) gives

$$\frac{\Delta F}{F} \approx \frac{1}{100}\left[\frac{s x_1}{s^2 + x_1 s + x_2}\right] \tag{6.11}$$

Evaluating $\Delta F/F$ at $s = 2$ and $x_1 = x_2 = 1$, we have

$$\left.\frac{\Delta F}{F}\right|_{\substack{s=2\\ x_1=x_2=1}} \approx \frac{1}{100}\left[\frac{(2)(1)}{(2)^2 + (1)(2) + (1)}\right] = \frac{2}{700} \tag{6.12}$$

This result has been obtained by using the first-order sensitivity function in equation (6.8). The *exact* fractional change $\Delta F/F$ due to the perturbation of x_1 may, of course, be calculated directly from equation (6.9); prior to the perturbation of x_1, F is

$$\left. F\right|_{\substack{s=2\\ x_1=x_2=1}} = (2)^2 + (1)(2) + (1) = 7 \tag{6.13}$$

and, after the perturbation to $x_1 = 1.01$, we have the perturbed value of F:

$$\left. F\right|_{\substack{s=2\\ x_1=1.01\\ x_2=1}} = (2)^2 + (1.01)(2) + (1) = 7.02 \tag{6.14}$$

Therefore, from equation (6.13) and (6.14), $\Delta F = 7.02 - 7 = 0.02$, so that

$$\left.\frac{\Delta F}{F}\right|_{\substack{s=2 \\ x_1=1\to1.01 \\ x_2=1}} = \frac{0.02}{7} = \frac{2}{700} \tag{6.15}$$

Clearly, in this particular numerical example, the first-order sensitivity, equation (6.7), provides *exactly* the correct answer, $\Delta F/F = \frac{2}{700}$. [The reason for this is apparent by inspection of equation (6.9); for the particular function $F(s, x_i) = s^2 + x_1 s + x_2$, it is readily shown that the higher-order sensitivity terms

$$\frac{x_1^n \partial^n F}{F \partial x_1^n} \tag{6.16}$$

are zero except when $n = 1$. Consequently, in equation (6.8), the approximation sign may be replaced by the equality sign.] We construct next an example in which the use of equation (6.8) involves a small error.

EXAMPLE 6.2: Repeat the previous example for the function

$$F \equiv F(s, x_1) = s^2 + x_1^2 s + 1 \tag{6.17}$$

SOLUTION: Proceeding as before,

$$S_{x_1}^F = \frac{2x_1^2 s}{s^2 + x_1^2 s + 1} \tag{6.18}$$

so that, by equation (6.8),

$$\frac{\Delta F}{F} \approx \frac{1}{100}\left(\frac{2x_1^2 s}{s^2 + x_1^2 s + 1}\right) \tag{6.19}$$

Evaluating $\Delta F/F$ at $s = 2$, $x_1 = x_2 = 1$ we have

$$\left.\frac{\Delta F}{F}\right|_{\substack{s=2 \\ x_1=x_2=1}} \approx \frac{1}{100}\left[\frac{(2)(2)}{(4) + (2) + (1)}\right] = \frac{4}{700} \tag{6.20}$$

The *exact* value of $\Delta F/F$ is obtained as follows:

$$F\Big|_{\substack{s=2 \\ x_1=x_2=1}} = (4) + (2) + (1) = 7$$

$$F\Big|_{\substack{s=2 \\ x_1=1.01 \\ x_2=1}} = (4) + (2)(1.0201) + (1) = 7.0402$$

giving $\Delta F = 7.0402 - 7 = 0.0402$ and

$$\left.\frac{\Delta F}{F}\right|_{\substack{s=2\\ x_1=1\to 1.01\\ x_2=1}} = \frac{0.0402}{7} = \frac{4.02}{700} \tag{6.21}$$

which is in excess of the result predicted in equation (6.20) by $\frac{1}{2}$%. The reader may wish to account for this error by evaluating the second-order sensitivity term

$$\frac{1}{2!}\frac{x_i^2\partial^2 F}{F\partial^2 x_i}\left(\frac{\Delta x_i}{x_i}\right)^2 \tag{6.22}$$

that is neglected in the derivation of equation (6.8).

In most practical situations, the first-order sensitivity term $S_{x_i}^F$ may be used in equation (6.8) to provide an accurate estimate of the fractional change $\Delta F/F$ of a network function. The higher-order terms of the Taylor series in equation (6.4) are negligible unless $\partial F/\partial x_i$ is very close to zero and/or the perturbations $\Delta F/F$ and $\Delta x_i/x_i$ are too large.

6-3 ELEMENTARY RULES FOR SENSITIVITY FUNCTIONS

The reader may save a considerable amount of time and effort when evaluating sensitivity functions by verifying and remembering the six rules (or properties) shown in Table 6-1. The proofs of rules A, B, C, D, and E are straightforwardly obtained from the definition of $S_{x_i}^F$ in equation (6.7) and are requested in Problems 6.4 to 6.9. [Note that rule A may be obtained from rule B by setting $n = -1$ and rule E from rule D by setting $n = 2$ and $F_2 = K$.]

The chain rule property F in Table 6-1 may be used if the function F is itself a function of *real* functions $g_1(x_i), g_2(x_i), g_3(x_i), \ldots, g_n(x_i)$. A simple example of this is the frequency response function that is obtained by setting $s = j\omega$ so that in polar form

$$F \equiv F(j\omega, x_i) = |F(j\omega, x_i)|e^{-j[< F(j\omega, x_i)]} \tag{6.23}$$

Then, the real magnitude function $|F(j\omega, x_i)|$ is associated with $g_1(x_i)$ and the real phase function $< F(j\omega, x_i)$ with $g_2(x_i)$, so that we may write F in the form $F(g_1(x_i), g_2(x_i))$ and apply the chain rule.

The proof of the rule is straightforward:

$$F(g_1(x_i), g_2(x_i), \ldots, g_j(x_i) \ldots g_n(x_i)) \tag{6.24}$$

TABLE 6-1

[A] F-Inverse Property:

$$S_{x_i}^{F^{-1}} = -S_{x_i}^{F}$$

[B] F-Exponent Property:

$$S_{x_i}^{F^n} = n\, S_{x_i}^{F}, \qquad n \text{ real}$$

[C] F-Summation Property:

$$S_{x_i}^{F_1+F_2+\ldots+F_n} = \frac{F_1 S_{x_i}^{F_1} + F_2 S_{x_i}^{F_2} + \ldots + F_n S_{x_i}^{F_n}}{F_1 + F_2 + \ldots + F_n}$$

[D] F-Product Property:

$$S_{x_i}^{F_1F_2F_3\ldots F_n} = \sum_{j=1}^{n} S_{x_i}^{F_j}$$

[E] F-Scalar Property:

$$S_{x_i}^{KF} = S_{x_i}^{F}, \qquad K \text{ real and independent of } x_i$$

[F] Chain Rule Property: Given

$$F = F(g_1(x_i), g_2(x_i), \ldots, g_j(x_i), \ldots, g_n(x_i)),$$

then

$$S_{x_i}^{F} = \sum_{j=1}^{n} S_{g_j}^{F} S_{x_i}^{g_j}$$

leads to

$$\frac{\partial F}{\partial x_i} = \sum_{j=1}^{n} \frac{\partial F \partial g_j}{\partial g_j \partial x_i}$$

so that

$$S_{x_i}^{F} \equiv \frac{x_i \partial F}{F \partial x_i} = \sum_{j=1}^{n} \frac{x_i \partial F \partial g_j}{F \partial g_j \partial x_i}$$

$$= \sum_{j=1}^{n} \left[\frac{g_j \partial F}{F \partial g_j}\right]\left[\frac{x_i \partial g_j}{g_j \partial x_i}\right] = \sum_{j=1}^{n} S_{g_j}^{F} S_{x_i}^{g_j} \qquad (6.25)$$

EXAMPLE 6.3: Evaluate $S_{x_1}^F$, where

$$F \equiv \left(2x_1s^2 + \frac{s}{x_1x_2} + x_1^3\right)^{-1/4} \sin x_1 \tag{6.26}$$

This problem has been chosen to allow use of the rules A,B,C,D, and E. Let us define

$$P \equiv 2x_1s^2 + \frac{s}{x_1x_2} + x_1^3 \quad \text{and} \quad Q \equiv \sin x_1 \tag{6.27}$$

Therefore,

$$F = P^{-1/4}Q \tag{6.28}$$

Rules A, B, and D may be applied to equation (6.28) to obtain

$$S_{x_1}^F = -\tfrac{1}{4}S_{x_1}^P + S_{x_1}^Q \tag{6.29}$$

so it is then necessary that the functions $S_{x_1}^P$ and $S_{x_1}^Q$ be evaluated. Noting that

$$P = 2x_1s^2 + \frac{s}{x_1x_2} + x_1^3 \tag{6.30}$$

it follows by rule C that

$$S_{x_1}^P = \frac{\left[2x_1s^2(1) + \dfrac{s}{x_1x_2}(-1) + x_1^3(3)\right]}{\left[2x_1s^2 + \dfrac{s}{x_1x_2} + x_1^3\right]} \tag{6.31}$$

where the terms in the ( ) brackets correspond to the $S_{x_1}^{F_1}$, $S_{x_1}^{F_2}$, $S_{x_1}^{F_3}$ functions and where $F_1 \equiv 2x_1s^2$, $F_2 \equiv s/x_1x_2$ and $F_3 \equiv x_1^3$.
The sensitivity function $S_{x_i}^Q$ follows directly from $Q = \sin x_1$, so

$$\frac{\partial Q}{\partial x_1} = \cos x_1 \quad \text{and} \quad S_{x_1}^Q \equiv \frac{x_1 \partial Q}{Q\, \partial x_1} = \frac{\cos x_1}{\sin x_1} x_1 = x_1 \cot x_1 \tag{6.32}$$

Substituting equations (6.31) and (6.32) into equation (6.29) leads to the required result:

$$S_{x_1}^F = -\left\{\frac{\left[2x_1s^2 - \dfrac{s}{x_1x_2} + 3x_1^3\right]}{4\left[2x_1s^2 + \dfrac{s}{x_1x_2} + x_1^3\right]}\right\} + x_1 \cot x_1 \tag{6.33}$$

6-4 POLYNOMIAL SENSITIVITY

A rational network function $F(s, x_i)$ may be written in the form

$$F(s, x_i) = \frac{N(s, x_i)}{D(s, x_i)} = \frac{\sum_{j=0}^{m} a_j(x_i)s^j}{\sum_{k=0}^{n} b_k(x_i)s^k} \tag{6.34}$$

where the real coefficients $a_j(x_i)$ and $b_k(x_i)$ generally possess some functional dependence on the physical parameter x_i. In a particular network synthesis problem, the network function $F(s)$ is specified; the coefficients a_j, b_k are therefore specified. *The functional dependence of these coefficients on the physical parameters x_i will be determined by the particular network topology that is used to realize $F(s)$.*

It follows from rules A and D that

$$S_{x_i}^F = S_{x_i}^N - S_{x_i}^D \tag{6.35}$$

The sensitivities of the polynomials N and D to the physical parameters x_i are, therefore, of direct interest. Consider, for example, the function $S_{x_i}^D$. We have

$$D \equiv \sum_{k=0}^{n} b_k(x_i)s^k$$

from which it follows directly that

$$S_{x_i}^D = \frac{x_i \sum_{k=0}^{n} \frac{\partial b_k(x_i)}{\partial x_i} s^k}{D(s, x_i)} \tag{6.36}$$

The sensitivity function $S_{x_i}^D$ of a polynomial D is itself a real rational function of s and, by inspection of equation (6.36), it follows that the poles of $S_{x_i}^D$ are the zeros of D and, therefore, also the poles of the network function F. The function F is specified so that its pole locations are not dependent on the synthesis procedure or the topology. Clearly, *the choice of synthesis procedure and topology can only influence the zeros of $S_{x_i}^D$ via the numerator coefficients $x_i\, \partial b_k(x_i)/\partial x_i$*. The poles of $S_{x_i}^D$ are always the poles of F and are invariant.

Further discussion of this topic is simplified by assuming that the term $S_{x_i}^N$ is either zero or of negligibly small magnitude. That is, we *assume that the parameter* x_i *does not influence the numerator coefficients* a_j. In that case,

$$S_{x_i}^F \approx \frac{-x_i \sum_{k=0}^{n} \frac{\partial b_k(x_i)}{\partial x_i} s^k}{D(s, x_i)} \tag{6.37}$$

and the poles of F and the poles of $S_{x_i}^F$ are identical. Different topological implementations of F will result in different numerators in equation (6.37), and the magnitude of $S_{x_i}^F$ is clearly reduced in some region R of the s-plane if it can be arranged that this numerator possesses zeros in or close to this region R.

The possibility of manipulating the zeros of equation (6.37) in some useful way in order to suppress the magnitude of $S_{x_i}^F$ may be pursued by noting that $D(s, x_i)$ may always be written in the *bilinear form*,

$$D(s, x_i) = D_1(s) + x_i D_2(s) \tag{6.38}$$

where $D_1(s)$, $D_2(s)$ are independent of x_i and of degree less than or equal to n. Writing the kth coefficient of $D(s, x_i)$ in the corresponding bilinear form, we have

$$b_k(x_i) = b_{1k} + x_i b_{2k} \tag{6.39}$$

where

$$D(s, x_i) = \sum_{k=0}^{n} b_k(x_i) s^k$$

so that

$$\frac{\partial b_k}{\partial x_i} = b_{2k}$$

and equation (6.37) becomes

$$S_{x_i}^F \approx \frac{-x_i D_2(s)}{D(s, x_i)} \tag{6.40}$$

and the zeros of $S_{x_i}^F$ are then simply the s-plane roots of the equation

$$\boxed{D_2(s) = \sum_{k=0}^{n} b_{2k}s^k = 0} \tag{6.41}$$

To illustrate the possibility of suppressing $S_{x_i}^F$ to zero in some useful way by manipulation of zeros, suppose that we wish to synthesize a second-order lowpass network function of the form

$$F(s) = \frac{H_0}{b_2s^2 + b_1s + b_0} \tag{6.42}$$

and that *there exists a particular real nonzero radian frequency* ω_c *at which the frequency response function* $F(j\omega)$ *must have zero sensitivity,* $S_{x_i}^F = 0$, *to a particular physical parameter* x_i. Assuming that H_0 is independent of x_i, it is always possible to arrange $F(s, x_i)$ in the bilinear form:

$$F(s) = \frac{H_0}{\left[b_{12}s^2 + b_{11}s + b_{10}\right] + x_i\left[b_{22}s^2 + b_{21}s + b_{20}\right]} \tag{6.43}$$

where the b coefficients are independent of x_i. Then, by equation (6.41), the zeros of $S_{x_i}^F$ are the roots of the equation

$$b_{22}s^2 + b_{21}s + b_{20} = 0 \tag{6.44}$$

Now, these roots must be at $\pm j\omega_c$ if $S_{x_i}^F$ is to equal zero at $s = \pm j\omega_c$, implying that in equation (6.44)

$$b_{21} = 0 \quad \text{and} \quad b_{20} = b_{22}\omega_c^2 \tag{6.44A}$$

or, in terms of equation (6.43), that $F(s, x_i)$ must have the general form

$$F(s, x_i) = \frac{H_0}{\left[b_{12}s^2 + b_{11}s + b_{10}\right] + x_ib_{22}\left[s^2 + \omega_c^2\right]} \tag{6.45}$$

Unfortunately, practical RC-active realizations are difficult to realize in such a way that the transfer function is of the form given in equation (6.45) and, in general, there has not been any significant progress in the development of synthesis techniques for which $F(s)$ *and* $S_{x_i}^F$ are prescribed functions. The general approach has been to synthesize $F(s)$ and then compute the resultant sensitivity functions.

6-5 SENSITIVITY OF THE TRANSFER FUNCTION TO Q-FACTOR AND NATURAL FREQUENCY

A second-order *biquadratic* transfer function $H(s)$ may be defined in terms of the Q factor of the pair of poles Q_P, the Q-factor of the pair of zeros Q_Z, and the natural frequencies $\omega_{P,Z}$ of the poles and zeros, by the equation

$$H(s) \equiv H_0 \frac{s^2 + \left(\dfrac{\omega_Z}{Q_Z}\right)s + \omega_Z^2}{s^2 + \left(\dfrac{\omega_P}{Q_P}\right)s + \omega_P^2} \equiv H_0 \frac{N(s)}{D(s)} \tag{6.46}$$

For *complex* poles and zeros, these singularities must lie on circles of radii $\omega_{P,Z}$ at angles $\theta_{P,Z}$, where

$$\theta_{P,Z} = \text{arc sin} \frac{1}{2Q_{P,Z}} \tag{6.47}$$

as shown in Fig. 6.1(a). Biquadratic transfer functions of this type are commonly encountered, and the parameters H_0, $Q_{P,Z}$ and $\omega_{P,Z}$ are normally used to specify and describe these functions. Of course, these parameters are functions of the physical parameters x_i.

In the following analysis, the sensitivities of the transfer function $H(s)$ to the parameters $\omega_{P,Z}$ and $Q_{P,Z}$ are determined. It is then shown that, for many practical applications, the effect of perturbations $\Delta Q_{P,Z}$ of the Q-factors $Q_{P,Z}$ on $H(s)$ is negligible compared with the effect of perturbations $\Delta\omega_{P,Z}$ of the natural frequencies $\omega_{P,Z}$. It follows directly from equation (6.46) that

$$S_{Q_P}^{H(s)} = \frac{\omega_P}{Q_P}\left(\frac{s}{D(s)}\right) \quad \text{and} \quad S_{\omega_P}^{H(s)} = -\frac{\omega_P}{Q_P}\left(\frac{s + 2\omega_P Q_P}{D(s)}\right) \tag{6.48}$$

The poles of $S_{Q_P}^{H(s)}$ and $S_{\omega_P}^{H(s)}$ are, therefore, the poles of $H(s)$ and are shown in Figs. 6.1(b) and 6.1(c), respectively. The zero of $S_{\omega_P}^{H(s)}$ is at the origin whereas, for $Q_P \gg 1$, the zero of $S_{\omega_P}^{H(s)}$ is *far-removed* at $-2Q_P\omega_P$ on the

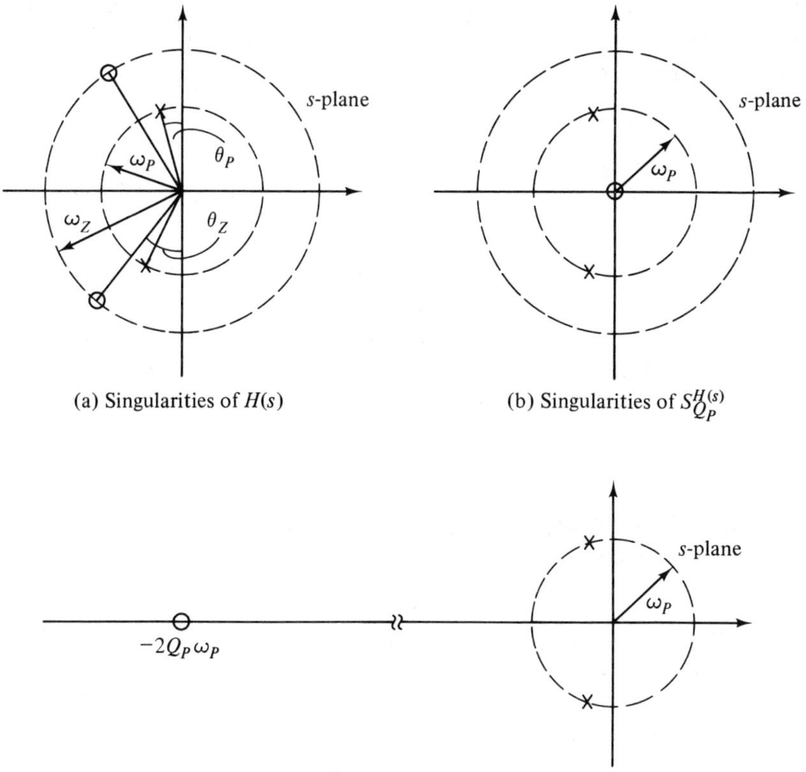

(a) Singularities of $H(s)$

(b) Singularities of $S_{Q_P}^{H(s)}$

(c) Singularities of $S_{\omega_P}^{H(s)}$

FIGURE 6.1

negative-real axis. Consequently, the magnitude of the function $S_{Q_P}^{H(s)}$ is much less than the magnitude of the function $S_{\omega_P}^{H(s)}$ at any point in the s-plane where $|s| \ll 2Q_P\omega_P$. This region of the s-plane includes that portion of the imaginary axis corresponding to real frequencies ω for which

$$\omega \ll 2Q_P\omega_P \tag{6.49}$$

Consequently, for frequency response sensitivities $S_{Q_P}^{H(j\omega)}$ and $S_{\omega_P}^{H(j\omega)}$ it follows that

$$|S_{Q_P}^{H(j\omega)}| \ll |S_{\omega_P}^{H(j\omega)}| \quad \text{if } \omega \ll 2Q_P\omega_P \text{ and } Q_P \gg 1 \tag{6.50}$$

An exactly similar expression applies to the numerator parameters Q_Z and ω_Z. The significance of this is that *in many high-Q circuits, it is possible to neglect the perturbations $\Delta Q_{P,Z}$ compared with the perturbations $\Delta\omega_{P,Z}$*

because of the relative insensitivity of $H(j\omega)$ to $\Delta Q_{P,Z}$. It is readily shown from equation (6.48) that we may neglect $\Delta Q_{P,Z}$ and only consider the effect on $H(j\omega)$ of $\Delta\omega_{P,Z}$ if

$$\boxed{\frac{\Delta Q_P}{Q_P} \ll 2Q_P\left(\frac{\Delta\omega_P}{\omega_P}\right)} \qquad \text{Condition for Neglecting } \Delta Q_P \tag{6.51}$$

where $Q_P \gg 1$ and $\omega \ll 2Q_P\omega_P$. Problem 6.22 requires numerical verification of this result. A similar expression to (6.51) applies to ΔQ_Z, Q_Z, $\Delta\omega_Z$, and ω_Z. In the next section, some general relationships are derived between sensitivity expressions involving the complex frequency response $H(j\omega)$ and the more useful real *gain* and *phase* functions, $|H(j\omega)|$ and $< H(j\omega)$.

6-6 SENSITIVITY AND THE STEADY-STATE FREQUENCY RESPONSE

So far, this discussion of sensitivity functions has implied that $S_{x_i}^{F(s)}$ is a real rational function of s. In many practical applications, the performance of a network function is specified only in terms of the behavior of that function on the imaginary s-plane axis $s = j\omega$. Writing the complex transfer function $H(j\omega)$ in polar form, we have

$$H(j\omega) = |H(j\omega)| \exp[j < H(j\omega)] \tag{6.52}$$

The term $|H(j\omega)|$ is defined as the *gain-frequency response* $G(\omega)$ and the term $< [H(j\omega)]$ as the *phase-frequency response* $\theta(\omega)$, so that

$$H(j\omega) = G(\omega) \exp[j\theta(\omega)] \tag{6.53}$$

where $G(\omega)$ and $\theta(\omega)$ are *real* functions of ω. The ease with which $G(\omega)$ and $\theta(\omega)$ may be measured under conditions of steady-state sinusoidal excitation suggests that the sensitivity functions $S_{x_i}^{G(\omega)}$ and $S_{x_i}^{\theta(\omega)}$ should be derived.

Assuming that $S_{x_i}^{H(j\omega)}$ has been obtained by substituting $s = j\omega$ into the expression for $S_{x_i}^{H(s)}$, then, by rule F in Table 6-1,

$$S_{x_i}^{H(j\omega)} = \left[S_{G(\omega)}^{H(j\omega)} S_{x_i}^{G(\omega)}\right] + \left[S_{\theta(\omega)}^{H(j\omega)} S_{x_i}^{\theta(\omega)}\right] \tag{6.54}$$

Now, it follows from equation (6.53) that

$$S_{G(\omega)}^{H(j\omega)} = 1 \tag{6.55}$$

and

$$S_{\theta(\omega)}^{H(j\omega)} \equiv \frac{\partial H(j\omega)\theta(\omega)}{\partial\theta(\omega)H(j\omega)} = \frac{G(\omega)\theta(\omega)}{H(j\omega)}\left[je^{j\theta(\omega)}\right] = j\theta(\omega) \tag{6.56}$$

Substituting equations (6.55) and (6.56) into equation (6.54), we have

$$S_{x_i}^{H(j\omega)} = S_{x_i}^{G(\omega)} + j\left[\theta(\omega)S_{x_i}^{\theta(\omega)}\right] \tag{6.56A}$$

Since $G(\omega)$ and $\theta(\omega)$ are *real* functions it follows that $S_{x_i}^{G(\omega)}$ and $S_{x_i}^{\theta(\omega)}$ are also *real* functions. Then, taking the real part of equation (6.56A), we have

$$\boxed{S_{x_i}^{G(\omega)} = \mathrm{Re}\left[S_{x_i}^{H(j\omega)}\right]} \qquad \text{GAIN SENSITIVITY} \tag{6.57}$$

and, taking the imaginary part of equation (6.56A), gives

$$\boxed{S_{x_i}^{\theta(\omega)} = \frac{1}{\theta(\omega)}\mathrm{Im}\left[S_{x_i}^{H(j\omega)}\right]} \qquad \text{PHASE SENSITIVITY} \tag{6.58}$$

Equations (6.57) and (6.58) are the expressions required for the gain and phase sensitivities. Alternative and equivalent forms of this result, which are often more convenient to use, are obtained by noting that if $S_{x_i}^{H(s)}$ is expressed as

$$S_{x_i}^{H(s)} \equiv \frac{N_e(s) + N_0(s)}{D_e(s) + D_0(s)} \tag{6.59}$$

where the subscripts e and o indicate the even and odd parts of the numerator and denominator, it follows that

$$S_{x_i}^{H(s)} = \frac{\left[N_e(s)D_e(s) - N_0(s)D_0(s)\right] + \left[N_0(s)D_e(s) - N_e(s)D_0(s)\right]}{D_e^2(s) - D_0^2(s)} \tag{6.60}$$

Then, substituting $s = j\omega$ into equation (6.60) and taking the real part we have, according to equation (6.57),

$$S_{x_i}^{G(\omega)} = \mathrm{Ev}\left\{ S_{x_i}^{H(s)}\Big|_{s=j\omega}\right\} = \left.\frac{N_e(s)D_e(s) - N_0(s)D_0(s)}{D_e^2(s) - D_0^2(s)}\right|_{s=j\omega} \tag{6.61}$$

and, taking the imaginary part,

$$S_{x_i}^{\theta(\omega)} = \mathrm{Od}\left\{ S_{x_i}^{H(s)}\Big|_{s=j\omega}\right\} = \frac{1}{j\theta(\omega)}\left[\frac{N_0(s)D_e(s) + N_e(s)D_0(s)}{D_e^2(s) - D_0^2(s)}\right]\Bigg|_{s=j\omega} \tag{6.62}$$

EXAMPLE 6.4: SENSITIVITIES OF GAIN $G(\omega)$ to Q_P AND ω_P: Use the preceeding results to determine the gain sensitivities $S_{Q_P}^{G(\omega)}$ and $S_{\omega_P}^{G(\omega)}$ for the second order *biquadratic* transfer function of equation (6.46).

SOLUTION: It follows directly from equation (6.48) that we may write

$$S_{Q_P}^{H(s)} = \frac{\dfrac{\omega_P s}{Q_P}}{s^2 + \dfrac{\omega_P s}{Q_P} + \omega_P^2} \tag{6.63}$$

so that the even and odd parts of the numerator and denominator of equation (6.63) are

$$\left.\begin{aligned} N_e(s) &= 0, & N_0(s) &= \frac{\omega_P s}{Q_P} \\ D_e(s) &= (s^2 + \omega_P^2), & D_0(s) &= \frac{\omega_P s}{Q_P} \end{aligned}\right\} \tag{6.64}$$

Substituting equation (6.64) into equation (6.61), we have

$$S_{Q_P}^{G(\omega)} = \left.\frac{(0)(s^2 + \omega_P^2) - \left(\dfrac{\omega_P s}{Q_P}\right)\left(\dfrac{\omega_P s}{Q_P}\right)}{(s^2 + \omega_P^2)^2 - \left(\dfrac{\omega_P s}{Q_P}\right)^2}\right|_{s=j\omega}$$

which may be rearranged to obtain

$$S_{Q_P}^{G(\omega)} = \frac{1}{1 - Q_P^2\left(\dfrac{s}{\omega_P} + \dfrac{\omega_P}{s}\right)^2}\Bigg|_{s=j\omega} = \frac{1}{1 + Q_P^2\left(\dfrac{\omega}{\omega_P} - \dfrac{\omega_P}{\omega}\right)^2} \tag{6.65}$$

This result is interesting because the *gain* $G(\omega)$ is quite insensitive to Q_P with $0 \leqslant S_{Q_P}^{G(\omega)} \leqslant 1$.

Repetition of the analysis above to obtain $S_{\omega_P}^{G(\omega)}$ leads to a quite different result: from equation (6.48),

$$S_{\omega_P}^{H(s)} = \frac{\dfrac{-\omega_P s}{Q_P} - 2\omega_P^2}{s^2 + \dfrac{\omega_P s}{Q_P} + \omega_P^2} \tag{6.66}$$

so that

$$\left.\begin{aligned} N_e(s) &= -2\omega_P^2, \qquad N_0(s) = -\frac{\omega_P s}{Q_P} \\ D_e(s) &= (s^2 + \omega_P^2), \qquad D_0(s) = \frac{\omega_P s}{Q_P} \end{aligned}\right\} \tag{6.66A}$$

Substituting equations (6.66A) into equation (6.61) and making use of the result in equation (6.65) we obtain, after some manipulation,

$$S_{\omega_P}^{G(\omega)} = -S_{Q_P}^{G(\omega)} + \left[\frac{2Q_P^2\dfrac{\omega_P}{\omega}\left(\dfrac{\omega}{\omega_P} - \dfrac{\omega_P}{\omega}\right)}{1 + Q_P^2\left(\dfrac{\omega}{\omega_P} - \dfrac{\omega_P}{\omega}\right)^2}\right] \tag{6.67}$$

The term in the [ ] brackets usually dominates equation (6.67) over the range of ω that is of interest. For example, equations (6.65) and (6.67) are plotted in Fig. 6.2 for $Q_P = 10$ and $\omega_P = 1$. Note the high sensitivity to ω_P in the region $\omega \approx \omega_P$. [In Problem 6.25, the reader is asked to prove that the maximum and minimum of $S_{\omega_P}^{G(\omega)}$ are of approximate magnitude Q_P and that they occur at approximately the frequencies $\omega_P(1 \pm 1/2Q_P)$.] Clearly, $G(\omega)$ is far more sensitive to ω_P than to Q_P if $Q_P \gg 1$. The extremum values of $S_{\omega_P}^{G(\omega)}$ are dependent on Q_P, whereas the maximum value of $\left|S_{Q_P}^{G(\omega)}\right|$ is unity.

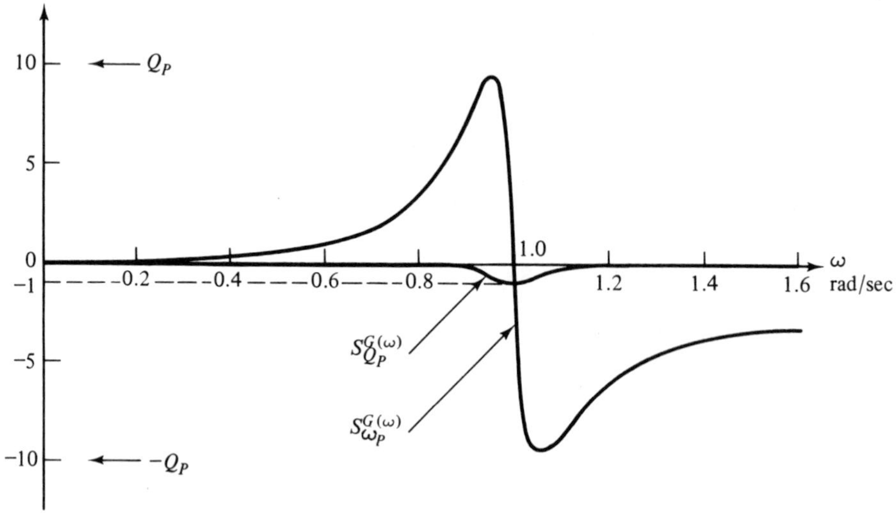

FIGURE 6.2 *Sensitivities of biquadratic gain function G(ω) to ω_P and Q_P.*

6-7 MULTIPARAMETER SENSITIVITY

In this section, consideration is given to the fact that *more than one* parameter x_i is likely to vary in a given circuit. A designer might wish to ensure that a particular network function $F(s)$ remains within certain bounds when *all* of the parameters x_i assume their worstcase values insofar as the resultant deviation of the transfer function is concerned.

Suppose that there exists n parameters $x_1, x_2, \ldots, x_i, \ldots, x_n$ that have small fractional perturbations $\Delta x_1/x_1, \Delta x_2/x_2, \ldots, \Delta x_i/x_i, \ldots, \Delta x_n/x_n$ from their nominal values. Then, to a first-order approximation, it may be shown that for arbitrarily small Δx_i the *total* change in the network function $\Delta F(s, x_i) \equiv \Delta F$ is given by

$$\frac{\Delta F}{F} \approx \sum_{i=1}^{n} S_{x_i}^{F} \frac{\Delta x_i}{x_i} \tag{6.68}$$

If the fractional parameter perturbations are *uniform* over all n elements; that is, if

$$\frac{\Delta x_1}{x_1} \equiv \frac{\Delta x_2}{x_2} \cdots \equiv \frac{\Delta x_i}{x_i} \cdots \equiv \frac{\Delta x_n}{x_n} \equiv \epsilon \tag{6.69}$$

then

$$\frac{\Delta F}{F} \approx \epsilon \sum_{i=1}^{n} S_{x_i}^{F} \tag{6.70}$$

where the *multiparameter sensitivity* of this rational function of s is now defined by

$$\boxed{MS_{x_i}^F \equiv \sum_{i=1}^{n} S_{x_i}^F} \quad \text{MULTIPARAMETER SENSITIVITY} \tag{6.71}$$

so that equation (6.70) becomes

$$\boxed{\frac{\Delta F}{F} \approx \epsilon MS_{x_i}^F} \tag{6.72}$$

Note that both $\Delta F/F$ and $MS_{x_i}^F$ are also rational functions of s.

EXAMPLE 6.5: Given the network transfer function

$$F \equiv F(s) = \frac{1}{1 + sC_1R_1 + \mu} \tag{6.73}$$

determine the multiparameter sensitivity function $MS_{x_i}^F$, where the parameters C_1, R_1, and μ form the set of perturbed parameters x_1, x_2, and x_3. Use the result to estimate $\Delta F(j\omega)/F(j\omega)$ at $\omega = 2$ and for $C_1 = R_1 = \mu = 1$ under uniform perturbations of these three parameters by $+1\%$ ($\epsilon = +0.01$).

SOLUTION: By direct analysis of equation (6.73),

$$S_{R_1}^F = S_{C_1}^F = -\frac{sC_1R_1}{F}, \qquad S_\mu^F = -\frac{\mu}{F} \tag{6.74}$$

Adding these single-element sensitivities according to equation (6.71), we have

$$MS_{R_1C_1\mu}^F = -\frac{(\mu + 2sC_1R_1)}{(1 + \mu) + sC_1R_1} \tag{6.75}$$

where the lower subscript ($R_1C_1\mu$) describes the parameters over which the summation is performed. Substituting equation (6.75) into equation (6.72) and letting $s = j\omega$ gives the fraction change $\Delta F(j\omega)/F(j\omega)$ as

$$\frac{\Delta F(j\omega)}{F(j\omega)} \approx -\epsilon\left[\frac{\mu + 2j\omega C_1R_1}{(1 + \mu) + j\omega C_1R_1}\right] \tag{6.76}$$

Substituting numerical values, we have

$$\frac{\Delta F(j2)}{F(j2)} \approx -(0.01)\left[\frac{1+j4}{2+j2}\right] = -0.01457\underline{/30.96^\circ} \tag{6.77}$$

The reader is requested to compare this answer with the *exact* result in Problem 6-27. Note that the fractional change $\Delta F(j2)/F(j2)$ of the *complex* number $F(j2)$ is itself a complex number. In many applications, the *real* function $|F(j\omega)|$ is of more interest because it is readily measurable. The following example describes the method used to derive $\Delta|F(j\omega)|/|F(j\omega)|$.

EXAMPLE 6.6: Repeat the previous example, but this time derive the fraction deviation $\Delta|F(j\omega)|/|F(j\omega)|$ of the *real gain* function $|F(j\omega)|$.

SOLUTION: The first step is to derive $S_{x_i}^{|F(j\omega)|}$ for $x_i = R_1$, C_1, and μ. This may be done by taking the real parts of equation (6.74), using the results in equations (6.57) and (6.61): Thus, from equation (6.74),

$$S_{R_1}^{|F(j\omega)|} = \text{Re}\left[S_{R_1}^{F(j\omega)}\right] = \left.\frac{-(-sC_1R_1)(sC_1R_1)}{(1+\mu)^2-(sC_1R_1)^2}\right|_{s=j\omega} \tag{6.78}$$

where the terms in the ( ) brackets correspond to the even and odd parts of the equation $-sC_1R_1/F$ as required by equation (6.61). It also follows from equation (6.78) that

$$S_{R_1}^{|F(j\omega)|} = S_{C_1}^{|F(j\omega)|} \tag{6.79}$$

Now, taking the real part of S_μ^F in equation (6.74), we obtain

$$S_{\mu}^{|F(j\omega)|} = \text{Re}\left[S_{\mu}^{F(j\omega)}\right] = \left.\frac{-(\mu)(1+\mu)}{(1+\mu)^2-(sC_1R_1)^2}\right|_{s=j\omega} \tag{6.80}$$

The single-element sensitivities in the above three equations are added and substituted into equation (6.72); thus,

$$\frac{\Delta|F(j\omega)|}{|F(j\omega)|} \approx \epsilon\left.\left[\frac{-\mu(1+\mu)+2s^2C_1^2R_1^2}{(1+\mu)^2-s^2C_1^2R_1^2}\right]\right|_{s=j\omega} \tag{6.81}$$

Therefore,

$$\boxed{\frac{\Delta|F(j\omega)|}{|F(j\omega)|} \approx -\epsilon\left[\frac{\mu(1+\mu)+2\omega^2C_1^2R_1^2}{(1+\mu)^2+\omega^2C_1^2R_1^2}\right]} \tag{6.82}$$

[Note that, as might be expected from an inspection of equation (6.73), the fractional change expressed in equation (6.82) is negative for all positive ϵ, μ,

C_1, and R_1 and for all ω.] Substituting numerical values into equation (6.82), we have

$$\frac{\Delta|F(j\omega)|}{|F(j\omega)|} \approx -\frac{1}{100}\left(\frac{2+8}{4+4}\right) = -\frac{1}{80} \tag{6.83}$$

The reader is requested to obtain the *exact* answer in Problem 6-28.

6-8 WORSTCASE SENSITIVITY

The multiparameter sensitivity criterion of equation (6.71) is of restricted usefulness because it only allows $\Delta F/F$ to be derived if the perturbations are *uniform*, as described in equation (6.69). Nevertheless, the *concept* is usefully extended in this section to allow the circuit designer to predict the worstcase fractional deviation of a *real* function [for example, $|F(j\omega)|$] due to perturbations of elements x_i in some range such that

$$\left|\frac{\Delta x_i}{x_i}\right| \leqslant \epsilon_{\max}, \qquad \forall\, i \tag{6.84}$$

The situation described by equation (6.84) might correspond to the set of resistors for which the tolerance specification is $\pm 1\%$; then $\epsilon_{\max} = 0.01$. If the fractional deviations of the parameters x_i are *nonuniform but bounded* by equation (6.84) and of values ϵ_i then, by equation (6.8),

$$\frac{\Delta F}{F} \approx \sum_{i=1}^{n} \epsilon_i S_{x_i}^F, \qquad |\epsilon_i| \leqslant \epsilon_{\max} \tag{6.85}$$

If F is a complex function, as is generally the case, then the terms $\epsilon_i S_{x_i}^F$ are also generally complex functions. The vector addition that is implied by equation (6.85) is shown in Fig. 6.3(a) and (b) for any particular value of s and for $n = 3$. It follows from the triangle inequality that any summation of a set of complex numbers c_i has the property that

$$\left|\sum_{i=1}^{n} c_i\right| \leqslant \sum_{i=1}^{n} |c_i| \tag{6.86}$$

so that, letting

$$c_i \equiv \epsilon_i S_{x_i}^F \tag{6.87}$$

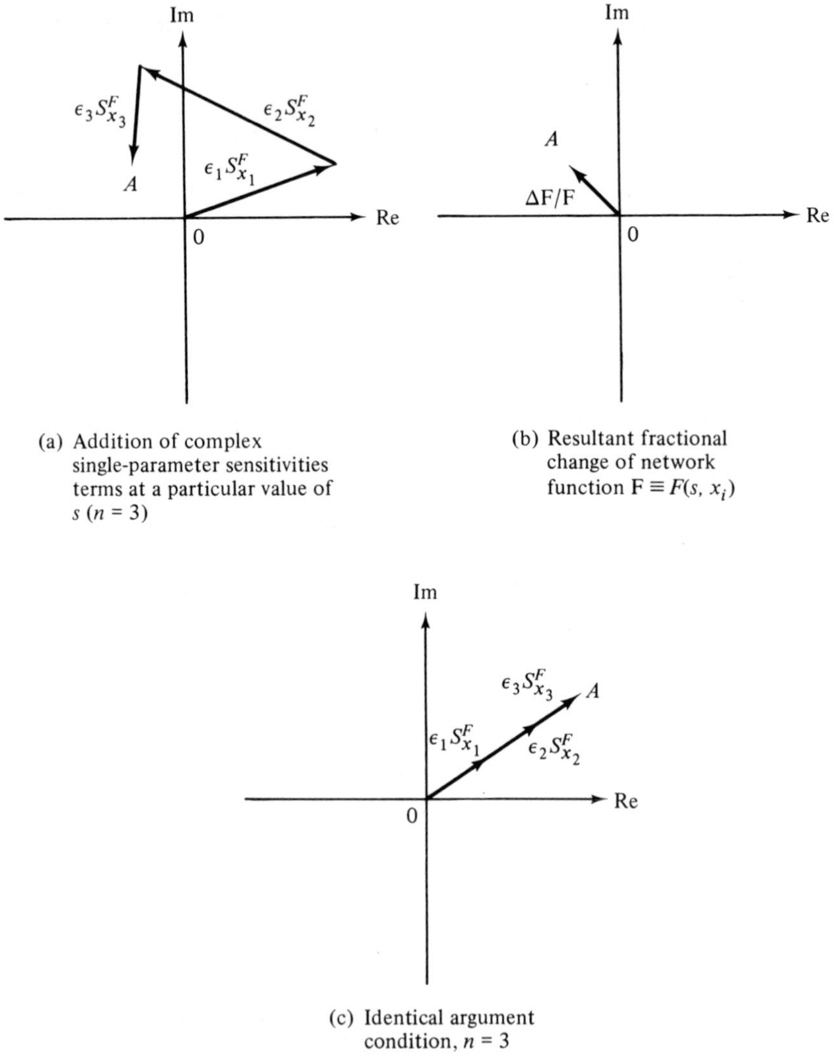

(a) Addition of complex single-parameter sensitivities terms at a particular value of s ($n = 3$)

(b) Resultant fractional change of network function $F \equiv F(s, x_i)$

(c) Identical argument condition, $n = 3$

FIGURE 6.3

it follows from equations (6.85), (6.86), and (6.87) that

$$\left|\frac{\Delta F}{F}\right| \leqslant \sum_{i=1}^{n} \left|\epsilon_i S^F_{x_i}\right| \qquad \text{(6.88)}$$

MAGNITUDE OF FRACTIONAL CHANGE IN F DUE TO NONUNIFORM ϵ_i

If and only if the arguments of the complex numbers c_i are identical over all i does it follow that

$$\left|\sum_{i=1}^{n} c_i\right| = \sum_{i=1}^{n} |c_i|$$

and, therefore,

$$\left|\frac{\Delta F}{F}\right| = \sum_{i=1}^{n} |\epsilon_i S_{x_i}^F| \tag{6.89}$$

if the arguments $< [\epsilon_i S_{x_i}^F]$ are identical over all i. The special situation described by equation (6.89) is shown in Fig. 6.3(c).

The result in equation (6.88) is usually not of direct usefulness, because we do not generally *know* the magnitudes of each ϵ_i term; we only know that $|\epsilon_i| \leqslant \epsilon_{\max}$. For *real* functions, such as $F \equiv |H(s, x_i)|$ or $< [H(s, x_i)]$, it is possible to derive an important worstcase bound on $|\Delta F/F|$. If F is *real*, then each of the terms $\epsilon_i S_{x_i}^F$ in equation (6.88) is real. The situation is shown in Fig. 6.4(a) for $n = 3$. It then follows that, since $|\epsilon_i| \leqslant \epsilon_{\max}$ and $\epsilon_i S_{x_i}^F$ are real,

$$|\epsilon_i S_{x_i}^F| \leqslant \epsilon_{\max} |S_{x_i}^F|, \qquad \forall i \text{ for real } F \tag{6.90}$$

or

$$\sum_{i=1}^{n} |\epsilon_i S_{x_i}^F| \leqslant \epsilon_{\max} \sum_{i=1}^{n} |S_{x_i}^F| \qquad \text{for real } F \tag{6.91}$$

Combining equations (6.88) and (6.91) provides the required result, which is

$$\left|\frac{\Delta F}{F}\right| \leqslant \epsilon_{\max} \sum_{i=1}^{n} |S_{x_i}^F|, \qquad F \text{ real} \tag{6.92}$$

Defining worstcase sensitivity as

$$\boxed{WS_{x_i}^F \equiv \sum_{i=1}^{n} |S_{x_i}^F|} \tag{6.93}$$

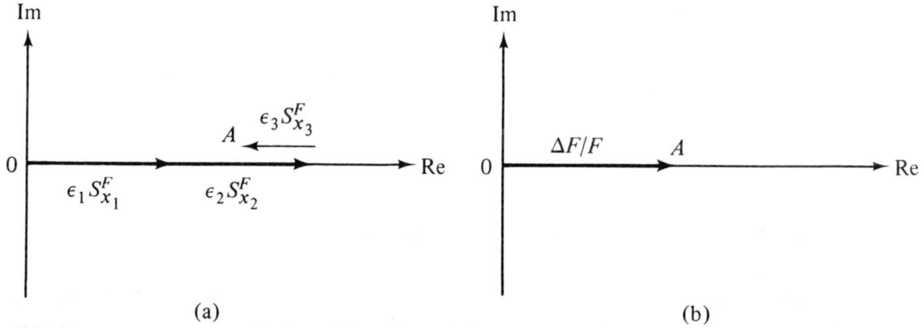

FIGURE 6.4 *(a) Addition of real single-element sensitivities terms at a particular value of s (n = 3); (b) resultant fractional change of real network function F (e.g., |F(s, x_i)|).*

allows equation (6.92) to be written as

$$\boxed{\left|\frac{\Delta F}{F}\right| \leqslant \epsilon_{\max} WS_{x_i}^{F}} \qquad F \text{ real} \quad \text{WORSTCASE DEVIATION} \tag{6.94}$$

Knowing the single-element sensitivities and the tolerances $\epsilon_{\max}$, equation (6.94) *allows the designer to calculate an upper bound on the fractional error* $|\Delta F/F|$ *for any real network function* F.

It is unlikely that at any particular value of s the fractional deviation $|\Delta F/F|$ will *equal* the upper bound of equation (6.94). For this to be the case, it is necessary that

$$\left.\begin{array}{ll} \text{(i)} & |\epsilon_i| = \epsilon_{\max}, \qquad \forall i \\ \text{and} \quad \text{(ii)} & \text{sign}\left[\epsilon_i S_{x_i}^{F}\right] \text{ are identical}, \qquad \forall i \end{array}\right\} \tag{6.95}$$

Condition (i) requires that *all* perturbations cause the x_i to take on their extremum values and condition (ii) states that the signs of the terms $\epsilon_i S_{x_i}^{F}$ are the same for all n elements. The reader is asked to prove (in Problem 6.29) that if condition (i) is satisfied, the probability that condition (ii) is also satisfied at any particular value of s is $2^{(1-n)}$. Thus, *for large* n, it is improbable that the upper bound in equation (6.94) is attained at any particular value of s. Nevertheless, if the designer requires a *guarantee* that the final design is within a given specification, then the worstcase-sensitivity result in equation (6.94) should be used.

EXAMPLE 6.7: Derive $WS_{R_1C_1\mu}^{F}$ for the situation described in Example 6.5.

SOLUTION: It is only necessary to add the magnitudes of the three single-element sensitivities in equations (6.78), (6.79) and (6.80) to obtain

$$WS_{R_1C_1\mu}^{F=|F(j\omega)|} = |S_{R_1}^{F}| + |S_{C_1}^{F}| + |S_{\mu}^{F}| = \frac{\mu(1+\mu) + 2\omega^2C_1^2R_1^2}{(1+\mu)^2 + \omega^2C_1^2R^2} \tag{6.96}$$

By comparison of equation (6.96) with the answer we have obtained in equation (6.82), $WS_{x_i}^{F} = |MS_{x_i}^{F}|$ is readily verifiable. This is, of course, to be expected because the single-element sensitivities in equations (6.78), (6.79), and (6.80) are *all* negative.

6-8.1 Lower Bound Worstcase Sensitivity

Suppose that a designer is faced with the situation that the element tolerances $\epsilon_{\max}$ and the single-element sensitivities $|S_{x_i}^{F}|$ lead, via equation (6.94), to a worstcase deviation $|\Delta F/F|$ which is unacceptably large. The next logical step is to search for some other synthesis of the function F that hopefully will lead to a *different circuit* for which the sensitivities $|S_{x_i}^F|$ are "sufficiently" low-valued. This is a topic of major research interest in the synthesis of active networks. In the next section, a *lower-bound* is established for the worstcase-sensitivity function $WS_{x_i}^F$ that is valid over all the passive RC elements. This lower-bound is *invariant* in the sense that it may be guaranteed that, for the particular network function that is of interest, it is *not possible* to find a network with a worstcase-sensitivity $WS_{x_i}^F$ that is better (less) than the invariant lower-bound that is to be derived here. Consequently, the designer may decide that, for the particular circuit under consideration over the particular region of the s-plane that is of concern, the worstcase sensitivity is sufficiently close to the invariant lower-bound that it is *not worth the effort of trying to find another circuit with lower sensitivities* $|S_{x_i}^F|$. Alternatively, the calculated worstcase sensitivity may be in excess of the lower-bound by several orders of magnitude, in which case a *search for another circuit might be highly rewarding*. In other words, a lower-bound on $WS_{x_i}^F$ (and, therefore, on $|\Delta F/F|$) is a useful "benchmark" for comparing circuits and synthesis techniques. In order to derive this lower-bound we need to establish some results on the invariance properties of multiparameter sensitivities.

6-9 MULTIPARAMETER INVARIANCE AND LOWER-BOUND WORSTCASE SENSITIVITY

In this section, an RC-active network is considered as an interconnection of Rs, Cs, and nullors. Consequently, if the actual network contains such building-blocks as controlled sources, inductors, gyrators and transformers, for example, it is necessary to implement each of these building-blocks using RC-nullor networks and to include these resistors, capacitors and nullors as part of the overall network. In Fig. 6.5(a), for example, the noninverting vcvs is implemented as indicated in Fig. 6.5(b) and, therefore, the following discussion of *passive-sensitivities* refers to the complete RC-nullor network and includes the effects of perturbations of the resistors R_1 and R_2 that are used to make the vcvs subnetwork.

Let $F(s, x_i)$ be a *dimensionless* network function; a voltage or current transform transfer function, for example. If the network is an interconnection of Rs, Cs, and nullors, then the parameters x_i are simply the values, R_i and C_i, of the passive elements. It is assumed that the network contains n_R resistors and n_C capacitors. The network function may be written $F(R_i, 1/sC_i)$, since it is a function of the individual transform impedances of each of the elements. Now, since $F(R_i, 1/sC_i)$ is dimensionless, *impedance-scaling* all of these transform impedances by a real scalar k does not change the transfer function; that is

$$F\left(R_i, \frac{1}{sC_i}\right) = F\left(kR_i, \frac{k}{sC_i}\right) \tag{6.97}$$

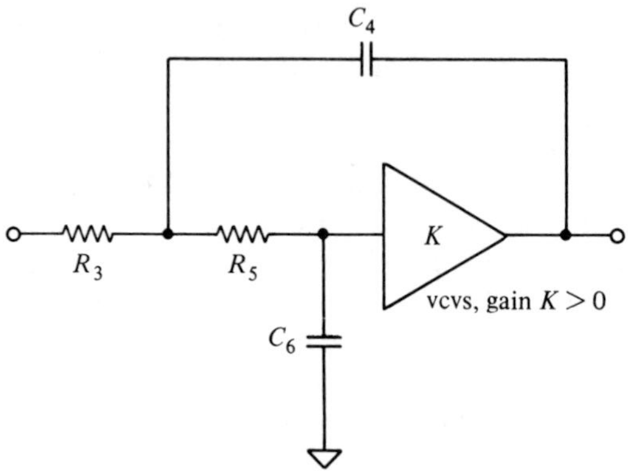

(a) An RC-OP AMP circuit

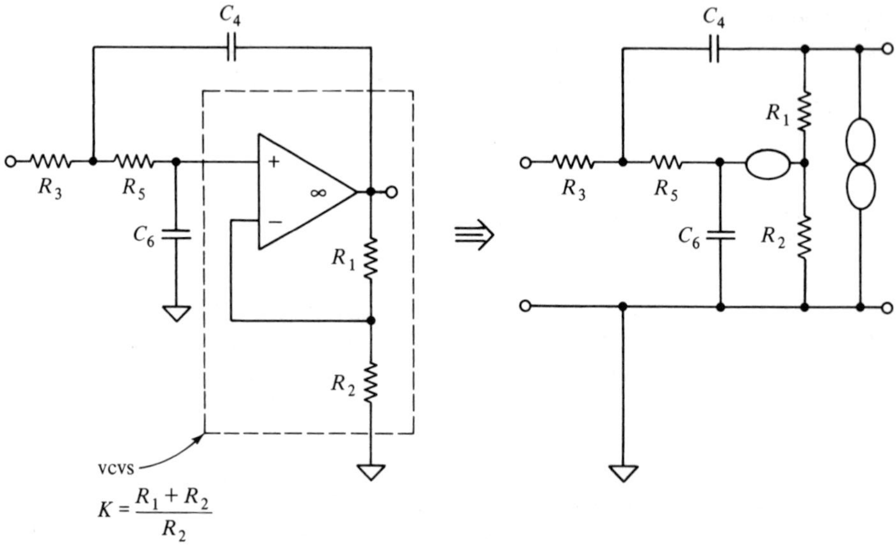

(b) The equivalent RC–nullor circuit

FIGURE 6.5

Differentiating equation (6.97) with respect to k using the chain rule for partial differentiation, we obtain

$$0 = \frac{\partial F}{\partial [kR_1]} R_1 + \frac{\partial F}{\partial [kR_2]} R_2 + \ldots + \frac{\partial F}{\partial [kR_{n_R}]} R_{n_R}$$

$$+ \frac{\partial F}{\partial \left[\dfrac{k}{sC_1} \right] sC_1} + \frac{\partial F}{\partial \left[\dfrac{k}{sC_2} \right] sC_2} + \ldots + \frac{\partial F}{\partial \left[\dfrac{k}{sC_{n_C}} \right] sC_{n_C}}$$

or

$$\sum_{i=1}^{n_R} \frac{\partial F}{\partial[kR_i]} R_i + \sum_{i=1}^{n_C} \frac{\partial F}{\partial\left[\frac{k}{sC_i}\right] sC_i} = 0 \tag{6.98}$$

Then, dividing by F and setting k to unity,

$$\sum_{i=1}^{n_R} S_{R_i}^F + \sum_{i=1}^{n_C} S_{1/sC_i}^F = 0 \tag{6.99}$$

However, since s is a constant throughout the differentiation and using $S_x^F = -S_{1/x}^F$ it follows from equation (6.99) that

$$\sum_{i=1}^{n_R} S_{R_i}^F - \sum_{i=1}^{n_C} S_{C_i}^F = 0$$

or

$$\boxed{MS_R^F - MS_C^F = 0} \quad \text{MULTIPARAMETER INVARIANCE PROPERTY NO. 1} \tag{6.100}$$

A *physical interpretation* of this invariance property is that if all n_R resistors are uniformly perturbed by the fraction $+\epsilon_i$ and all n_C capacitors by the fraction $-\epsilon_i$, then the resultant change of the dimensionless network function F is zero for all s.

A second invariance property is obtained by multiplying all n_C capacitances C_i by the real scalar quantity k, in which case we obtain exactly the same transfer function that is obtained by multiplying the complex operator s by k. That is,

$$F(R_i, kC_i, s) = F(R_i, C_i, ks) \tag{6.101}$$

Differentiating both sides with respect to k and dividing by F, we have

$$\sum_{i=1}^{n_C} \frac{1}{F} \frac{\partial F}{\partial[kC_i]} C_i = \frac{1}{F}\left[\frac{\partial F}{\partial[ks]} s\right] = \frac{1}{k} \cdot \frac{\partial[\ln F]}{\partial[\ln(s)]}$$

Setting $k = 1$ and using equation (6.100) provides the second invariance property,

$$\boxed{MS_C^F = MS_R^F = \frac{\partial[\ln F]}{\partial[\ln s]}} \quad \text{INVARIANCE PROPERTY NO. 2} \tag{6.102}$$

The right-hand side of this equation is *invariant* in the sense that it is a function of F and *not* a function of the topology that is used to realize F.

6-9.1 Sensitivity Invariance and Frequency Response

We now consider the special case where F is the complex frequency response transfer function $H(j\omega) \equiv G(\omega)\exp[j\theta(\omega)]$. Letting $F \equiv H(s)$ in equation (6.102) and evaluating on the imaginary axis ($s = j\omega$), we have

$$MS_C^{H(j\omega)} = MS_R^{H(j\omega)} = \frac{\partial[\ln H(j\omega)]}{\partial[\ln(j\omega)]} = S_\omega^{H(j\omega)} \tag{6.103}$$

Furthermore, letting x_i equal ω in equation (6.56A),

$$S_\omega^{H(j\omega)} = S_\omega^{G(\omega)} + j\theta(\omega)S_\omega^{\theta(\omega)} \tag{6.104}$$

where $G(\omega)$ and $\theta(\omega)$ are the *real* gain and phase functions, respectively. Eliminating $S_\omega^{H(j\omega)}$ from the equations (6.103) and (6.104) and taking the real part of the result,

$$\boxed{\mathrm{Re}\left[MS_C^{H(j\omega)}\right] = \mathrm{Re}\left[MS_R^{H(j\omega)}\right] = S_\omega^{G(\omega)}} \tag{6.105}$$

Taking the imaginary part of equations (6.103) and (6.104),

$$\boxed{\mathrm{Im}\left[MS_C^{H(j\omega)}\right] = \mathrm{Im}\left[MS_R^{H(j\omega)}\right] = \theta(\omega)S_\omega^{\theta(\omega)}} \tag{6.106}$$

Alternative and more useful interpretations of equations (6.105) and (6.106) are obtained by noting that

$$\mathrm{Re}\left[MS_{x_i}^{H(j\omega)}\right] \equiv \mathrm{Re}\sum_{i=1}^{n} S_{x_i}^{H(j\omega)} = \sum_{i=1}^{n} \mathrm{Re}\left[S_{x_i}^{H(j\omega)}\right]$$

$$= \sum_{i=1}^{n} S_{x_i}^{G(\omega)} = MS_{x_i}^{G(\omega)} \tag{6.107}$$

and, similarly,

$$\mathrm{Im}\left[MS_{x_i}^{H(j\omega)}\right] = MS_{x_i}^{\theta(\omega)}\theta(\omega) \tag{6.108}$$

Using equations (6.107) and (6.108) in (6.105) and (6.106), we obtain

$$\boxed{\begin{aligned} MS_C^{G(\omega)} &= MS_R^{G(\omega)} = S_\omega^{G(\omega)} = \frac{\partial[\ln G(\omega)]}{\partial[\ln \omega]} \\ MS_C^{\theta(\omega)} &= MS_R^{\theta(\omega)} = S_\omega^{\theta(\omega)} = \frac{\partial[\ln \theta(\omega)]}{\partial[\ln \omega]} \end{aligned}} \quad \text{MULTIPARAMETER SENSITIVITY INVARIANTS} \tag{6.109}$$

These two equations are the required sensitivity invariants for the gain function $G(\omega)$ and phase function $\theta(\omega)$. Note that the multiparameter sensitivities are equal to the logarithmic frequency derivatives of $G(\omega)$ and $\theta(\omega)$. It is of considerable practical consequence to observe that there exists this inherent relationship between the rates of change of $G(\omega)$ and $\theta(\omega)$ with frequency ω and the multiparameters sensitivities. Clearly, from equation (6.109), *highly selective gain and/or phase characteristics imply high-valued multiparameter sensitivities. It is, therefore, expected that the sensitivity (or tolerance) problem becomes most acute for high-quality high-order filters in the region of ω where $G(\omega)$ and $\theta(\omega)$ are changing most rapidly.*

6-9.2 Lower-Bound Worstcase Sensitivity and Frequency Response

In this section, the invariance results in equations (6.109) are used to derive a *physically meaningful* lower-bound on the worstcase fractional deviations $|\Delta G(\omega)/G(\omega)|$ and $|\Delta\theta(\omega)/\theta(\omega)|$.

First, it is observed that if there exists a set of n real variables $r_1, r_2, \ldots, r_i, \ldots, r_n$ which are constrained so that they have a constant sum, that is

$$\sum_{i=1}^{n} r_i \equiv K, \qquad K \text{ constant} \tag{6.110}$$

Then it is readily shown that

$$\sum_{i=1}^{n} |r_i| \geqslant |K| \tag{6.111}$$

and, further, that

$$\left.\begin{aligned} &\sum_{i=1}^{n} |r_i| = |K| \\ &\text{if and only if} \quad \text{(i)} \quad r_i \geqslant 0, \quad \forall i \\ &\qquad\qquad\text{or} \quad \text{(ii)} \quad r_i \leqslant 0, \quad \forall i \end{aligned}\right\} \tag{6.112}$$

If we let

$$r_i \equiv S_{C_i}^{G(\omega)} \tag{6.113}$$

so that, from equations (6.113) and (6.110), it follows that the sum K is simply

$$K = \frac{\partial[\ln G(\omega)]}{\partial[\ln \omega]} \tag{6.114}$$

and, by equation (6.111),

$$\sum_{i=1}^{n_C} |S_{C_i}^{G(\omega)}| \equiv WS_C^{G(\omega)} \geqslant \left| \frac{\partial[\ln G(\omega)]}{\partial[\ln \omega]} \right| \tag{6.115}$$

Similarly,

$$WS_R^{G(\omega)} \geqslant \left| \frac{\partial[\ln G(\omega)]}{\partial[\ln \omega]} \right| \tag{6.116}$$

$$WS_C^{\theta(\omega)} \geqslant \left| \frac{\partial[\ln \theta(\omega)]}{\partial[\ln \omega]} \right| \tag{6.117}$$

and

$$WS_R^{\theta(\omega)} \geqslant \left| \frac{\partial[\ln \theta(\omega)]}{\partial[\ln \omega]} \right| \tag{6.118}$$

The right-hand sides of the four equations above are the lower-bound worstcase sensitivities; they are also [see equations (6.109)] the magnitudes of the multiparameter sensitivities $|MS_{RC}^{G(\omega)}|$ and they are *invariant*. We will use the notation $LW_{RC}^{G(\omega)}$ to describe the right sides of equations (6.115) and (6.116) so that $LWS_{RC}^{G(\omega)} = |MS_{RC}^{G(\omega)}|$.

Different circuit implementations of $G(\omega)$ generally possess different single-element sensitivities $S_{R_i,C_i}^{G(\omega)}$, but it can be guaranteed that there *does not exist* a circuit implementation of $H(j\omega)$ for which the actual worstcase sensitivity $WS_{RC}^{G(\omega)}$ is less than $LWS_{RC}^{G(\omega)} \equiv |MS_{RC}^{G(\omega)}|$. Therefore, it is convenient to use $LWS_{RC}^{G(\omega)}$ as a "benchmark" for the comparison of the passive RC sensitivities of the different synthesis techniques.

It *is* possible for the worstcase sensitivities WS in equations (6.115) through (6.118) to equal exactly the lower bounds on the right-hand sides. In this case, equation (6.112) must be valid. That is, either *all* the single-element sensitivities are non-negative or *all* the single-element sensitivities are non-positive.

6-9.3 An Illustrative Example of *MS*, *LWS*, and *WS*

The preceding analysis of sensitivity functions may be summarized by means of the numerical example shown in Fig. 6.6(a). The individual complex sensitivities $S_{x_i}^{H(s)}$ for a three-element network are assumed and shown in Fig. 6.6(a) for a particular point on the imaginary axis. (Any point in the s-plane could be used, but the *frequency response* sensitivity invariants apply on the imaginary axis.) The single-element sensitivities are shown for a particular value of ω as $(1 + j)$, $(-2 + j)$, $(-1 - 3j)$, so that the *complex* multiparameter sensitivity $MS_x^{H(j\omega)}$ is the vector sum $(-2 - j)$, as shown in Fig. 6.6(a). The real parts of each of the vectors in Fig. 6.5(b) give the sensitivities of the gain function $G(\omega)$ to x_i, as shown in Fig. 6.6(b) and Fig. 6.6(c). The multiparameter sensitivity $MS_x^{G(\omega)}$ is given by $(1 + (-2) + (-1)) = -2$ so that the lower-bound worstcase sensitivity *for this function* $H(j\omega)$ is simply given by

$$LWS_x^{G(\omega)} = |MS^{G(\omega)}| = |-2| = 2$$

whereas the actual worstcase sensitivity for *the particular topology that has the particular single-element sensitivities* of Fig. 6.6(a) is given by

$$WS_x^{G(\omega)} \equiv |S_{x_1}^{G(\omega)}| + |S_{x_2}^{G(\omega)}| + |S_{x_3}^{G(\omega)}| = 1 + 2 + 1 = 4$$

Thus, if $\epsilon_{\max}$ is 2%, then the worstcase deviation $|\Delta G(\omega)/G(\omega)|$ at this value of ω is $(2 \times 4) = 8\%$, and the lower-bound worstcase deviation (for which there *may* exist a corresponding circuit implementation) is $(2 \times 2) = 4\%$. The designer must decide whether to try a different circuit with the objective of coming closer to the lower-bound worstcase deviation of 4%. There is no guarantee that this is possible, because there is no guarantee that there exists a realization of $H(s)$ that has the lower-bound performance, $WS_x^{G(\omega)} = LWS_x^{G(\omega)}$.

EXAMPLE 6.8: Derive the worstcase fractional deviation $|\Delta G(\omega)/G(\omega)|$ of the gain function $G(\omega)$ for the damped integrator circuit shown in Fig. 6.7. Compare the result with the lower-bound worstcase deviation. Assume that R_1, R_2, and C_2 have tolerances of $\pm 1\%$; that is, $\epsilon_{\max} \equiv 0.01$.

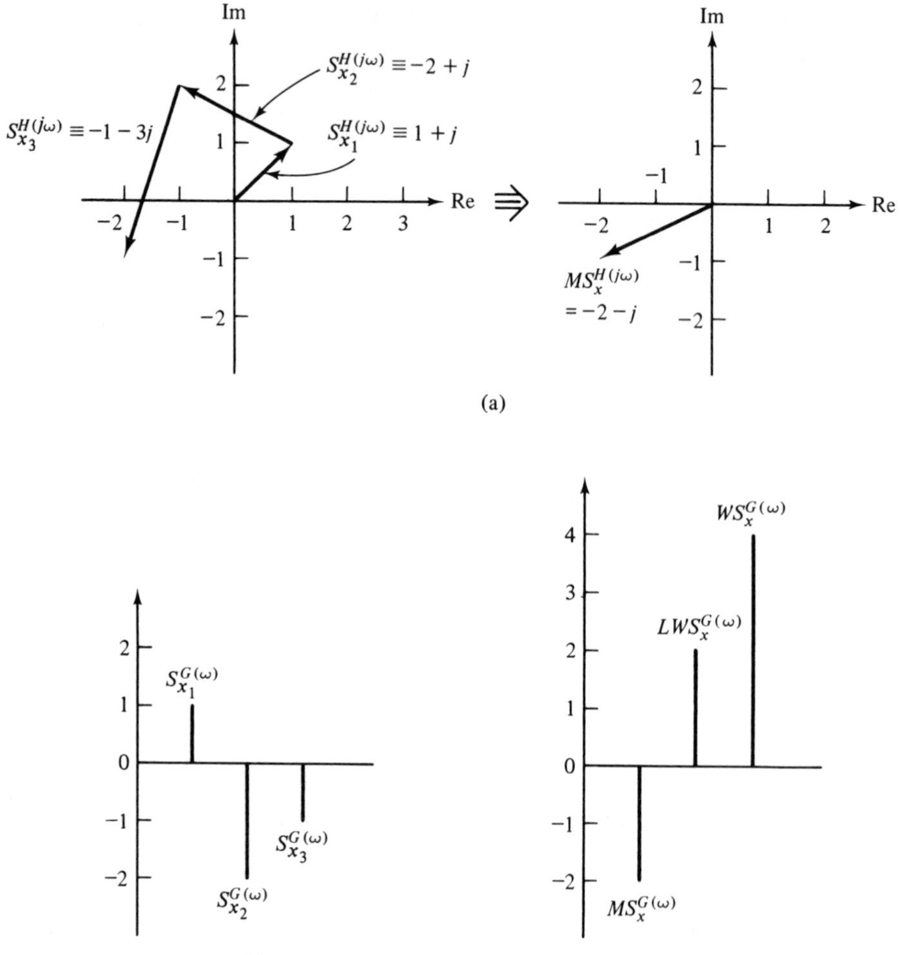

FIGURE 6.6 *(a) Complex multiparameter sensitivity for a particular $s = j\omega$; (b) single-element real sensitivities $S_{x_i}^{G(\omega)}$; (c) real multiparameter-worstcase- and lower-bound worstcase sensitivities.*

SOLUTION: Direct analysis of the circuit gives the complex frequency response transfer function

$$H(j\omega) \equiv \frac{V_2(j\omega)}{V_1(j\omega)} = \frac{-R_2}{R_1[1 + j\omega C_2 R_2]} \tag{6.119}$$

and the gain function

$$G(\omega) \equiv |H(j\omega)| = \frac{R_2}{R_1\left[1 + \omega^2 C_2^2 R_2^2\right]^{1/2}} \tag{6.120}$$

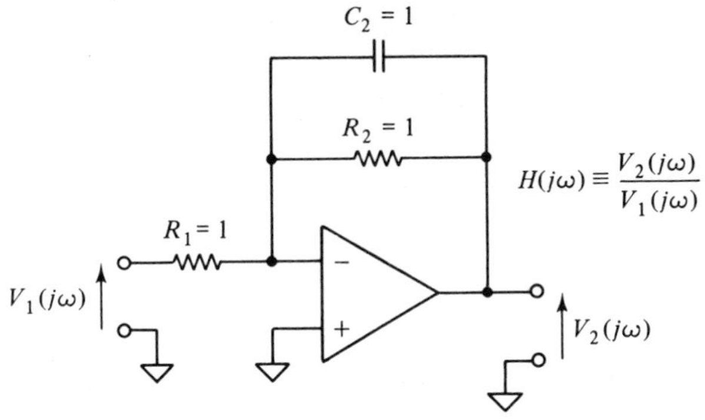

FIGURE 6.7

Direct analysis of this equation provides the single-element sensitivities

$$\left.\begin{aligned} S_{R_1}^{G(\omega)} &= -1 \\ S_{R_2}^{G(\omega)} &= \frac{1}{1 + \omega^2 C_2^2 R_2^2} \\ S_{C_2}^{G(\omega)} &= \frac{-\omega^2 C_2^2 R_2^2}{1 + \omega^2 C_2^2 R_2^2} \end{aligned}\right\} \tag{6.121}$$

Adding the moduli of equations (6.121), we have

$$WS_{RC}^{G(\omega)} = 2$$

Thus, although $H(j\omega)$ and $G(\omega)$ are functions of ω, in this particular case the worstcase sensitivity is *constant* for all ω. Since $\epsilon_{max} = 0.01$, then, by equation (6.94),

$$\left| \frac{\Delta G(\omega)}{G(\omega)} \right| \leqslant 0.01 \times 2 = \underline{\underline{0.02}} \tag{6.122}$$

corresponding to a worstcase deviation of 2% at all ω.

The *lower-bound* worstcase deviation is calculated directly from the transfer function $G(\omega)$. Substituting $C_2 = R_2 = R_1 = 1$, as given in Fig. 6.7, into (6.120) we have

$$G(\omega) = (1 + \omega^2)^{-1/2} \tag{6.123}$$

Now, from equation 6.115 and 6.116,

$$LWS_R^{G(\omega)} + LWS_C^{G(\omega)} = 2\left| \frac{\partial[\ln G(\omega)]}{\partial[\ln \omega]} \right| = 2|S_\omega^{G(\omega)}| \tag{6.124}$$

The sensitivity term $S_{\omega}^{G(\omega)}$ is calculated directly from equation (6.123) as

$$S_{\omega}^{G(\omega)} = -\frac{\omega^2}{1 + \omega^2} \tag{6.125}$$

so that, from equations (6.115) and (6.116) the worstcase sensitivity to *all RC* elements is

$$LWS_{RC}^{G(\omega)} = 2|S_{\omega}^{G(\omega)}| = \frac{2\omega^2}{1 + \omega^2} \tag{6.126}$$

and the corresponding lower-bound deviation is, for $\epsilon_{max} = 0.01$, simply

$$\left|\frac{\Delta G(\omega)}{G(\omega)}\right|_{\substack{\text{lower-bound}\\\text{worstcase}}} = \frac{0.02\omega^2}{1 + \omega^2} \tag{6.127}$$

The *actual* worstcase deviation of 2% [equation (6.122)] only coincides with the theoretical lower bound [equation (6.127)] at $\omega = \infty$; there is no known *systematic* method for finding another circuit implementation of $H(j\omega)$ with a corresponding $WS_{RC}^{G(\omega)}$ that is closer to the lower bound.

EXAMPLE 6.9: Prove that the network shown in Fig. 6.8 exhibits lower-bound worstcase sensitivity behavior for $G(\omega)$ and $\theta(\omega)$ over all ω, C_1 and R_1.

SOLUTION: The voltage transfer function is

$$H(j\omega) = \frac{1}{1 + j\omega C_1 R_1} \tag{6.128}$$

and, therefore,

$$G(\omega) = \left(1 + \omega^2 C_1^2 R_1^2\right)^{-1/2} \tag{6.129}$$

and

$$\theta(\omega) = -\arctan(\omega C_1 R_1) \tag{6.130}$$

By direct analysis of equation (6.128)

$$S_{C_1}^{G(\omega)} = S_{R_1}^{G(\omega)} = -\frac{\omega^2 C_1^2 R_1^2}{1 + \omega^2 C_1^2 R_1^2} \tag{6.131}$$

All (both) the single-element sensitivities *have the same sign* over all ω, C_1, and R_1 and therefore (by the argument on page 217) the function $WS_{RC}^{G(\omega)}$ is equal to the lower bound $LWS_{RC}^{G(\omega)}$ over all ω, C_1, and R_1. An exactly similar argument is used to show that $WS_{RC}^{\theta(\omega)} = LWS_{RC}^{\theta(\omega)}$ because, from equation (6.130), the single-element sensitivities $S_{R_1}^{\theta(\omega)}$ and $S_{C_1}^{\theta(\omega)}$ *have the same sign*.

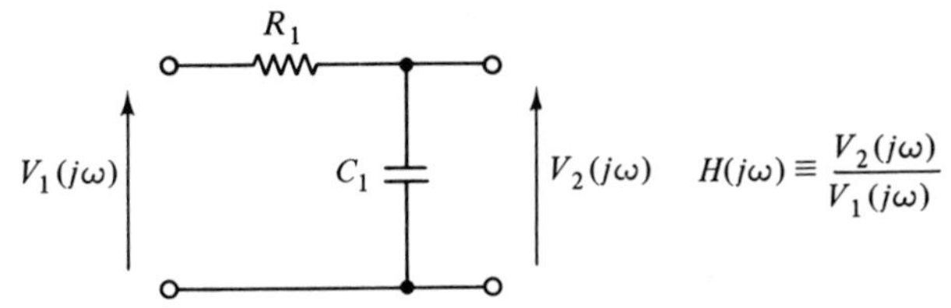

FIGURE 6.8

An *alternative* proof of the above is obtained by calculating the lower-bound worstcase sensitivity *directly* from equation (6.103), thus

$$LWS_{RC}^{G(\omega)} = |MS_{RC}^{G(\omega)}| = 2|S_{\omega}^{G(\omega)}| = +\frac{2\omega^2 C_1^2 R_1^2}{1 + \omega^2 C_1^2 R_1^2} \tag{6.132}$$

and the *actual* worstcase sensitivity, from addition of the moduli of equations (6.131), is

$$WS_{RC}^{G(\omega)} \equiv |S_{R_1}^{G(\omega)}| + |S_{C_1}^{G(\omega)}| = \frac{2\omega^2 C_1^2 R_1^2}{1 + \omega^2 C_1^2 R_1^2} \tag{6.133}$$

It follows by comparison of equations (6.132) and (6.133) that $WS_{RC}^{G(\omega)} = LWS_{RC}^{G(\omega)}$ over all ω, C_1, and R_1. Thus, the proof has been obtained by direct calculation of $WS_{RC}^{G(\omega)}$ and $LWS_{RC}^{G(\omega)}$.

6-10 THE SUM-OF-SQUARES OR SCHOEFFLER CRITERION

It has been noted previously that the worstcase-sensitivity criterion is pessimistic in the sense that it predicts the absolute upper bound on $|\Delta F/F|$ under conditions where *all* parameters x_i take on their appropriate extremum values $x_{iN}(1 \pm \epsilon_{max})$ corresponding to conditions (i) and (ii) of equation (6.95). The probability of this occurring becomes increasingly remote as the number of parameters n increases. The worstcase deviation also is less probable if the single-element sensitivities $|S_{x_i}^F|$ do not contain any obviously dominant terms. For example, the sensitivities $|S_{x_i}^F|$ are given for $n = 5$ in Fig. 6.9(a), where the $|S_{x_5}^F|$ term is clearly dominant. On the other hand, in Fig. 6.9(b) the terms $|S_{x_i}^F|$ are equally distributed over all five elements. For both Fig. 6.9(a) and Fig. 6.9(b), $WS_{x_i}^F = 20$. Now, *if* $|\epsilon_i| = \epsilon_{max}$ *for all five elements*, the probability that the worstcase deviation of $20\epsilon_{max}$ will occur is the same for both Fig. 6.9(a) and Fig. 6.9(b), requiring that the signs of $\epsilon_i S_{x_i}^F$ are the same over all five elements.

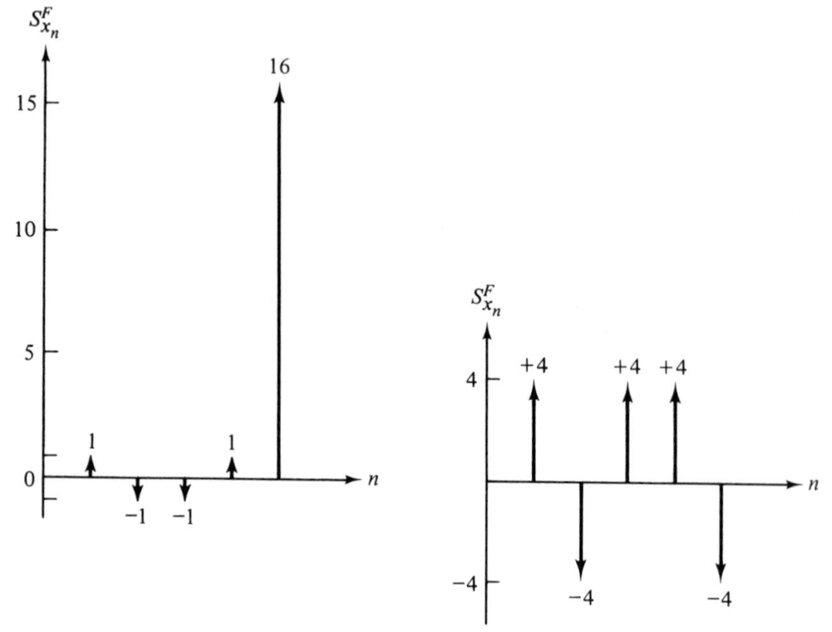

(a) Nonuniform distribution

(b) Uniform magnitude distribution

FIGURE 6.9

However, if the signs of these terms are purely random, then with $|\epsilon_i| = \epsilon_{max}$, the *actual* deviations are given by the following 2^5 possibilities:

$$\frac{\Delta F}{F} = \epsilon_{max}\left[\pm 4 \pm 4 \pm 4 \pm 4 \pm 4\right], \qquad \text{Fig. 6.1(b)} \qquad (6.134)$$

and

$$\frac{\Delta F}{F} = \epsilon_{max}\left[\pm 1 \pm 1 \pm 1 \pm 1 \pm 16\right], \qquad \text{Fig. 6.1(a)} \qquad (6.134A)$$

With random choices of polarities, the probability that $|\Delta F/F|$ exceeds $12\epsilon_{max}$, for example, is easily shown to be $\frac{1}{8}$ for equation (6.134), whereas the probability that $|\Delta F/F|$ *exceeds* $12\epsilon_{max}$ in equation (6.135) is $\frac{15}{16}$. This rather simple example therefore demonstrates that the probability of $|\Delta F/F|$ *approaching* worstcase is far greater if one of the single-element sensitivities is dominant over the others.

The sum-of-squares, or Schoeffler sensitivity criterion, is defined by

$$\boxed{\mathbb{S}^F_x \equiv \sum_{i=1}^{n} |S^F_{x_i}|^2} \qquad \textsc{Schoeffler Criterion} \qquad (6.135)$$

and *is a measure of the uniformity* of the $|S_{x_i}^F|$ terms as well as of their magnitudes. Thus, in the previously considered example,

$$\mathbb{S}_x^{G(\omega)} = 5 \times (4)^2 = 80, \qquad \text{Fig. 6.9(b)}$$

and

$$\mathbb{S}_x^{G(\omega)} = \left[4 \times (1)^2\right] + \left[(16)^2\right] = 260, \qquad \text{Fig. 6.9(a)}$$

The Schoeffler criterion does not provide a measure of $WS_x^{G(\omega)}$ but, in the case above, indicates that the situation in Fig. 6.9(b) is preferable to that in Fig. 6.9(a) on the basis of the lower value of $\mathbb{S}_x^{G(\omega)}$, indicating a more uniform distribution of single-element sensitivities.

6-10.1 Schoeffler-Sensitivity Invariants

The invariance properties of the Schoeffler criterion are now described for *real* functions F. Once more, we make use of an elementary property of a set of real variables $r_1, r_2, \ldots, r_i, \ldots, r_n$ that have a constant sum K. Thus, given that

$$\sum_{i=1}^{n} r_i \equiv K, \qquad K \text{ constant} \tag{6.136}$$

it may be shown that the sum S, where

$$S \equiv \sum_{i=1}^{n} |r_i|^2, \tag{6.137}$$

has *minimum value* where

$$|r_i| = \left|\frac{K}{n}\right|, \qquad \forall i \tag{6.138}$$

Thus, substituting equation (6.138) into equation (6.137) gives the minimum value of S as

$$S_{\min} = n\left|\frac{K}{n}\right|^2 = \frac{|K|^2}{n} \tag{6.139}$$

We now use this result to obtain the minimum value of $\mathbb{S}_x^F$, where the single-element sensitivities $S_{x_i}^F$ are associated with r_i and the sum K with $MS_{x_i}^F$ as required by equation (6.137). Then, $\mathbb{S}_x^F$ is, *by definition*, equal to S

in equation (6.137) so that, according to equation (6.139),

$$\mathbb{S}_x^F \geqslant \frac{|K|^2}{n} = \frac{1}{n}|MS_x^F|^2 \qquad \text{LOWER-BOUND ON } \mathbb{S}_x^F \tag{6.140}$$

This is the required lower bound on $\mathbb{S}_x^F$; it is important to realize that it is invariant *for a specified F and a specified number of elements n.* The lower bound of $\mathbb{S}_x^F$ decreases as n increases for a particular F.

An upper bound on $\mathbb{S}_x^F$ is derived by making use of the inequality

$$\left(\sum_{i=1}^{n} |r_i|\right)^2 \geqslant \sum_{i=1}^{n} |r_i|^2 \tag{6.141}$$

so that, substituting $r_i = S_{x_i}^F$ leads immediately to

$$\mathbb{S}_{x_i}^F \leqslant \left[WS_{x_i}^F\right]^2 \qquad \text{UPPER BOUND ON } \mathbb{S}_{x_i}^F \tag{6.142}$$

Combining equations (6.140) and (6.142), we have

$$\frac{1}{n}|MS_x^F|^2 \leqslant \mathbb{S}_x^F \leqslant [WS_x^F]^2 \tag{6.143}$$

If $\mathbb{S}_{x_i}^F$ is close to the upper bound then the single-element sensitivities contain *one* dominant term, implying that $\epsilon_{\max} WS_x^F$ is a meaningful estimate of circuit performance; if $\mathbb{S}_x^F$ is close to the lower bound and $n \gg 1$, then the single-element sensitivities do not contain a dominant term $S_{x_i}^F$ and $\epsilon_{\max} WS_x^F$ is a highly pessimistic estimate of circuit performance. Circuits in this category are compared more meaningfully by using $\mathbb{S}_x^F$.

6-10.2 Frequency Response Invariants of $\mathbb{S}_{RC}^{H(j\omega)}$

If $F \equiv G(\omega)\exp[j\theta(\omega)]$, then it follows from equation (6.103) that

$$MS_{RC}^{G(\omega)} \equiv MS_R^{G(\omega)} + MS_C^{G(\omega)} = 2S_\omega^{G(\omega)} \tag{6.144}$$

and, similarly,

$$MS_{RC}^{\theta(\omega)} = 2S_\omega^{\theta(\omega)} \tag{6.145}$$

Using the equations (6.144) and (6.145) in equation (6.140),

$$\mathbb{S}_{RC}^{G(\omega)} \geqslant \frac{4}{n}|S_{\omega}^{G(\omega)}|^2 \quad \text{and} \quad \mathbb{S}_{RC}^{\theta(\omega)} \geqslant \frac{4}{n}|S_{\omega}^{\theta(\omega)}|^2 \qquad \text{SCHOEFFLER SENSITIVITY INVARIANTS} \tag{6.146}$$

6-11 SENSITIVITY TO IDEALLY-ZERO OR IDEALLY-INFINITE PARAMETERS

In many practical situations, the parameter x_i is ideally-zero or ideally-infinite. For example, if x_i is the resistance r_x in the hybrid-π transistor model of Fig. 4.6(a) then its *ideal* value is zero. If x_i is the gain bandwidth product B of an OP AMP then its *ideal* value is infinity. In a practical application, it is useful to know the fractional charge $\Delta F/F$ of a network function due to the departure of r_x from zero, for example, or due to the finiteness of B, for example.

A mathematical difficulty arises if one attempts to compute

$$\frac{\Delta F}{F} = S_{x_i}^F\left(\frac{\Delta x_i}{x_i}\right)$$

when x_i is either equal to zero or infinity because $\Delta x_i/x_i$ is either infinite or zero, respectively. This difficulty is resolved by noting that a nominal value of x_i equal to zero in equation (6.2) implies that

$$F(s, x_i) - F(s, 0) = \sum_{n=1}^{\infty} \frac{x_i^n \partial^n F(s, x_i)}{n! \partial^n x_i} \tag{6.147}$$

so

$$\frac{F(s, x_i) - F(s, 0)}{F(s, x_i)} = \sum_{n=1}^{\infty} \frac{x_i^n \partial^n F(s, x_i)}{F(s, x_i) n! \partial^n x_i}$$

Then, letting x_i tend to zero in the equation above, we have

$$\frac{\Delta F(s, x_i)}{F(s, x_i)} \approx \frac{x_i \partial F(s, x_i)}{F(s, x_i) \partial x_i}, \qquad x_i \to 0$$

or

$$\boxed{\frac{\Delta F}{F} \approx S_{x_i}^F, \qquad x_i \to 0}$$

FRACTIONAL CHANGE DUE TO AN IDEALLY-ZERO PARAMETER (6.148)

The approximation above becomes *exact* if

$$\frac{\partial^n F(s, x_i)}{\partial^n x_i} = 0 \qquad \text{for } n > 1 \tag{6.149}$$

Ideally-infinite parameters: Suppose x_i is ideally-infinite, as in the case of OP AMP bandwidth B. Then x_i^{-1} is ideally-zero and the analysis above may be applied to a Taylor series expansion in x_i^{-1} giving, from equation (6.148),

$$\frac{\Delta F}{F} \approx S_{1/x_i}^F, \qquad \frac{1}{x_i} \to 0$$

or

$$\boxed{\frac{\Delta F}{F} \approx -S_{x_i}^F, \qquad x_i \to \infty}$$

FRACTIONAL CHANGE DUE TO AN IDEALLY-INFINITE PARAMETER (6.150)

The approximation above becomes *exact* if equation (6.149) is valid.

EXAMPLE 6.10: It is found in section 7-4.2 that a particular Sallen-Key lowpass filter has an active-sensitivity performance

$$S_B^{H_b(s)} = \frac{K^2\alpha_1}{B}\left[\frac{s^2}{s^2 + \dfrac{\omega_P s}{Q_P} + \omega_P^2}\right] \tag{6.151}$$

where $H_b(s)$ is the transfer function, B is the gain bandwidth product of the OP AMP, and K, α_1, Q_P, and ω_P are constants. Derive expressions for

(i) the fractional deviation of the gain function $G(\omega)$ due to the *finiteness* of B;

and

(ii) the fractional deviation of $G(\omega)$ due to small perturbations of B.

SOLUTION: (i) Defining

$$t_{HP}(s) \equiv \frac{s^2}{s^2 + \dfrac{\omega_P s}{Q_P} + \omega_P^2} \tag{6.152}$$

in (6.151) and taking the real part as required by equation (6.105), we have

$$S_B^{G(\omega)} = \frac{K^2\alpha_1}{B} \operatorname{Re}[t_{HP}(j\omega)] \tag{6.153}$$

Using equations (6.150) and (6.153) with $F \equiv G(\omega)$ and $B \equiv x_i$,

$$\boxed{\frac{\Delta G(\omega)}{G(\omega)} \approx -\frac{K^2\alpha_1}{B} \operatorname{Re}[t_{HP}(j\omega)] \qquad B \to \infty} \tag{6.154}$$

Employing equation (6.152) to expand $\operatorname{Re}[t_{HP}(j\omega)]$ in equation (6.154),

$$\boxed{\frac{\Delta G(\omega)}{G(\omega)} \approx -\frac{K^2\alpha_1}{B} \operatorname{Re}\left[\frac{1}{1 - \left(\dfrac{\omega_P}{\omega}\right)^2 - j\left(\dfrac{\omega_P}{Q_P\omega}\right)}\right], \qquad B \to \infty} \tag{6.155}$$

This is the fractional deviation in the gain $G(\omega)$ due to the *finiteness* of B.

(ii) Given that B is *finite* and has a small fractional perturbation $\Delta B/B$ about the nominal finite value, then $\Delta B/B$ causes a fractional charge $\Delta G(\omega)/G(\omega)$, as in the conventional expression

$$\boxed{\frac{\Delta G(\omega)}{G(\omega)} \approx S_B^{G(\omega)}\left(\frac{\Delta B}{B}\right), \qquad \frac{\Delta B}{B} \to 0}$$

where $S_B^{G(\omega)}$ is given in equation (6.153).

6-12 STATISTICAL MEASURES OF SENSITIVITY

The WS_x^F and $\mathbb{S}_x^F$ criteria have the *advantage* that they may be expressed algebraically and compared with the 'benchmark' lower-bound sensitivity expressions. They have the *disadvantage* of being overly simplistic in that no allowance is made for the statistical distribution of ϵ_i over a large number of the resistors and capacitors. Techniques have been developed for analyzing a given network with perturbations ϵ_i that possess a specified probability distribution. The majority of these techniques involve a computer-aided simulation of the network to calculate an error-distribution curve for $|\Delta G(\omega)/G(\omega)|$. This curve provides a criterion of circuit performance or, alternatively, a weighted area under the curve might be used. Such methods are extremely useful if an accurate prediction of the probability of $|\Delta G(\omega)/G(\omega)|$ lying within acceptable limits is essential; for example, large-volume manufacturers of a particular microcircuit active filter might choose a *statistical* approach to the analysis of tolerances. For the purposes of developing an *understanding* of the sensitivity behavior of the different synthesis techniques and corresponding structures, computer-aided statistical methods are of limited usefulness.

6-13 SUMMARY

The sensitivity of the network function $F(s)$ to real parameters x_i is of fundamental importance in the synthesis and design of active networks. The first-order sensitivity function $S_{x_i}^{F(s)}$ is generally a useful criterion for estimating the fractional change of $F(s)$ that results from a small fractional change of x_i. There are simple rules (Table 6-1) that enable one to readily derive algebraic functions $S_{x_i}^{F(s)}$.

If $F(s)$ is a polynomial, then $S_{x_i}^{F(s)}$ is generally a ratio of two polynomials, and the poles of $S_{x_i}^{F(s)}$ are also the zeros of $F(s)$. Since $F(s)$ is usually invariant, in the sense that it is a specified design criterion, it follows that the poles of $S_{x_i}^{F(s)}$ are, in fact, invariant. Different topological implementations of $F(s)$ will generally possess $S_{x_i}^{F(s)}$ functions that differ in their zero locations but not in their pole locations.

PROBLEMS

6-1. Given that

$$F(s, x_i) \equiv 3s^2 + x_1 s + x_2$$

derive $S_{x_2}^F$ and thereby estimate the fractional change in F at $s=4$, $x_1=2$, $x_2=1$ due to a -1% change in x_2. Use the formula

$$\frac{\Delta F}{F} \approx S_{x_2}^F\left(\frac{\Delta x_2}{x_2}\right)$$

6-2. Calculate $F(s, x_i)$ at the values $s=4$, $x_1=2$, $x_2=1$ and again at the values $s=4$, $x_1=2$, $x_2=0.99$ corresponding to the -1% change of x_2 described in Problem 6.1. Compare the *actual* change in $F(s, x_i)$ with that obtained in Problem 6.1.

6-3. Repeat Problems 6.1 and 6.2 but for the function

$$F(s, x_i) \equiv 3s^2 + x_1 s + x_2^3$$

6-4. Prove the inverse property

$$S_{x_i}^{F^{-1}} = -S_{x_i}^{F}$$

6-5. Prove the exponent property

$$S_{x_i}^{F^n} = nS_{x_i}^{F}$$

6-6. Prove the summation property

$$S_{x_i}^{F_1+F_2+F_3+\cdots+F_n} = \frac{F_1 S_{x_i}^{F_1} + F_2 S_{x_i}^{F_2} + \cdots + F_n S_{x_i}^{F_n}}{F_1 + F_2 + F_3 + \cdots + F_n}$$

6-7. Prove the product property

$$S_{x_i}^{F_1 F_2 F_3 \cdots F_n} = \sum_{j=1}^{n} S_{x_i}^{F_j}$$

6-8. Prove the scalar property

$$S_{x_i}^{KF} = S_{x_i}^{F}, \qquad K \text{ real and independent of } x_i$$

6-9. Prove the chain rule property given in Table 6-1.

6-10. Given

$$F = \frac{1}{3s^2 + 2s + x_2}$$

derive $S_{x_2}^{F}$ by using the inverse property.

6-11. Given

$$F = \left(3s^2 + 2s + x_2\right)^{-5/2}$$

derive $S_{x_2}^{F}$ by using the exponent property.

6-12. Given

$$F = 23\left(3s^2 + 2s + x_2\right)^{-5/2}$$

derive $S_{x_2}^{F}$ by using the scalar property.

6-13. Given

$$F \equiv 5x^3 + \sin x + 25\sqrt{x}$$

Use the summation property to derive S_x^F.

6-14. Given

$$F \equiv (5x^3)(\sin x)(25\sqrt{x}\,)$$

use the product property to derive S_x^F.

6-15. Derive S_x^F for the following functions:

(a) $F \equiv 2 + 3jx$

(b) $F \equiv [2 + 5\omega^2 x + 7\omega^4 x] + j[\omega^3 x + 4\omega^5 x]$

(c) $F \equiv \dfrac{2+3jx}{1+4jx}$

(d) $F \equiv \left|\dfrac{2+3jx}{1+4jx}\right|$

6-16. Given that

$$F \equiv (x+2)^{1/2} \sin x$$

let $g_1(x) \equiv x + 2$ and $g_2 \equiv x$. Then, derive S_x^F by using the chain rule.

6-17. Repeat Example 6.3 in the text of this chapter by letting

$$g_1(x) \equiv 2x_1 s^2 + \frac{s}{x_1 x_2} + x_1^3 \quad \text{and} \quad g_2(x) \equiv x_1$$

so that

$$F = [g_1(x)]^{1/4} \sin[g_2(x)]$$

Show that your answer agrees with equation (6.33).

6-18. Given

$$Q(s, x) \equiv (x+1)s^2 + xs + 2$$

use equation (6.36) to derive the function S_x^Q and sketch a pole-zero diagram of the function S_x^Q for $x = 1$. Confirm that equation (6.41) gives the zeros of S_x^Q for this case.

6-19. Given

$$H(s) = \frac{1}{s^2 + 0.1s + 1 + 2x(s^2+1)}$$

show that

$$S_x^{H(s)} = 0 \qquad \text{at } s = \pm j$$

6-20. Prove equations (6.48) from equation (6.40) and plot pole-zero diagrams for $S_{Q_p}^{H(s)}$ and $S_{\omega_p}^{H(s)}$, given that $Q_p \equiv 10$ and $\omega_p \equiv 1$ rad/sec.

6-21. Use the pole-zero diagrams in Problem 6-20 to sketch the functions $|S_{Q_P}^{H(j\omega)}|$ and $|S_{\omega_P}^{H(j\omega)}|$ and thereby verify equation (6.50).

6-22. Given

$$H(s)=\frac{1}{s^2+\dfrac{\omega_P s}{Q_P}+\omega_P^2}, \qquad \omega_P=1, \qquad Q_P=10$$

and small perturbations of ω_P and Q_P,

$$\frac{\Delta\omega_P}{\omega_P}=0.01, \qquad \frac{\Delta Q_P}{Q_P}=0.01$$

show that
(a) equation (6.51) is valid and
(b) that the perturbation $\Delta Q_P/Q_P$ has a negligible effect on $H(j\omega)$ compared with the perturbation $\Delta\omega_P/\omega_P$.

6-23. Given

$$S_{x_i}^{H(s)}=\frac{2s^2+s+3}{s^2+5s+4}$$

write expressions for $N_e(s)$, $N_0(s)$, $D_e(s)$, and $D_0(s)$ as required by equation (6.59). Then, use equations (6.61) and (6.62) to derive the gain sensitivity function $S_{x_i}^{G(\omega)}$ and the phase sensitivity function $S_{x_i}^{\theta(\omega)}$ as functions of frequency ω.

Write a polynomial $H(s, x_i)$ for which the above expression $S_{x_i}^{H(s)}$ *is* the corresponding sensitivity function. [*Hint*: Use equations (6.41) and (6.39).]

6-24. Example 6.4 in the text is a derivation of the *gain* sensitivities $S_{\omega_P}^{G(\omega)}$ and $S_{\omega_P}^{Q_P}$ for the normalized bandpass function. Repeat Example 6.4 for the *phase* sensitivities $S_{\omega_P}^{\theta(\omega)}$ and $S_{Q_P}^{\theta(\omega)}$. Sketch these phase sensitivity functions for $\omega_P=1$ and $Q_P=10$ and compare with Fig. 6.2.

6-25. The expression for $S_{\omega_P}^{G(\omega)}$ in equation (6.67) is important because many second-order RC-active circuits are highly sensitive to the perturbation $\Delta\omega_P/\omega_P$. Prove that, for $Q_P \gg 1$, $S_{\omega_P}^{G(\omega)}$ has maxima and minima of approximate magnitudes Q_P and $-Q_P$, respectively, occurring at approximately

$$\omega_P=\omega_P\left(1-\frac{1}{2Q_P}\right) \quad \text{and} \quad \omega_P=\omega_P\left(1+\frac{1}{2Q_P}\right)$$

respectively.

6-26. For the normalized bandpass function of equation (6.63), prove that

$$S_{\omega_P}^{H(s)} = -S_{\omega}^{H(s)}$$

and, therefore, that

$$S_{\omega_P}^{G(\omega)} = -S_{\omega}^{G(\omega)} = -\frac{\partial \ln G(\omega)}{\partial \ln \omega}$$

Comment on the significance of this result.

6-27. In Example 6.5, with

$$F(s) = \frac{1}{1 + sC_1R_1 + \mu}$$

it is shown that *uniform* small fractional perturbations ϵ of the parameters C_1, R_1, and μ give the multiparameter sensitivity in equation (6.75) and the fractional change in $F(j\omega)$ in equation (6.76). Verify equation (6.77) by substituting numerical values directly into equation (6.73) to obtain the *actual* change $\Delta F(j2)/F(j2)$ caused by $+1\%$ perturbations of C_1, R_1, and μ.

6-28. Verify equation (6.83) by substituting numerical values directly into equation (6.73) to obtain the *actual* change $|\Delta F(j2)/F(j2)|$ caused by $+1\%$ perturbations of C_1, R_1, and μ.

6-29. Prove that

$$MS_{x_i}^{F} = WS_{x_i}^{F}$$

if and only if

$$\text{sign}\left[S_{x_i}^{F} \right]$$

are identical over all i.

6-30. At a particular frequency $s = j\omega_1$, the single-element sensitivities of the gain function $G(\omega_1)$ of an RC-active network are given by

i	1	2	3	4	5	6
$S_{x_i}^{G(\omega_1)}$	-1	1	-15	2	-2	1

where $x_1, x_2, \ldots, x_6$ are the only RC elements in the filter.

(a) Derive $MS_{RC}^{G(\omega_1)}$, $WS_{RC}^{G(\omega_1)}$ and $\mathbb{S}_{RC}^{G(\omega_i)}$.

(b) What is the value of the logarithmic derivative term

$$\frac{\partial \ln G(\omega)}{\partial \ln \omega}$$

at $\omega = \omega_1$?

(c) Given that

$$-0.02 \leqslant \epsilon_i \leqslant +0.02, \qquad i=1, 2, \ldots, 6$$

and

$$\epsilon_i \equiv \frac{\Delta x_i}{x_i},$$

derive an upper bound on the value of $\Delta G(\omega_1)/G(\omega_1)$.

(d) If $\epsilon_i = -0.01$ for $i = 1, 2, \ldots, 6$, derive the corresponding $\Delta G(\omega_1)/G(\omega_1)$.

(e) Derive the upper and lower bounds of the inequality in equation (6.143) for this particular numerical example. Is $\mathbb{S}_{RC}^{G(\omega_1)}$ close to the upper bouna, $[WS_{RC}^{G(\omega_1)}]^2$, or to the lower bound, $\frac{1}{n}|MS_{RC}^{G(\omega_1)}|^2$? Comment on the significance of this result.

6-31. A particular *RC*-active circuit has a voltage transfer function [see equation (8.6)], given by

$$H(s)=\left(\frac{1}{R_0R_1C_1C_2}\right)\left[\frac{1}{s^2+s\left(\dfrac{1}{R_5C_1}\right)+\left(\dfrac{R_4}{R_1R_2R_3C_1C_2}\right)}\right]$$

Assume that a *nominal* design corresponds to

$$R_0=R_1=R_2=R_3=R_4=1$$
$$C_1=C_2=1$$
$$R_5=10$$

(a) Derive the single-element sensitivities

$$S_{R_i}^{H(s)}, \qquad i=1, 2, 3, 4, 5$$

and

$$S_{C_i}^{H(s)}, \qquad i=1, 2$$

(b) From the above, prove that equation (6.100) is valid for this particular circuit.

(c) From the expression above for $H(s)$, derive

$$\frac{\partial[\ln H(s)]}{\partial[\ln s]} \equiv S_s^{F(s)}$$

and thereby prove that equation (6.102) is valid for this particular circuit.

6-32. The *group delay* $\tau(\omega)$ of a steady-state sinusoidal frequency response function $G(\omega)\exp[j\theta(\omega)]$ is defined by

$$\tau(\omega) \equiv \frac{\partial\theta(\omega)}{\partial\omega}$$

Use equations (6.109), (6.117), and (6.118) to prove that
(a)

$$\theta\left[MS_{R,C}^{\theta(\omega)}\right] = \omega\tau(\omega)$$

and
(b)

$$|\theta|WS_{R,C}^{\theta(\omega)} \geq |\omega\tau(\omega)|$$

6-33. For the transfer function $H(s)$ and the nominal design given in Problem 6-31, derive the invariant lowerbound on the Schoeffler multiparameter sensitivity criterion $\mathbb{S}_{RC}^{G(\omega)}$. That is, evaluate the right-hand side of equation (6.146) for this particular case. [*Hint*: To evaluate the $S_{\omega}^{G(\omega)}$ term it may be useful to combine the result in equation (6.67) with the result of Problem 6-26.]

6-34. A particular *RC*-active filter has an active-sensitivity function of the form

$$S_B^{H(s)} = \frac{K_0}{B}\left[\frac{s\omega_P}{s^2 + \dfrac{s\omega_P}{Q_P} + \omega_P^2}\right]$$

where K_0, ω_P and Q_P are constants and B is the gain bandwidth product of the OP AMP. Derive an expression for the fractional deviations of the gain $G(\omega)$ and the phase $\theta(\omega)$ due to the *finiteness* of B. Sketch $\Delta G(\omega)/G(\omega)$ for $\omega_P = 1$, $Q_P = 10$, and K_0/B normalized to unity.

6-35. Given

$$t_{LP}(s) \equiv \frac{1}{D(s)}, \qquad t_{BP}(s) \equiv \frac{\omega_P s}{D(s)}, \qquad t_{HP}(s) \equiv \frac{s^2}{D(s)}$$

and

$$D(s) \equiv s^2 + \frac{\omega_P s}{Q_P} + \omega_P^2$$

prove the following:

(i) $S_{\omega_P}^{t_{LP}(s)} = 2t_{HP}(s) + \dfrac{1}{Q_P} t_{BP}(s)$

(ii) $S_{\omega_P}^{t_{BP}(s)} = t_{HP}(s) - t_{LP}(s)$

(iii) $S_{\omega_P}^{t_{HP}(s)} = -2t_{LP}(s) - \dfrac{1}{Q_P} t_{BP}(s)$

(iv) $S_{Q_P}^{t_{LP}(s)} = S_{Q_P}^{t_{BP}(s)} = S_{Q_P}^{t_{HP}(s)} = t_{BP}(s)$

6-36. An *RC*-active filter has an active-sensitivity function

$$S_B^{H(s)} = \frac{K_0}{B} t_{\mathrm{HP}}(s), \qquad K_0 \text{ constant}$$

where $H(s) = K_1 t_{\mathrm{LP}}(s)$, K_1 constant. Use the results in Problem 6-35 to derive the effective perturbations $\Delta Q_P/Q_P$ and $\Delta\omega_P/\omega_P$ that cause the same perturbation $\Delta H(s)$ that is caused by the effect of finite bandwidth B. Assume K_0 and K_1 are independent of B. [*Hint*: Use the results in Problem 6.35 to derive $\Delta H(s)/H(s)$ due to $\Delta Q_P/Q_P$ and $\Delta\omega_P/\omega_P$, and then equate this to $-S_B^{H(s)}$, which is the change due to finite B; solve for $\Delta Q_P/Q_P$ and $\Delta\omega_P/\omega_P$.]

BIBLIOGRAPHY

BLOSTEIN, M. L., "Some Bounds on Sensitivity in RLC Networks," *Proc. 1st Allerton Conf. Circuit Systems Theory*, p. 488–501, 1963.

MITRA, S. K., *Analysis and Synthesis of Linear Active Networks*, Wiley 1969, chapter 5.

7

DECOMPOSITIONS, STRUCTURES, AND SENSITIVITY

7-1 INTRODUCTION

In this and subsequent chapters, the topics of RC-active circuit *synthesis* and *design* are presented. In previous chapters the topics of *analysis* and *modeling* are emphasized. The reader should, at this point, be ready to tackle the problems involved in designing RC-active networks that have specific prescribed transform transfer functions $H(s)$. The problem that confronts the designer is often specified in terms of a prescribed rational transform transfer function of the form

$$H(s) = \frac{\sum_{i=0}^{m} a_i s^i}{\sum_{i=0}^{n} b_i s^i} \equiv \frac{N(s)}{D(s)}, \qquad m \leqslant n \tag{7.1}$$

In many cases, the function $H(s)$ has been derived from specified constraints on the magnitude steady-state frequency response $G(\omega) \equiv |H(j\omega)|$. There are many filter design and synthesis textbooks that provide the designer with tabulated coefficients a_i, b_i that correspond to lowpass, bandpass, and highpass functions $G(\omega)$ with varying degrees of passband flatness, transition region selectivity, and stopband attenuation (see Daniels and Zvrev in references at the end of this chapter). For example, the widely used Butterworth and Chebychev lowpass functions are tabulated in most passive filter design textbooks and the coefficients are presented in Tables 7-1 and 7-2 for the normalized case corresponding to a cut-off frequency of 1 radian per second and for orders $n = 1$ to 7.

It is assumed that the transfer function coefficients are known and that the designer is faced with the problems of selecting and designing an appropriate RC-active implementation of $H(s)$. *A major criterion that determines the selection of a given type of circuit is the sensitivity performance of the circuit to expected perturbations of such parameters as the RC elements and the open-loop gains A_0 and gain-bandwidth products (GBPs) B of the OP AMPs.* In this chapter, some widely used network structures are described and their sensitivity performance is analyzed in terms of their

TABLE 7-1

Butterworth low pass; $\omega_0 = 1$.

n	b_1	b_2	b_3	b_4	b_5	b_6	b_7
1	1						
2	$\sqrt{2}$	1					
3	2	2	1				
4	2.613	3.414	2.613	1			
5	3.236	5.236	5.236	3.236	1		
6	3.864	7.464	9.141	7.464	3.864	1	
7	4.494	10.103	14.606	14.606	10.103	4.494	1

Chebychev low pass; $\omega_0 = 1$.
Passband ripple = 1.0 db.

n	b_1	b_2	b_3	b_4	b_5	b_6	b_7
1	0.509						
2	0.996	0.907					
3	2.521	2.012	2.035				
4	2.694	5.275	3.457	3.628			
5	4.726	7.933	13.750	7.627	8.142		
6	4.456	13.632	17.446	28.021	13.471	14.512	
7	6.958	17.867	44.210	46.531	70.867	30.063	32.566

$$H(s) = \frac{a_0}{\sum_{i=0}^{n} b_i s^i}, \qquad a_0 = b_0 = 1$$

signal-flow graphs. Some insight is gained by comparing the sensitivity behavior of these structures with the theoretical lower bounds that have been derived in the previous chapter.

7-2 THE DIRECT-FORM STRUCTURE

The direct-form (DF) structure is characterized by a one-to-one correspondence between the coefficients a_i, b_i of the transfer function $H(s)$ and the transmittances of the DF signal-flow graph. For example, the DF signal-flow graph shown in Fig. 7.1 is an interconnection of inverting-integrators $-s^{-1}$ and the transmittances $\pm a_i$, $\pm b_i$. The transmittances a_n, $-a_{n-1}, \ldots, \pm a_0$ are known as the *feedforward transmittances* and the transmittances b_{n-1}, $-b_{n-2}, \ldots, \pm b_0$ as the *feedback transmittances.* Direct analysis of this signal-flow graph for the node variable $V_n(s)$ at node n indicates that

$$V_n(s) = V_1(s) - b_{n-1}s^{-1}V_n(s)$$
$$- b_{n-2}s^{-2}V_n(s) \cdots - b_1 s^{-n+1}V_n(s) - b_0 s^{-n}V_n(s)$$

Therefore,

$$\frac{V_n(s)}{V_1(s)} = \frac{s^n}{s^n + b_{n-1}s^{n-1} + b_{n-2}s^{n-2} \cdots b_1 s + b_0} \tag{7.2}$$

By direct analysis of the feedforward part of the network shown in Fig. 7.1 we have

$$V_2(s) = a_n V_n(s) + a_{n-1}s^{-1}V_n(s) + a_{n-2}s^{-2}V_n(s)$$
$$\cdots + a_1 s^{-n+1}V_n(s) + a_0 s^{-n}V_n(s) \tag{7.3}$$

Eliminating $V_n(s)$ from equations (7.2) and (7.3) leads directly to the required input-output transfer function

$$H(s) \equiv \frac{V_2(s)}{V_1(s)} = \sum_{i=0}^{n} \frac{a_i s^i}{b_i s^i}, \qquad b_n = 1 \tag{7.4}$$

which is clearly equated to the expression in equation (7.1) by ensuring that

$$a_i = 0, \qquad (n + 1) \leqslant i \leqslant m \tag{7.5}$$

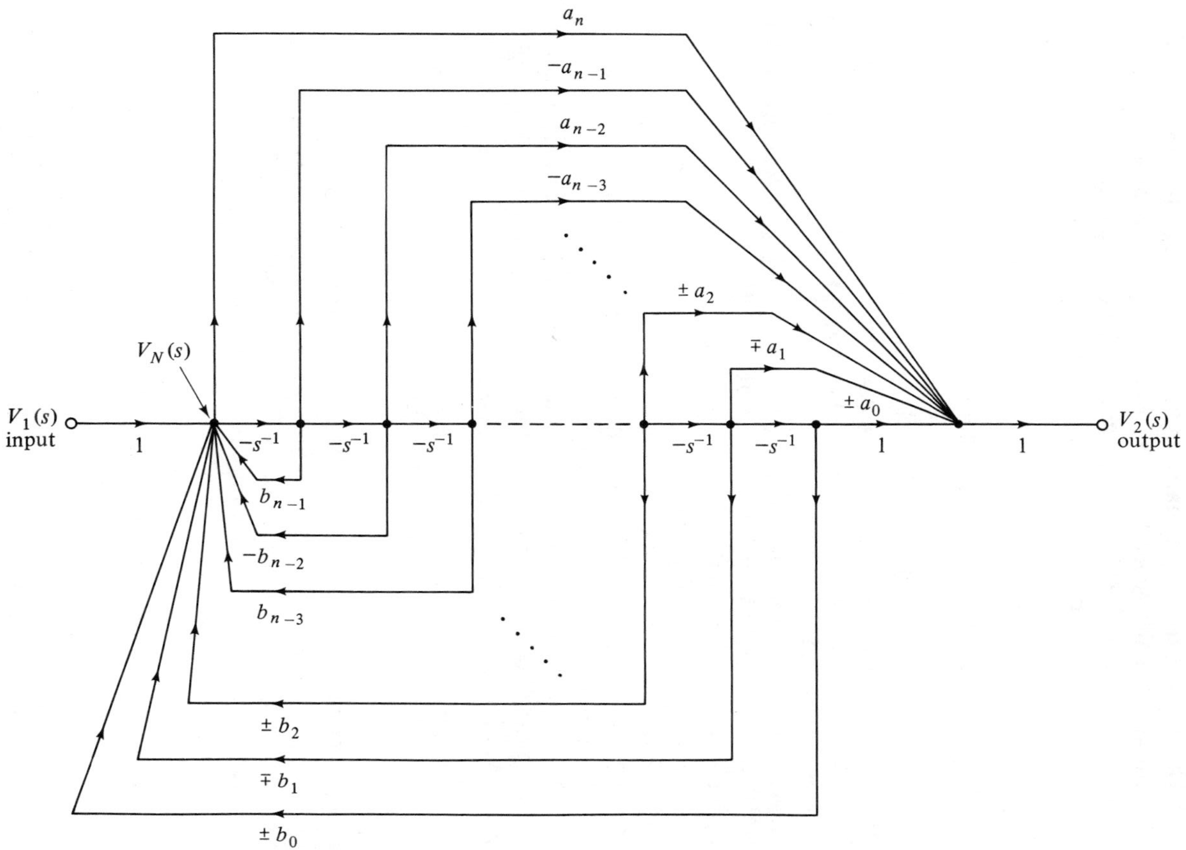

FIGURE 7.1 *Direct-form signal-flow graph using inverting-integrators.*

There are various alternative forms of DF signal-flow graph implementations of the general transfer function $H(s)$. For example, noninverting integrators $+s^{-1}$ may be employed or differentiators $\pm s$. The reader is referred to Problem 7-1 for further study of these alternative DF structures.

7-2.1 RC-Active Direct-Form Circuits

It follows by inspection of Fig. 7.1 that the *RC*-active implementation of this particular DF signal-flow graph requires circuitry to perform inverting-integration and summation. The single OP AMP circuits shown in Figs. 5.5(b) and 5.7 are used extensively for this purpose. A fifth-order implementation of the DF structure is shown in Fig. 7.2(a), where OP AMPs 1 and 2 perform inverting-summation; OP AMPs 3, 4, 5, 6, and 7 perform inverting-integration; and OP AMPs 8, 9, 10, 11, 12, and 13 provide inverting-amplification in *alternate* feedforward and feedback paths. These inverting-amplifiers provide polarity reversal of the corresponding feedforward and feedback paths to correspond to a realization of $H(s)$ for which all coefficients a_i and b_i are positive.

It is readily established that *all stable* transfer functions $H(s)$ require inverting-amplifiers in *alternate* feedback paths because stability implies that the denominator $D(s)$ of $H(s)$, where

$$D(s) = \sum_{i=0}^{n} b_i s^i \tag{7.6}$$

is a polynomial with n left-half s-plane roots; such a polynomial $D(s)$ is known as a *strictly Hurwitz polynomial* and it has the property that all the coefficients b_i are of *the same sign*. Thus, by inspection of Fig. 7.1, it is necessary to provide inverting-amplifiers in alternate feedback paths b_i. In Fig. 7.2(a), for example, OP AMPs 11, 12, and 13 ensure that the coefficients b_4, b_2, and b_0 have the same sign as coefficients b_5, b_3, and b_1.

It is not possible to make general statements with regard to the requirement for inverting-amplifiers in the *feedforward* paths. In Fig. 7.2(a), inverting-amplifiers are used in alternate feedforward paths to correspond to the case for which all coefficients a_i of $H(s)$ have the same sign; in practice, some of the a_i may be negative; it then becomes necessary to add or remove inverting-amplifiers from the corresponding feedforward paths. For example, if it is required that Fig. 7.2(a) be modified so that a_3 and a_2 are both negative, then the inverting-amplifier must be removed from the a_3 feedforward path and an inverting-amplifier must be introduced into the a_2 feedforward path.

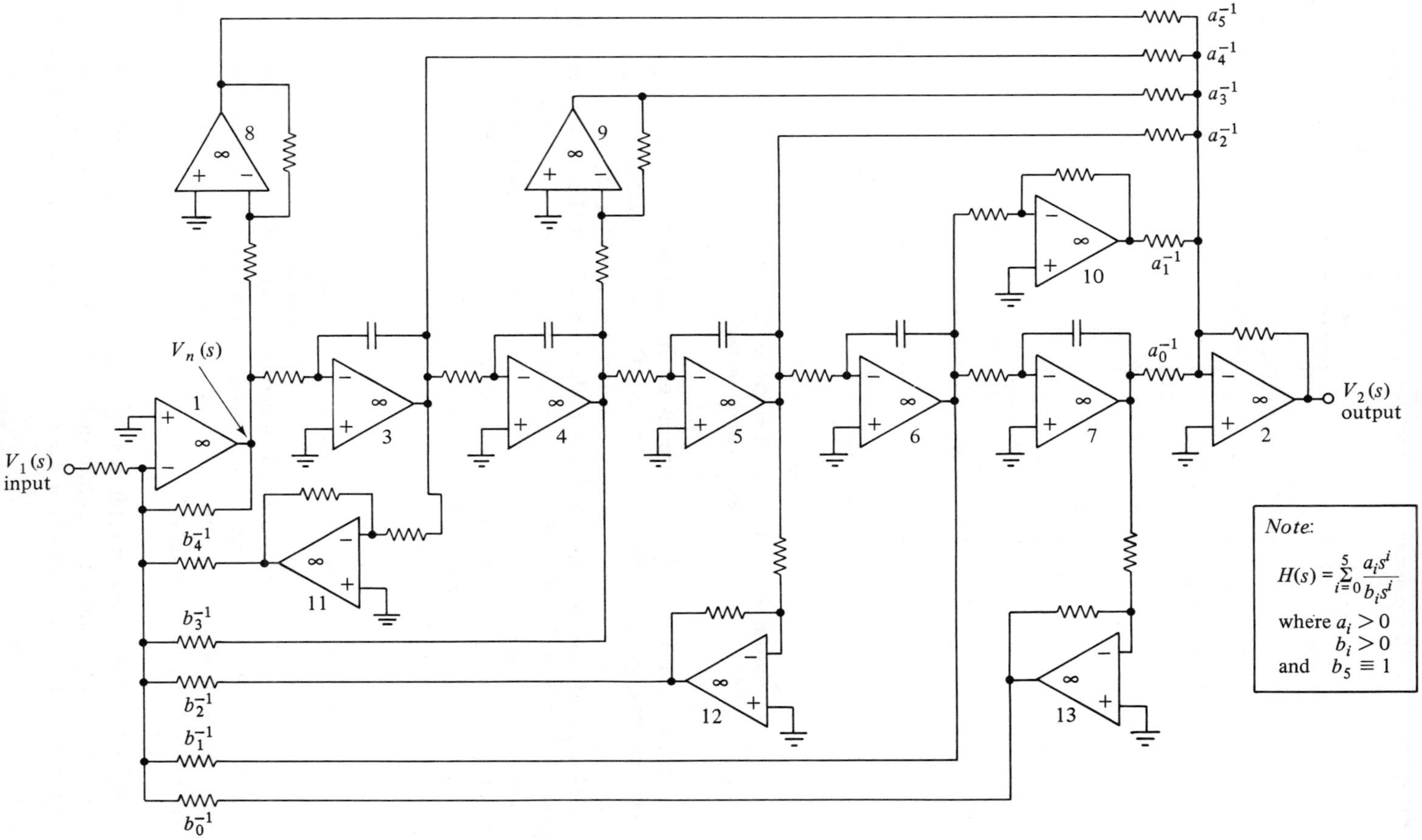

FIGURE 7.2 *(a) Direct form n = 5 inverting-integrator circuit C = R = 1 unless shown otherwise;*

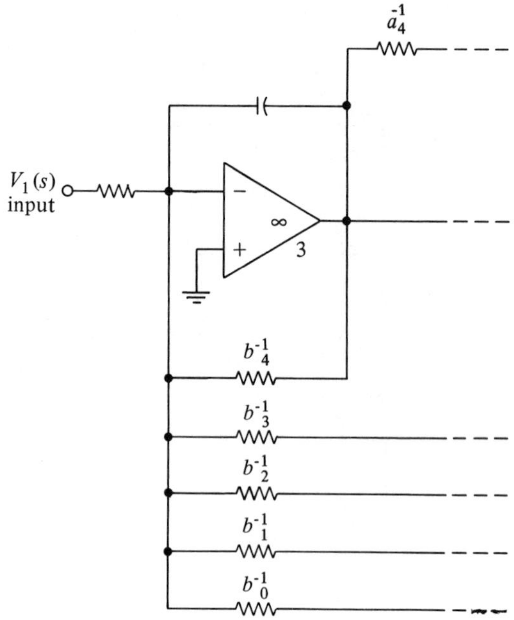

FIGURE 7.2 *(b)Elimination of OP AMPs 1 and 11 for $m < n$; $a_5 = 0$. Resistor values shown.*

In historical terms, the DF circuits of the type shown in Fig. 7.2(a) were the first types of *RC*-active OP AMP networks used for the implementation of voltage transfer functions; the inverting-integrator, -amplifier and -summer are the basic building-blocks of the *analog computer*. For this reason the DF structure is employed widely in simulation studies that use the analog computer. The number of OP AMPs required depends on the degree of the numerator m and denominator n of $H(s)$ and on the number of numerator coefficients a_i of unlike polarity. It is always possible to realize $H(s)$ with n_A OP AMPs where

$$n_A \leqslant 2n + 4 \tag{7.7}$$

and where n is the degree of denominator of $H(s)$. With modern microelectronic technology, it is often economically feasible to use as many as $(2n + 4)$ OP AMPs for the implementation of an nth order transfer function $H(s)$. The DF structure is seldom used to implement *high-order* functions because the structure possesses inherent sensitivity limitations. These limitations are considered in section 7-2.2.

Simplified DF *RC*-active circuit for $m < n$: Note that the OP AMP implementation of the DF structure that is shown in Fig. 7.2(a) for $n = 5$ may be simplified significantly if the degree m of the numerator $N(s)$ of

$H(s)$ is less than the degree n of the denominator $D(s)$ of $H(s)$. In this case, the node $V_n(s)$ of Fig. 7.1 [and voltage $V_n(s)$ in Fig. 7.2(a)] are not explicitly required so that OP AMPs 1 and 11 may be eliminated and the input circuitry of Fig. 7.2(a) modified, as shown in Fig. 7.2(b). Thus, for $m < n$, the total number of OP AMPs required for an inverting-integrator DF implementation is not more than $(2n + 1)$.

7-2.2 Sensitivity Performance of the DF Structure

The analysis in this section involves the calculation of the sensitivity of the transfer function $H(s)$ to the transmittances a_i and b_i of the DF structure. It follows directly from equation (7.1) and the definition of the sensitivity function $S_{x_i}^{H(s)}$ that

$$S_{a_i}^{H(s)} = \frac{a_i s^i}{N(s)} \tag{7.8}$$

$$S_{b_i}^{H(s)} = \frac{-b_i s^i}{D(s)} \tag{7.9}$$

where $H(s) = N(s)/D(s)$. In many applications, the magnitudes of the sensitivity functions in equations (7.8) and (7.9) are *much* greater than unity and intolerably large-valued. For simplicity, we pursue here the case for which $H(s)$ does not possess any finite zeros (that is, $N(s)$ is *unity*). Then, we may neglect equation (7.8) and concentrate our attention on equation (7.9), from which we may write

$$S_{b_i}^{H(s)} = -\frac{b_i s^i}{D(s)} = -b_i s^i H(s), \qquad N(s) \equiv 1 \tag{7.10}$$

Taking the real part, according to equation (6.57), gives the gain sensitivity to the b_i coefficient as

$$S_{b_i}^{G(\omega)} = \mathrm{Re}\left[S_{b_i}^{H(j\omega)}\right] = -b_i \,\mathrm{Re}\left[(j\omega)^i H(j\omega)\right] \tag{7.11}$$

This is the required expression for the gain sensitivity. Note the dependence on the magnitude of the coefficient b_i. It is this dependence on the coefficient b_i that essentially accounts for the poor sensitivity performance of the DF structure. A glance at Table 7-1 reveals that the b_i coefficients increase rapidly with order n for the Butterworth and Chebychev functions, implying a corresponding increase with n of the sensitivities $S_{b_i}^{G(\omega)}$.

The term Re $[(j\omega)^i H(j\omega)]$ in equation (7.11) has magnitude less than unity throughout the passband (where $0 \leqslant \omega \leqslant 1$), although it may be shown to exhibit a maxima of approximately unity near the bandedge ($\omega \simeq 1$) for "brick-wall"-type gain functions $G(\omega)$. The reader should check this statement for one of the functions specified in Table 7-1.

EXAMPLE 7.1: Coefficient Sensitivity Calculation: Chebychev n = 7. The result in equation (7.11) is now demonstrated for the $n = 7$ Chebychev lowpass 1.0 db passband ripple transfer function in Table 7.1 that is given by

$$H(s) = \frac{1}{\begin{bmatrix} [32.566]s^7 + [30.063]s^6 + [70.867]s^5 + [46.531]s^4 + [44.210]s^3 \\ + [17.867]s^2 + [6.958]s + 1 \end{bmatrix}} \tag{7.12}$$

where the corresponding magnitude response $G(\omega) \equiv |H(j\omega)|$ is shown in Fig. 7.3.

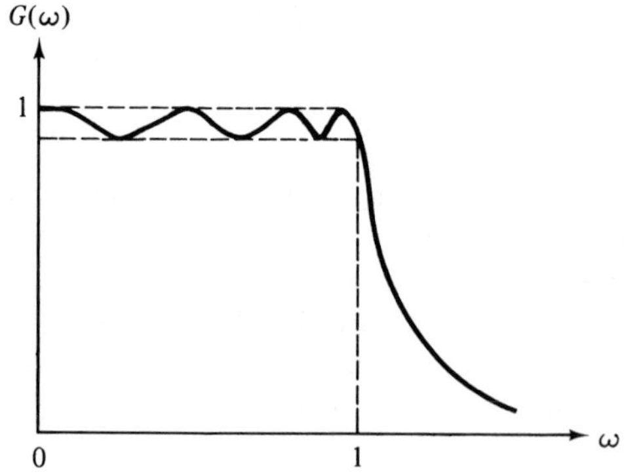

FIGURE 7.3 *Chebychev n = 7 magnitude response G(ω); passband ripple = 1.0 db.*

SOLUTION: Consider now the calculation of the sensitivities of $H(s)$ and $G(\omega)$ to one of the coefficient transmittances b_i. For example, we choose the transmittance b_5 so that, from equations (7.9) and (7.12),

$$S_{b_5}^{H(s)} = \frac{[-70.867]s^5}{\begin{bmatrix} [32.566]s^7 + [30.063]s^6 + [70.867]s^5 + [46.531]s^4 + [44.210]s^3 \\ + [17.867]s^2 + [6.958]s + 1 \end{bmatrix}} \tag{7.13}$$

The pole-zero diagrams for $H(s)$ and $S_{b_5}^{H(s)}$ are shown in Fig. 7.4. The corresponding pole-zero diagrams for the coefficient transmittances $b_{1,2,3,4,6,7}$ only differ from Fig. 7.4 by the *number of zeros at the origin*. The

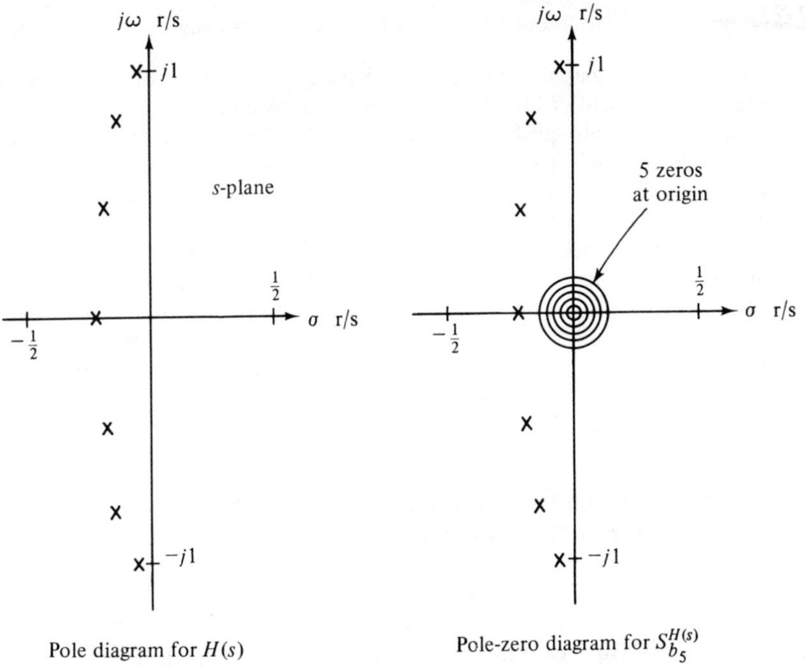

FIGURE 7.4 *Pole-zero diagrams for Chevychev lowpass function $H(s)$ and $S_{b_5}^{H(s)}$; $n = 7$; 1.0 db passband ripple.*

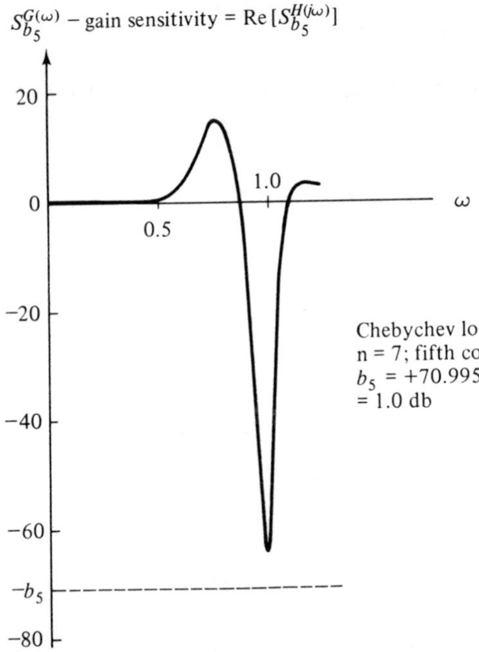

FIGURE 7.5 *A coefficient sensitivity for the direct-form structure, $S_{b_5}^{G(\omega)}$.*

sensitivity of the magnitude response (or gain) $G(\omega)$ to b_5 is obtained from equations (6.57) and (7.13); that is, $s = j\omega$ is substituted into equation (7.13) and then $S_{b_5}^{G(\omega)}$ is obtained as the real part of $S_{b_5}^{H(j\omega)}$. The result is shown in Fig. 7.5, from which it is observed that the five zeros at the s-plane origin clearly contribute to the excellent low-valued gain sensitivity $S_{b_5}^{G(\omega)}$ in the region $\omega < \frac{1}{2}$. However, in the region $\frac{1}{2} \leqslant \omega \leqslant \frac{6}{5}$ it is observed from Fig. 7.5 that the gain sensitivity $S_{b_5}^{G(\omega)}$ exhibits a peak-to-peak excursion from approximately $+15$ to -65. Consequently, a small fractional positive error ϵ_5 in b_5 will cause $G(\omega)$ to be in error by approximately $15\epsilon_5\%$ at $\omega \simeq 0.75$ and approximately $-65\epsilon_5\%$ at $\omega \simeq 1.0$. Clearly, the practical implication of this result is that the coefficient b_5 (and therefore conductance b_5 of Fig. 7.2) must be accurate to better than $\frac{1}{10}\%$ if the response $G(\omega)$ is to in any way resemble that of Fig. 7.3.

7-2.3 Bandedge DF Transmittance Sensitivities for the Butterworth Transfer Function

In this section, it is shown that the sensitivity of $G(\omega)$ to the transmittances a_i, b_i at the cut-off (or bandedge) frequency $\omega = 1$ may be easily derived for the Butterworth transfer function $H(s)$. It follows from the definition of the Butterworth function $H(s)$ that

$$|H(s)^2| = H(s)H(-s) \equiv \frac{1}{1 + (-1)^n s^{2n}} \qquad \text{BUTTERWORTH CASE} \tag{7.14}$$

It may be shown from equation (7.14) that the bandedge value $H(j1)$ of the nth-order Butterworth function $H(s)$ is

$$H(j1) = \frac{1}{\sqrt{2}} e^{-j\pi n/4} \tag{7.15}$$

so that the magnitude frequency response at the bandedge is

$$G(1) \equiv |H(j1)| = \frac{1}{\sqrt{2}} \tag{7.16}$$

We wish to determine the sensitivity of $G(1)$ to the transmittances b_i of the DF structure; thus, substituting equation (7.15) into equation (7.11) and setting ω to unity, we have

$$S_{b_i}^{G(1)} = -\frac{b_i}{\sqrt{2}} \operatorname{Re}\left[e^{j\frac{\pi}{4}(2i-n)} \right]$$

or

$$S_{b_i}^{G(1)} = -\frac{b_i}{\sqrt{2}}\cos\left[\frac{\pi}{4}(2i-n)\right] \qquad \text{BANDEDGE SENSITIVITY AT CUT-OFF; BUTTERWORTH CASE, DF STRUCTURE, ORDER } n \tag{7.17}$$

which may be expressed in the form

$$\left.\begin{aligned}
S_{b_i}^{G(1)} &= -\frac{b_i}{\sqrt{2}} && \text{for} \quad (2i-n) = 8k \\
S_{b_i}^{G(1)} &= -\frac{b_i}{2} && \text{for} \quad (2i-n) = 8k \pm 1 \\
S_{b_i}^{G(1)} &= 0 && \text{for} \quad (2i-n) = 8k \pm 2 \\
S_{b_i}^{G(1)} &= +\frac{b_i}{2} && \text{for} \quad (2i-n) = 8k \pm 3 \\
S_{b_i}^{G(1)} &= +\frac{b_i}{\sqrt{2}} && \text{for} \quad (2i-n) = 8k \pm 4
\end{aligned}\right\} \; k \text{ integer} \tag{7.18}$$

It follows directly from equation (7.17) that the worstcase sensitivity of $G(1)$ to the b_i coefficients is

$$WS_{b_i}^{G(1)} = \frac{1}{\sqrt{2}} \sum_{i=0}^{n} \left| b_i \cos\left[\frac{\pi}{4}(2i-n)\right]\right| \tag{7.19}$$

Thus, if the tolerance specification on b_i is such that

$$(1-\epsilon_{\max})b_{iN} \leqslant b_i \leqslant (1+\epsilon_{\max})b_{iN}$$

where b_{iN} is the nominal (exact) value of b_i, then by the result obtained in equation (6.94) we may write the *worstcase* fractional deviation of $G(1)$ as

$$\left.\frac{\Delta G(1)}{G(1)}\right|_{\text{Worstcase}} = \frac{\epsilon_{\max}}{\sqrt{2}} \sum_{i=0}^{n} \left| b_i \cos\left[\frac{\pi}{4}(2i-n)\right]\right| \tag{7.20}$$

EXAMPLE 7.2: Determine the worstcase fractional deviation of $G(1)$ for the 7th-order DF Butterworth lowpass implementation due to coefficient tolerances given by $\epsilon_{\max} = 0.001$ (that is, the b_i have 0.1% tolerance specifications).

SOLUTION: Substituting the b_i coefficients from Table 7-1 for $n = 7$ into equation (7.20) and using $\epsilon_{max} = 0.001$, we obtain

$$\left.\frac{\Delta G(1)}{G(1)}\right|_{\text{Worstcase}} = 0.0302 \qquad (7.21)$$

implying that $\pm 0.1\%$ tolerances for b_i coefficients permit a worstcase deviation in $G(1)$ of 3.02%; this is $WS_{b_i}^{G(1)} = 30.2$.

This example illustrates the requirement for high accuracy in the transmittance coefficients b_i of the DF structure and implies that the conductances b_i of the corresponding OP AMP implementation of Fig. 7.2 must exhibit comparable accuracies. It is primarily because of this high-sensitivity characteristic of DF circuits that "analog-computer type" DF RC-active networks are not recommended for the implementation of high-order ($n \gg 1$) transfer functions.

Remarks. The DF RC-active circuits have the advantage that they are easily designed, since it is only necessary to determine the circuit conductances a_i, b_i by direct inspection of the rational function $H(s)$. Unfortunately, the high-valued sensitivities $|S_{b_i}^{H(s)}|$ and $|S_{b_i}^{G(\omega)}|$ inhibit the usefulness of DF circuits, because these sensitivities are generally proportional to the order n of the transfer function.

In the remainder of this chapter, we consider other structures that have sensitivity performance superior to that of the DF structure and have, therefore, been used in the design of practical *high-quality* RC-active circuits.

7-3 CASCADED BIQUADRATIC DIRECT-FORM STRUCTURES

The transmittance-sensitivity problem that is associated with the high-order DF structure may be alleviated by employing a cascade connection of biquadratic transfer functions $H_b(s)$ to realize the higher-order transfer function $H(s)$, as indicated in Fig. 7.6. In general,

$$H_b(s) \equiv \frac{a_{2b}s^2 + a_{1b}s + a_{0b}}{b_{2b}s^2 + b_{1b}s + b_{0b}} \qquad (7.22)$$

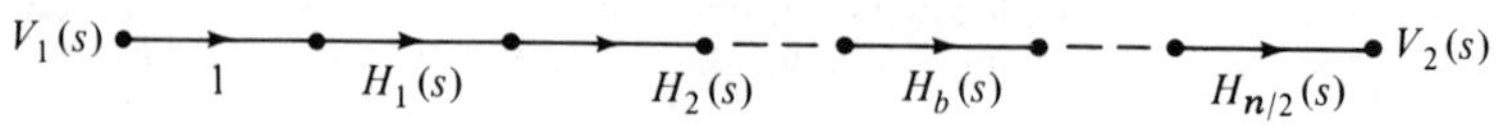

FIGURE 7.6 *Cascaded biquadratic structure.*

and the overall higher-order transfer function of *maximum* order n, obtained by the cascade connection of biquadratic sections with transfer functions $H_1(s), H_2(s), \ldots, H_b(s), \ldots, H_{n/2}(s)$, is

$$H(s) = \prod_{b=1}^{n/2} H_b(s) = \prod_{b=1}^{n/2} \frac{a_{2b}s^2 + a_{1b}s + a_{0b}}{b_{2b}s^2 + b_{1b}s + b_{0b}} \tag{7.23}$$

7-3.1 The Fully Biquadratic Direct-Form Structure BIQUAD DF

The structure shown in Fig. 7.7 is the $n = 2$ version of the DF structure in Fig. 7.1. It is described here as *fully* biquadratic because *all* of the coefficients of both the numerator $(a_{2b}s^2 + a_{1b}s + a_{0b})$ and denominator $(b_{2b}s^2 + b_{1b}s + b_{0b})$ are implemented so that a one-to-one correspondence exists between coefficients a_{ib}, b_{ib} and transmittances of the signal-flow graph.

An inverting-integrator OP AMP implementation of the BIQUAD DF structure is shown in Fig. 7.8 for the case where all six coefficients a_{ib}, b_{ib} $(i = 0, 1, 2)$ are positive. Note that six OP AMPs are required.

It is possible to reduce the number of OP AMPs required for the implementation of a biquadratic function $H_b(s)$ by suitably decomposing the numerator $N_b(s)$ and/or denominator $D_b(s)$. In the next section, we consider structures that have DF denominators but do *not* have DF numerators. These structures are employed to implement *RC*-active biquadratic sections requiring four (or less) OP AMPs.

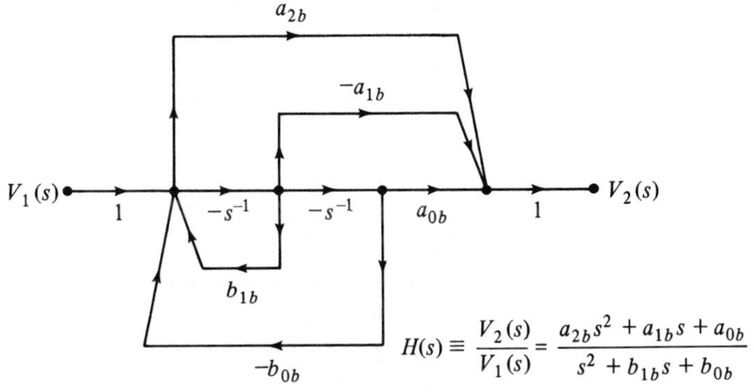

FIGURE 7.7 *Fully biquadratic direct-form structure-"BIQUAD DF."*

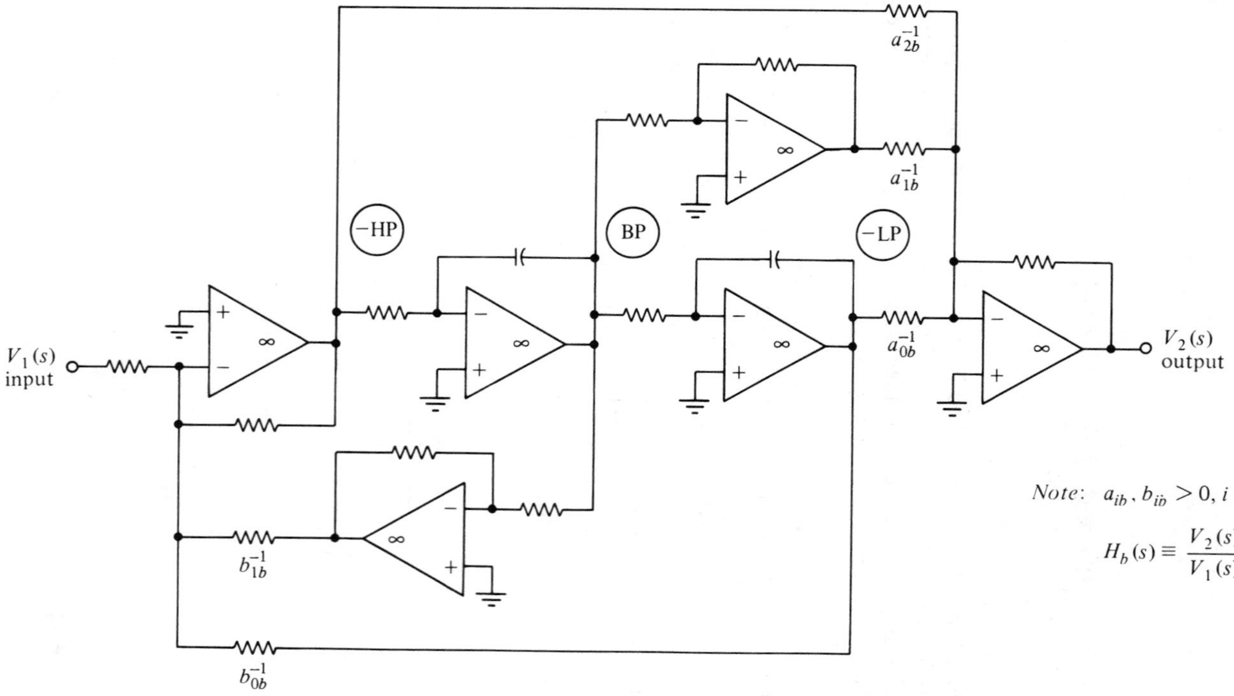

Note: $a_{ib}, b_{ib} > 0, i = 0, 1, 2$

$$H_b(s) \equiv \frac{V_2(s)}{V_1(s)} = \frac{a_{2b}s^2 + a_{1b}s + a_{0b}}{s^2 + b_{1b}s + b_{0b}}$$

FIGURE 7.8 *BIQUAD DF inverting-integrator implementation.*

7-3.2 Biquadratic Structures with Direct-Form Denominators and Decomposed Numerators

The number of OP AMPs required to realize a biquadratic section may be reduced by forming the numerator somewhat differently than for the BIQUAD DF structure previously discussed. The detailed *RC*-active *OP AMP circuit implementations* of these BIQUAD structures are considered in chapter 8; only the signal-flow graphs and their sensitivity properties are explained here.

We define the lowpass quadratic function LP, bandpass quadratic function BP and highpass quadratic function HP in terms of the equation

$$T(s) = \frac{H_0 s^x}{b_2 s^2 + b_1 s + b_0}, \qquad H_0 \text{ constant} \tag{7.24}$$

so that $T(s)$ is defined as an LP function for $x = 0$, a BP function for $x = 1$, and an HP function for $x = 2$.

The LBI/DF BIQUAD structure: The mnemonic LBI/DF implies that the corresponding biquadratic transfer function $H_b(s)$ possesses a numerator polynomial that is formed by the summation of the lowpass function *L*P, bandpass function *B*P, and *i*nput function $V_1(s)$, and that the denominator is a DF implementation. The LBI/DF BIQUAD signal-flow graph is shown in Fig. 7.9(a), where the LP, $-$LP, and BP outputs are indicated. The feedforward transmittances $\alpha_2, \alpha_1, \alpha_0$ are used to sum the input, $-$BP, and $-$LP output waveforms to realize the output $V_2(s)$ where

$$V_2(s) = \underbrace{\alpha_2 V_1(s)}_{\text{input term}} + \underbrace{\alpha_1 \left[\frac{-sV_1(s)}{s^2 + b_{1b}s + b_{0b}}\right]}_{-\text{BP term}} + \underbrace{\alpha_0 \left[\frac{-b_{0b}V_1(s)}{s^2 + b_{1b}s + b_{0b}}\right]}_{-\text{LP term}} \tag{7.25}$$

giving

$$H_b(s) \equiv \frac{V_2(s)}{V_1(s)} = \frac{\alpha_2 s^2 + [\alpha_2 b_{1b} - \alpha_1]s + [\alpha_2 - \alpha_0]b_{0b}}{s^2 + b_{1b}s + b_{0b}} \tag{7.26}$$

Clearly, the denominator coefficients imply a DF denominator because b_{1b} and b_{0b} appear explicitly in the signal-flow graph as transmittances. The numerator of equation (7.26) is related to the numerator polynomial coefficients a_{2b}, a_{1b} and a_{0b} of equation (7.22) via the decompositions

$$\boxed{\begin{aligned} a_{2b} &= \alpha_2 \\ a_{1b} &= \alpha_2 b_{1b} - \alpha_1 \\ a_{0b} &= [\alpha_2 - \alpha_0]b_{0b} \end{aligned}} \qquad \text{NUMERATOR DECOMPOSITION FOR LBI/DF BIQUAD} \tag{7.27}$$

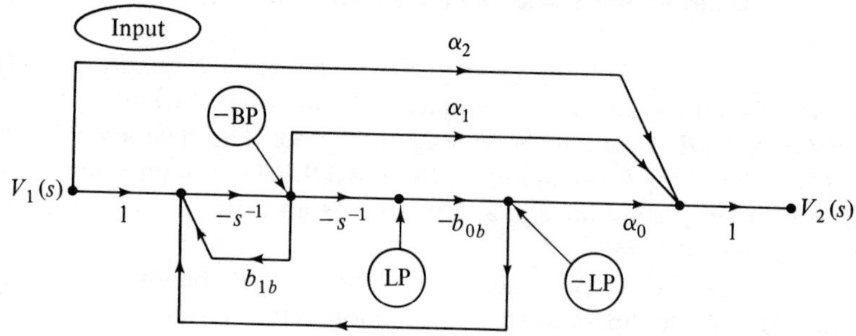

(a) Biquad LBI/DF

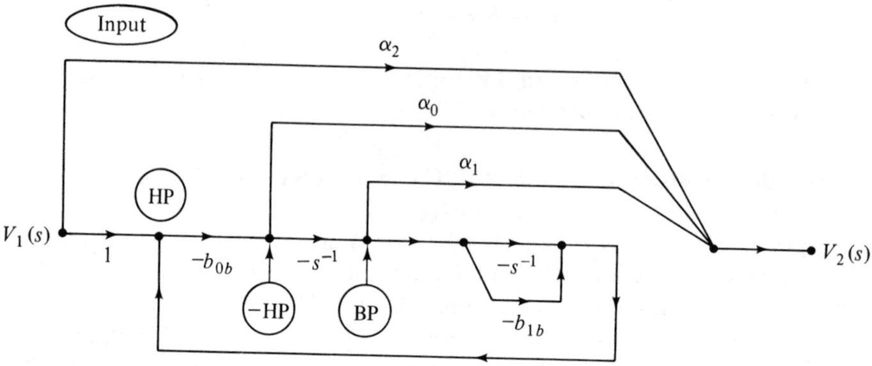

(b) Biquad BHI/DF

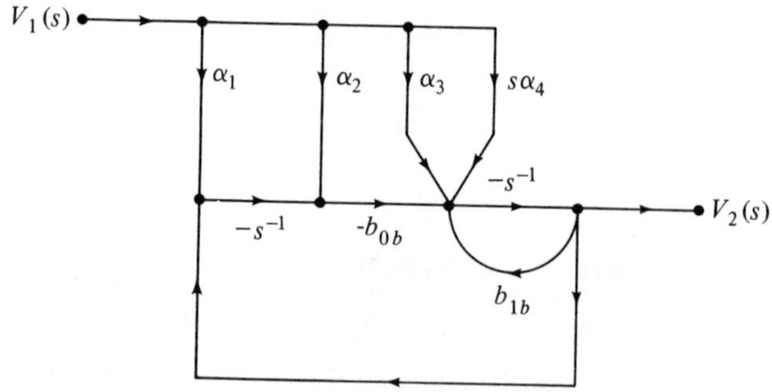

(c) Biquad MI/DF

FIGURE 7.9

Note that for any particular set of b_0, b_1 it is always possible to choose α_1, α_2, and α_3 to realize any particular set of a_{2b}, a_{1b}, a_{0b}. By appropriate choices of b_0, b_1, α_0, α_2 it is possible to realize any stable biquadratic transfer function $H_b(s)$. Finally, and most important, it will be found in chapter 8 that this structure may be implemented with only four OP AMPs.

The BHI/DF BIQUAD structure: As implied by the mnemonic BHI/DF, this structure forms the numerator by summation of the input with the BP and HP functions. The BHI/DF BIQUAD structure is diagrammed in Fig. 7.9(b), from which it may be shown that the functions HP, −HP, and BP are available directly for feedforward summation with the input waveform $V_1(s)$. Direct analysis of Fig. 7.9(b) proceeds in a similar way to that performed in equation (7.25) and leads to

$$H_b(s) \equiv \frac{V_2(s)}{V_1(s)} = \frac{[\alpha_2 - \alpha_0 b_{0b}]s^2 + [\alpha_2 b_{1b} b_{0b} + \alpha_1 b_{0b}]s + \alpha_2 b_{0b}}{s^2 + b_{1b} b_{0b} s + b_{0b}} \tag{7.28}$$

corresponding to the numerator decomposition

$$\boxed{\begin{aligned} a_{2b} &= \alpha_2 - \alpha_0 b_{0b} \\ a_{1b} &= \alpha_2 b_{1b} b_{0b} + \alpha_1 b_{0b} \\ a_{0b} &= \alpha_2 b_{0b} \end{aligned}} \qquad \text{NUMERATOR DECOMPOSITION FOR BHI/DF BIQUAD} \tag{7.29}$$

This structure is useful because it may be realized with four OP AMPs by the methods to be described in chapter 8; it differs from the LBI/DF BIQUAD in that the functions ±HP are available and *not* ±LP.

The MI/DF BIQUAD structure: The mnemonic MI implies that the numerator polynomial of $H_b(s)$ is realized by employing *m*ultiple-*i*nput paths instead of a single output summation node. The MI/DF structure is shown in Fig. 7.9(c), from which it is observed that the input $V_1(s)$ is summed into three nodes of the signal-flow graph via the transmittances α_1, α_2, and $(\alpha_3 + s\alpha_4)$. Direct analysis of this structure is requested in Problem 7-10 and leads to

$$H_b(s) \equiv \frac{V_2(s)}{V_1(s)} = -\left[\frac{\alpha_4 s^2 + (\alpha_3 - \alpha_2 b_0)s + \alpha_1 b_{0b}}{s^2 + b_{1b} s + b_{0b}}\right] \tag{7.30}$$

corresponding to the numerator decomposition

$$\boxed{\begin{aligned} a_{2b} &= -\alpha_4 \\ a_{1b} &= \alpha_2 b_{0b} - \alpha_3 \\ a_{0b} &= -\alpha_1 b_{0b} \end{aligned}} \qquad \text{NUMERATOR DECOMPOSITION FOR MI/DF BIQUAD} \tag{7.31}$$

The important feature of the MI/DF BIQUAD structure becomes apparent on consideration of the corresponding OP AMP implementation in chapter 8, where it is found that only *three* OP AMPs are required to implement this BIQUAD.

7-3.3 Sensitivity Performance of Cascaded Biquadratic DF Structures

In this section, it is shown that *cascading* biquadratic DF sections $H_b(s)$ to realize a higher-order transfer function $H(s)$ generally results in a structure with a sensitivity performance superior to that of the corresponding DF structure of Fig. 7.1. The analysis is for a cascade connection of DF BIQUADs of the types shown in Fig. 7.7 and Fig. 7.8.

It follows directly from equations (7.22), (7.23), (7.8), and (7.9) that

$$S_{a_{kb}}^{H(s)} = S_{a_{kb}}^{H_b(s)} = +\left(\frac{a_{kb}s^k}{a_{2b}s^2 + a_{1b}s + a_{0b}}\right), \qquad k = 0, 1, 2 \tag{7.32}$$

and, similarly,

$$S_{b_{kb}}^{H(s)} = S_{b_{kb}}^{H_b(s)} = -\left(\frac{b_{kb}s^k}{b_{2b}s^2 + b_{1b}s + b_{0b}}\right), \qquad k = 0, 1, 2 \tag{7.33}$$

Equations (7.32) and (7.33) are exact expressions for the sensitivities of the overall transfer function $H(s)$ to any of the transmittances a_{kb}, b_{kb} of the bth DF BIQUAD of Fig. 7.6, where $k = 0, 1, 2$.

The results in equations (7.32) and (7.33) may be simplified by writing $H_b(s)$ in terms of Q_P, Q_Z, ω_P, and ω_Z so that, by comparing equations (7.22) and (6.46), it follows that

$$H_b(s) = \frac{a_{2b}}{b_{2b}} \cdot \frac{s^2 + \dfrac{\omega_Z s}{Q_Z} + \omega_Z^2}{s^2 + \dfrac{\omega_P s}{Q_P} + \omega_P^2} \tag{7.34}$$

where

$$\left.\begin{aligned} \frac{\omega_Z}{Q_Z} &= \frac{a_{1b}}{a_{2b}}, & \frac{\omega_P}{Q_P} &= \frac{b_{1b}}{b_{2b}} \\ \omega_Z^2 &= \frac{a_{0b}}{a_{2b}}, & \omega_P^2 &= \frac{b_{0b}}{b_{2b}} \end{aligned}\right\} \tag{7.35}$$

Then, substituting equations (7.35) into equations (7.32) and (7.33) we obtain the following sensitivities:

$$S_{a_{0b}}^{H_b(s)} = +\left[\frac{\omega_Z^2}{s^2 + \dfrac{\omega_Z s}{Q_Z} + \omega_Z^2}\right], \qquad S_{b_{0b}}^{H_b(s)} = -\left[\frac{\omega_P^2}{s^2 + \dfrac{\omega_P s}{Q_P} + \omega_P^2}\right]$$

$$S_{a_{1b}}^{H_b(s)} = +\frac{1}{Q_Z}\left[\frac{\omega_Z s}{s^2 + \dfrac{\omega_Z s}{Q_Z} + \omega_Z^2}\right], \qquad S_{b_{1b}}^{H_b(s)} = -\frac{1}{Q_P}\left[\frac{\omega_P s}{s^2 + \dfrac{\omega_P s}{Q_P} + \omega_P^2}\right]$$

$$S_{a_{2b}}^{H_b(s)} = +\left[\frac{s^2}{s^2 + \dfrac{\omega_Z s}{Q_Z} + \omega_Z^2}\right], \qquad S_{b_{2b}}^{H_b(s)} = -\left[\frac{s^2}{s^2 + \dfrac{\omega_P s}{Q_P} + \omega_P^2}\right]$$

(7.36)

These equations are important and will be used to derive the major sensitivity equations for a variety of biquadratic *RC*-active filters. The terms in brackets are called herein the *normalized quadratic functions*; they will be written as

$$t_{LP}(s) \equiv \frac{\omega_P^2}{D(s)}, \qquad t_{BP} \equiv \frac{\omega_P s}{D(s)}$$

$$t_{HP}(s) \equiv \frac{s^2}{D(s)}$$

NORMALIZED QUADRATIC FUNCTIONS (7.37)

where $D(s) \equiv s^2 + \dfrac{\omega_P s}{Q_P} + \omega_P^2$ in the case of equations containing ω_P and Q_P. The *normalized quadratic functions* $t_{LP}(s)$, $t_{BP}(s)$, and $t_{HP}(s)$ are most useful in the sensitivity analysis of biquadratic circuits.

Sensitivity of the gain function $G_b(\omega) \equiv |H_b(j\omega)|$: To determine the sensitivity of the gain function $G_b(\omega)$ to the coefficients a_{kb} and b_{kb} it is necessary, according to equation (6.57), to set $s = j\omega$ in equation (7.36) and take the real part. For example, in the case of the sensitivities to b_{0b}, b_{1b}, b_{2b} in equation (7.36) we have

$$S_{b_{0b}}^{G_b(\omega)} = -\mathrm{Re}\left[t_{\mathrm{LP}}(j\omega)\right], \quad S_{b_{1b}}^{G_b(\omega)} = -\frac{1}{Q_P}\mathrm{Re}\left[t_{\mathrm{BP}}(j\omega)\right]$$

and

$$S_{b_{2b}}^{G_b(\omega)} = -\mathrm{Re}\left[t_{\mathrm{HP}}(j\omega)\right] \tag{7.38}$$

The expressions Re $[t_{\mathrm{LP,\,BP,\,HP}}(j\omega)]$ in equation (7.38) occur often in the sensitivity analysis of biquadratic circuits. They are sketched in Fig. 7.10 for $\omega_P = 1$ rad/sec. and $Q_P = 10$. Note that Re $[t_{\mathrm{LP}}(j\omega)]$ and Re $[t_{\mathrm{HP}}(j\omega)]$ exhibit extrema of approximately $\pm\frac{1}{2}Q_P$ at the "half-power" frequencies $\omega_{1,2} \equiv \omega_P\left[1 \pm \frac{1}{2}Q_P^{-1}\right]$ and that Re $[t_{\mathrm{BP}}(j\omega)]$ exhibits a maxima of approximately Q_P at the frequency ω_P. It may be shown that these three functions always behave in this way for all $Q_P \gg 1$.

It is important to note that for $Q_P \gg 1$ the function $S_{b_{1b}}^{G_b(\omega)}$ in equation (7.38) will be bounded between 0 and -1, whereas the functions $S_{b_{0b}}^{G_b(\omega)}$ and $S_{b_{2b}}^{G_b(\omega)}$ have extremum values of $\pm\frac{1}{2}Q_P$. Consequently, $G_b(\omega)$ is *more sensitive to* b_{1b} (by a factor of up to about $\frac{1}{2}Q_P$) than it is sensitive to either b_{0b} or b_{2b}.

EXAMPLE 7.3: *Lowpass Chebychev Function H(s), n = 7, 1.0 db Passband Ripple.* The lowpass Chebychev function $H(s)$ in equation (7.12) may be realized as a cascade of (bi)quadratic functions $H_b(s)$ by factorizing the denominator, thus

$$H(s) = \underbrace{\frac{1}{[4.868146]s + 1}}_{H_1(s)} \times \underbrace{\frac{1}{[4.339354]s^2 + [1.606203]s + 1}}_{H_2(s)}$$
$$\times \underbrace{\frac{1}{[1.530343]s^2 + [0.391997]s + 1}}_{H_3(s)} \times \underbrace{\frac{1}{[1.007384]s^2 + [0.0920944]s + 1}}_{H_4(s)} \tag{7.39}$$

The corresponding coefficients b_{kb} are shown in Table 7-2. *Note that these coefficients are generally much smaller than the coefficients b_i in the denominator of the overall transfer function $H(s)$*, equation (7.12). For this reason, we might expect that the sensitivities $S_{b_{kb}}^{H(s)}$ will be considerably smaller in magnitude than is obtained for the DF implementation. This may easily be verified by substituting the coefficients from Table 7-2 into equation (7.33). For example, let us derive the sensitivity of the overall transfer function $H(s)$

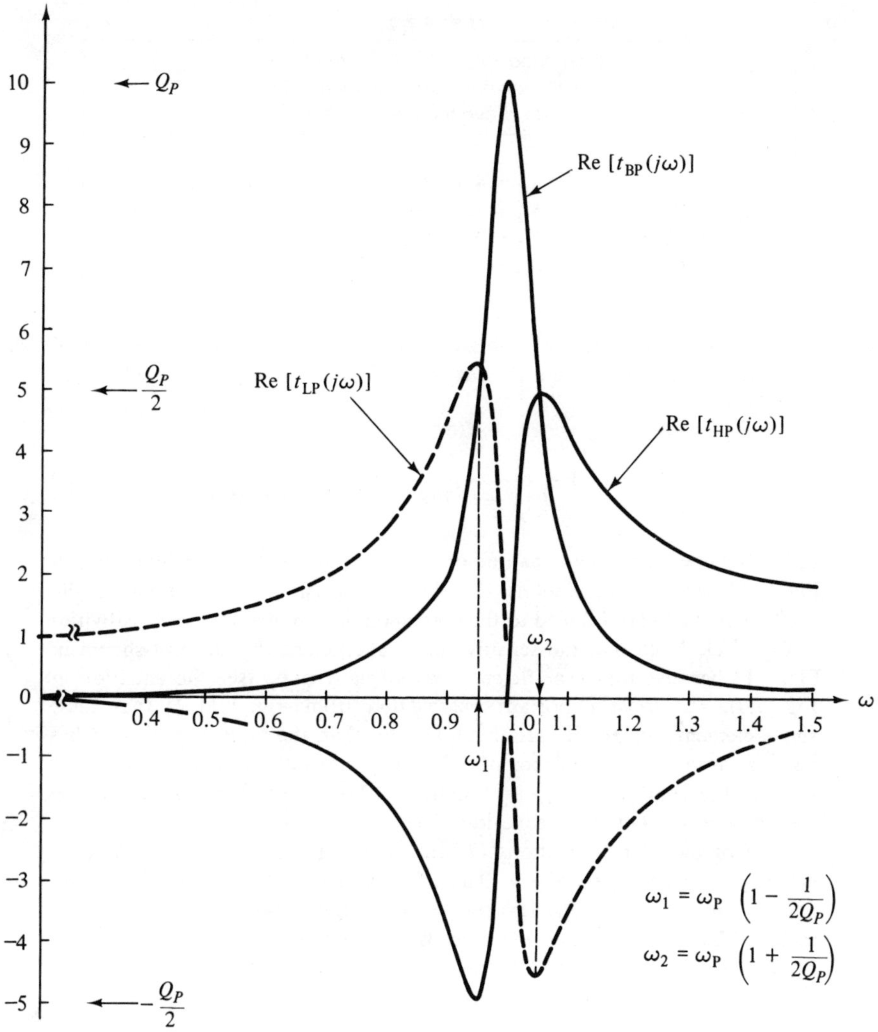

FIGURE 7.10 *Real parts of normalized quadratic functions for $Q_P = 10$, $\omega_P = 1$ rad/sec.*

to the coefficient $b_{14} = 0.0920944$ in the fourth section $H_4(s)$ of the cascade implementation implied by equation (7.39). Then, from equation (7.33), and with $k = 1$ and $b = 4$, we have

$$S_{b_{14}}^{H(s)} = -\frac{b_{14}s}{b_{24}s^2 + b_{14}s + b_{04}}$$

$$= -\frac{[0.0920944]s}{[1.007384]s^2 + [0.0920944]s + 1} \tag{7.40}$$

TABLE 7-2

Cascaded-BIQUAD DF coefficients for the lowpass Chebychev n = 7 1.0 db ripple transfer function.

			Stage
$b_{21} = 0,$	$b_{11} = 4.868146,$	$b_{01} = 1$	1
$b_{22} = 4.339354,$	$b_{12} = 1.606203,$	$b_{02} = 1$	2
$b_{23} = 1.530343,$	$b_{13} = 0.391997,$	$b_{03} = 1$	3
$b_{24} = 1.007384,$	$b_{14} = 0.0920944,$	$b_{04} = 1$	4

The gain sensitivity $S_{b_{14}}^{G(\omega)}$ is obtained from equation (7.40) by means of equation (6.57). Thus, taking the real part of $S_{b_{14}}^{H(j\omega)}$ in equation (7.40) leads directly to

$$S_{b_{14}}^{G(\omega)} = -\text{Re}\left[\frac{[0.092044]j\omega}{[1-\omega^2 1.007384] + j\omega[0.092044]}\right] \tag{7.41}$$

which is shown graphically as one of the single-element sensitivity curves in Fig. 7.11. Similar calculations for the gain sensitivities to seven of the coefficients in Table 7-2 lead to the corresponding single-element sensitivities in Fig. 7.11. Note that the sensitivities to coefficients b_{2b} are not shown in Fig. 7.11 because these coefficients are normally unity (see the equation in Fig. 7.7). The most important observation from Fig. 7.11 is that these single-element sensitivities are much smaller than the typical result that has been derived in Fig. 7.5 for the DF implementation. This is primarily because the coefficients b_{kb} in equation (7.33) are smaller than the corresponding coefficients b_i in equation (7.9).

As predicted by equations (7.38), the sensitivities $S_{b_{12}}^{G_b(\omega)}$, $S_{b_{13}}^{G_b(\omega)}$, $S_{b_{14}}^{G_b(\omega)}$ exhibit extremum values of -1. Thus, the tolerance requirements on b_{12}, b_{13}, and b_{14} are a factor of Q_{Pb} less severe, where Q_{Pb} is the Q factor of the bth section, than the tolerance requirements on the corresponding coefficients b_{02}, b_{03}, b_{04}. In agreement with equations (7.38), it is observed that the peak magnitudes of $S_{b_{03}}^{G_b(\omega)}$ and $S_{b_{04}}^{G_b(\omega)}$ approximate $\frac{1}{2}Q_{P3}$ and $\frac{1}{2}Q_{P4}$, respectively.

7-3.4 Worstcase-Sensitivity Comparison Between the Cascaded Biquadratic Stucture and Direct-Form Structure

The superiority of the cascaded-biquadratic structure compared with the DF structure may be inferred by comparing the worstcase gain sensitivities according to the definition of worstcase sensitivity that is given in equation (6.93), where $F \equiv G(\omega)$ and $x_i \equiv b_i$ for the DF case and $x_i \equiv b_{bk}$ for the cascaded-biquadratic case. Then, by direct definition of

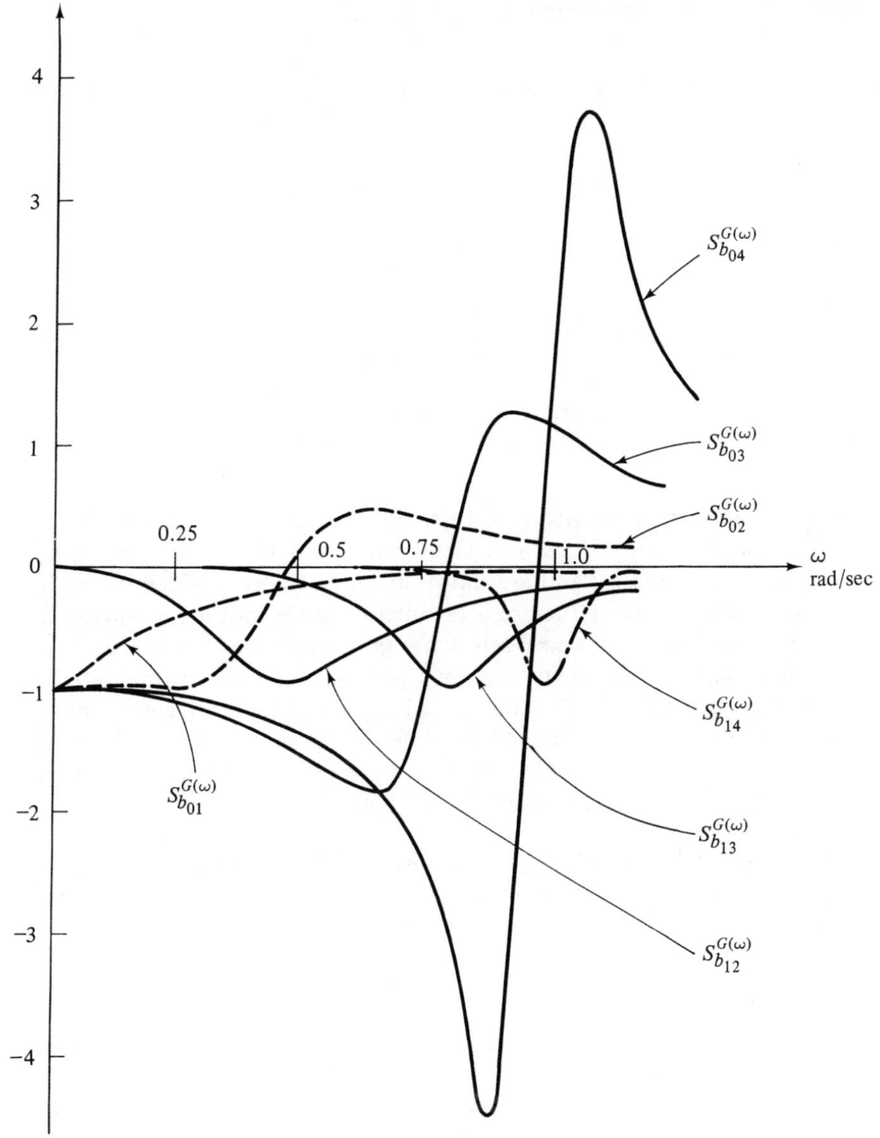

FIGURE 7.11 *Coefficient sensitivities. Example 7.3: Chebychev lowpass n = 7. Cascaded biquads.*

the worstcase sensitivity function,*

$$WS_{b_i}^{G(\omega)} = \sum_{i=0}^{n-1} \left| \mathrm{Re} \frac{-b_i s^i}{s^n + b_{n-1}s^{n-1} + \ldots + b_0} \right| \tag{7.42}$$

DIRECT FORM:
Equation (7.12), for example

and

$$WS_{b_k}^{G(\omega)} = \sum_{b=1}^{n/2} \sum_{k=0}^{1} \left| \mathrm{Re} \frac{-b_{bk}s^k}{s^2 + b_{b1}s + b_{b0}} \right| \tag{7.43}$$

CASCADED BIQUAD:
Equation (7.23), for example

These worstcase sensitivity expressions are evaluated from the b_i coefficients in Table 7-1 ($n = 7$, Chebychev) and the corresponding b_{bk} coefficients in Table 7-2 and are shown in Fig. 7.12, from which it may be concluded that the DF structure is far inferior to the cascaded-biquadratic structures insofar as worstcase-sensitivity performance is concerned.

Note that, for simplicity, we have pursued a numerical example for which the numerator coefficients a_i and a_{bk} are absent. Of course, equations (7.8) and (7.32) can be analyzed in a similar way to derive additional terms in equations (7.42) and (7.43) that represent the additional contributions of a_i and a_{bk} to the worstcase sensitivities.

BIQUAD sensitivities in terms of Q-factor and natural frequency: Recall that the biquadratic function $H_b(s)$ is often characterized [equation (6.46)] in terms of Q-factors $Q_{P,Z}$ and natural frequencies $\omega_{P,Z}$ in the form

$$H_b(s) = H_{0b} \left\{ \frac{s^2 + \dfrac{\omega_Z}{Q_Z}s + \omega_Z^2}{s^2 + \dfrac{\omega_P}{Q_P}s + \omega_P^2} \right\} \tag{7.44}$$

If we choose to describe $H_b(s)$ in this way, then it is a simple matter to

*Note that the coefficients b_n in the DF case and b_{b2} in the BIQUAD DF case are normalized to unity here to correspond to the inverting-integrator implementations of Figs. 7.1 and Fig. 7.7, respectively. Conductances corresponding to b_n and b_{b2} do not exist in the corresponding OP AMP circuits and, therefore, these coefficients are omitted from the worstcase-sensitivity calculations.

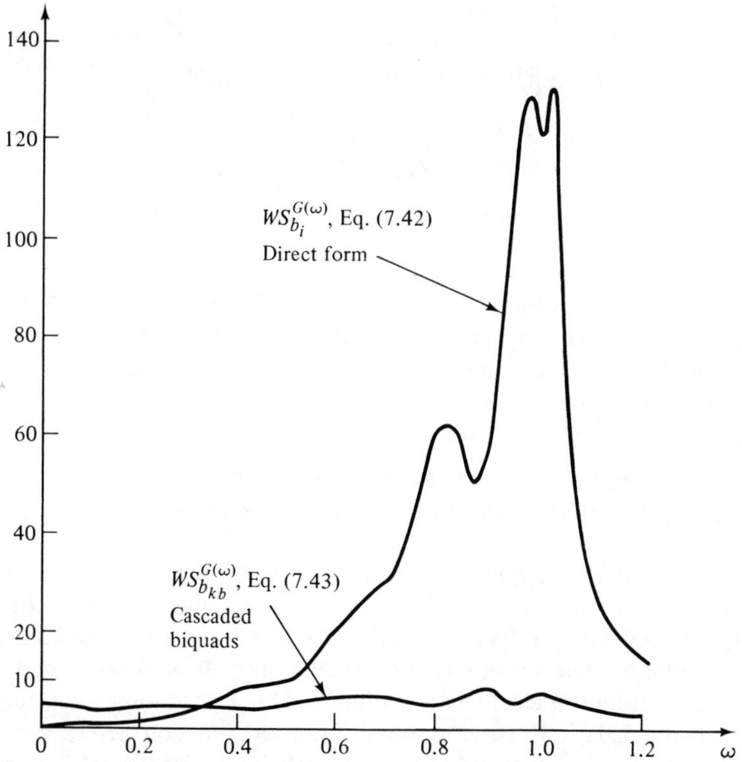

FIGURE 7.12 *Worstcase-sensitivity comparison between direct form and cascaded-biquadratic structures; seventh-order lowpass chebychev; 1.0 db passband ripple.*

express $Q_{P,Z}$ and $\omega_{P,Z}$ in terms of the transfer function coefficients by comparing equations (6.46) and (7.22), from which it follows that

$$H_{0b} = \frac{a_{2b}}{b_{2b}} \tag{7.45}$$

$$Q_P = \frac{1}{b_{1b}} \sqrt{b_{0b} b_{2b}}\,, \qquad \omega_P = \sqrt{\frac{b_{0b}}{b_{2b}}} \tag{7.46}$$

$$Q_Z = \frac{1}{a_{1b}} \sqrt{a_{0b} a_{2b}}\,, \qquad \omega_Z = \sqrt{\frac{a_{0b}}{a_{2b}}} \tag{7.47}$$

Therefore, from equations (7.45), (7.46), and (7.47), we may write $Q_{P,Z}$ sensitivities and $\omega_{P,Z}$ sensitivities in terms of the coefficient sensitivities as follows:

$$\left.\begin{aligned} S_{Q_P}^{H_b(s)} &= \tfrac{1}{2}S_{b_{0b}}^{H_b(s)} + \tfrac{1}{2}S_{b_{2b}}^{H_b(s)} - S_{b_{1b}}^{H_b(s)} \\ S_{\omega_P}^{H_b(s)} &= \tfrac{1}{2}S_{b_{0b}}^{H_b(s)} - \tfrac{1}{2}S_{b_{2b}}^{H_b(s)} \\ S_{Q_Z}^{H_b(s)} &= \tfrac{1}{2}S_{a_{0b}}^{H_b(s)} + \tfrac{1}{2}S_{a_{2b}}^{H_b(s)} - S_{a_{1b}}^{H_b(s)} \\ S_{\omega_Z}^{H_b(s)} &= \tfrac{1}{2}S_{a_{0b}}^{H_b(s)} - \tfrac{1}{2}S_{a_{2b}}^{H_b(s)} \end{aligned}\right\} \tag{7.48}$$

Note that the biquadratic structures BIQUAD DF, LBI/DF, BHI/DF, and MI/DF all have $b_{2b} \equiv 1$ so that the equations (7.45) and (7.46) may be simplified, and the terms $S_{b_{2b}}^{H_b(s)}$ in equations (7.48) may be set to zero for these particular structures.

7-3.5 On the Sensitivity Performance of Biquadratic Structures with Decomposed Numerators

Three types of biquadratic structures with decomposed numerators are introduced in section 7-3.2; they are described as the LBI/DF, BHI/DF, and MI/DF types, according to the way in which the numerator is decomposed. The sensitivity performance for all of these structures is determined from the results in equations (7.38). For example, consider the case of the LBI/DF structure with transfer function $H_b(s)$ given by equation (7.26). Suppose we require the sensitivity function $S_{\alpha_1}^{H_b(s)}$, where $\pm\,\alpha_1$ is one of the feedforward transmittances in Fig. 7.9(a). Then, it follows from equation (7.27) that

$$S_{\alpha_1}^{a_{1b}} = \frac{-\alpha_1}{a_{1b}}$$

Since none of the coefficients b_{0b}, b_{1b}, a_{2b}, and a_{0b} depend on α_1, we may write, using $S_{a_{1b}}^{H_b(s)}$ from equations (7.36), that

$$S_{\alpha_1}^{H_b(s)} = S_{a_{1b}}^{H_b(s)} S_{\alpha_1}^{a_{1b}} = -\frac{\alpha_1}{Q_Z a_{1b}}\left[\frac{\omega_Z s}{s^2 + \dfrac{\omega_Z s}{Q_Z} + \omega_Z^2}\right] \tag{7.49}$$

Writing Q_Z in terms of the numerator coefficients of $H_b(s)$ as $Q_Z = \dfrac{1}{a_{1b}}\sqrt{a_{0b}a_{2b}}$ and substituting into equation (7.49), we have

$$S_{\alpha_1}^{H_b(s)} = -\frac{\alpha_1}{\sqrt{a_{0b}a_{2b}}}\, t_{\mathrm{BP}}(s)$$

Taking the real part, we obtain

$$S_{\alpha_1}^{G_b(\omega)} = -\frac{\alpha_1}{\sqrt{a_{0b}a_{2b}}}\mathrm{Re}[\,t_{\mathrm{BP}}(j\omega)\,]$$

so that

$$\left|S_{\alpha_1}^{G_b(\omega)}\right|_{\max} \simeq \frac{\alpha_1}{\sqrt{a_{0b}a_{2b}}}|Q_Z| \tag{7.50}$$

The designer obtains $|Q_Z| \gg 1$ by letting $\alpha_1 \to \alpha_2 b_{1b}$, in which case the sensitivity expression in equation (7.50) is *approximately proportional to* Q_Z. This result is typical of the LBI/DF, BHI/DF, and MI/DF structures; that is, the placement of complex zeros of $H_b(s)$ close to the imaginary axis ($|Q_Z| \gg 1$) results in a high sensitivity of the gain function $G_b(\omega)$ to such transmittances as α_1 in equation (7.26) and $\alpha_{2,3}$ in equation (7.30).

7-4 BIQUADRATIC STRUCTURES EMPLOYING ONE FEEDBACK PATH

Previously considered structures require more than one feedback path and correspond to circuits that require more than one amplifier. In this section, we consider structures that have only one feedback path; they often require only one amplifier and for this reason correspond to a class of widely used *RC*-active circuits. It will be found that some of these structures have the disadvantage, compared with the DF BIQUAD structure, of poor sensitivity performance under conditions of high Q-factors $Q_{P,Z}$.

Consider the structure in Fig. 7.13(a), where N_{RC} is a passive *RC* network with accessible terminals 1, 2, and 3 and a connection to ground as shown. The amplifier $\pm K$ is considered here to be a voltage-controlled voltage source with *ideal* gain $\pm K$. There are many *RC*-active circuits that may be represented by means of the structure shown in Fig. 7.13(a) and, for this reason, it is useful to examine some of the fundamental properties of the transfer function $H_b(s)$, where

$$H_b(s) \equiv \frac{V_2(s)}{V_1(s)} = \pm K\frac{V_3(s)}{V_1(s)} \tag{7.51}$$

First, we review some of the pertinent properties of the open-circuit voltage transfer functions of the passive *RC* network N_{RC}. It is well known that any open-circuit voltage transfer function, such as $T_{31}(s)$ or $T_{32}(s)$ as defined in Fig. 7.13(b), of a passive (transformerless) *RC* network has the

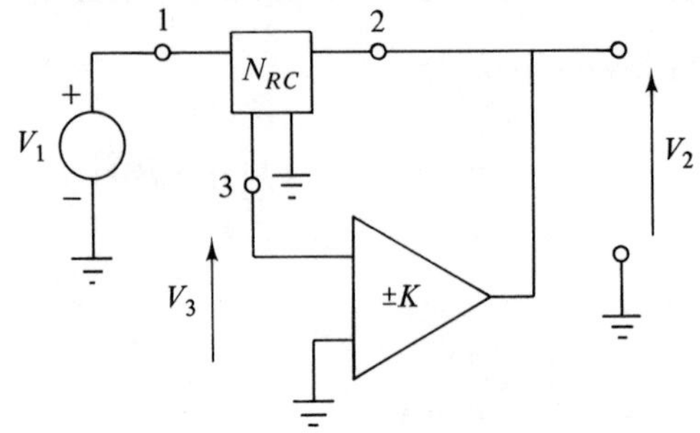

(a) General structure with one feedback path, $K \neq 0$.

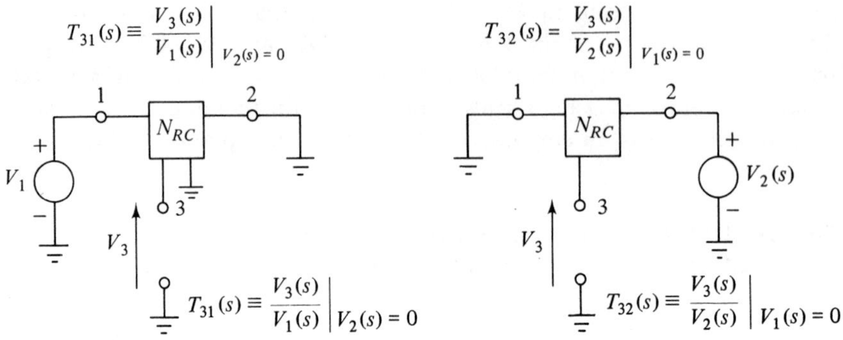

(b) Definitions of $T_{31}(s)$ and $T_{32}(s)$.

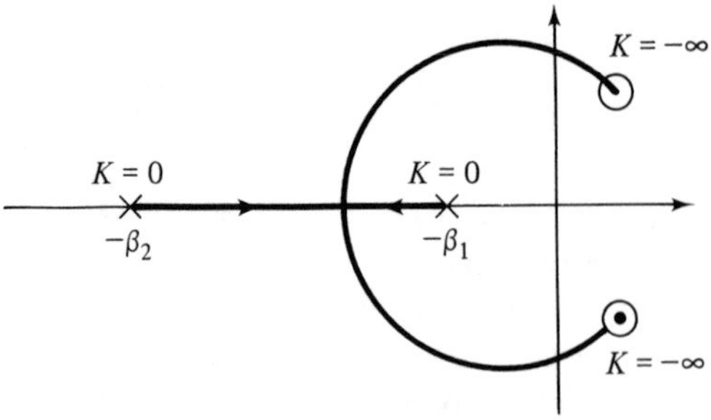

(c) Root locus diagram for poles of $H_b(s)$ versus gain K.

FIGURE 7.13

following properties:

(i) the poles are simple and on the negative real axis,

(ii) the zeros may be anywhere except on the positive real axis,

(iii) the magnitude responses of all open-circuit voltage transfer functions $|T(j\omega)|$ are everywhere less than or equal to unity for *ladder type* RC networks.

Of course, the poles of $T_{31}(s)$ and $T_{32}(s)$ shown in Fig. 7.13(b) are identical, and we shall assume that the network N_{RC} is of second order (it contains two capacitors), so that we may write

$$T_{31}(s) = \frac{a_2 s^2 + a_1 s + a_0}{(s + \beta_1)(s + \beta_2)} \quad \text{and} \quad T_{32}(s) = \frac{\alpha_2 s^2 + \alpha_1 s + \alpha_0}{(s + \beta_1)(s + \beta_2)} \tag{7.52}$$

where β_1 and β_2 are real and positive because of property (i) above. It is sometimes intuitively useful to express the denominators of equations (7.52) in terms of the Q-factor Q_N and natural frequency ω_N of the passive network N; thus,

$$(s + \beta_1)(s + \beta_2) \equiv s^2 + \frac{\omega_N s}{Q_N} + \omega_N^2 \tag{7.53}$$

so that

$$\omega_N = \sqrt{\beta_1 \beta_2} \tag{7.54}$$

and

PASSIVE RC NETWORK N_{RC}

$$Q_N = \frac{1}{\sqrt{\dfrac{\beta_1}{\beta_2}} + \sqrt{\dfrac{\beta_2}{\beta_1}}} \tag{7.55}$$

It follows from equation (7.55) that the maximum value of Q_N is equal to $\frac{1}{2}$ and occurs where β_1 is equal to β_2; we have, then,

$$\boxed{0 < Q_N < \tfrac{1}{2}} \tag{7.56}$$

BOUND ON Q-FACTOR OF PASSIVE NETWORK

(Of course, it is this very constraint that prevents us from using *passive* RC networks to realize highly selective magnitude transfer functions.)

Analysis of the general structure: It follows directly by superposition and from Fig. 7.13(a) that

$$V_3(s) = T_{31}(s)V_1(s) + T_{32}(s)V_2(s) \tag{7.57}$$

Furthermore,

$$\pm KV_3(s) = V_2(s) \tag{7.58}$$

so that, by eliminating $V_3(s)$ from equations (7.57) and (7.58),

$$\boxed{H_b(s) \equiv \frac{V_2(s)}{V_1(s)} = \frac{\pm KT_{31}(s)}{1 \mp KT_{32}(s)}} \qquad \text{GENERAL TRANSFER FUNCTION} \tag{7.59}$$

We now assume that N_{RC} is a second-order (transformerless) passive *RC* network with open-circuit voltage transfer functions provided by equations (7.52). By substituting equation (7.52) into equation (7.59),

$$\boxed{H_b(s) = \frac{\pm K\left[a_2s^2 + a_1s + a_0\right]}{(s + \beta_1)(s + \beta_2) \mp K(\alpha_2s^2 + \alpha_1s + \alpha_0)}} \qquad \text{GENERAL BIQUADRATIC TRANSFER FUNCTION} \tag{7.60}$$

Note that the zeros of $H_b(s)$ are the zeros of $T_{31}(s)$, whereas the poles of $H_b(s)$ depend on the poles of the network N_{RC}, the zeros of $T_{32}(s)$, and the amplifier gain $\pm K$.

It is sometimes convenient to write the transfer function $H_b(s)$ directly in terms of the denominator coefficients b_2, b_1, and b_0 so that

$$H_b(s) = \pm K\left[\frac{a_2s^2 + a_1s + a_0}{b_2s^2 + b_1s + b_0}\right] \tag{7.61}$$

where, by comparison with equation (7.60),

$$b_0 = (\beta_1\beta_2) \mp K\alpha_0, \qquad b_1 = (\beta_1 + \beta_2) \mp K\alpha_1, \qquad b_2 = 1 \mp K\alpha_2 \tag{7.62}$$

Alternatively, if $H_b(s)$ is expressed in terms of Q-factors and natural frequencies in the form

$$H_b(s) \equiv H_{0b}\left[\frac{s^2 + \dfrac{\omega_Z s}{Q_Z} + \omega_Z^2}{s^2 + \dfrac{\omega_P s}{Q_P} + \omega_P^2}\right] \tag{7.63}$$

then, by comparison with equations (7.62) and (7.61), it follows that

$$\left.\begin{aligned} H_{0b} = \frac{\pm K a_2}{1 \mp K\alpha_2}, \quad \omega_Z &= \sqrt{\frac{a_0}{a_2}}, \quad Q_Z = \frac{1}{a_1}\sqrt{a_0 a_2} \\ \omega_P &= \sqrt{\frac{b_0}{b_2}}, \quad Q_P = \frac{1}{b_1}\sqrt{b_0 b_2} \end{aligned}\right\} \tag{7.64}$$

and, of particular importance to this discussion,

$$\omega_P^2 = \frac{\omega_N^2 \mp K\alpha_0}{1 \mp K\alpha_2} \tag{7.65}$$

$$Q_P = Q_N \frac{\left[1 \mp \dfrac{K\alpha_0}{\omega_N^2}\right]^{1/2} [1 \mp K\alpha_2]^{1/2}}{\left[1 \mp \dfrac{K\alpha_1 Q_N}{\omega_N}\right]} \tag{7.66}$$

This expression is important because it reveals that the Q-factor Q_P of the overall transfer function $H_b(s)$ may be increased to an arbitrarily large magnitude by appropriate selection of the gain $\pm K$. Clearly, this could be achieved in a variety of ways; for example, by selecting a network N_{RC} for which $\alpha_0 = \alpha_2 = 0$ and using a positive gain $+K$. In this case, $\omega_P = \omega_N$ and $Q_P = Q_N[1 - K\alpha_1 Q_N/\omega_N]^{-1}$, where Q_P may be made arbitrarily large by letting $+K$ approach the value $\omega_N/\alpha_1 Q_N$. Of course, inspection of equation (7.66) reveals other possibilities for increasing Q_P to an arbitrarily large value; selecting a negative gain $-K$ and α_1 as zero is another possibility. All of these possibilities for increasing Q_P by appropriately constraining α_1, α_2, α_0, Q_N, ω_N, and $\pm K$ have been studied by circuit designers, and they lead to a wide variety of quite different RC-active circuit topologies.

Some insight into the effect of the amplifier gain $\pm K$ on the Q-factor Q_P and natural frequency ω_P is possible by considering the root locus for the denominator of equation (7.60) as a function of K. At $K = 0$ the roots of the denominator are at $-\beta_1$ and $-\beta_2$, which are the poles of the passive network N_{RC}. At $K = -\infty$ the roots of the denominator are at the roots of $\alpha_2 s^2 + \alpha_1 s + \alpha_0$, which are the zeros of $T_{32}(s)$. These two extremum negative values for K are shown in Fig. 7.13(c) for the case of two complex conjugate right-hand plane zeros of $T_{32}(s)$. It may be proven that the root locus is either on a circle or on the real axis between $-\beta_2$ and $-\beta_1$, in the manner shown in Fig. 7.13(c).

7-4.1 Sensitivity to Amplifier Imperfections

The amplifier $\pm K$ is often a single OP AMP circuit of the inverting type $-K$, as shown in Fig. 5.5(a) or the noninverting type $+K$ (Fig. 5.8). The nonideal gain $k(s)$ of these amplifiers may be written as a function of the *ideal* gain K, the dc open-loop gain A_0 and the gain-bandwidth product B of the OP AMP. For example, equations (5.22) and (5.36) describe the nonideal gain $k(s, A_0, B, K)$ for the inverting and noninverting single OP AMP amplifiers, respectively. In this section, it is shown that the sensitivity of the transfer function $H_b(s)$ to the nonideal behavior of the OPAMP is a problem that is particularly troublesome with the single-amplifier biquadratic structure of Fig. 7.13(a).

Writing $\pm k(s)$ as the corresponding *nonideal* gain function in equation (7.60) and using

$$(s) \equiv \frac{\partial H_b(s)}{\partial k(s)} \frac{k(s)}{H_b(s)}$$

gives

$$S_{k(s)}^{H_b(s)} = \left[\frac{\pm k(s)\left(\alpha_2 s^2 + \alpha_1 s + \alpha_0\right)}{(s + \beta_1)(s + \beta_2) \mp k(s)\left(\alpha_2 s^2 + \alpha_1 s + \alpha_0\right)} \right] \pm 1 \qquad (7.67)$$

For satisfactory operation of the structure, we assume that

$$k(s) \simeq K \qquad (7.68)$$

in the region of the s-plane that is of interest, so that equation (7.67) simpliflies to

$$S_{k(s)}^{H_b(s)} \simeq \left[\frac{\pm K\left(\alpha_2 s^2 + \alpha_1 s + \alpha_0\right)}{(s + \beta_1)(s + \beta_2) \mp K\left(\alpha_2 s^2 + \alpha_1 s + \alpha_0\right)} \right] \pm 1 \qquad (7.69)$$

Substituting equations (7.62) into equation (7.69) and using equations

(7.65) and (7.66), we obtain

$$S_{k(s)}^{H_b(s)} \simeq \left[\frac{\pm K}{1 \mp K\alpha_2}\right]\left[\frac{\alpha_2 s^2 + \alpha_1 s + \alpha_0}{s^2 + \dfrac{\omega_P s}{Q_P} + \omega_P^2}\right] \pm 1 \tag{7.70}$$

This is the required expression for the sensitivity of the transfer function to the gain function $k(s)$, where $k(s) \simeq K$. The denominator polynomial $\left[s^2 + \dfrac{\omega_P s}{Q_P} + \omega_P^2\right]$ is invariant in the sense that it is specified in the required transfer function $H_b(s)$; the designer can, therefore, only control the complex sensitivity function of equation (7.70) by making a suitable choice for $\pm K$, α_0, α_1, and α_2. This possibility is considered in the following section, where structures that use $+K$ amplifiers (noninverting) are classified as type I and those that use $-K$ (inverting) amplifiers as type II, which leads to the two kinds of circuits shown in Table 7-3.

The *noninverting single-OP AMP amplifier* circuit of Fig. 5.8 has a nonideal gain $k(s) \equiv V_2(s)/V_1(s)$ that is given by equation (5.35). Using equation (5.1) for $A(s)$ in equation (5.35), we get

$$\frac{1}{k(s)} = \frac{1}{K} + \frac{1}{A_0} + \frac{s}{B} \tag{7.71}$$

where the OP AMP gain has been assumed to be *dominant pole* as in equation (5.1). By direct analysis of equation (7.71), we find that

$$S_{A_0}^{k(s)} = \frac{k(s)}{A_0} \simeq \frac{K}{A_0}$$

and

$$S_{B}^{k(s)} = \frac{sk(s)}{B} \simeq \frac{sK}{B}$$

NONINVERTING-AMPLIFIER GAIN SENSITIVITIES: Fig. 5.8 (7.72)

where it is assumed that $k(s) \simeq K$ over the region of the s-plane that is of interest.

Similarly, the inverting single OP AMP amplifier circuit shown in Fig. 5.5(a) has a nonideal gain that is provided by equation (5.22), which may be written in the form

$$\frac{1}{k(s)} = \frac{1}{K} + \frac{(1+K)}{KA_0} + \frac{(1+K)s}{KB} \tag{7.73}$$

TABLE 7-3

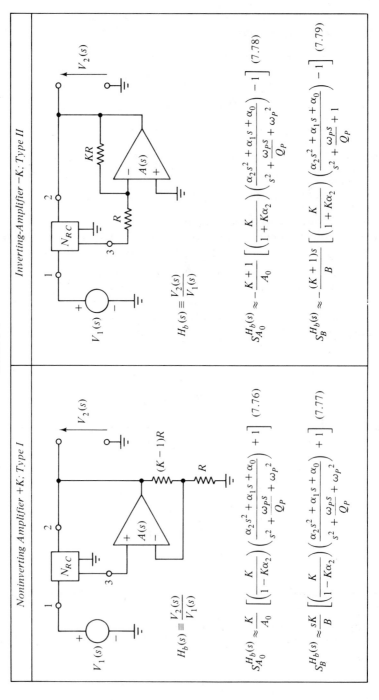

Noninverting Amplifier +K; Type I	*Inverting-Amplifier −K; Type II*
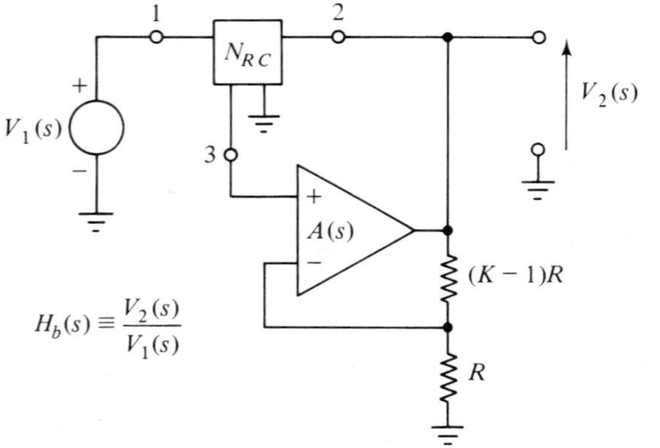	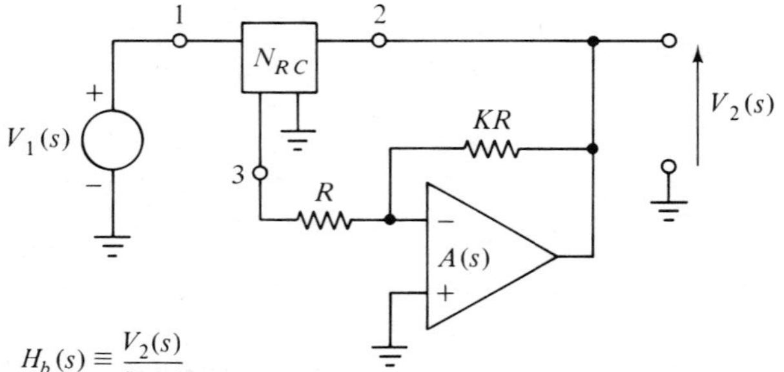
$H_b(s) \equiv \frac{V_2(s)}{V_1(s)}$	$H_b(s) \equiv \frac{V_2(s)}{V_1(s)}$
$S_{A_0}^{H_b(s)} \approx \frac{K}{A_0}\left[\left(\frac{K}{1-K\alpha_2}\right)\left(\frac{\alpha_2 s^2+\alpha_1 s+\alpha_0}{s^2+\frac{\omega_P s}{Q_P}+\omega_P^2}\right)+1\right]$ (7.76)	$S_{A_0}^{H_b(s)} \approx -\frac{K+1}{A_0}\left[\left(\frac{K}{1+K\alpha_2}\right)\left(\frac{\alpha_2 s^2+\alpha_1 s+\alpha_0}{s^2+\frac{\omega_P s}{Q_P}+\omega_P^2}\right)-1\right]$ (7.78)
$S_{B}^{H_b(s)} \approx \frac{sK}{B}\left[\left(\frac{K}{1-K\alpha_2}\right)\left(\frac{\alpha_2 s^2+\alpha_1 s+\alpha_0}{s^2+\frac{\omega_P s}{Q_P}+\omega_P^2}\right)+1\right]$ (7.77)	$S_{B}^{H_b(s)} \approx -\frac{(K+1)s}{B}\left[\left(\frac{K}{1+K\alpha_2}\right)\left(\frac{\alpha_2 s^2+\alpha_1 s+\alpha_0}{s^2+\frac{\omega_P s}{Q_P}+1}\right)-1\right]$ (7.79)

which leads directly to

$$S_{A_0}^{k(s)} = \frac{(K+1)k(s)}{KA_0} \simeq \frac{(K+1)}{A_0}$$

and

$$S_{B}^{k(s)} = \frac{(K+1)k(s)s}{KB} \simeq \frac{(K+1)s}{B}$$

INVERTING-AMPLIFIER GAIN SENSITIVITIES: Fig. 5.5(a) (7.74)

where we again assume $k(s) \simeq K$ over the region of s that is of interest. The inverting amplifier shown in Table 7-3 has an input resistance R that terminates the network N_{RC} at terminal 3; it is necessary to include this resistance R in the calculation of $T_{32}(s)$ in equation (7.52). It follows that α_2, α_1, and α_0 are generally dependent on R.

The major practical observation derivable from equations (7.72) and (7.74) is that $k(s)$ *becomes increasingly sensitive to* A_0 *and* B *as the ideal gain* K *is increased.* This is a serious design problem in the case of type II structures because it will be found that K is generally very large ($> 4Q_P^2$), thereby implying that $k(s)$ is highly dependent on A_0 and B. The sensitivity of the transfer function $H_b(s)$ to A_0 and B is

$$S_{A_0,\,B}^{H_b(s)} = S_{k(s)}^{H_b(s)} S_{A_0,\,B}^{k(s)} \tag{7.75}$$

Thus, by substituting equations (7.72) or (7.74) and (7.70) into (7.75) we obtain the sensitivity expressions for $S_{A_0,\,B}^{H_b(s)}$ that are shown in Table 7-3. [In many practical situations, the $+1$ terms in equations (7.76) to (7.79) in Table 7-3 may be neglected.]

7-4.2 Type I Noninverting Amplifier (+ K) Biquadratic Structures

The first category of single-amplifier structures corresponds to the case for which the network N_{RC} has

$$\alpha_0 \equiv \alpha_2 \equiv 0 \qquad \text{TYPE I} \tag{7.80}$$

so that, by inspection of equation (7.52), it is clear that $T_{32}(s)$ *is a bandpass function*:

$$T_{32}(s) = \frac{\alpha_1 s}{(s+\beta_1)(s+\beta_2)} \qquad \text{BANDPASS } T_{32} \tag{7.81}$$

Although $T_{32}(s)$ is a bandpass function, the overall transfer function $H_b(s)$ is *not* necessarily bandpass because its zeros depend on $T_{31}(s)$ via the coefficients a_2, a_1, and a_0 [see equation (7.60)].

The constraints in equation (7.80) allow the previously described expressions for $H_b(s)$, ω_P, Q_P, $S_{A_0}^{H_b(s)}$ and $S_B^{H_b(s)}$ to be simplified. Thus, substituting equation (7.80) into equations (7.60), (7.65), (7.66), (7.76), and (7.77) provides the corresponding equations supplied in Table 7-4 for the type I structure. Since $Q_N \leqslant \frac{1}{2}$ and we generally require $Q_P > \frac{1}{2}$, then by inspection of the equation for Q_P in Table 7-4 it is clear that K must be positive. That is, a *type I structure requires a noninverting amplifier* $+K$.

The root locus diagram for the denominator of $H_b(s)$ is shown in Fig. 7.14(a); for complex poles of $H_b(s)$, the pole positions lie on a circle with their center at the origin and radius $\omega_P = \omega_N$. As K is increased, the poles

TABLE 7-4 *Type I +K noninverting gain with bandpass $T_{32}(s)$.*

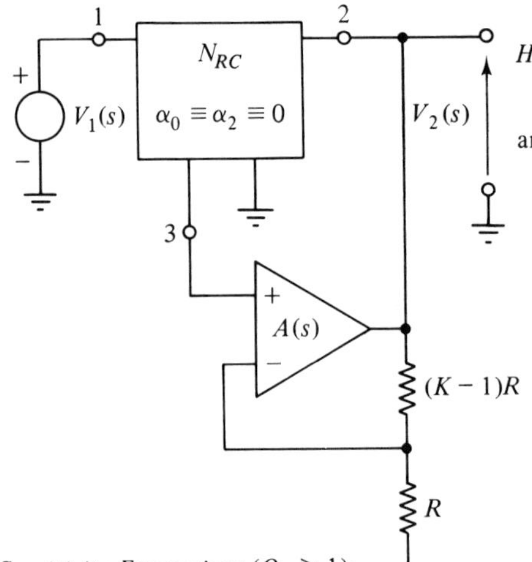

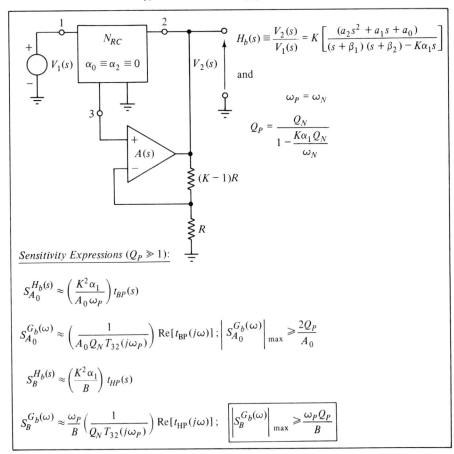

$$H_b(s) \equiv \frac{V_2(s)}{V_1(s)} = K\left[\frac{(a_2 s^2 + a_1 s + a_0)}{(s+\beta_1)(s+\beta_2) - K\alpha_1 s}\right]$$

and

$$\omega_P = \omega_N$$

$$Q_P = \frac{Q_N}{1 - \dfrac{K\alpha_1 Q_N}{\omega_N}}$$

Sensitivity Expressions ($Q_P \gg 1$):

$$S_{A_0}^{H_b(s)} \approx \left(\frac{K^2\alpha_1}{A_0\omega_P}\right) t_{BP}(s)$$

$$S_{A_0}^{G_b(\omega)} \approx \left(\frac{1}{A_0 Q_N T_{32}(j\omega_P)}\right) \mathrm{Re}[t_{\mathrm{BP}}(j\omega)]\,; \quad \left|S_{A_0}^{G_b(\omega)}\right|_{\max} \geqslant \frac{2Q_P}{A_0}$$

$$S_B^{H_b(s)} \approx \left(\frac{K^2\alpha_1}{B}\right) t_{HP}(s)$$

$$S_B^{G_b(\omega)} \approx \frac{\omega_P}{B}\left(\frac{1}{Q_N T_{32}(j\omega_P)}\right) \mathrm{Re}[t_{\mathrm{HP}}(j\omega)]\,; \quad \boxed{\left|S_B^{G_b(\omega)}\right|_{\max} \geqslant \frac{\omega_P Q_P}{B}}$$

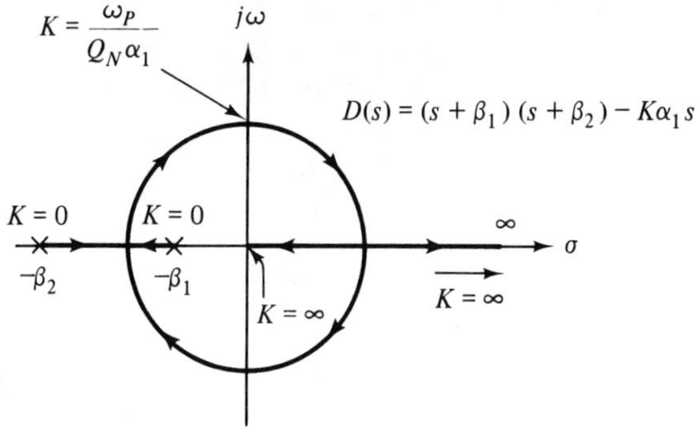

(a) Type I structure: root locus

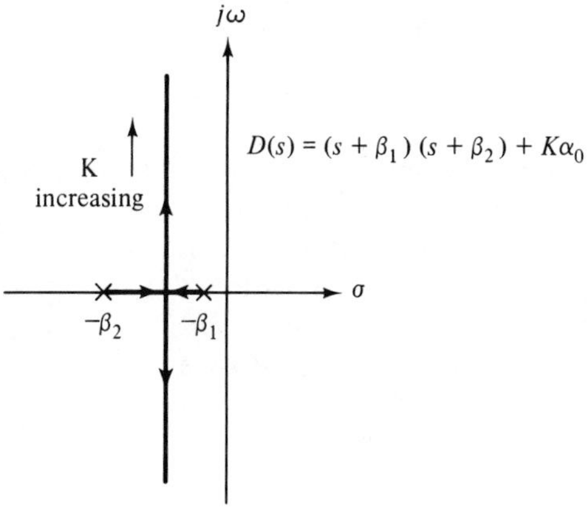

(b) Type II A structure: root locus

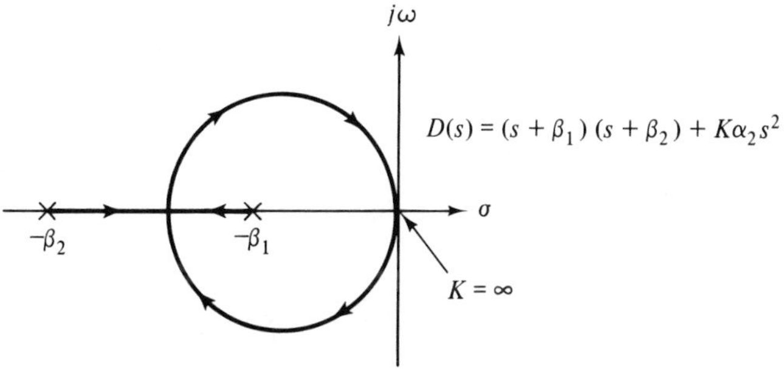

(c) Type II B structure: root locus

FIGURE 7.14

migrate toward the $j\omega$-axis, and they lie on the imaginary axis when $Q_P = \infty$ and $K = \omega_N/Q_N\alpha_1 = \omega_P/Q_N\alpha_1$.

It is noted from Table 7-4 that if $Q_P \gg 1$, then

$$K \simeq \frac{\omega_P}{Q_N\alpha_1}, \qquad Q_P \gg 1 \tag{7.82}$$

Furthermore, by inspection of equations (7.52) and (7.53) it is clear that

$$T_{32}(j\omega_N) = |T_{32}(j\omega_N)| = \frac{Q_N\alpha_1}{\omega_N} \tag{7.83}$$

Therefore, from equations (7.82) and (7.83) we have

$$\boxed{KT_{32}(j\omega_N) \simeq 1 \qquad \text{for } Q_P \gg 1} \tag{7.84}$$

By property (iii), $|T_{32}(j\omega)| \leqslant 1$ over all ω, so that equation (7.84) implies that $K \geqslant 1$. Furthermore, the *minimum* value of K that may be selected is also limited by equation (7.84) because, for $Q_P \gg 1$, reducing K downward toward unity clearly implies increasing $T_{32}(j\omega_P)$ upward toward its upper bound of unity; unfortunately, as $T_{32}(j\omega_P) \to 1$ the *spread* in RC-element values (for example, $C_{\max} : C_{\min}$) becomes intolerably large.

The sensitivity of the gain function $G_b(\omega)$ to amplifier imperfections is obtained by setting $s = j\omega$ and taking the real part of the $S_{A_0,\,B}^{H_b(s)}$ expressions in Table 7-4, so that, for $Q_P \gg 1$,

$$S_{A_0}^{G_b(\omega)} \simeq +\frac{K^2\alpha_1}{A_0\omega_p}\operatorname{Re}\left[t_{\mathrm{BP}}(j\omega)\right], \qquad S_B^{G_b(\omega)} \simeq +\frac{K^2\alpha_1}{B}\operatorname{Re}\left[t_{\mathrm{HP}}(j\omega)\right] \tag{7.85}$$

Further insight into the nature of these sensitivity functions is possible by using the equations in Table 7-4 to eliminate K and α_1 from equations (7.85); thus, we note from Table 7-4 that

$$\frac{K\alpha_1}{\omega_P} = \left(1 - \frac{Q_N}{Q_P}\right)\frac{1}{Q_N} \tag{7.86}$$

and

$$K = \left(1 - \frac{Q_N}{Q_P}\right)\frac{1}{T_{32}(j\omega_P)} \tag{7.87}$$

Substituting equations (7.86) and (7.87) into (7.85), we have

$$S_{A_0}^{G_b(\omega)} = \frac{1}{A_0}\left[\frac{\left(1 - \dfrac{Q_N}{Q_p}\right)^2}{Q_N T_{32}(j\omega_p)}\right] \mathrm{Re}\left[t_{\mathrm{BP}}(j\omega)\right]$$

and, for $Q_P \gg 1$,

$$S_{A_0}^{G_b(\omega)} \simeq \frac{1}{A_0}\left[\frac{1}{Q_N T_{32}(j\omega_p)}\right] \mathrm{Re}\left[t_{\mathrm{BP}}(j\omega)\right] \tag{7.88}$$

Similarly,

$$S_{B}^{G_b(\omega)} = \frac{\omega_p}{B}\left[\frac{\left(1 - \dfrac{Q_N}{Q_p}\right)^2}{Q_N T_{32}(j\omega_p)}\right] \mathrm{Re}\left[t_{\mathrm{HP}}(j\omega)\right]$$

and, for $Q_P \gg 1$,

$$S_{B}^{G_b(\omega)} \simeq \frac{\omega_p}{B}\left[\frac{1}{Q_N T_{32}(j\omega_p)}\right] \mathrm{Re}\left[t_{\mathrm{HP}}(j\omega)\right] \tag{7.89}$$

This result is important because it describes a fundamental limitation of type I single-amplifier networks. The only means by which equations *(7.88) and (7.89)* may be controlled by the designer is via the terms $Q_N T_{32}(j\omega_p)$. Clearly, ω_p, B, A_0, Re $[t_{\mathrm{BP}}(j\omega)]$ and Re $[t_{\mathrm{HP}}(j\omega)]$ are directly or indirectly specified and beyond the control of the designer. *Minimization of equations* (7.88) *and* (7.89) *is therefore achieved by maximizing* $Q_N T_{32}(j\omega_p)$. In chapter 8, we are concerned with the techniques for achieving this maximization. The functions $S_{A_0, B}^{G_b(\omega)}$ are sketched in Fig. 7.15. Note that

$$\begin{aligned} \left|S_{A_0}^{G_b(\omega)}\right|_{\max} &\geqslant \frac{2Q_p}{A_0} && \text{at } \omega \simeq \omega_p \\ \left|S_{B}^{G_b(\omega)}\right|_{\max} &\geqslant \frac{\omega_p Q_p}{B} && \text{at } \omega_{1,2} \simeq \omega_p\left(1 \pm \frac{1}{2Q_p}\right) \end{aligned} \tag{7.90}$$

where the equality is only approached as $Q_N \to \frac{1}{2}$ and $T_{32}(j\omega_p) \to 1$.

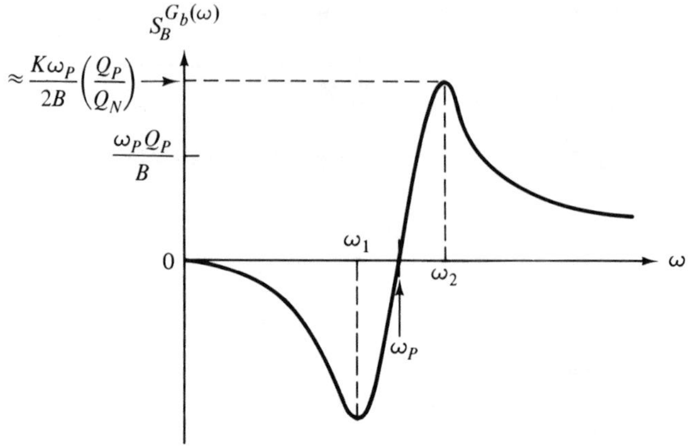

(a) Gain sensitivity of type I noninverting amplifier circuit to B: $Q_P >> 1$.

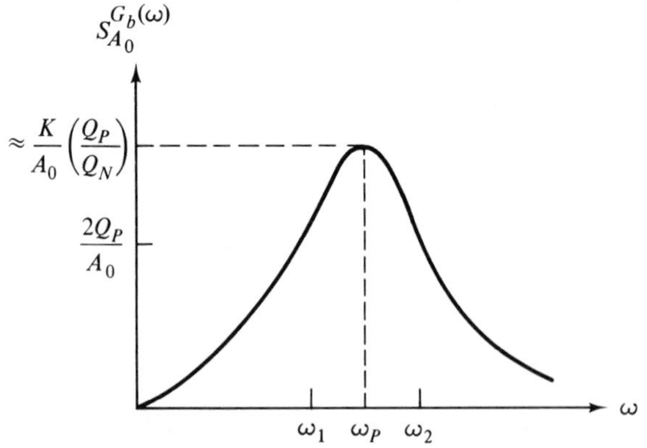

(b) Gain sensitivity of type I noninverting amplifier circuit to A_0, $Q_P >> 1$.

FIGURE 7.15

7-4.3 Type II Inverting Amplifier ($-K$) Biquadratic Structures

There are essentially two simple categories of single feedback path structures that require *inverting* amplifiers. They are defined as follows:

$$\text{Type IIA:} \qquad \alpha_1 \equiv \alpha_2 \equiv 0 \tag{7.91}$$

$$\text{Type IIB:} \qquad \alpha_0 \equiv \alpha_1 \equiv 0 \tag{7.92}$$

so that we have a *lowpass function* $T_{32}(s)$ *for type IIA and a highpass function* $T_{32}(s)$ *for type IIB*.

Type IIA structures: $\alpha_1 \equiv \alpha_2 \equiv 0$, lowpass $T_{32}(s)$: the pertinent expressions for the type IIA circuits are obtained by substituting equation (7.91) into equations (7.61), (7.65), (7.66), (7.78), and (7.79). Thus, we have

$$H_b(s) = \frac{-K(a_2 s^2 + a_1 s + a_0)}{(s+\beta_1)(s+\beta_2) + K\alpha_0} \qquad \text{TYPE II STRUCTURE: TABLE 7-3} \tag{7.93}$$

$$\omega_P = \omega_N\left(1 + \frac{K\alpha_0}{\omega_N^2}\right)^{1/2}, \qquad Q_P = Q_N\left(1 + \frac{K\alpha_0}{\omega_N^2}\right)^{1/2} \tag{7.94}$$

$$S_{A_0}^{H_b(s)} \approx -\left[\frac{K(K+1)\alpha_0}{A_0\omega_p^2}\right] t_{\text{LP}}(s) \tag{7.95}$$

and

$$S_B^{H_b(s)} \approx -\left[\frac{K(K+1)\alpha_0}{B\omega_p}\right] t_{\text{BP}}(s) \tag{7.96}$$

A root locus diagram for the type IIA structure is shown in Fig. 7.14(b). Note that the locus is on a vertical line in the left-half plane for the case of complex roots.

Some serious practical limitations of this structure are now derived from the preceding equations. First, we note from equation (7.52) that

$$T_{32}(0) = \frac{\alpha_0}{\omega_N^2} \leqslant 1 \tag{7.97}$$

and is not greater than unity by property (iii) of an RC network. Then, using equation (7.94) for Q_P and (7.56), we have the important result that

$$KT_{32}(0) > 4Q_P^2 - 1 \tag{7.98}$$

so that, by equation (7.97),

$$K > 4Q_P^2 - 1 \tag{7.99}$$

Since it is always true that $Q_P > \frac{1}{2}$ for complex poles then, by equation (7.99), $K > 0$, which implies that a noninverting amplifier is *required* for the type IIA structure. It is important to note that, for $Q_P \gg 1$,

$$\boxed{K > 4Q_P^2} \tag{7.100}$$

Clearly, very large (often impractically large) values of gain K are required for even quite moderate Q-factors Q_P. For example, a Q_P of 10 requires a value of K that is in excess of 400.

Useful insight into the practical limitations of this structure is obtained by using equations (7.94), (7.97), and (7.100) to eliminate K and α_0 from equations (7.95) and (7.96). Thus, for $Q_P \gg 1$,

$$S_B^{H_b(s)} \approx -\left[\frac{\omega_P Q_P^2}{BQ_N^2 T_{32}(0)}\right] t_{\mathrm{BP}}(s) \tag{7.101}$$

and

$$S_{A_0}^{H_b(s)} \approx -\left[\frac{Q_P^2}{A_0 Q_N^2 T_{32}(0)}\right] t_{\mathrm{HP}}(s) \tag{7.102}$$

Setting $s = j\omega$ and taking the real parts of these equations, we obtain

$$\boxed{\begin{aligned} S_B^{G_b(\omega)} &\simeq \left[\frac{-1}{Q_N^2 T_{32}(0)}\right]\left(\frac{\omega_P Q_P^2}{B}\right) \mathrm{Re}[t_{\mathrm{BP}}(j\omega)] \\ &\text{and} \\ S_{A_0}^{G_b(\omega)} &\simeq \left[\frac{-1}{Q_N^2 T_{32}(0)}\right]\left(\frac{Q_P^2}{A_0}\right) \mathrm{Re}[t_{\mathrm{LP}}(j\omega)] \end{aligned}} \tag{7.103}$$

for $Q_P \gg 1$. It follows from equations (7.97), (7.56), and (7.103) that

$$\left|S_B^{G_b(\omega)}\right| \geqslant \frac{4\omega_P Q_P^2}{B}\left|\text{Re}\left[t_{\text{BP}}(j\omega)\right]\right| \tag{7.104}$$

and

$$\left|S_{A_0}^{G_b(\omega)}\right| \geqslant \frac{4Q_P^2}{A_0}\left|\text{Re}\left[t_{\text{LP}}(j\omega)\right]\right| \tag{7.105}$$

so that we may use the equations above with the diagrams in Fig. 7.10 to write the lower bounds on the maximum sensitivities as

$$\left|S_B^{G_b(\omega)}\right|_{\max} \geqslant \frac{4\omega_P Q_P^3}{B} \qquad \text{at } \omega = \omega_P \tag{7.106}$$

$$\left|S_{A_0}^{G_b(\omega)}\right|_{\max} \geqslant \frac{2Q_P^3}{A_0} \qquad \text{at } \omega = \omega_P\left(1 \pm \frac{1}{2Q_p}\right) \tag{7.107}$$

for $Q_P \gg 1$.

This dependence of the gain $G_b(\omega)$ on Q-factor Q_P *virtually eliminates type IIA structures from consideration except at very low frequencies and very low Q-factors* Q_P. The reader may wish to try Problem 7.16 to be convinced of the severity of the limitations imposed by equations (7.106) and (7.107).

Type IIB structures: $\alpha_0 \equiv \alpha_1 \equiv 0$ highpass $T_{32}(s)$: the pertinent expressions for the type IIB circuits are obtained by substituting equation (7.92) into equations (7.61), (7.65), (7.66), (7.78), and (7.79):

$$H_b(s) = \frac{-K\left(a_2 s^2 + a_1 s + a_0\right)}{(s+\beta_1)(s+\beta_2) + K\alpha_2 s^2} \qquad \begin{array}{l}\text{TYPE II}\\ \text{STRUCTURE:}\\ \text{TABLE 7-3}\end{array} \tag{7.108}$$

$$\omega_P = \omega_N\left(\frac{1}{1+K\alpha_2}\right)^{1/2}, \qquad Q_P = Q_N(1+K\alpha_2)^{1/2} \tag{7.109}$$

$$S_{A_0}^{H_b(s)} \approx -\left[\frac{K(K+1)\alpha_2}{A_0(1+K\alpha_2)}\right] t_{\text{HP}}(s) \tag{7.110}$$

and

$$S_B^{H_b(s)} \approx -\left[\frac{K(K+1)\alpha_2}{B(1+K\alpha_2)}\right]\left(\frac{s^2}{\omega_P}\right)t_{\rm BP}(s) \tag{7.111}$$

The root locus diagram for the denominator of the transfer function $H_b(s)$ is shown in Fig. 7.14(c). Note that the locus is a circle for complex roots [poles of $H_b(s)$] and that the high-Q_P region is near the origin in this diagram.

It is now possible to derive some serious practical limitations of the type IIB structure. From equation (7.52) we have

$$T_{32}(\infty) = \alpha_2 \leqslant 1 \tag{7.112}$$

which is not greater than unity by property (iii) of an *RC* network. Then, using equations (7.109), (7.112), and (7.56), it follows that

$$KT_{32}(\infty) > 4Q_P^2 - 1 \tag{7.113}$$

so that, by equation (7.112),

$$K > 4Q_P^2 - 1 \tag{7.114}$$

Again we note that since $Q_P > \frac{1}{2}$, we have $K > 0$ and, therefore, a *noninverting* amplifier is required; for $Q_P \gg 1$,

$$K > 4Q_P^2 \tag{7.115}$$

Using equations (7.109), (7.112), and (7.115) to eliminate K and α_2 from equations (7.110) and (7.111) we have, for $Q_P \gg 1$,

$$S_B^{H_b(s)} \simeq \left[\frac{-1}{Q_N^2 T_{32}(\infty)}\right]\left(\frac{s^2 Q_P^2}{\omega_P B}\right)t_{\rm BP}(s) \tag{7.116}$$

and

$$S_{A_0}^{H_b(s)} \simeq \left[\frac{-1}{Q_N^2 T_{32}(\infty)}\right]\left(\frac{Q_P^2}{A_0}\right)t_{\rm HP}(s) \tag{7.117}$$

Setting $s = j\omega$ and taking the real parts of the above equations we have, for $Q_P \gg 1$,

$$S_B^{G_b(\omega)} \simeq \left[\frac{1}{Q_N^2 T_{32}(\infty)}\right]\left(\frac{\omega^2 Q_P^2}{\omega_P B}\right) \operatorname{Re}[t_{BP}(j\omega)]$$

and

$$S_{A_0}^{G_b(\omega)} \simeq \left[\frac{-1}{Q_N^2 T_{32}(\infty)}\right]\left(\frac{Q_P^2}{A_0}\right) \operatorname{Re}[t_{HP}(j\omega)] \tag{7.118}$$

Note that the magnitudes of these sensitivities are minimized by designing with $Q_N \to \frac{1}{2}$ and $T_{32}(\infty) \to 1$; also, it follows from equations (7.118), (7.56), and (7.112) that

$$\left|S_B^{G_b(\omega)}\right|_{\max} \geqslant \frac{4\omega^2 Q_P^2}{B\omega_P} \left|\operatorname{Re}[t_{BP}(j\omega)]\right|_{\max} \tag{7.119}$$

and

$$\left|S_{A_0}^{G_b(\omega)}\right|_{\max} \geqslant \frac{4Q_P^2}{A_0} \left|\operatorname{Re}[t_{HP}(j\omega)]\right|_{\max} \tag{7.120}$$

Furthermore, it may be shown from these equations and Fig. 7.10 that equations (7.106) and (7.107) also apply to the type IIB structure. This dependence of $G_b(\omega)$ on Q_P^3 makes this structure undesirable except at low Q-factors Q_P and at low frequencies.

7-4.4 Summary

The general structure in Fig. 7.13(a) consists of a single VCVS with gain $\pm K$ and a three-terminal grounded (transformerless) RC network N_{RC}. The transfer function $H_b(s)$ is characterized by the poles β_1 and β_2 of N_{RC}, the zeros of $T_{31}(s)$, the zeros of $T_{32}(s)$, and the polarity and magnitude of the amplifier gain $\pm K$, as described by equation (7.60). The classification of these types of structures is given in terms of the coefficients α_2, α_1, and α_0 of the function $T_{32}(s)$. Thus, a type I network corresponds to $\alpha_2 \equiv \alpha_0 \equiv 0$, so that $T_{32}(s)$ is bandpass and it turns out that the amplifier gain must be chosen as positive $+K$ for $Q_P > \frac{1}{2}$. It is found that type I networks possess sensitivities of $G_b(\omega)$ to A_0 and B that are greater than $2Q_P/A_0$ and $\omega_P Q_P/B$, respectively, and that these sensitivities are minimized by maximizing the quantity $Q_N T_{32}(j\omega_P)$.

Type II circuits are generally inferior to type I circuits because $H_b(s)$ and $G_b(\omega)$ are far more sensitive to perturbations of A_0 and B. This is primarily because the magnitude of the inverting gain $-K$ must be greater than $4Q_P^2$ which, in turn, implies a high sensitivity of $k(s)$, $H_b(s)$, and $G_b(\omega)$ to A_0 and B. Minimization of these sensitivities to A_0 and B is achieved by maximizing $Q_N^2 T_{32}(0)$ for type IIA circuits and by maximizing $Q_N^2 T_{32}(\infty)$ for type IIB circuits. The performance of these type II circuits is summarized by equations (7.106) and (7.107), implying a dependence of $G_b(\omega)$ on B and A_0 that is at least a factor of Q_P^2 in excess of the corresponding result [equation (7.90)] for type I circuits.

It will be found in chapter 8 that maximizing $Q_N^2 T_{32}(0)$, $Q_N T_{32}(j\omega_P)$, and $Q_N^2 T_{32}(\infty)$ usually implies a severe increase in the ratio between the maximum and minimum values of the *RC* elements. This large *RC* "element-spread" is often a practical fabrication problem.

7-5 RESISTIVELY TERMINATED LOSSLESS TWO-PORT STRUCTURES

It is convenient to commence this discussion of resistively terminated lossless two-port structures by referring to the diagram in Fig. 7.16(a). The two-port network N is by definition *lossless*; thus, according to the definition of losslessness that is given in section 2-2, the total energy E_T delivered to network N is zero, so that

$$E_T(\infty) \equiv \int_{-\infty}^{\infty} \left[v_1(t) i_1(t) + v_2(t) i_2(t) \right] dt = 0 \quad \text{LOSSLESS 2-PORT} \tag{7.121}$$

over all possible square integrable waveforms $v_1(t)$, $i_1(t)$, $v_2(t)$, and $i_2(t)$. In general, the network N contains lossless elements such as $\pm L$, $\pm C$, transformers, and gyrators.

We define the input voltage as $e_1(t)$ with transform $E_1(s)$. Furthermore, we note that the source resistance R_s implies that the maximum available power $P_{\max}$ that can be delivered to the input port 1, with input impedance $Z_1(j\omega)$, is

$$P_{\max} = \frac{|E_1|^2}{4R_s} \tag{7.122}$$

where $|E_1|$ refers to a sinusoidal excitation source. The reader is requested to prove in Problem 7-17 that the power delivered to the input port $Z_1(j\omega)$ only equals the maximum $P_{\max}$, if $Z_1(j\omega)$ is equal to R_s.

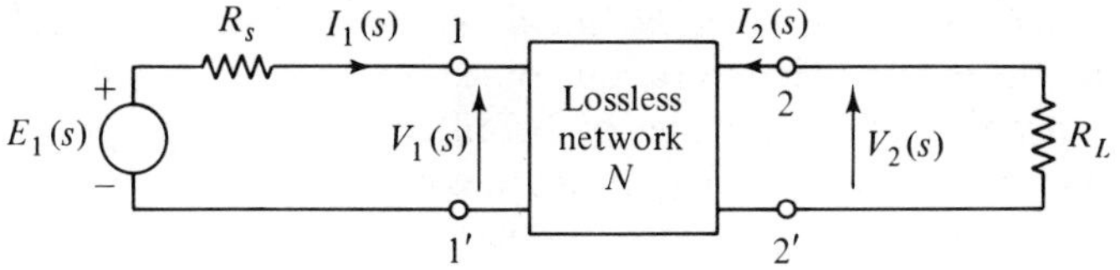

(a) Resistively terminated lossless two port.

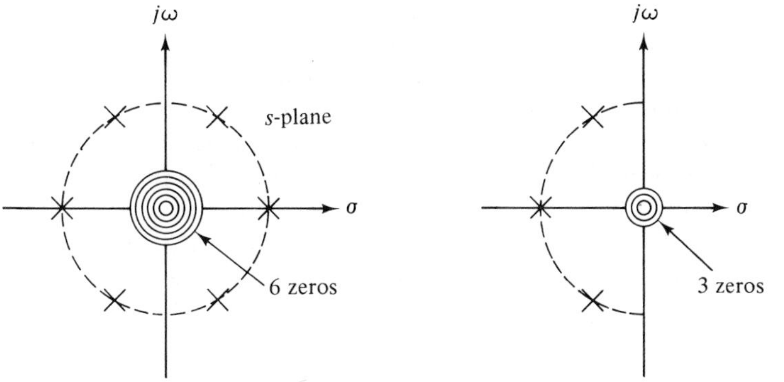

Pole-zero diagram for $|\rho_1(s)|^2$

Pole-zero diagram for $\rho_1(s)$

(b)

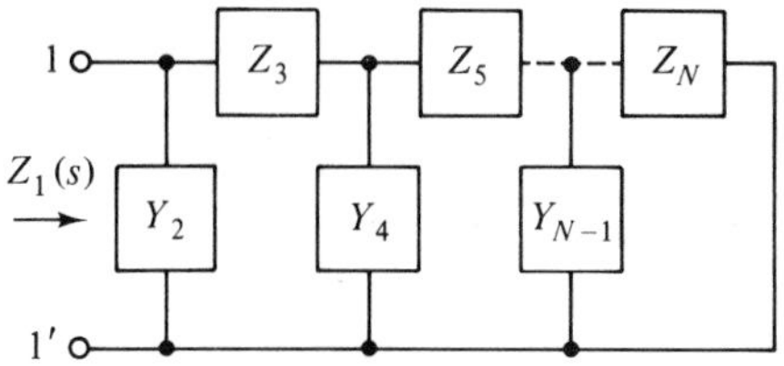

(c) Ladder topology.

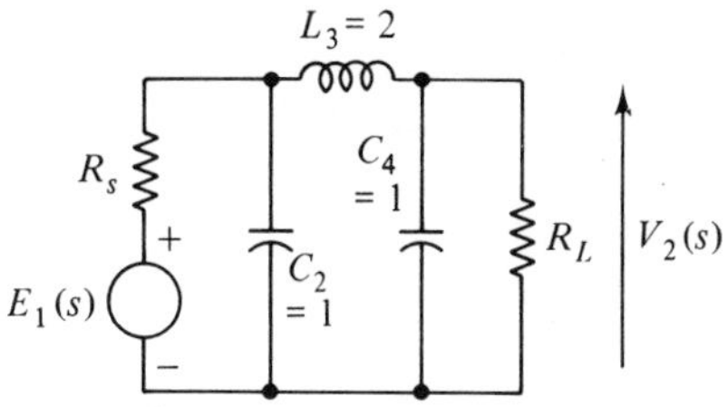

(d) Butterworth, $n = 3$, equiterminated.

FIGURE 7.16

Passive *LC* filter synthesis and design are concerned with finding a circuit having a required input-output function $H(s) \equiv V_2(s)/E_1(s)$. These topics are well documented and have led to successful techniques for implementing highly selective "brick-wall"-type magnitude functions $G(\omega) \equiv |H(j\omega)|$. In this book we are *not* primarily concerned with the synthesis and design of passive *LC* filters; however, *it must be recognized that LC filters may be designed in such a way that the magnitude function $G(\omega)$ is* *extraordinarily insensitive to perturbations of the LC elements and to the terminating resistances*. Therefore, our objective here is limited to providing a brief review of the principles of LC filter design and to providing an explanation for the low-sensitivity properties of these LC filters. Following the description of the sensitivity properties of lossless two-ports, the signal-flow graph representations of lossless *ladder* two-ports are described.

7-5.1 Derivation of the Feldtkeller Energy Equation

Consider the behavior of the network shown in Fig. 7.16(a) under conditions of steady-state sinusoidal excitation. All of the allowable elements of the lossless two-port network N absorb zero net energy over any one period of the steady-state sinusoidal excitation. Consequently, the average power absorbed by the network N under steady-state sinusoidal conditions is zero. Therefore, it follows that the average power $P_1(\omega)$ delivered to port 1 [via the input impedance $Z_1(j\omega)$] must equal the average power $P_2(\omega)$ delivered to the load resistance R_L. Writing

$$Z_1(j\omega) = R_1(\omega) + jX(\omega) \tag{7.123}$$

it follows that

$$P_1(\omega) = |I_1(j\omega)|^2 R_1(\omega) = \frac{|E_1(j\omega)|^2 R_1(\omega)}{|R_s + Z_1(j\omega)|^2} \tag{7.124}$$

and

$$P_2(\omega) = \frac{|V_2(j\omega)|^2}{R_L} \tag{7.125}$$

Thus, due to the losslessness of N, we may use $P_1(\omega) = P_2(\omega)$ in equations (7.124) and (7.125) to obtain

$$\frac{P_2(\omega)}{|E_1(j\omega)|^2/4R_s} = \frac{4R_s R_1}{|R_s + Z_1(j\omega)|^2} \tag{7.126}$$

This expression is simplified in two ways. First, by noting that $|E_1(j\omega)|^2/4R_s$ is equal to $P_{\max}$ from equation (7.122). Second, by noting that

$$4R_s R_1 \doteq |R_s + Z_1(j\omega)|^2 - |R_s - Z_1(j\omega)|^2 \tag{7.127}$$

By substituting equations (7.127) and (7.122) into (7.126),

$$\frac{P_2(\omega)}{P_{\max}} = 1 - \left| \frac{R_s - Z_1(j\omega)}{R_s + Z_1(j\omega)} \right|^2 \tag{7.128}$$

This is the required result, and we shall express it in the terms that are used by *LC* filter designers. Thus, we note that $P_2(\omega)/P_{\max}$ may be defined in terms of $V_2(j\omega)$ and $E_1(j\omega)$ as

$$\frac{P_2}{P_{\max}} = \frac{4R_s|V_2(j\omega)|^2}{R_L|E_1(j\omega)|^2} = \frac{4R_s}{R_L}|H(j\omega)|^2 \tag{7.129}$$

where $H(s)$ is the input-output transform function given by $H(s) \equiv V_2(s)/E_1(s)$. The *transmission coefficient* $t(s)$ is defined by the relationship

$$t(s) \equiv 2\sqrt{\frac{R_s}{R_L}}\; H(s) \tag{7.130}$$

so that, by comparison with equation (7.129),

$$|t(j\omega)|^2 = \frac{P_2(\omega)}{P_{\max}} \tag{7.131}$$

It is important to note that $P_2(\omega) \leqslant P_{\max}$, so that $|t(j\omega)|^2$ is bounded by the relationship

$$0 \leqslant |t(j\omega)|^2 \leqslant 1 \tag{7.132}$$

We define the *input reflection coefficient* $\rho_1(s)$ as

$$\rho_1(s) \equiv \frac{R_s - Z_1(s)}{R_s + Z_1(s)} \tag{7.133}$$

so that equations (7.128), (7.131) and (7.133) provide the Feldtkeller equation

$$|t(j\omega)|^2 = 1 - |\rho_1(j\omega)|^2 \quad \text{FELDTKELLER EQUATION} \tag{7.134}$$

This result expresses the fact that there exists a balance of average (normalized) power at port 1 of the network N; the sum of the normalized transmitted power $|t(j\omega)|^2$ and the normalized reflected power $|\rho_1(j\omega)|^2$ at port 1 is unity. The Feldtkeller equation is important for two immediate purposes: it may be used to design LC filters from a specified magnitude function $G(\omega)$ and, secondly, it may be used to explain the important low-sensitivity properties of LC filters. It follows directly from equation (7.134) that

$$|t(s)|^2 + |\rho_1(s)|^2 = 1$$

and

$$t(s)t(-s) + \rho_1(s)\rho_1(-s) = 1 \tag{7.135}$$

7-5.2 Review of the *LC*-Filter Synthesis Procedure

The usual design problem is to realize a prescribed transfer function $H(s)$. The step-by-step procedure may be divided into two parts: first, the determination of $Z_1(s)$ for specific terminations R_s and R_L; second, the *synthesis* of a particular network N with termination R_L and input impedance $Z_1(s)$. The procedure is summarized below:

(i) Determine $|t(s)|^2$, using equation (7.130).

(ii) Determine $\rho_1(s)$, using equation (7.135).

(iii) Determine $Z_1(s)$, using equation (7.133).

(iv) Synthesize a lossless two-port network N that is terminated at port 2 in R_L and with the required input impedance $Z_1(s)$.

Step (iv) is a nontrivial exercise in network synthesis that is beyond the scope of this text. However, the four steps above are illustrated in the following example.

Synthesis of a Butterworth lowpass *LC* filter, $n = 3$: In this example, the principles involved in the design of a resistively terminated LC filter are reviewed for the Butterworth $n = 3$ transfer function. The required nth-order transfer function $H(j\omega)$ is as follows:

$$|H(j\omega)|^2 \equiv \frac{H_0}{1 + \omega^{2n}}, \qquad H_0 \text{ constant} \tag{7.136}$$

from which the normalized result in equation (7.14) follows by substituting $\omega = s/j$. Thus,

$$|H(s)|^2 = H(s)H(-s) = \frac{H_0}{1 + (-1)^n s^{2n}} \tag{7.137}$$

We *choose* to synthesize an LC filter that implements equation (7.136) under the equiterminated condition, $R_s \equiv R_L \equiv 1\ \Omega$. Then, it follows from equations (7.136) and (7.130) that

$$|t(j\omega)|^2 = \frac{4H_0}{1+\omega^{2n}} \tag{7.138}$$

Since $|t(j\omega)|^2$ is bounded by equation (7.132), we select $H_0 = \frac{1}{4}$ so that this bound is exactly satisfied over all ω. Thus, substituting $H_0 = \frac{1}{4}$ and $\omega = s/j$ into equation (7.138) gives the left-hand side of the Feldtkeller equation as

$$|t(s)|^2 \equiv t(s)t(-s) = \frac{1}{1-s^6} \tag{7.139}$$

We have completed part (i) of the four-step procedure. Next, we use the Feldtkeller equation to find a suitable input reflection coefficient $\rho_1(s)$. Thus, by substituting equation (7.139) into (7.135),

$$|\rho_1(s)|^2 = 1 - \frac{1}{1-s^6} = \frac{-s^6}{1-s^6} \tag{7.140}$$

We must now find a suitable $\rho_1(s)$ from equation (7.140). The pole-zero diagram of $|\rho_1(s)|^2$ is shown in Fig. 7.16(b); since $|\rho_1(s)|^2$ is equal to $\rho_1(s)\rho_1(-s)$ it follows that the poles and zeros of $\rho_1(s)$ are the negatives of the poles and zeros of $\rho_1(-s)$. Since the Feldtkeller equation implies that the poles of $\rho_1(s)$ are the poles of $t_1(s)$, then for stability we must select the poles of $\rho_1(s)$ to be the three left-half plane poles of $|\rho_1(s)|^2$. Therefore, the poles of $\rho_1(s)$ are at -1, $-\frac{1}{2} \pm j\sqrt{3}/2$, and then

$$\rho_1(s) = \frac{s^3}{(s+1)\left(s+\frac{1}{2}+j\frac{\sqrt{3}}{2}\right)\left(s+\frac{1}{2}-j\frac{\sqrt{3}}{2}\right)} \tag{7.141}$$

or

$$\rho_1(s) = \frac{s^3}{s^3+2s^2+2s+1} \tag{7.142}$$

We have completed step (ii) of the design procedure. [Note that the zeros of $|\rho_1(s)|^2$ are generally not constrained to be at the origin as in this particular example; so, since stability is *not* a problem, there is generally not any need to assign left-half plane zeros to $\rho_1(s)$ and right-half plane zeros to $\rho_1(-s)$. There is, therefore, *generally* some choice here as to *which* zeros of $|\rho_1(s)|^2$ are assigned to $\rho_1(s)$, and it may be shown that this choice has some consequence on the sensitivity behavior of the resultant *LC* filter.]

The driving-point impedance $Z_1(s)$ is now found by substituting equation (7.142) and $R_s = 1$ into equation (7.133) to obtain

$$Z_1(s) = \frac{1 - \rho_1(s)}{1 + \rho_1(s)} = \frac{2s^2 + 2s + 1}{2s^3 + 2s^2 + 2s + 1} \tag{7.143}$$

Step (iii) of the procedure is now complete, and the final step is to synthesize a lossless two-port N that possesses the driving-point impedance $Z_1(s)$ when terminated at port 2 with $R_L = 1\ \Omega$. We do not pursue the general problem of the $Z_1(s)$ driving-point synthesis but simply observe that the ladder topology shown in Fig. 7.12(c) possesses a driving-point impedance that is given by

$$Z_1(s) = \cfrac{1}{Y_2 + \cfrac{1}{Z_3 + \cfrac{1}{Y_4 + \dots \dots \cfrac{1}{R_L}}}} \tag{7.144}$$

This expansion for the driving-point impedance of a ladder topology is important in the synthesis of *LC* filters; it is known as a *Cauer form.* We may perform a continued fraction expansion of $Z_1(s)$ in equation (7.143), to obtain

$$Z_1(s) = \cfrac{1}{s + \cfrac{1}{2s + \cfrac{1}{s + \cfrac{1}{1}}}} \tag{7.145}$$

so that, by comparison of equations (7.144) and (7.145) it is evident that we may select

$$Y_2 = s, \quad Z_3 = 2s, \quad Y_4 = s, \quad R_L = 1$$

and the resultant equiterminated lowpass Butterworth $n = 3$ *LC* ladder filter is that shown in Fig. 7.12(d). This completes the design example.

7-5.3 Sensitivity Performance of Resistively Terminated Lossless Two-Ports

In this section, it is shown that resistively terminated lossless two-port networks may be designed to yield magnitude functions $G(\omega)$ that are highly insensitive to the parameters x_i of the two-port network at all passband frequencies if $|t(j\omega)|$ is approximately unity throughout the passband.

It is required that N be *lossless*; since *active* elements such as $-C$ and $-L$ may also be lossless, there is no requirement that all the elements within the two-port be *passive*. For the purposes of this discussion, we require that the parameters x_i of the lossless network N satisfy the Feldtkeller equation in the sense that

$$|t(j\omega, x_i)|^2 = 1 - |\rho_1(j\omega, x_i)|^2, \qquad \forall\, x_i \tag{7.146}$$

In the case of a passive LC network N, the parameters x_i are the values of the LC elements. [It is important to note that in subsequent chapters we consider RC-active two-ports N for which equation (7.146) is satisfied for some but not all of the RC elements employed to synthesize the two-port.] Writing $|t| \equiv |t(j\omega, x_i)|$ and $|\rho_1| \equiv |\rho_1(j\omega, x_i)|$ and differentiating equation (7.146) with respect to x_i, we find that

$$2|t|\frac{\partial|t|}{\partial x_i} = -2|\rho_1|\frac{\partial|\rho_1|}{\partial x_i} \tag{7.147}$$

so that

$$S_{x_i}^{|t|} \equiv \frac{\partial|t|x_i}{\partial x_i|t|} = -\frac{|\rho_1|\partial|\rho_1|}{|t|^2\partial x_i}\cdot x_i$$

which we write in the form

$$S_{x_i}^{|t|} = S_{x_i}^{G(\omega)} = -\frac{x_i}{|t|}\left(\frac{|\rho_1|}{|t|}\right)\left(\frac{\partial|\rho_1|}{\partial x_i}\right) \tag{7.148}$$

where we equate $S_{x_i}^{G(\omega)}$ to $S_{x_i}^{|t|}$ due to the proportionality between $t(s)$ and $H(s)$ that is implicit in equation (7.130). Consider now the sensitivity function in equation (7.148) in the passband at some frequency ω_1 where $|t(j\omega, x_i)|$ is at the upper bound of unity. Then, by the Feldtkeller equation, $|\rho_1(j\omega_1, x_i)|$ is zero. We sketch a typical relationship between $|\rho_1(j\omega, x_{iN})|$ and ω in Fig. 7.17(a), where x_{iN} is the nominal design value of the parameter x_i. Consider now that x_i is perturbed by some arbitrarily small amount Δx_i from the nominal value x_{iN}. Then, since $|\rho_1(j\omega, x_i)|$ is a continuous non-negative function of the parameters x_i it follows that the

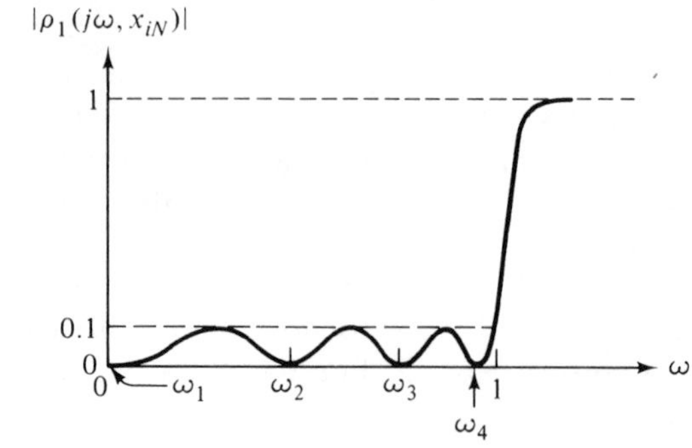

(a) Typical relationship between input reflection coefficient and frequency.

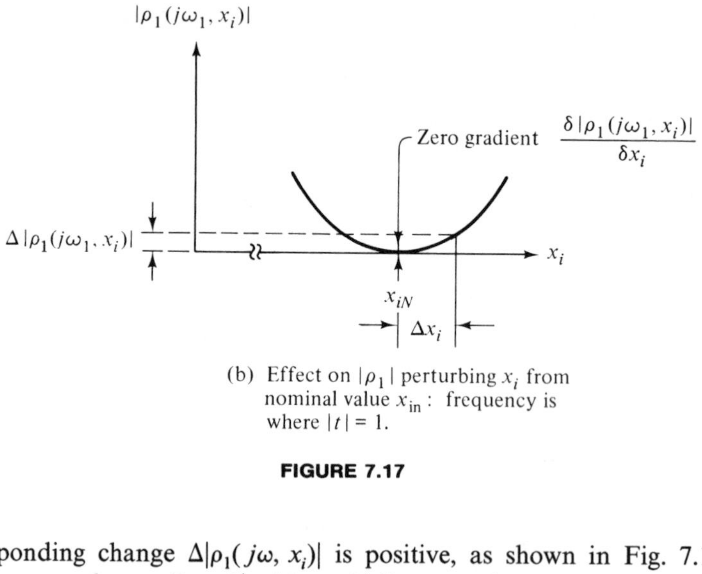

(b) Effect on $|\rho_1|$ perturbing x_i from nominal value x_{in}: frequency is where $|t| = 1$.

FIGURE 7.17

corresponding change $\Delta|\rho_1(j\omega, x_i)|$ is positive, as shown in Fig. 7.17(b). Furthermore, the gradient $\partial|\rho_1(j\omega_1, x_i)|/\partial x_i$ must be zero; if this were not true, it would always be possible to find some arbitrarily small $\pm\Delta x_i$ that would cause $|\rho_1(j\omega_1, x_i)|$ to be negative and thereby contradict the inherent positivity of this function. Thus, at all frequencies $\omega_1, \omega_2, \omega_3, \ldots$, etc., where $|t(j\omega_{1,2,3}, x_i)|$ are unity, the function $\partial|\rho|/\partial x_i$ is zero and, therefore, by equation (7.148), $S_{x_i}^{G(\omega_1)} = S_{x_i}^{G(\omega_2)} = S_{x_i}^{G(\omega_3)} = \ldots = 0$. It also follows that the term $[|\rho_1|/|t|]$ in equation (7.148) is much less than unity throughout the passband region because, by definition of *passband*, $|t| \cong 1$ (and $|\rho_1| \ll 1$). Thus, both terms within the parentheses of equation (7.148) are zero at the frequencies where $|t| = 1$, and we can expect $S_{x_i}^{G(\omega)}$ to be low-valued throughout the passband.

EXAMPLE 7.4: Chebychev Lowpass LC Filter, $n = 7$, 1.0 dB Passband Ripple. For purposes of comparison with previous sensitivity calculations that have been performed for the DF structure and cascaded-BIQUAD DF structure, we return to the Chebychev lowpass $n = 7$ filter that is specified in Table 7-1. A suitable *LC* filter realization of this filter that corresponds to a passband for which $|t(j\omega)|$ has maximum values of unity is given in Fig. 7.18(a). The frequencies ω_1, ω_2, ω_3, and ω_4, where $|t(j\omega)|$ is unity, are given by

$$\omega_1 = 0, \quad \omega_2 = 0.42, \quad \omega_3 = 0.78, \quad \omega_4 = 0.97 \text{ rad/sec} \qquad (7.149)$$

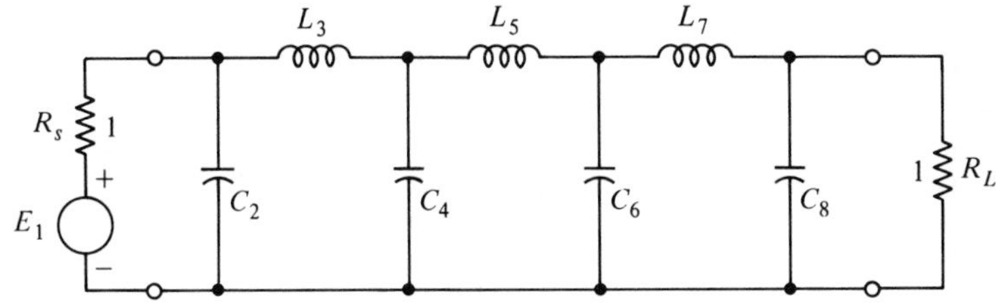

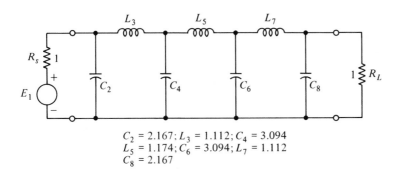

$C_2 = 2.167; L_3 = 1.112; C_4 = 3.094$
$L_5 = 1.174; C_6 = 3.094; L_7 = 1.112$
$C_8 = 2.167$

(a) Lowpass Chebychev, $n = 7$, 1.0 dB bandpass ripple.

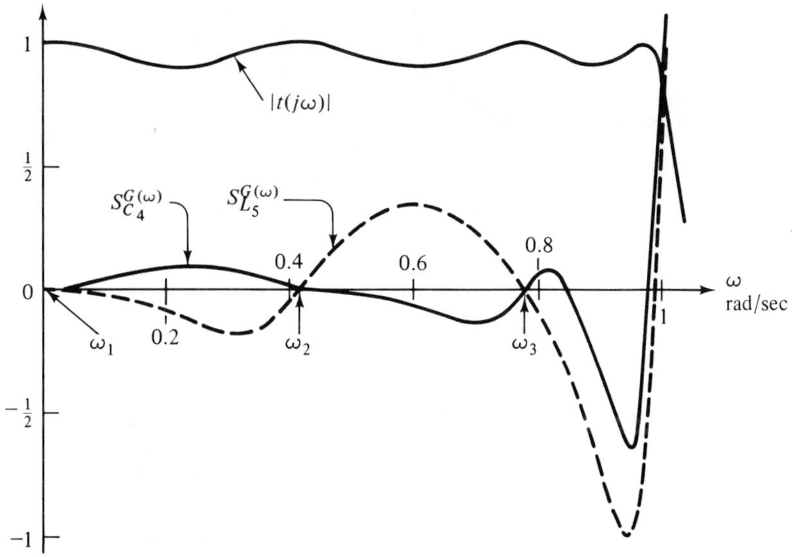

(b) Single-element sensitivities and $|t(j\omega)|$ for an LC Chebychev filter.

FIGURE 7.18

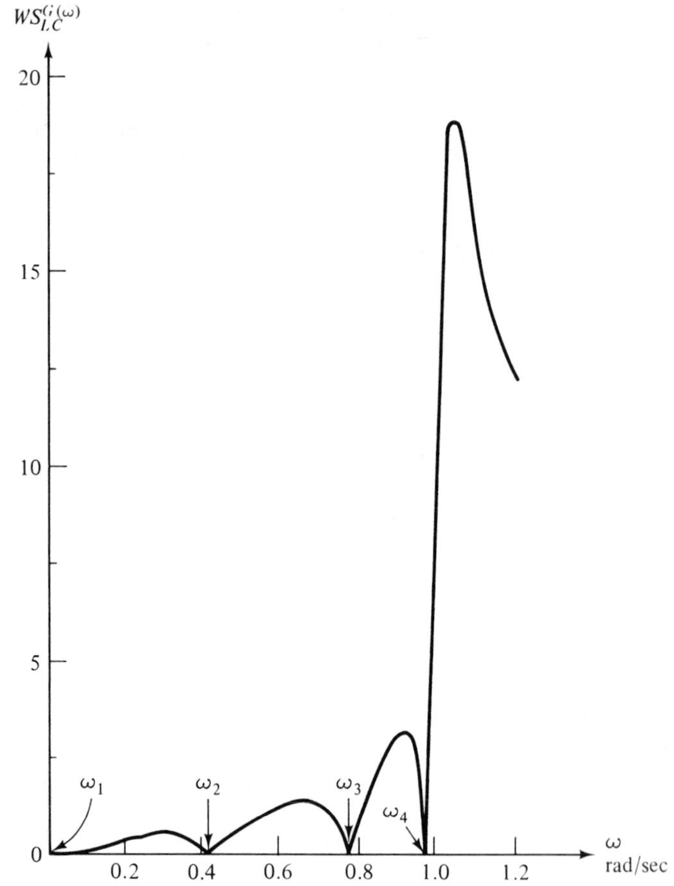

(c) Worstcase sensitivity to LC elements.

FIGURE 7.18 *(Continued).*

The function $|t(j\omega)|$ is shown in Fig. 7.18(b), and the sensitivity functions $S_{C_4, L_5}^{G(\omega)}$ are also shown on the same diagram. Note that wherever $|t(j\omega)|$ is unity, the sensitivities $S_{C_4, L_5}^{G(\omega)}$ are zero. Furthermore, the maximum values of $|S_{C_4, L_5}^{G(\omega)}|$ are less than unity and are typical of the seven single-element LC sensitivities for this network N. This result should be compared with the typical DF transmittance-conductance sensitivity shown in Fig. 7.5 for the same transfer function.

The worstcase sensitivity function obtained by summing the seven LC single-element sensitivities $WS_{LC}^{G(\omega)}$ is shown in Fig. 7.18(c) for this particular numerical example. By comparison with Fig. 7.12 it follows that the LC filter of Fig. 7.18(a) is far less sensitive to its LC elements than is either the DF structure or the cascaded-BIQUAD DF structure to the corresponding transmittances-conductances. Note that the $WS_{LC}^{G(\omega)}$ curve of Fig. 7.18(c) is low-valued throughout the passband and zero at the frequencies ω_1, ω_2, ω_3, and ω_4 where, of course, all the single-element sensitivities are also zero due to the fact that $|t|$ is unity at these frequencies.

Sensitivity to the resistive terminations: The reader may have noticed that we have not thus far considered the sensitivity performance of the resistively terminated LC ladder filter with respect to the resistive terminations R_s and R_L. It may be shown, and is stated here without proof, that

$$|S_{R_s, R_L}^{G(\omega)}| \leqslant 1, \qquad \forall\, \omega \tag{7.150}$$

That is, the sensitivity of the gain function $G(\omega)$ to the termination resistance R_s or R_L is everywhere less than unity.

7-5.4 Signal-Flow Graph Representations of Resistively Terminated *LC* Ladder Filters

Consider the general M branch ladder topology shown in Fig. 7.19 in terms of parallel impedances Z_i and series admittances Y_{i+1}, with the voltages V_i and currents I_{i+1} chosen with the alternating polarities indicated in the diagram. Then, for the individual branches of the ladder topology, we may write the following steady-state equations:

$$\left.\begin{array}{ll}
\text{Branch 1:} & -E_1Y_1 = I_1 - V_2Y_1 \\
\text{Branch 2:} & 0 = V_2 + [I_1 + I_3]Z_2 \\
\text{Branch 3:} & 0 = I_3 - [V_2 + V_4]Y_3 \\
\vdots & \vdots \\
\text{Branch } i: & 0 = V_i + [I_{i-1} + I_{i+1}]Z_i \\
\text{Branch } i+1: & 0 = I_{i+1} - [V_i + V_{i+2}]Y_{i+1} \\
\text{Branch } i+2: & 0 = V_{i+2} + [I_{i+1} + I_{i+3}]Z_{i+2} \\
\text{Branch } i+3: & 0 = I_{i+3} - [V_{i+2} + V_{i+4}]Y_{i+3} \\
\vdots & \vdots \\
\text{Branch } M-1: & 0 = I_{M-1} - [V_{M-2} + V_M]Y_{M-1} \\
\text{Branch } M: & 0 = V_M + [I_{M-1}]Z_M
\end{array}\right\} \tag{7.151}$$

This set of M equilibrium equations is equivalent to (and completely describes) the ladder topology of Fig. 7.19. The matrix formulation of these equations is also shown in Fig. 7.19.

The electrical network of Fig. 7.19, the set of equations in (7.151), and the signal-flow graph that we are about to describe are simply different

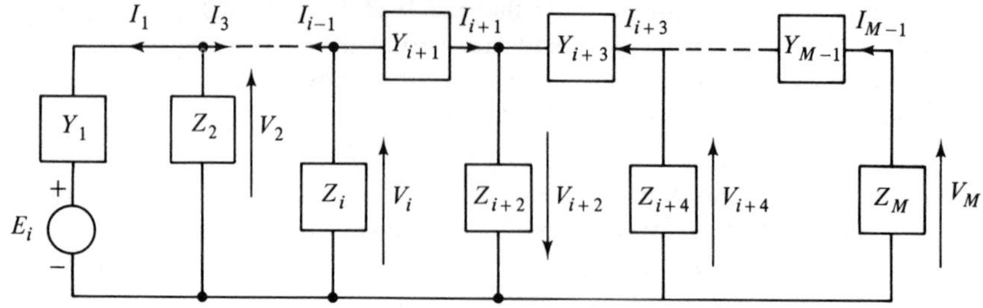

(a)

$$
\begin{bmatrix} -E_1 Y_1 \\ 0 \\ 0 \\ \vdots \\ 0 \\ 0 \\ \vdots \\ 0 \\ 0 \end{bmatrix}
=
\begin{bmatrix}
1 & Y_1 & 0 & 0 & \cdots & & & & & 0 \\
Z_2 & 1 & Z_2 & 0 & \cdots & & & & & 0 \\
0 & -Y_3 & 1 & -Y_3 & 0 & \cdots & & & & 0 \\
\vdots & & & & \ddots & & & & & \vdots \\
0 & & 0 & Z_i & 1 & Z_i & 0 & \cdots & & 0 \\
0 & \cdots & & 0 & -Y_{i+1} & 1 & -Y_{i+1} & 0 & \cdots & 0 \\
\vdots & & & & & & & & & \vdots \\
0 & \cdots & & & 0 & 0 & -Y_{M-1} & 1 & & -Y_{M-1} \\
0 & \cdots & & & & 0 & 0 & Z_M & & 1
\end{bmatrix}
\begin{bmatrix} I_1 \\ V_2 \\ I_3 \\ \vdots \\ V_i \\ I_{i+1} \\ \vdots \\ I_{M-1} \\ V_M \end{bmatrix}
$$

(b)

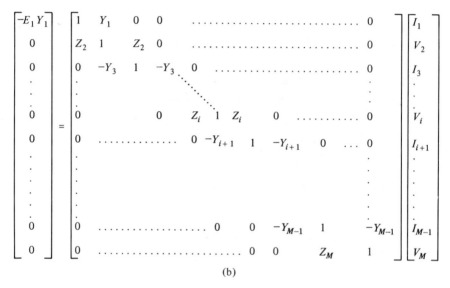

FIGURE 7.19 *Ladder topology and corresponding matrix description.*

ways of describing the same system. *The signal-flow graph description is useful because it allows RC-active implementations of these ladder filters to be described in a straightforward way.* Consider the signal-flow graph descriptions of the equations that describe the ith and $(i + 1)$th branches in (7.151); thus, in Fig. 7.20(a) we *choose* to represent the signals I_{i-1} and I_{i+1} as inputs for the ith branch equations and the signals V_i and V_{i+2} as inputs for the $(i + 1)$th branch equation. The output signals are V_i and I_{i+1}, respectively, and the nonunity transmittances are $-Z_i$ and $+Y_{i+1}$, respectively. Similarly, the signal-flow graphs for the input branch equation and output branch equation are shown in Fig. 7.20(a). The *complete* signal-flow graph representation of Fig. 7.19 and equations (7.151) is, therefore, obtained by appropriately interconnecting the individual two-input/single-output branch signal-flow graphs to result in Fig. 7.20(b). It is

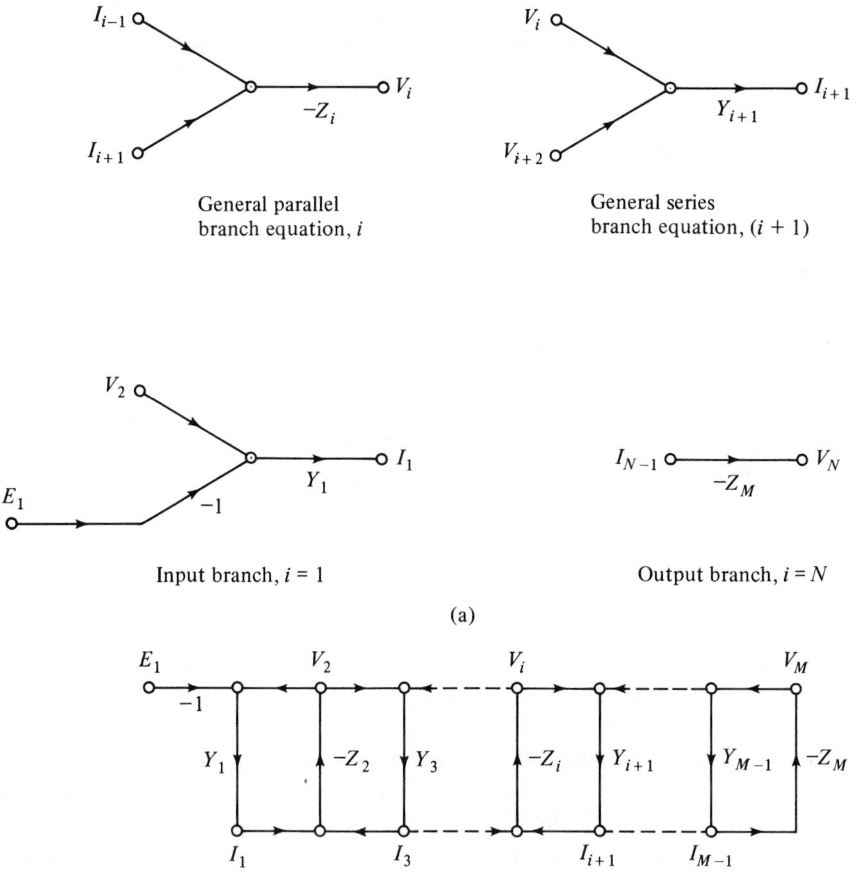

(b) A signal-flow graph description of the ladder topology in Fig. 7.19.

FIGURE 7.20

now pertinent to consider the particular signal-flow graphs that correspond to the widely employed lowpass and bandpass *LC* filter topologies.

Lowpass *LC*-filter signal-flow graphs: The most straightforward application of the signal-flow graph in Fig. 7.20(b) is the integrator simulation of the type of lowpass *LC* filter that is shown in Fig. 7.18(a). That is,

$$\left.\begin{aligned} &Y_1 \equiv G_1, \quad Z_2 \equiv \frac{1}{sC_2}, \quad Y_3 \equiv \frac{1}{sL_3}, \quad \ldots, \quad Z_i \equiv \frac{1}{sC_i}, \\ &Y_{i+1} = \frac{1}{sL_{i+1}}, \quad Y_{M-1} \equiv \frac{1}{sL_{M-1}}, \quad Z_M \equiv R_M \end{aligned}\right\} \tag{7.152}$$

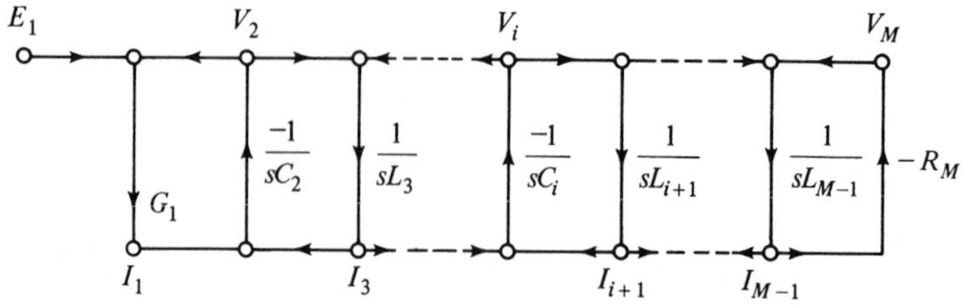

FIGURE 7.21 *Integrator signal-flow graph model of the LC lowpass filter. Figure 7.18(a).*

The corresponding signal-flow graph is shown in Fig. 7.21, from which it is observed that the implementation requires noninverting and inverting integrators to realize the transmittances $Y_{i+1} = 1/sL_{i+1}$ and $-Z_i = -1/sC_i$, respectively. Of course, the sensitivities of the transfer function $H(s)$ to the terms $1/L_{i+1}$ and $-1/C_i$ in the integrating transmittances of the signal-flow graph are identical to that of the LC-filter transfer $H(s)$ to the LC elements L_{i+1} and C_i. That is, $S_{L_{i+1}}^{H(s)}$ and $S_{C_i}^{H(s)}$ are the same for both the LC electrical filter and the corresponding signal-flow graph.

It is important, however, to realize that *all* of the *LC electrical filter* parameters $C_2, L_3, \ldots, C_i, L_{i+1}, \ldots, L_{M-1}$, which we denote x_i, satisfy the Feldtkeller equation (7.146) and lead to the low sensitivities $S_{x_i}^{G(\omega)}$ throughout the passband, where $|t(j\omega)| \simeq 1$.

Bandpass *LC*-filter signal-flow graphs: There is a variety of LC-filter topologies that are used to realize *bandpass* magnitude functions $G(\omega)$. We consider here the particularly straightforward case for which all the finite zeros of $H(s)$ are at the origin in the s-plane. Such ladder filters consist of the series LC branches Y_{i+1} and parallel LC branches Z_i shown in Fig. 7.22(a). The corresponding expressions for Y_{i+1} and Z_i are shown in Fig. 7.22(a), from which it follows (by comparison with Fig. 7.20) that we may write signal-flow graph models for these ladder branches, as shown in Fig. 7.22(b) and (c). The signal-flow graph for the complete bandpass LC ladder filter is therefore given by substituting Figs. 7.22(b) and (c) into Fig. 7.20(b), thereby resulting in the required complete graph in Fig. 7.22(d).

There is a large number of methods that may be used to implement RC-active versions of Fig. 7.22(d). We consider some of these methods in chapter 9; it may be obvious to the reader that we can use *inverting integrators* to implement the transmittances $-1/sC_i$ and $-1/sL_i$. Similarly, *noninverting integrators* can be used to implement the transmittances $+1/sC_i$ and $+1/sL_i$. Thus, if we allow one OP AMP for each inverting integrator and two OP AMPs for each noninverting integrator then an nth order RC-active version of the bandpass SFG in Fig. 7.22(d) *requires* $3n$ *OP AMPs*.

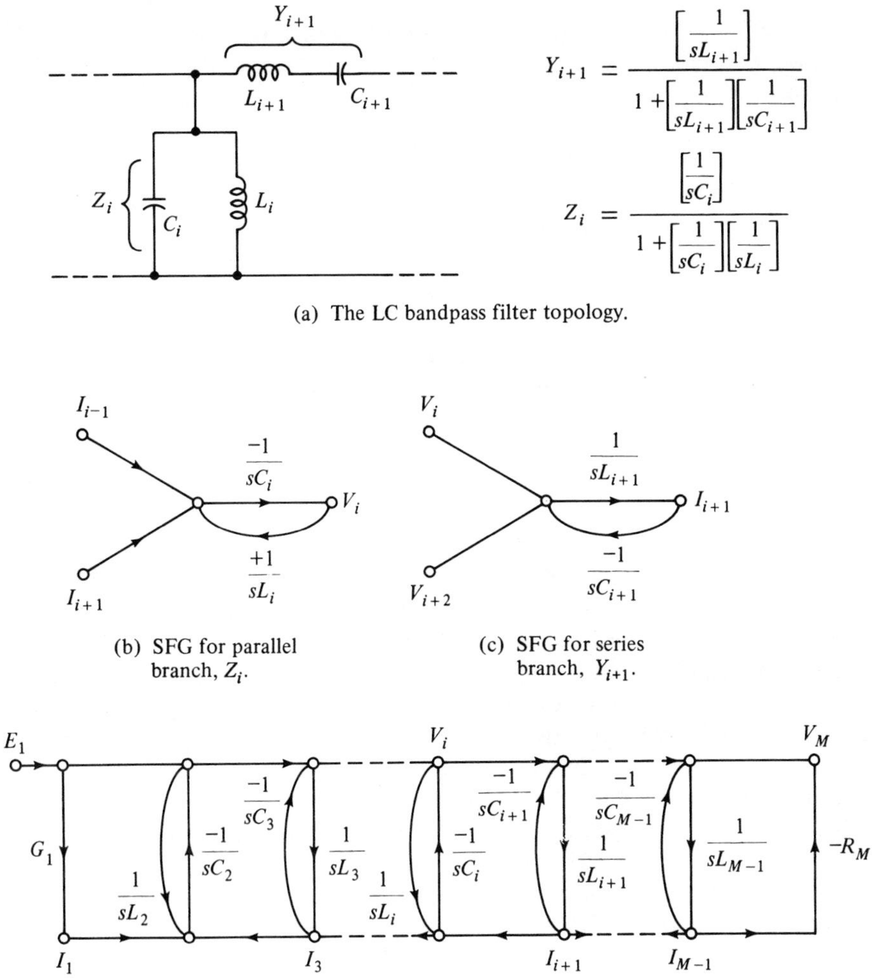

$$Y_{i+1} = \frac{\left[\frac{1}{sL_{i+1}}\right]}{1 + \left[\frac{1}{sL_{i+1}}\right]\left[\frac{1}{sC_{i+1}}\right]}$$

$$Z_i = \frac{\left[\frac{1}{sC_i}\right]}{1 + \left[\frac{1}{sC_i}\right]\left[\frac{1}{sL_i}\right]}$$

(a) The LC bandpass filter topology.

(b) SFG for parallel branch, Z_i.

(c) SFG for series branch, Y_{i+1}.

(d) SFG for the LC bandpass filter; suitable for R-C active implementation using integrators.

FIGURE 7.22

7-5.5 Summary of the Performance of *LC* Filters

It is possible to design a *bandpass LC ladder* filter directly from a lowpass *LC* prototype ladder filter. The design procedure involves the lowpass-to-bandpass transformation and is considered in some detail in chapter 9.

The *losslessness* of the *LC* two-port may be used to develop the Feldtkeller equation (7.134) from which it is possible to synthesize a given $G(\omega)$ in such a way that $G(\omega)$ is highly insensitive to the *LCR*-element values throughout the passband. This is acheived by ensuring that $|t(j\omega)|$ is close to the upper bound of unity throughout the passband.

By means of a Chebychev $n = 7$ numerical example, it is found that the passband sensitivity performance with respect to *LC*-element values is far superior to that of the DF BIQUAD implementation.

7-6 SUMMARY

In this chapter, we have described some of the most important network structures and the sensitivity behavior of these structures with respect to critical parameters has been analyzed. The direct-form (DF) structure is found to be highly sensitive to the feedback transmittances b_i, for example. Cascaded biquadratic direct-form structures (DF BIQUADs) alleviate this sensitivity problem by essentially decoupling the stages so that the overall sensitivity performance (in the worstcase sense) depends only on that of the individual biquadratic sections. The sensitivity functions associated with individual biquadratic sections are often much smaller in magnitude than that of the overall DF structure.

Various decompositions of the biquadratic function are possible. The LBI/DF, HBI/DF and MI/DF possess DF denominators and are shown, in chapter 9, to be particularly economical in terms of the *number* of OP AMPs required. At the expense of an increase in the sensitivity functions, it is possible to decompose the denominator of a biquadratic function so that only one feedback loop is required (as opposed to two feedback loops for DF denominators). Three types of single feedback path biquadratic structures are considered, two of which possess sensitivities $S^{G_b(\omega)}_{A_0, B}$ that are at least proportional to Q_p^3 and are, therefore, not recommended. However, the type I structure with a bandpass function $T_{32}(s)$ is useful and will be shown to lead to acceptable biquadratic *RC*-active circuits, provided that some care is taken to control the quantity $Q_N T_{32}(j\omega_P)$.

In the subsequent chapters of this book, we essentially incorporate the electronic building-blocks that are described in chapters 2, 3, 4, and 5 into the structures that are described in this chapter and thereby learn to design *RC*-active filter circuits.

PROBLEMS

7-1. The direct-form signal-flow graph employs inverting-integrator transmittances $-s^{-1}$. Replace each transmittance $-s^{-1}$ by a *noninverting-integrator* transmittance $+s^{-1}$ (Fig. 7.1) and thereby prove that general Nth-order transfer functions $H(s)$ may be realized in this way. Derive the values of the remaining transmittances in terms of a_i and b_i.

Repeat the above for noninverting differentiators $+s$ and inverting differentiators $-s$.

7-2. Design a direct-form inverting-integrator version of Fig. 7.2(a) to realize the transfer function

$$H(s)=\frac{s^2+\dfrac{\omega_Z s}{Q_Z}+\omega_Z^2}{s^2+\dfrac{\omega_P s}{Q_P}+\omega_P^2}, \qquad \begin{array}{l}\omega_{Z,P}>0\\ Q_{Z,P}>0\end{array}$$

expressing, where necessary, the conductances a_0, a_1, a_2, b_0, b_1, and b_2 in terms of ω_Z, ω_P, Q_P, and Q_Z. (All unmarked RC elements in Fig. 7.2(a) have unity value.)

7-3. Modify the design in Problem 7.2 by the addition of an extra inverting amplifier to realize negative values of Q_Z.

7-4. Use the scheme shown in Fig. 7.2(b) to eliminate an operational amplifier from Fig. 7.2(a) and thereby realize the transfer function

$$H(s)=\frac{s^2+3}{s^3+2s^2+9s+7}$$

using inverting-integrators.

7-5. Derive expressions for the $-$HP, $+$BP, and $-$LP outputs of the direct-form biquadratic structure BIQUAD DF shown in Fig. 7.8.

7-6. Derive expressions for the $-$BP, $+$LP, and $-$LP outputs of the signal-flow graph in Fig. 7.9(a).

7-7. Derive expressions for the $+$HP, $-$HP, and $+$BP outputs of the signal-flow graph in Fig. 7.9(b).

7-8. The LBI/DF signal-flow graph in Fig. 7.9(a) is used to realize

$$H_b(s)=5\left(\frac{s^2+0.1s+2}{s^2+0.05s+1}\right)$$

Derive the numerical values of the transmittances $+\alpha_2$, $+b_{1b}$, $+b_{0b}$, $-\alpha_1$, and $-\alpha_0$ required in Fig. 7.9(a).

7-9. Repeat Problem 7.8 for the BHI/DF signal-flow graph in Fig. 7.9(b).

7-10. Analyze the MI/DF signal-flow graph in Fig. 7.9(c) to prove that equation (7.30) is valid. Select transmittances $-b_{0b}$, b_{1b}, α_1, α_2, α_3, and α_4 to give the biquadratic transfer function having $\omega_P = 1$, $Q_P = 10$, $\omega_Z = 2$, $Q_Z = \infty$ and a dc gain $H(0)$ of 100.

7-11. Prove equations (7.72) from equation (7.71) for the noninverting single-OP AMP amplifier circuit shown in Fig. 5.8.

At a closed-loop voltage gain K equal to 100 and at a frequency of 10 kHz, the gain-bandwidth product B of the OP AMP increases from 1 MHz to 1.1 MHz. Use equation (7.72) to estimate the fractional change in the phase function $\angle[K(j\omega)]$ of the amplifier. [*Hint*: You can apply the result in equation (6.57) to equation (7.72) by letting $H(j\omega)$ to be equal to $K(j\omega)$].

7-12. Derive equation (7.73) from equation (5.22) and thereby prove equations (7.74) for the inverting amplifier of Fig. 5.5(a).

7-13. Verify that equation (7.101) is valid for the type IIA noninverting amplifier $+K$ structure and that equation (7.118) is valid for the type IIB noninverting amplifier $+K$ structure.

7-14. A type I noninverting amplifier structure, as described in section 7-4.2, is designed using a passive network N_{RC} for which

$$\omega_N = 10^4 \text{ rad/sec}$$

$$Q_N = \frac{1}{4}, \qquad \alpha_1 = 2\times 10^4 \text{ rad/sec}$$

Select the value of K necessary to realize

$$\omega_P = 10^4 \text{ rad/sec}, \qquad Q_P = 25$$

7-15. For the numerical design of the type I structure in Problem 7.14, the sensitivity of the transfer function $H_b(s)$ to the gain-bandwidth product B is given by

$$S_B^{H_b(s)} \approx \frac{K^2\alpha_1}{B}\left[\frac{s^2}{s^2 + \dfrac{\omega_P s}{Q_P} + \omega_P^2}\right], \qquad Q_p \gg 1$$

as shown in Table 7.4. Use the *numerical values* of Problem 7.14 to sketch the gain sensitivity function $S_B^{G_b(\omega)}$. [*Hint*: Take the real part of $S_B^{H_b(s)}$ and use equation (6.56A) and Fig. 7.10 to obtain this sketch].
What is the approximate value of $|S_B^{G_b(\omega)}|_{\max}$ for this numerical example? If $B \approx 2\pi\times 10^7$ rad/sec and $\Delta B/B = 0.01$, what is the corresponding approximate maximum fractional variation of $G_b(\omega)$? Comment on the significance of this result.

7-16. A type IIA inverting-amplifier structure, of the type discussed in section 7-4.3, is used to implement a transfer function $H_b(s)$ having

$$Q_P = 25$$

$$\omega_P = 10^4 \text{ rad/sec}$$

Assuming that $\Delta B/B = 0.01$, $Q_N = \frac{1}{4}$, and $T_{32}(0) = \frac{1}{2}$, use equation (7.103) to derive $S_B^{G_b(\omega)}$ for this example. Then, estimate the minimum value of B necessary to ensure that over all ω the above $\Delta B/B$ perturbation changes $G_b(\omega)$ by less than 10%. Comment on the significance of this result.

7-17. Prove that the power delivered to port 1 of the resistively terminated lossless two-port shown in Fig. 7.16(a) equals $P_{\max}$ if and only if the input impedance $Z_1(j\omega)$ at port 1 equals the source resistance R_s.

BIBLIOGRAPHY

BUDAK, A., *Passive and Active Network Analysis and Synthesis*, Houghton Mifflin, 1974, chapters 17, 10, 11, 12, 20.

DANIELS, R. W. *Approximation Methods for Electronic Filter Design*, McGraw-Hill, New York, 1974.

GIRLING, F. E. J., and E. F. GOOD, "Active Filters 7 and 8: The Two Integrator Loop," *Wireless World*, vol. 76, pp. 117–119 , March 1970, and pp. 134–139, April 1970.

GIRLING, F. E. J., and E. F. GOOD, "Active Filters 12 and 13: The Leapfrog or Active Ladder Synthesis; Applications of the Active Ladder Synthesis," *Wireless World*, vol. 76, p. 341–345, July 1970 and pp. 445–450, Sept. 1970.

TEMES, G. C., and S. K. MITRA, *Modern Filter Theory and Design*, Wiley, 1973, chapters 2 and 3.

ZVEREV, A. I. *Handbook of Filter Synthesis*, Wiley, New York, 1967.

8

BIQUADRATIC FILTER SYNTHESIS AND DESIGN

8-1 INTRODUCTION

The topic of biquadratic *RC*-active filter design is important because a standard and widely used method of implementing high-order transfer functions is by cascading biquadratic sections. The sensitivity performance of *RC*-active filters that are realized by this method is described in chapter 6, where it is found that the worstcase passive (*RC*) sensitivities are inherently higher throughout the passband than is obtained using equiterminated active ladder filters. Nevertheless, cascaded biquadratic *RC*-active filters have several advantages over active ladders: they often allow independent tuning of individual pairs of complex poles or zeros and they are well suited to a modular approach to the implementation of higher-order filters.

The number of different *RC*-active circuit topologies that exists for realizing quadratic and biquadratic transfer functions is enormous. Thousands of papers have been published on the topic, and there is not presently any unanimity among circuit theorists and designers concerning which types of circuits are best for specific applications. With this in mind, the reader should consider the material in this chapter (and the related sensitivity analyses of chapter 6) as an introduction to the more important *classes* of biquadratic and quadratic circuits. They are:

(i) Two integrator loop circuits (TILs),

(ii) Generalized immittance converter circuits (GICs),

(iii) Single-amplifier type I circuits (Sallen-Key).

These three classes of circuits represent hundreds of different individual RC-active topologies, some of which are analyzed here in depth.

Some important general sensitivity expressions are developed in section 8-2 for the biquadratic case; the sensitivity of the transfer function to a parameter x_i is related to the sensitivities $S_{x_i}^{\omega_{Z,P}}$ and $S_{x_i}^{Q_{P,Z}}$. It is important to realize that the designer is primarily interested in the accuracy of the *transfer function* and *not* primarily interested in the accuracies of $\omega_{Z,P}$ and $Q_{P,Z}$. Therefore, the emphasis is always on the calculation of $S_{x_i}^{H_b(s)}$ or $S_{x_i}^{G_b(\omega)}$ and not on the calculation of $\Delta Q_P/Q_P$ and $\Delta\omega_P/\omega_P$. Some simple conditions [equations (8.20)] are derived that ensure that an RC-active quadratic circuit is *near-optimum* insofar as worstcase passive sensitivity is concerned. A large number of quadratic circuits, including many that are analyzed in this chapter, are readily shown to be near-optimum in this sense. The TIL circuit is shown to be an example of a circuit with near-optimum worstcase passive sensitivity performance.

The Schoeffler sensitivity criteria $\mathbb{S}_{R,C}^{G_b(\omega)}$ have been defined in chapter 6; in section 8-2, it is shown that it is straightforward to estimate $\mathbb{S}_{R,C}^{G_b(\omega)}$ if certain commonly satisfied conditions [equations (8.25)] are met. Exact single-parameter sensitivities $S_{x_i}^{G_b(\omega)}$ are derived in the form

$$S_{x_i}^{G_b(\omega)} = q_1 + q_2\,\mathrm{Re}\left[t_{\mathrm{BP}}(j\omega)\right] + q_3\,\mathrm{Re}\left[t_{\mathrm{HP}}(j\omega)\right]$$

where $q_{1,2,3}$ are real constants and $\mathrm{Re}[t_{\mathrm{BP}}(j\omega)]$ and $\mathrm{Re}[t_{\mathrm{HP}}(j\omega)]$ are the normalized Q_p-dependent curves shown in Fig. 7.10.

Section 8-3 is devoted entirely to the classification, analysis, and modeling of TIL biquadratic and quadratic circuits. Particular attention is given to the analysis of high-frequency effects due to the finite OP AMP gain bandwidth products B. Recent results on the superior high-frequency performance that is achieved with type 1B noninverting integrators are described in some detail.

Section 8-4 includes the classification, analysis, and modeling of GIC quadratic and biquadratic circuits. Simulated-inductance circuits and FDNR circuits are analyzed and the nonideal performance is derived. It is found that GIC circuits compare favorably with TIL circuits in that the worstcase passive sensitivity is near-optimum and the active sensitivities are low-valued.

The type I [bandpass feedback $T_{32}(s)$] single-amplifier $(+K)$ circuits are the subject of section 8-5. A detailed study of the Sallen-Key lowpass structure is provided; the passive sensitivity is shown to be near-optimum if the circuit is carefully designed, but active sensitivity and large element spread are shown to be the disadvantages of this circuit under high-Q_P conditions. By careful design, active sensitivity may be minimized.

The type II single-amplifier $(-K)$ circuits are omitted because (as shown in chapter 6) the active sensitivities are much more serious than is obtained with type I circuits and, therefore, they are significantly less useful from a practical viewpoint.

8-2 ON THE SENSITIVITY PERFORMANCE OF BIQUADRATIC CIRCUITS

In this section, some general comments and results are presented that will enable the passive-sensitivity performance of biquadratic high-Q_P circuits to be derived. The circuit designer is, of course, interested in the accuracy with which the biquadratic transfer function $H_b(s)$ may be achieved. If we write $H_b(s)$ in the forms

$$H_b(s) = \frac{a_{2b}s^2 + a_{1b}s + a_{0b}}{b_{2b}s^2 + b_{1b}s + b_{0b}} \quad \text{or} \quad H_0\left[\frac{s^2 + \left(\dfrac{\omega_Z}{Q_Z}\right)s + \omega_Z^2}{s^2 + \left(\dfrac{\omega_P}{Q_P}\right)s + \omega_P^2}\right] \tag{8.1}$$

then it follows directly by the chain rule property in Table 6-1 that we may write

$$S_{x_i}^{H_b(s)} = S_{x_i}^{H_0} + \left[S_{\omega_p}^{H_b(s)}\right]S_{x_i}^{\omega_p} + \left[S_{\omega_Z}^{H_b(s)}\right]S_{x_i}^{\omega_Z} + \left[S_{Q_p}^{H_b(s)}\right]S_{x_i}^{Q_p} + \left[S_{Q_Z}^{H_b(s)}\right]S_{x_i}^{Q_Z} \tag{8.2}$$

where x_i is some real parameter, such as the value of an R or C element, upon which $H_b(s)$ is dependent. Now, the terms $S_{\omega_p}^{H_b(s)}$, $S_{\omega_Z}^{H_b(s)}$, $S_{Q_p}^{H_b(s)}$, and $S_{Q_Z}^{H_b(s)}$ are *invariant* in the sense that they depend only on the required function $H_b(s)$ and not on the particular RC-active circuit that is employed. On the other hand, the terms $S_{x_i}^{H_0}$, $S_{x_i}^{\omega_p}$, $S_{x_i}^{\omega_Z}$, $S_{x_i}^{Q_p}$, $S_{x_i}^{Q_Z}$ depend on the *choice* of RC-active circuit and, therefore, are at the influence of the designer. It is sometimes convenient to compare biquadratic circuits on the basis of these five circuit-dependent sensitivity terms. The role of these five terms in equation (8.2) may be thought of as circuit-dependent *real* weighting coefficients; the terms in the brackets [ ] are invariant. Taking the real part of equation (8.2), we have

$$S_{x_i}^{G_b(\omega)} = S_{x_i}^{H_0} + \left[S_{\omega_p}^{G_b(\omega)}\right]S_{x_i}^{\omega_p} + \left[S_{\omega_Z}^{G_b(\omega)}\right]S_{x_i}^{\omega_Z} + \left[S_{Q_P}^{G_b(\omega)}\right]S_{x_i}^{Q_P} + \left[S_{Q_Z}^{G_b(\omega)}\right]S_{x_i}^{Q_Z} \tag{8.3}$$

where $G_b(\omega)$ is the gain function $|H_b(j\omega)|$. The *invariant* terms $S_{\omega_{P,Z}}^{G_b(\omega)}$ and $S_{Q_{P,Z}}^{G_b(\omega)}$ are shown in Fig. 6.2 (for ω_P and Q_P); the remaining terms in equation (8.3) depend on the particular circuit and are constants if x_i is an R or C element.

8-2.1 The High-Q Approximation

In many practical applications, Q_P and/or $|Q_Z|$ are much greater than unity. Consequently, the terms $[S_{\omega_p}^{G_b(\omega)}]$ and $[S_{\omega_Z}^{G_b(\omega)}]$ in equation (8.3) exhibit large peak-to-peak variations from $-Q_{P,Z}$ to $+Q_{P,Z}$ and therefore dominate the terms $[S_{Q_p}^{G_b(\omega)}]$ and $[S_{Q_Z}^{G_b(\omega)}]$ which, according to Fig. 6.2, have magnitudes that do not exceed unity. Thus, if

$$\left|S_{x_i}^{Q_{P,Z}}\right| \ll \left|Q_{P,Z}S_{x_i}^{\omega_{P,Z}}\right| \tag{8.4}$$

then equation (8.3) is given approximately by

$$\boxed{S_{x_i}^{G_b(\omega)} \approx S_{x_i}^{H_0} + S_{x_i}^{\omega_P}\left[S_{\omega_P}^{G_b(\omega)}\right] + S_{x_i}^{\omega_Z}\left[S_{\omega_Z}^{G_b(\omega)}\right]} \tag{8.5}$$

It is important to note that this approximation may only be used if inequality (8.4) is valid.

A quadratic example: the lowpass two-integrator loop: Consider the second-order lowpass two-integrator loop *RC*-active filter shown in Fig. 8.1. This circuit is an implementation of the signal-flow graph shown in Fig. 7.7; by direct analysis, we find that

$$H_b(s) \equiv \frac{V_2(s)}{V_1(s)} = \left(\frac{1}{R_0R_2C_1C_2}\right)\left[\frac{1}{s^2 + s\left(\dfrac{1}{R_QC_1}\right) + \left(\dfrac{R_4}{R_1R_2R_3C_1C_2}\right)}\right] \tag{8.6}$$

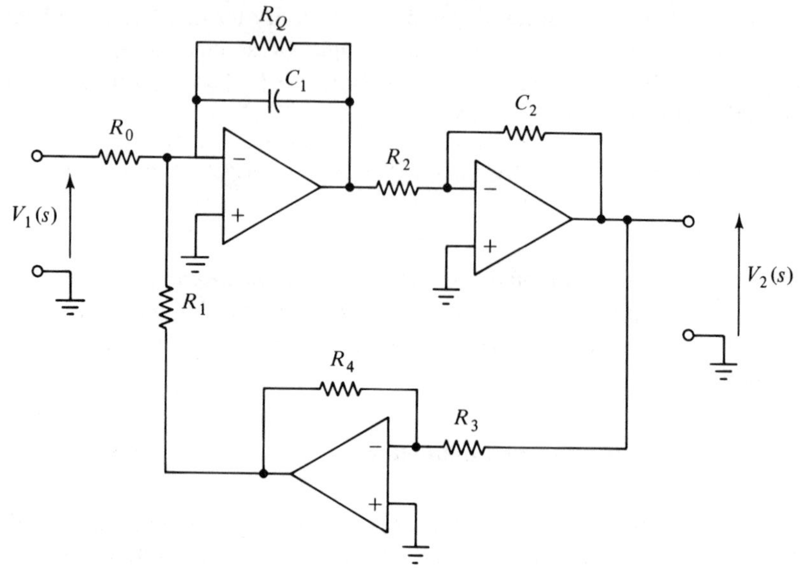

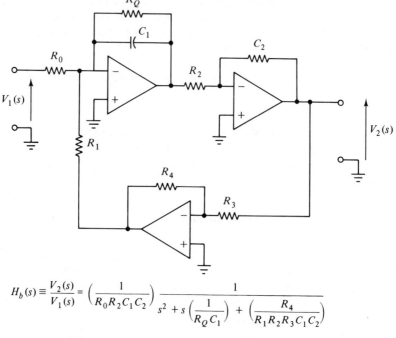

$$H_b(s) \equiv \frac{V_2(s)}{V_1(s)} = \left(\frac{1}{R_0R_2C_1C_2}\right)\frac{1}{s^2 + s\left(\dfrac{1}{R_QC_1}\right) + \left(\dfrac{R_4}{R_1R_2R_3C_1C_2}\right)}$$

FIGURE 8.1 *The lowpass two-integrator loop circuit.*

It follows from equations (8.1) and (8.6) that

$$\left.\begin{aligned} H_0 = (R_0R_2C_1C_2)^{-1}, \qquad \omega_P &= \left(\frac{R_4}{R_1R_2R_3C_1C_2}\right)^{1/2} \\ Q_P = R_QC_1&\left(\frac{R_4}{R_1R_2R_3C_1C_2}\right)^{-1/2} \end{aligned}\right\} \tag{8.7}$$

Having obtained expressions for H_0, ω_P, and Q_P it is a straightforward calculation to determine each of the single-element sensitivities $S_{x_i}^{H_0}$, $S_{x_i}^{Q_P}$, $S_{x_i}^{\omega_P}$, where x_i are the RC-element values. These are tabulated in Table 8-1. It follows from inspection of Table 8-1 that if $Q_P \gg 1$, then equation (8.4) and therefore equation (8.5) are valid for the six elements R_1, R_2, R_3, R_4, C_1, and C_2. Using the data in Table 8-1 with equation (8.5) and omitting the terms involving ω_Z indicates that

$$S_{R_2}^{G_b(\omega)} \approx S_{C_1}^{G_b(\omega)} \approx S_{C_2}^{G_b(\omega)} \approx -1 - \tfrac{1}{2}\left[S_{\omega_p}^{G_b(\omega)}\right] \tag{8.8}$$

TABLE 8-1

S_{x_i} \ x_i	R_1	R_2	R_3	R_4	R_Q	C_1	C_2	R_0
$S_{x_i}^{H_0}$	0	-1	0	0	0	-1	-1	-1
$S_{x_i}^{\omega_P}$	$-\frac{1}{2}$	$-\frac{1}{2}$	$-\frac{1}{2}$	$\frac{1}{2}$	0	$-\frac{1}{2}$	$-\frac{1}{2}$	0
$S_{x_i}^{\omega_Z}$	—	—	—	—	—	—	—	—
$S_{x_i}^{Q_P}$	$-\frac{1}{2}$	$-\frac{1}{2}$	$-\frac{1}{2}$	$\frac{1}{2}$	1	$\frac{1}{2}$	$-\frac{1}{2}$	0
$S_{x_i}^{Q_Z}$	—	—	—	—	—	—	—	—

and

$$S_{R_1}^{G_b(\omega)} \approx S_{R_3}^{G_b(\omega)} \approx -S_{R_4}^{G_b(\omega)} \approx -\tfrac{1}{2}\left[S_{\omega_p}^{G_b(\omega)}\right] \tag{8.9}$$

The curve for $S_{\omega_p}^{G_b(\omega)}$ in Fig. 6.2 may then be used to obtain the corresponding passive-sensitivity curves shown in Fig. 8.2. Note that these curves have maxima and minima of approximately $Q_P/2$ at $\omega \approx \omega_p\left(1 \pm \dfrac{1}{2Q_P}\right)$.

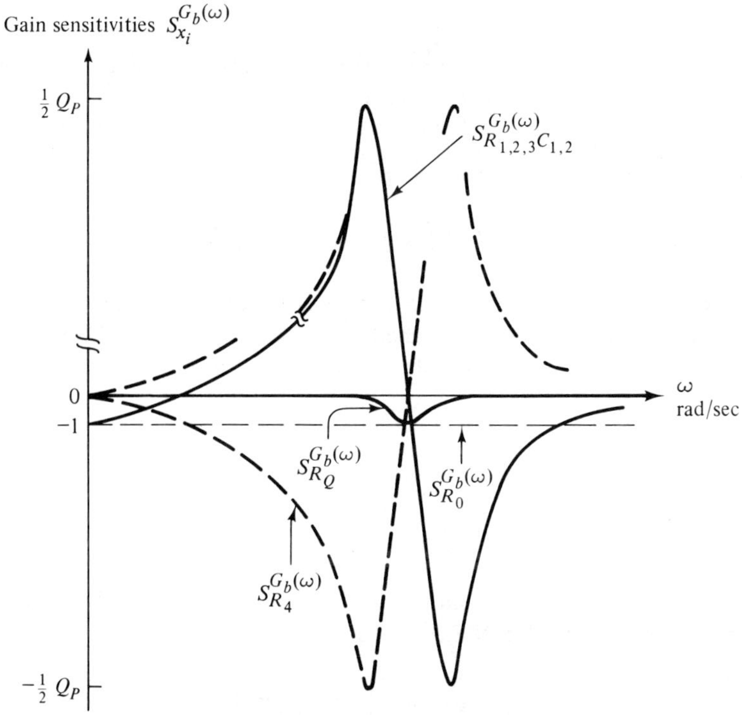

FIGURE 8.2 *Single-element sensitivities of the lowpass two-integrator loop, $Q_p \gg 1$.*

The sensitivity of $G_b(\omega)$ to the remaining elements R_0 and R_Q are calculated directly from equation (8.3) and Table 8-1, to obtain

$$S_{R_0}^{G_b(\omega)} = -1, \qquad S_{R_Q}^{G_b(\omega)} = S_{Q_p}^{G_b(\omega)} \tag{8.10}$$

The curve for $S_{Q_p}^{G_b(\omega)}$ in Fig. 6.2 is then used to obtain the sensitivity of $G_b(\omega)$ to the element R_Q as shown in Fig. 8.2. Note that $G_b(\omega)$ is relatively insensitive to R_Q and R_0. This completes our example showing the use of equations (8.3) and (8.5) to estimate the passive sensitivities.

8-2.1 Near-Optimum Passive-Sensitivity Performance

In this subsection, we derive simple conditions on the sensitivities $S_{x_i}^{\omega_p}$, $S_{x_i}^{Q_p}$, and $S_{x_i}^{H_0}$ that ensure the sensitivity performance of *quadratic RC-active* filters is close to the optimum insofar as worstcase-sensitivity performance and Schoeffler-sensitivity performance are concerned. The analysis is restricted to quadratic lowpass, bandpass, and highpass functions of the form

$$H_b(s) = \frac{H_0 s^k}{s^2 + \left(\dfrac{\omega_P}{Q_P}\right)s + \omega_P^2}, \qquad k = 0,\ 1, \text{ or } 2 \tag{8.11}$$

for which equation (8.3) simplifies to

$$S_{x_i}^{G_b(\omega)} = S_{x_i}^{H_0} + \left[S_{\omega_p}^{G_b(\omega)}\right]S_{x_i}^{\omega_p} + \left[S_{Q_p}^{G_b(\omega)}\right]S_{x_i}^{Q_p} \tag{8.12}$$

The terms involving ω_Z and Q_Z have been omitted because $H_b(s)$ does not have nonzero zeros. It follows immediately from equation (8.12) that if the parameters x_i are, for example, the resistors or capacitors of the network, then

$$WS_{R,C}^{G_b(\omega)} \equiv \sum_{i=1}^{n_{R,C}} \left|S_{R_i,C_i}^{G_b(\omega)}\right| \leqslant WS_{R,C}^{H_0} + WS_{R,C}^{\omega_p}\left|S_{\omega_p}^{G_b(\omega)}\right| + WS_{R,C}^{Q_p}\left|S_{Q_p}^{G_b(\omega)}\right| \tag{8.13}$$

Now, it is recalled from equations (6.115), (6.116), and (6.124) that the invariant lower bound for $WS_{R,C}^{G_b(\omega)}$ is as follows:

$$LWS_{R,C}^{G_b(\omega)} = \left|\frac{\partial \ln G_b(\omega)}{\partial \ln \omega}\right| \tag{8.14}$$

where the notation R, C implies all the Rs *or* all the Cs are included in the summation so that, from (8.13) and (8.14), we have

$$\left|\frac{\partial \ln G_b(\omega)}{\partial \ln \omega}\right| \leqslant WS_{R,C}^{G_b(\omega)} \leqslant WS_{R,C}^{H_0} + WS_{R,C}^{\omega_p}\left|S_{\omega_p}^{G_b(\omega)}\right| + WS_{R,C}^{Q_p}\left|S_{Q_p}^{G_b(\omega)}\right| \tag{8.15}$$

It follows from equation (8.11) that

$$S_{\omega_p}^{G_b(\omega)} = -S_{\omega}^{G_b(\omega)} + 2 - k, \qquad k = 0, 1, 2 \tag{8.16}$$

so that equations (8.15) and (8.16) provide the following upper and lower bounds on $WS_{R,C}^{G_b(\omega)}$:

$$\begin{aligned}\left|S_{\omega}^{G_b(\omega)}\right| &\leqslant WS_{R,C}^{G_b(\omega)} \\ &\leqslant WS_{R,C}^{H_0} + WS_{R,C}^{\omega_P}\left[\left|S_{\omega}^{G_b(\omega)}\right| + (2-k)\right] + WS_{R,C}^{Q_P}\left|S_{Q_p}^{G_b(\omega)}\right|\end{aligned} \tag{8.17}$$

We now inspect the right-hand side of equation (8.17) and ask ourselves the question as to which terms might conceivably be minimized so that the right-hand side approaches the theoretical lower bound of $|S_{\omega}^{G_b(\omega)}|$? Since all summed terms on the right-hand side are non-negative, the theoretical lower bound is achieved only if

$$\boxed{\begin{aligned}WS_{R,C}^{H_0} &= k - 2,\\ WS_{R,C}^{Q_P} &= 0,\\ WS_{RC}^{\omega_P} &= 1\end{aligned}} \qquad \text{CONDITIONS FOR OPTIMAL WORSTCASE SENSITIVITY TO THE RESISTORS OR CAPACITORS} \tag{8.18}$$

in which case equation (8.17) simply becomes

$$WS_{R,C}^{G_b(\omega)} = \left|S_{\omega}^{G_b(\omega)}\right| \equiv LWS_{R,C}^{G_b(\omega)} \tag{8.19}$$

A circuit will *approach* lower-bound worstcase-sensitivity performance if $WS_{R,C}^{\omega_p} \approx 1$ and the terms containing H_0 and Q_P in equation (8.17) are negligibly small. Since $|S_{\omega}^{G_b(\omega)}|$ has a peak magnitude of Q_P and $|S_{Q_p}^{G_b(\omega)}|$ has a peak magnitude of 1, then *near-optimum* worstcase sensitivity is achieved if

$WS_{R,C}^{H_0} \ll Q_p\, WS_{R,C}^{\omega_p}$ $WS_{R,C}^{Q_p} \ll Q_p\, WS_{R,C}^{\omega_p}$ $WS_{R,C}^{\omega_p} \approx 1$	CONDITIONS FOR NEAR-OPTIMUM WORSTCASE SENSITIVITY TO THE RESISTORS OR THE CAPACITORS	(8.20)

It is stressed that many circuits satisfy these three conditions and are, therefore, described as *near-optimum in the worstcase sense.*

A typical example: the Lowpass Two-integrator Loop: We now reconsider Fig. 8.1 and the corresponding expressions in equation (8.7). According to the single-element sensitivities of Table 8-1, equations (8.7) indicate that

$WS_R^{H_0} = W_C^{H_0} = 1 + 1 = 2$ $WS_R^{\omega_p} = 2, \quad WS_C^{\omega_p} = 1$ $WS_R^{Q_p} = 3, \quad WS_C^{Q_p} = 1$	WORSTCASE-SENSITIVITIES FROM TABLE 8-1	(8.21)

If $Q_P \gg 1$, then it follows directly from equations (8.21) that the conditions for near-optimality in (8.20) are satisfied for the capacitors. Thus,

$$WS_C^{G_b(\omega)} \approx WS_C^{\omega_p}|S_\omega^{G_b(\omega)}| = |S_\omega^{G_b(\omega)}| \tag{8.22}$$

which is, of course, the optimal (lower bound) for $WS_C^{G_b(\omega)}$ that has been derived in equation (6.116). The worstcase sensitivity $WS_R^{G_b(\omega)}$ is not quite as close to lower bound, because the first two conditions in equation (8.20) are met but the third condition, $WS_R^{\omega_p} \approx 1$, is *not* met. In fact, $WS_R^{\omega_p} = 2$ and we have, therefore,

$$WS_R^{G_b(\omega)} \approx WS_R^{\omega_p}|S_\omega^{G_b(\omega)}| = 2|S_\omega^{G_b(\omega)}| \tag{8.23}$$

which is twice the lower bound in equation (6.116).

Estimation of the Schoeffler sensitivity $\mathbb{S}_{R,C}^{G_b(\omega)}$: It is often possible to estimate the Schoeffler-sensitivity criterion. The *exact* expression for $\mathbb{S}_{R,C}^{G_b(\omega)}$, for example, is obtained from equation (8.3) as

$$\mathbb{S}_{R,C}^{G_b(\omega)} \equiv \sum_{i=1}^{n_{R,C}} \left|S_{R_i,C_i}^{G_b(\omega)}\right|^2 = \sum_{i=1}^{n_{R,C}} \left|S_{R_1,C_i}^{H_0} + S_{R_i,C_i}^{\omega_p}\left[S_{\omega_p}^{G_b(\omega)}\right] + S_{R_i,C_i}^{Q_p}\left[S_{Q_p}^{G_b(\omega)}\right]\right|^2 \tag{8.24}$$

For $Q_P \gg 1$, the term $[S_{\omega}^{G_b(\omega)}]$ exhibits a peak-to-peak variation from approximately $-Q_P$ to $+Q_P$. Thus, for all resistors R_i and capacitors C_i that contribute to ω_P [such as R_1, R_2, R_3, R_4, C_1, and C_2 in equation (8.7)], we can expect that the term $S_{R_i, C_i}^{\omega_p}[S_{\omega_p}^{G_b(\omega)}]$ will play a dominant role in equation (8.24). For many of the most important quadratic circuits, it can be shown readily that

$$\boxed{\begin{aligned} &\left|S_{R_i, C_i}^{H_0}\right| \ll Q_P \\ &\left|S_{R_i, C_i}^{Q_P}\right| \ll Q_P \end{aligned}} \qquad \begin{aligned} &i = 1, 2, \ldots, n_{R, C} \\ &i = 1, 2, \ldots, n_{R, C} \end{aligned} \tag{8.25}$$

which is sufficient to ensure that equation (8.24) may be approximated by

$$\mathbb{S}_{R, C}^{G_b(\omega)} \equiv \sum_{i=1}^{n_{R, C}} \left|S_{R_i, C_i}^{G_b(\omega)}\right|^2 \approx \sum_{i=1}^{n_{r, c}} \left|S_{R_i, C_i}^{\omega_p}\left[S_{\omega}^{G_b(\omega)}\right]^2\right| \quad i = 1, 2, \ldots, n_{r, c}$$

where $R_1, R_2, \ldots, R_{n_r}$ or $C_1, C_2, \ldots, C_{n_c}$ *are the resistors or capacitors that contribute to the expression for* ω_p. (For example, n_R is equal to 6, and n_r is equal to 4 for the circuit shown in Fig. 8.1). The equation above may be written in the form

$$\boxed{\mathbb{S}_{R, C}^{G_b(\omega)} \approx \left|\frac{\partial \ln G_b(\omega)}{\partial \ln \omega}\right|^2 \sum_{i=1}^{n_{r, c}} \left|S_{R_i C_i}^{\omega_p}\right|^2} \tag{8.26}$$

ESTIMATE OF SCHOEFFLER SENSITIVITY UNDER CONDITIONS OF EQUATION (8.25)

whereas the theoretical lower bound on $\mathbb{S}_{R, C}^{G_b(\omega)}$ is obtained directly from equation (6.140) by letting F be defined as $G_b(\omega)$ and n as the total number of resistors or capacitors $n_{R, C}$, so that

$$L\mathbb{S}_{R, C}^{G_b(\omega)} = \frac{1}{n_{R, C}} \left|MS_{R,C}^{G_b(\omega)}\right|^2$$

which, by equation (6.109), is simply

$$\boxed{L\mathbb{S}_{R, C}^{G_b(\omega)} = \frac{1}{n_{R, C}} \left|\frac{\partial \ln G_b(\omega)}{\partial \ln \omega}\right|^2} \tag{8.27}$$

For example, the circuit shown in Fig. 8.1 corresponds to $n_R = 6$, $n_C = 2$, $n_r = 4$, $n_c = 2$. Then the single-element sensitivities in Table 8-1 indicate

$$\sum_{i=1}^{i=n_r=4} \left|S_{R_i}^{\omega_p}\right|^2 = \left(-\tfrac{1}{2}\right)^2 + \left(-\tfrac{1}{2}\right)^2 + \left(-\tfrac{1}{2}\right)^2 + \left(+\tfrac{1}{2}\right)^2 = 1$$

and

$$\sum_{i=1}^{i=n_c=2} \left|S_{C_i}^{\omega_p}\right|^2 = \left(-\tfrac{1}{2}\right)^2 + \left(-\tfrac{1}{2}\right)^2 = \tfrac{1}{2}$$

Substituting the equations above into equation (8.26), we have

$$\mathbb{S}_R^{G_b(\omega)} \approx \left|\frac{\partial \ln G_b(\omega)}{\partial \ln \omega}\right|^2, \qquad \mathbb{S}_C^{G_b(\omega)} \approx \frac{1}{2}\left|\frac{\partial \ln G_b(\omega)}{\partial \ln \omega}\right|^2 \tag{8.27A}$$

whereas, by equation (8.27), the lower bound sensitivities are

$$L\mathbb{S}_R^{G_b(\omega)} = \frac{1}{6}\left|\frac{\partial \ln G_b(\omega)}{\partial \ln \omega}\right|^2, \qquad L\mathbb{S}_C^{G_b(\omega)} \approx \frac{1}{2}\left|\frac{\partial \ln G_b(\omega)}{\partial \ln \omega}\right|^2$$

The interpretation of this result is that the Schoeffler sensitivities are approximately lower bound over the capacitors and approximately six times lower bound over the resistors for the circuit shown in Fig. 8.1.

It is important to observe that $L\mathbb{S}_{R,C}^{G_b(\omega)}$ *is a function of the number of resistors or capacitors* $n_{R,C}$ in the circuit; generally we find that the relative Schoeffler-sensitivity performances of *different* quadratic circuits may be realistically compared only if they contain approximately the same number $n_{R,C}$ of R, C elements.

8-2.2 Exact Calculation of Passive-Sensitivities $S_{xi}^{G_b(\omega)}$

In this subsection, consideration is given to the fact that the assumptions of high-Q (that is $Q_P \gg 1$, $|Q_Z| \gg 1$) are not always valid so that equations (8.4) and therefore (8.5) may not be used.

Whenever equation (8.4) is not a valid approximation, it is preferable to use the exact results for the coefficient sensitivities $S_{b_{ib},\,a_{ib}}^{G_b(\omega)}$ provided in equations (7.36) and (7.38). The *exact* transfer function $H_b(s)$ is usually available in the form

$$H_b(s) = \frac{a_{2b}s^2 + a_{1b}s + a_{0b}}{b_{2b}s^2 + b_{1b}s + b_{0b}} \tag{8.28}$$

where the coefficients a_{1b}, b_{ib} are known functions of a particular RC element x_i. Thus, it is straightforward to calculate the sensitivities $S_{x_i}^{a_{2b}}$, $S_{x_i}^{a_{1b}}, S_{x_i}^{a_{0b}}, S_{x_i}^{b_{2b}}, S_{x_i}^{b_{1b}} S_{x_i}^{b_{0b}}$, and use the chain rule to obtain

$$S_{x_i}^{G_b(\omega)} = \sum_{j=0}^{2} \left[S_{x_i}^{G_b(\omega)} S_{x_i}^{a_{jb}} + S_{b_{jb}}^{G_b(\omega)} S_{x_i}^{b_{jb}} \right] \tag{8.29}$$

The terms $S_{a_{jb}}^{G_b(\omega)}$ and $S_{b_{jb}}^{G_b(\omega)}$ are simply the real parts of $S_{a_{jb}}^{H_b(s)}$ and $S_{b_{jb}}^{H_b(s)}$ and are shown graphically in Fig. 7.10.

Example of an exact calculation of $S_{xi}^{G_b(\omega)}$: Reconsidering equation (8.6), we have by comparison equation (8.1)

$$b_{2b} = R_0 R_2 C_1 C_2, \qquad b_{1b} = \frac{R_0 R_2 C_2}{R_Q} \qquad b_{0b} = \frac{R_0 R_4}{R_1 R_3} \tag{8.30}$$

Suppose that we require the *exact* expression for $S_{R_1}^{G_b(\omega)}$, for example. Then equation (8.29) simplifies *drastically* to

$$S_{R_2}^{G_b(\omega)} = S_{b_{2b}}^{G_b(\omega)} S_{R_2}^{b_{2b}} + S_{b_{1b}}^{G_b(\omega)} S_{R_2}^{b_{1b}} \tag{8.31}$$

Using equations (7.38), (8.30), and (8.31) we have

$$\boxed{S_{R_2}^{G_b(\omega)} = -\mathrm{Re}[t_{\mathrm{HP}}(j\omega)] - \frac{1}{Q_P}\mathrm{Re}[t_{\mathrm{BP}}(j\omega)]} \tag{8.32}$$

where $S_{R_2}^{b_{2b}}$ and $S_{R_2}^{b_{1b}}$ are equal to unity. The function in (8.32) may be constructed graphically from Fig. 7.10.

In practical cases, this often tedious calculation of the *exact* sensitivities $S_{R_i, C_i}^{G_b(\omega)}$ is simply not necessary because *either* $|Q_{P,Z}| \gg 1$ and the high-Q approximations may be used *or*, if $|Q_{P,Z}|$ are not much greater than unity the passive sensitivities are so small that they may usually be neglected by comparison with the active sensitivities $S_B^{G_b(\omega)}$.

8-3 TWO-INTEGRATOR LOOP BIQUADRATIC CIRCUITS(TILs)

There is a variety of RC-active biquadratic circuits that are based on the use of two integrating OP AMP circuits. The reader may refer back to section 7-3 for a description and analysis of four distinct types of signal-flow graphs (SFGs) that may be used to realize biquadratic functions; they

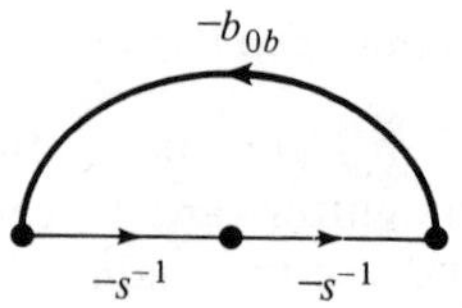

(a) The SFG for the two-integrator loop.

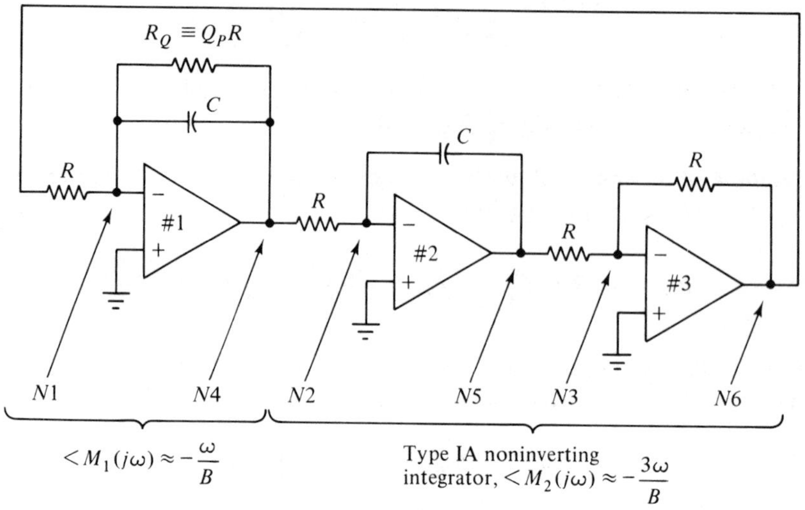

(b) The basic TIL-3A circuit: parallel damping resistor R_Q.

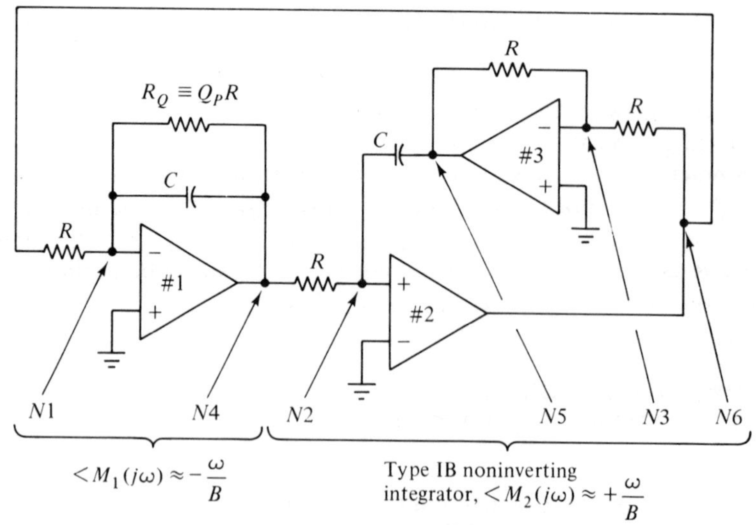

(c) The basic TIL-3B circuit: parallel damping resistor R_Q.

FIGURE 8.3

are the fully biquadratic direct-form structure BIQUAD DF, the LBI/DF, the HBI/DF, and the MI/DF structures. All four of these SFGs have direct-form denominators that are realized by a loop in the SFG of the type shown in Fig. 8.3(a), consisting of two inverting-integrating transmittances $-s^{-1}$ and an inverting transmittance $-b_{0b}$. The reader may inspect Figs. 7.7 and 7.9 to identify this two-integrator loop (TIL) in each of the SFGs. (Other types of TIL structures exist but are beyond the scope of this text).

Let us now consider the OP AMP building-block circuits that are available from sections 5-3 and 5-4 and that may be used to implement the TIL of Fig. 8.3(a). First, the inverting-amplifier and inverting-integrator of Fig. 5.5 may be used as shown in Fig. 8.3(b). This will be referred to as a TIL-3A circuit, because it employs *three* OP AMPs while using a type 1A noninverting configuration (see Table 5-2) to implement $b_{0b}s^{-1}$. Second, we could use the *ideally* equivalent TIL circuit shown in Fig. 8.3(c) to realize Fig. 8.3(a); this circuit is referred to as a TIL-3B circuit, because it uses *three* OP AMPs while using a type 1B noninverting configuration to implement $b_{0b}s^{-1}$. *The TIL-3A and TIL-3B circuits are the basic building-blocks for most of the important three- and four-amplifier RC-active biquadratic filters.* It is recalled from section 5-4.2 that the type 1A and 1B noninverting integrators are entirely equivalent insofar as ideal performance ($A_0 \rightarrow \infty$, $B \rightarrow \infty$) is concerned so that the TIL-3A and TIL-3B circuits are therefore *ideally equivalent*. The passive sensitivities $S_{R_i, C_i}^{H(s)}$, for example, are identical and so are the ideal transfer functions $H(s)$ for any particular input/output points. An important distinction between TIL-3A and TIL-3B is that the *nonideal* performance and, in particular, the dependence on gain bandwidth products B of the OP AMPs is significantly different. This is because the type 1A inverting-integrator is a *phase lag* type ($< M(j\omega) \approx -3\omega/B$), whereas the type 1B inverting-integrator is a *phase lead* type ($< M(j\omega) \approx +\omega/B$).

8-3.1 High-Frequency Behavior of TIL-3A

The TIL-3A circuit is characterized by a high-frequency performance that, for $Q_P \gg 1$, is determined by the total loop error phase shift $\phi_L(\omega)$, which is simply defined in terms of the nonideal multiplicative functions (see Fig. 8.3(b)) as

$$\phi_L(\omega) \equiv [< M_1(j\omega) + < M_2(j\omega)] \quad \text{LOOP ERROR PHASE SHIFT} \tag{8.33}$$

which, for this circuit is simply

$$\boxed{\phi_L(\omega) \approx -\frac{\omega}{B} - \frac{3\omega}{B} \approx -\frac{4\omega}{B}} \quad \text{TIL-3A LOOP ERROR PHASE SHIFT} \tag{8.34}$$

In the following analysis, it will be shown that it is essentially this nonideal error $\phi_L(\omega)$ in loop phase shift that determines the nonideal effective Q-factor $\hat{Q}_P$. In fact, we shall show that the effective Q-factor $\hat{Q}_P$ at high frequencies is related to the ideal low-frequency ($B \to \infty$) Q-factor Q_P by the relationship

$$\boxed{\frac{1}{\hat{Q}_P} \approx \frac{1}{Q_P} - \phi_L(\omega)} \quad \text{EFFECTIVE } Q\text{-FACTOR } \hat{Q}_P \text{ AND LOOP PHASE SHIFT} \tag{8.35}$$

A *complete* analysis of the TIL-3A circuit shown in Fig. 8.3(b) requires that we introduce an excitation (in the form of summing resistors at nodes $N1$, $N2$, or $N3$) and also that we define an output node. However, it is always possible to write the steady-state response $H_b(j\omega)$ of this single-feedback loop circuit in the form

$$H_b(j\omega) = \frac{N(j\omega)}{1 - I_1(j\omega)I_2(j\omega)} \equiv \frac{N(j\omega)}{D(j\omega)} \tag{8.36}$$

where $I_1(j\omega)$ is the voltage transfer function of the inverting (damped) integrator, $I_2(j\omega)$ is that of the noninverting integrator, and $N(j\omega)$ is the feedforward transfer function between input and output. For example, an ideal bandpass response (BP) is realized [see Fig. 7.9(a)] by introducing the input at node $N1$ of OP AMP #1 and defining the output at node $N4$ of OP AMP #1 in which case $N(j\omega)$ is simply equal to $I_1(j\omega)$. *The essential aspects of the nonideal behavior of the TIL circuits are revealed by deriving the high-frequency behavior of the denominator $D(j\omega)$ in terms of an effective Q-factor $\hat{Q}_P$ and effective natural frequency $\hat{\omega}_P$.*

The nonideal voltage transfer function of the damped integrator $I_1(j\omega)$ is now derived. Substituting equation (5.1) into equation (5.20) and choosing $Y_2(s) = sC + GQ_P^{-1}$ and $Y_1(s) = G$, corresponding to the *damped* integrator in Fig. 8.3(b), we have

$$H(s) \equiv I_1(s) = -\left(\frac{sC}{G} + \frac{1}{Q_P}\right)\left\{1 + \frac{1}{A_0} + \frac{s}{B_1}\left[1 + \left(\frac{sC}{G} + \frac{1}{Q_P}\right)^{-1}\right]\right\}^{-1}$$

which may be simplified to

$$\boxed{I_1(j\omega) \approx -\frac{\omega_P}{j\omega}\left[1 - \frac{\omega_P}{B_1} - j\left(\frac{\omega}{B_1} - \frac{\omega_P}{\omega Q_P}\right)\right]} \tag{8.37}$$

with the assumptions

$$Q_P \gg 1, \quad \omega \gg \frac{\omega_P}{Q_P}, \quad A_0 \gg 1, \quad \omega^2 \gg \frac{\omega_P B_1}{A_0}, \quad \omega_P \equiv \frac{G}{C} \tag{8.38}$$

Similarly, equation (5.20) gives

$$I_2(j\omega) \approx \underbrace{\left\{-\frac{\omega_P}{j\omega}\left[\left(1 - \frac{\omega_P}{B_2}\right) - \frac{j\omega}{B_2}\right]\right\}}_{\text{inverting-integrator}} \times \underbrace{\left\{-1\left[1 - j\frac{2\omega}{B_3}\right]\right\}}_{\text{inverter}} \tag{8.39}$$

for the non-inverting type 1A integrator. Substituting equations (8.37) and (8.39) into (8.36) and assuming $\omega / B_{1,2,3} \ll 1$, we obtain

$$D(j\omega) = 1 - \left(\frac{\omega_P}{\omega}\right)^2\left[\left(1 - \frac{\omega_P}{B_1} - \frac{\omega_P}{B_2}\right) + j\left(\frac{\omega_P}{\omega Q_P} - \frac{\omega}{B_1} - \frac{\omega}{B_2} - \frac{2\omega}{B_3}\right)\right] \tag{8.40}$$

Consider now the *ideal* behavior of $D(j\omega)$ corresponding to $B_{1,2,3} = \infty$ in equation (8.40); thus,

$$D(j\omega)_{\text{Ideal}} = 1 - \left(\frac{\omega_P}{\omega}\right)^2\left(1 + \frac{\omega_P}{\omega Q_P}\right) \tag{8.41}$$

We now note that $D(j\omega)$ has the same form as $D(j\omega)_{\text{Ideal}}$ if we define the effective natural frequency $\hat{\omega}_P$ and the effective Q-factor $\hat{Q}_P$ as

$$\boxed{\begin{aligned} \hat{\omega}_P &\equiv \omega_P\left(1 - \frac{\omega_P}{B_1} - \frac{\omega_P}{B_2}\right)^{1/2} \\ \hat{Q}_P &\equiv Q_P\left(\frac{\hat{\omega}_P}{\omega_P}\right)^3\left[1 - \frac{\omega^2 Q_P}{\omega_P}\left(\frac{1}{B_1} + \frac{1}{B_2} + \frac{2}{B_3}\right)\right]^{-1} \end{aligned}} \tag{8.42}$$

EFFECTIVE NATURAL FREQUENCY
EFFECTIVE Q-FACTOR

Substituting equations (8.42) into (8.40), we have

$$D(j\omega) = 1 - \left(\frac{\hat{\omega}_P}{\omega}\right)^2\left(1 + j\frac{\hat{\omega}_P}{\omega\hat{Q}_P}\right) \tag{8.43}$$

Comparing equations (8.41) and (8.43) it is observed that the terms $\hat{\omega}_P$ and $\hat{Q}_P$ may simply replace ω_P and Q_P, respectively, in (8.41) to describe the nonideal behavior of $D(j\omega)$. Since it is always true that

$$\omega_P \ll B_{1,2,3} \tag{8.44}$$

and usually that

$$B_1 \approx B_2 \approx B_3 \equiv B \tag{8.45}$$

we use the simplified expressions obtained from (8.42) as

$$\hat{\omega}_P \approx \omega_P\left(1 - \frac{\omega_P}{B}\right) \tag{8.46}$$

and

$$\hat{Q}_P \approx Q_P\left(1 - \frac{4\omega Q_P}{B}\right)^{-1} \tag{8.47}$$

APPROXIMATE EFFECTIVE Q AND ω_P; $\omega \ll B$

which can be written in terms of fractional changes as

$$\frac{\Delta Q_P}{Q_P} \equiv \frac{\hat{Q}_P - Q_P}{Q_P} = \frac{4\omega Q_P}{B - 4\omega Q_P} \tag{8.48}$$

and

$$\frac{\Delta\omega_P}{\omega_P} \equiv \frac{\hat{\omega}_P - \omega_P}{\omega_P} = -\frac{\omega_P}{B} \tag{8.49}$$

FRACTIONAL CHANGES IN Q-FACTOR AND ω_P; TIL-3A

If

$$\omega \ll \frac{B}{4Q_P} \quad \text{then} \quad \frac{\Delta Q_P}{Q_P} \approx \frac{4\omega Q_P}{B} \tag{8.50}$$

Some remarks on the behavior of TIL-3A at high frequencies are now pertinent. First, it follows immediately from equations (8.47) and (8.34) that we have proven that equation (8.35) is valid. Furthermore, it is clear that *the nonideal loop phase error* $\phi_L(\omega)$ *is the cause of an increase in Q-factor* by a fractional amount of approximately $4\omega Q_P/B$, and there is a *decrease in natural frequency* ω_P by a fractional amount of approximately ω_P/B. How does this effect the gain function $G_b(\omega) \equiv |H_b(j\omega)|$? Of course, we have not yet determined the effect of B on the numerator $N(j\omega)$ of equation (8.36) and, therefore, the question cannot be answered completely unless $N(j\omega)$ is known; this requires that we specify the input and output connections of the specific TIL-3A circuit that is used and this, in turn, will depend on which one of the four SFGs (DF BIQUAD, LBI/DF, HBI/DF, or MI/DF) that is realized with this circuit. However, for all of these SFGs, the numerator $N(j\omega)$ is not nearly as dependent on B as is the denominator $D(j\omega)$ for all applications involving the realization of $\pm$LP, $\pm$ BP, and $\pm$HP transfer functions. The reader is asked to check this for the important case where $N(j\omega) = I_1(j\omega)$ in Problem 8.3. Thus, neglecting the small dependence of $N(j\omega)$ on B, equation (8.1) indicates

$$\frac{\Delta G_b(\omega)}{G_b(\omega)} \approx \left[S_{\omega_P}^{G_b(\omega)} \right] \frac{\Delta\omega_P}{\omega_P} + \left[S_{Q_P}^{G_b(\omega)} \right] \frac{\Delta Q_P}{Q_P} \tag{8.51}$$

where, of course, the terms in [ ] brackets are shown in Fig. 6.2 and are given explicitly in equations (6.65) and (6.67). The $S_{\omega_P}^{G_b(\omega)}$ term has a Q_P-dependent peak-to-peak variation with ω, whereas $|S_{Q_P}^{G_b(\omega)}|$ is not larger than unity. We have two terms contributing to $\Delta G_b(\omega)$ via the perturbations $\Delta\omega_P/\omega_P$ and $\Delta Q_P/Q_P$ so that from equations (8.49) and (8.50)

$$\frac{\Delta G_b(\omega)}{G_b(\omega)} \approx \underbrace{-\frac{\omega_P}{B}\left[S_{\omega_P}^{G_b(\omega)} \right]}_{\text{first term}} + \underbrace{\frac{4\omega Q_P}{B}\left[S_{Q_P}^{G_b(\omega)} \right]}_{\text{second term}} \tag{8.52}$$

The reader may visualize the variations of these terms with ω by inspecting Fig. 6.2. The perturbation $\Delta Q_P/Q_P$ contributes to $\Delta G_b(\omega)/G_b(\omega)$ via the second term in equation (8.52), which has an extremum value of approximately $4\omega_P Q_P/B$ occurring at $\omega = \omega_P$. The perturbation $\Delta\omega_P/\omega_P$ has a less serious effect on $\Delta G_b(\omega)/G_b(\omega)$, via the first term in equation (8.52), which has extremum values of $\pm\omega_P Q_P/B$ at $\omega \approx [1 \pm \omega_P/2Q_P]$.

A simple predistortion technique for TIL-3A: It is possible to predistort the initial design to compensate for the effect of B at high frequencies. Solving equations (8.46) and (8.47) for ω_P and Q_P gives the predistorted

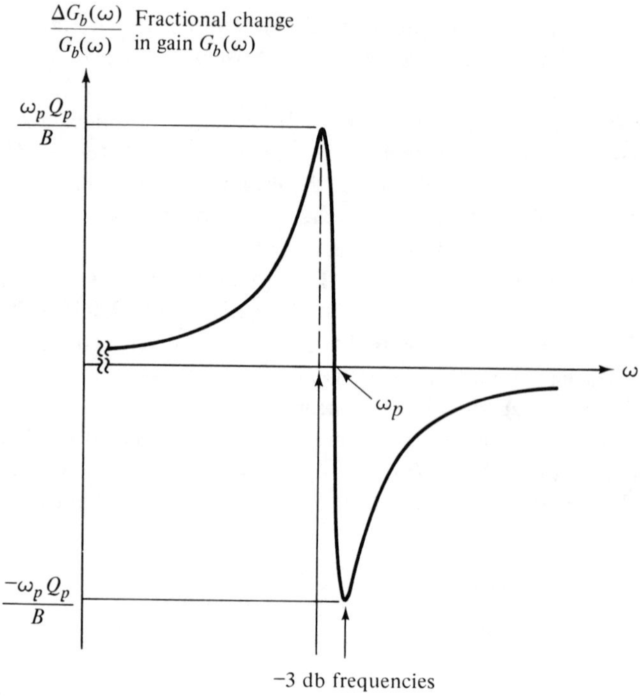

(a) Effect of bandwidth B on gain $G_b(\omega)$ for two-integrator loop TIL-3B.

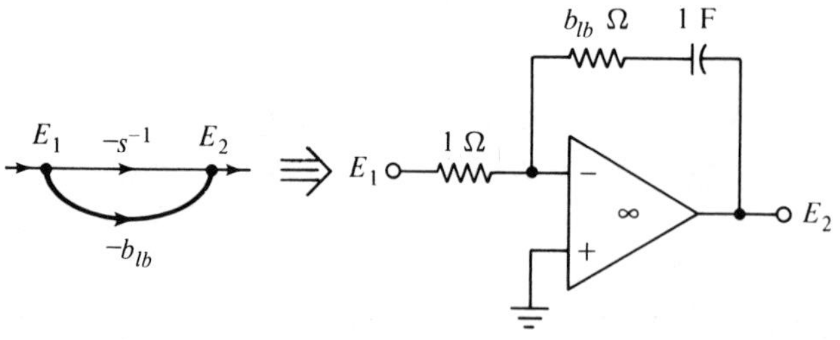

(b) Implementation of the feedforward transmittance $-b_{1b}$ using a series resistance b_{1b}.

FIGURE 8.4

initial design values as

$$\omega_P \approx \frac{B}{2}\left[1 - \left(1 - \frac{4\hat{\omega}_P}{B}\right)^{1/2}\right]$$

and

$$Q_P^{-1} \approx \hat{Q}_P^{-1} + \frac{4\hat{\omega}_P}{B}$$

PREDISTORTED INITIAL DESIGN VALUES FOR ω_P AND Q_P (8.52A)

Thus, the *ideal* design ($B_{1,2,3} = \infty$) is performed for the above values of ω_P and Q_P, *where* $\hat{\omega}_P$ *and* $\hat{Q}_P$ *are the actual required nonideal design values.* This gives predistorted values for the damping resistance R_Q and capacitances C as

$$R_Q = Q_P R = \frac{\hat{Q}_P R}{\left(1 + \dfrac{4\hat{\omega}_P \hat{Q}_P}{B}\right)} \tag{8.53}$$

and

$$C = \frac{1}{\omega_P R} = \frac{1}{\dfrac{BR}{2}\left[1 - \left(1 - \dfrac{4\hat{\omega}_P}{B}\right)^{1/2}\right]} \tag{8.54}$$

PREDISTORTED VALUES FOR R_Q AND C

For example, we design TIL-3A for a required nonideal (actual) $\hat{Q}_P$ and $\hat{\omega}_P$ given by

$$\hat{Q}_P \equiv 10, \qquad \hat{\omega}_P \equiv 10^4 \text{ rad/sec}$$

and we assume that we have

$$R \equiv 10^4\ \Omega, \qquad B \equiv 2 \times 10^6 \text{ rad/sec} \tag{8.55}$$

Substituting equations (8.55) into (8.53) and (8.54) gives

$$R_Q = \underline{\underline{83\tfrac{1}{3}\ \text{k}\Omega}}, \qquad C = \underline{\underline{1.005(05)\ \text{nF}}}$$

compared with ideal values ($B = \infty$) of $R_Q = 100$ kΩ, $C = 1$ nF.

8-3.2 High-Frequency Behavior of TIL-3B

The ideal behavior ($A_{01,2,3}$ and $B_{1,2,3} = \infty$) of TIL-3B is, of course, identical to that of TIL-3A. However, it is important to observe from Fig. 8.3(c) that the nonideal loop phase error $\phi_L(\omega)$ is, to a first approximation, equal to zero because

$$\phi_L(\omega) = [< M_1(j\omega) + < M_2(j\omega)] \approx -\frac{\omega}{B} + \frac{\omega}{B} = 0 \qquad (8.56)$$

where it will be recalled from the analysis in section 5-4.2 for the type 1A noninverting integrator that $M_2(j\omega) \approx +\omega/B$ if the two bandwidths [B_2 and B_3 in Fig. 8.3(c)] are both equal to B. *It is this important property of approximately zero loop phase error $\phi_L(\omega)$ that leads to the exceptionally accurate values of effective Q-factor $\hat{Q}_p$ that are possible with this circuit.*

Analysis of the high-frequency behavior of TIL-3B proceeds in a similar way to that described in section 8-3.1. In this case, $I_2(j\omega)$ is obtained for the type 1B noninverting integrator from equation (5.45) as

$$I_2(j\omega) \approx \frac{\omega_P}{j\omega}\left[\left(1 - \frac{\omega_P}{B_2}\right) - j\omega\left(\frac{1}{B_2} - \frac{2}{B_3}\right)\right] \qquad (8.57)$$

where RC $= \omega_P^{-1}$ and B_2 and B_3 correspond to OP AMPs #2 and #3, respectively, in Fig. 8.3(c). Of course, $I_1(j\omega)$ is given by equation (8.37), so that we use equations (8.37) and (8.57) in (8.36) to obtain

$$D(j\omega) \approx 1 - \left(\frac{\omega_P}{\omega}\right)^2\left[\left(1 - \frac{\omega_P}{B_1} - \frac{\omega_P}{B_2}\right) + j\left(\frac{\omega_P}{\omega Q_P} - \frac{\omega}{B_1} - \frac{\omega}{B_2} + \frac{2\omega}{B_3}\right)\right] \qquad (8.58)$$

Using the same technique as outlined in section 8-3.1, we compare equations (8.58) and (8.41) to determine the effective natural frequency $\hat{\omega}_P$ and effective Q-factor $\hat{Q}_P$ as

$$\hat{\omega}_P = \omega_P\left(1 - \frac{\omega_P}{B_1} - \frac{\omega_P}{B_2}\right)^{1/2}$$

and

$$\hat{Q}_P = Q_P\left(\frac{\hat{\omega}_P}{\omega_P}\right)^3\left[1 - \frac{Q_P\omega^2}{\omega_P}\left(\frac{1}{B_1} + \frac{1}{B_2} - \frac{2}{B_3}\right)\right]^{-1}$$

$\hat{Q}_P$ AND $\hat{\omega}_P$ TIL-3B (8.59)

If we use OP AMPs on the same semiconductor chip (for example, as part of a quad-op amp single package), then $B_1 \approx B_2 \approx B_3 \equiv B$, and we have

$$\hat{\omega}_P \approx \omega_P\left(1 - \frac{\omega_P}{B}\right) \quad \text{and} \quad \hat{Q}_P \approx Q_P\left(1 - \frac{3\omega_P}{B}\right) \qquad \hat{Q}_P \text{ AND } \omega_P \text{ EQUAL Bs TIL-3B} \tag{8.60}$$

implying that

$$\frac{\Delta\omega_P}{\omega_P} \approx -\frac{\omega_P}{B} \qquad \frac{\Delta Q_P}{Q_P} \approx -\frac{3\omega_P}{B} \qquad \text{TIL-3B EQUAL Bs} \tag{8.61}$$

Comparing equations (8.61) with (8.49) and (8.50) reveals that $\Delta\omega_P/\omega_P$ is approximately the same for both TIL-3B and TIL-3A. However, $\Delta Q_P/Q_P$ *is significantly less for TIL-3B than for TIL-3A* if $Q_P \gg 1$. Substituting equations (8.61) into (8.51) and neglecting the much smaller term containing $\Delta Q_P/Q_P$, we find that

$$\frac{\Delta G_b(\omega)}{G_b(\omega)} \approx \left[S_{\omega_p}^{G_b(\omega)}\right]\left(-\frac{\omega_P}{B}\right) \qquad \text{TIL-3B EQUAL Bs} \tag{8.62}$$

which is shown in Fig. 8.4(a). It is possible to predistort the initial TIL-3B circuit to compensate for the above gain error. The downward shift of ω_p that is implied by equation (8.62) may be cancelled by predistorting the capacitors C as shown in equation (8.54). It is generally not necessary to predistort R_Q.

The TIL-3B circuit is similar to the coupled-GIC circuit (Tables 5-6, 5-7, and 5-9) in the sense that it may be designed to be self-compensating, or *self-enhancing*; that is, the effect of B on the Q-factor is virtually negligible if matched GBPs are used.

8-3.3 Biquadratic Three- and Four-Amplifier Circuits

In this section, some practical three- and four-amplifier biquadratic RC-active filter circuits are described. These circuits are implementations

TABLE 8-2

The LBI/DF biquadratic circuit

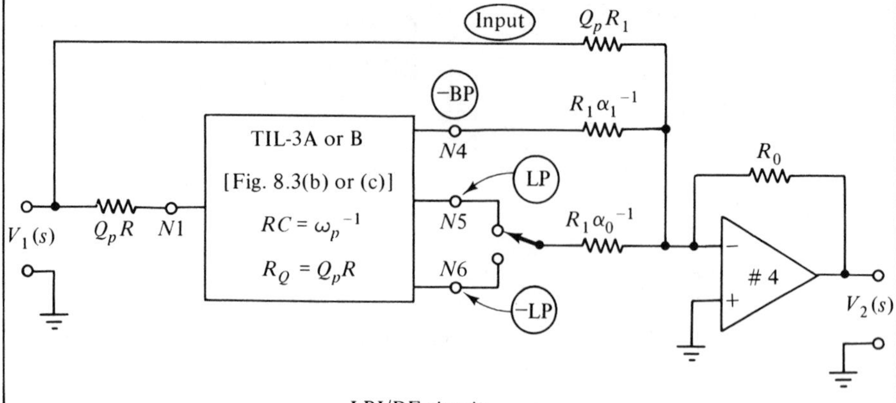

LBI/DF circuit

Transfer function:

$$H_b(s) \equiv \frac{V_2(s)}{V_1(s)} = H_0 \frac{s^2 + \left[\frac{\omega_z s}{Q_z}\right] + \omega_z^2}{s^2 + \left[\frac{\omega_p s}{Q_p}\right] + \omega_p^2}$$

where

$$H_0 = -\frac{R_0}{R_1} \cdot \frac{1}{Q_p}; \qquad \omega_z^2 = \omega_p^2 [1 \mp \alpha_0], \pm \text{LP}$$

and

$$\left[\frac{\omega_z}{Q_z}\right] = \omega_p [Q_p^{-1} - \alpha_1]$$

Q_p-Enhancement condition: Use TIL-3B, $B_1 = B_2 = B_3$

$$\frac{\Delta\omega_p}{\omega_p} \approx -\frac{\omega_p}{B}; \qquad WS_R^{G_b(\omega)} \approx 2 \times LWS_R^{G_b(\omega)}$$

$$WS_C^{G_b(\omega)} \approx LWS_C^{G_b(\omega)}$$

of the SFGs that are analyzed in section 7-3.2; they employ the TIL-3A or TIL-3B circuits as the essential building-blocks and are distinguished only by the ways in which input summing resistors are introduced at nodes $N1$, $N2$, and $N3$ of Figs. 8.3(b) or (c), and by the location of the output node.

The LBI/DF biquadratic circuit, Table 8-2: This circuit is an implementation of the SFG shown Fig. 7.9(a) and is shown in Table 8-2 along with a summary of the circuit behavior. The input signal, the negative bandpass signal ($-$BP), and either the positive lowpass ($+$LP) or negative lowpass signal ($-$LP) are summed by amplifier #4 to realize the general biquadratic function $H_b(s)$. Note that the input summing resistance has value $Q_P R$, where R is the value of the resistors inside the TIL-3A or TIL-3B circuit. The damped inverting-integrator inside the TIL circuit requires a damping resistor $R_Q = Q_P R$ in *parallel* with the capacitance C, as shown in Figs. 8.3(b) and (c). Circuit design is straightforward because H_0, ω_P, ω_Z, Q_P, and Q_Z are specified, and it follows directly from the expression for $H_b(s)$ in Table 8-2 that the circuit RC-element values are given by

$$RC = \omega_P^{-1} \tag{8.63}$$

$$R_Q = Q_P R \tag{8.64}$$

$$R_1\alpha_0^{-1} = \left[\left(\frac{\omega_Z}{\omega_P}\right)^2 - 1\right]^{-1} R_1 \tag{8.65}$$

$$R_1\alpha_1^{-1} = \left[1 - \frac{Q_P\omega_Z}{Q_Z\,\omega_P}\right]^{-1} Q_P R_1 \tag{8.66}$$

$$R_0 = -H_0 Q_P R_1 \tag{8.67}$$

RESISTOR DESIGN EQUATIONS FOR THE LBI/DF CIRCUIT, +LP OUTPUT

Some general observations concerning these design equations are worthwhile. The designer must specify resistors R, C, R_Q, $R_1\alpha_0^{-1}$, $R_1\alpha_1^{-1}$, $Q_P R_1$, and R_0; this involves selection of the *six* parameters R, C, α_0, α_1, R_1, and R_0 to determine the *five* design parameters H_0, Q_P, Q_Z, ω_P, and ω_Z. Consequently, the designer has the freedom to *specify* R and R_1, for example, or C and R_1 and then solve the five equations above for the remaining unknowns. Also note that equation (8.65) for $R_1\alpha_0^{-1}$ is negative if

$$\omega_P > \omega_Z \quad \text{CONDITION FOR USING } -\text{LP OUTPUT} \tag{8.68}$$

and equation (8.66) for $R_1\alpha_1^{-1}$ is negative if

$$Q_P\omega_Z > Q_Z\omega_P \quad \text{EXTRA INVERTER CONDITION} \tag{8.69}$$

The use of *negative* resistors for the feedforward paths containing $R_1\alpha_0^{-1}$ and $R_1\alpha_1^{-1}$ in the circuit of Table 8-2 is avoided by simply introducing an extra inverting-amplifier to create $+$BP from $-$BP if (8.69) is valid and by using the $-$LP output instead of the $+$LP output if (8.68) is valid. In most applications, equation (8.69) is *not* valid and, therefore, the extra (fifth) amplifier is *not* required.

EXAMPLE 8.1: Design an ideal LBI/DF biquadratic circuit as shown in Table 8-2 that provides the following desired transfer function parameters:

$$\left.\begin{aligned} &\omega_P = 10^4 \text{ rad/sec}, \qquad Q_Z = \infty \\ &\omega_Z = 2 \times 10^4 \text{ rad/sec}, \qquad Q_P = 10 \\ &H_0 = 0.1 \end{aligned}\right\} \tag{8.70}$$

with a TIL resistive level given by

$$\boxed{R = 10 \text{ k}\Omega}$$

SOLUTION: It follows from equations (8.63), (8.64), and (8.70) that

$$\boxed{C = 10 \text{ nF}, \qquad R_Q = 100 \text{ k}\Omega}$$

Then, from equations (8.67) and (8.70),

$$\frac{R_0}{R_1} = H_0 Q_P = 1$$

We may *select* R_1 as

$$\boxed{R_1 = 10 \text{ k}\Omega}$$

so that

$$\boxed{R_0 = 10 \text{ k}\Omega}$$

Finally, equations (8.65), (8.66), and (8.70) indicate that the feedforward summation resistors are

$$R_1\alpha_0^{-1} = \left[(2)^2 - 1\right]^{-1} \times 10^4 \ \Omega$$

and

$$R_1\alpha_1^{-1} = [1 - 0]^{-1} \times 10 \times 10^4 \ \Omega$$

which is

$$\boxed{R_1\alpha_0^{-1} = 3\tfrac{1}{3} \text{ k}\Omega, \qquad R_1\alpha_1^{-1} = 100 \text{ k}\Omega}$$

The complete circuit implementation is shown in Fig. 8.5(a), where TIL-3B is employed in order to take advantage of Q_P-enhancement ($\hat{Q}_P \approx Q_P$).

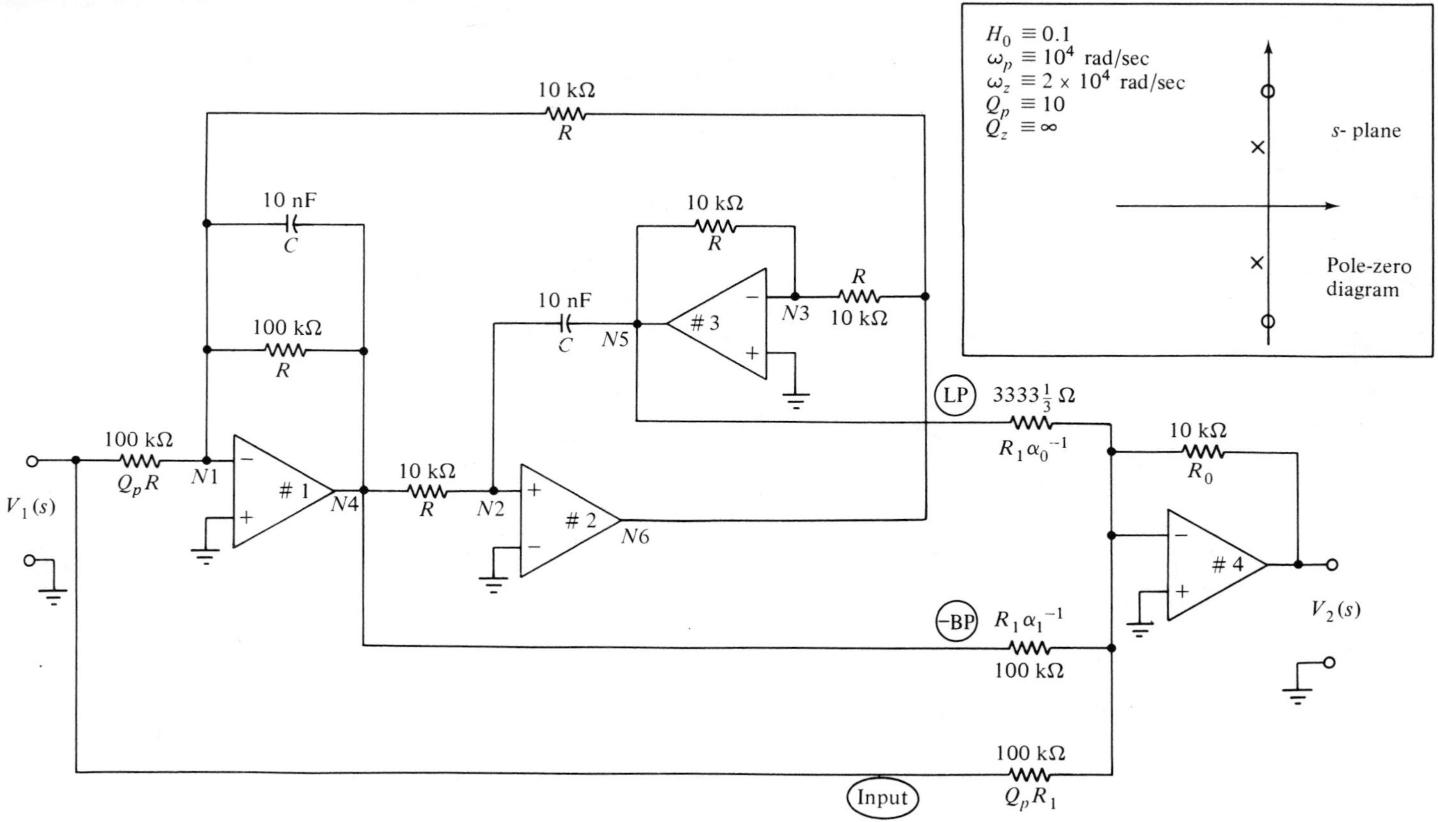

(a) Design example 8.1, LBI/DF. Uses TIL–3B, example 8.1

FIGURE 8.5

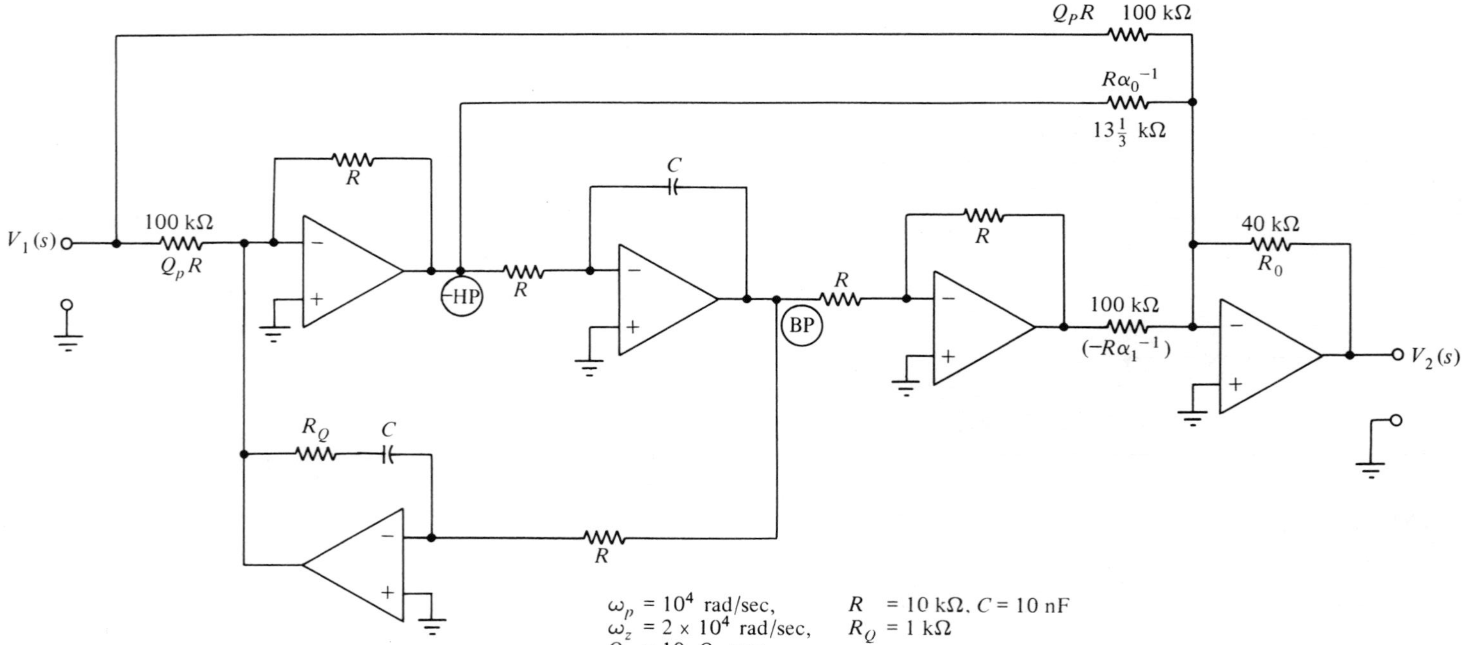

$\omega_p = 10^4$ rad/sec, $R = 10$ kΩ, $C = 10$ nF
$\omega_z = 2 \times 10^4$ rad/sec, $R_Q = 1$ kΩ
$Q_p = 10$, $Q_z = \infty$

(b) Design example 8.3 BHI/DF circuit.

FIGURE 8.5 *(Continued).*

The *nonideal performance* of this circuit is derived from the results in sections 8-2.1 and 8-3.1. Applying the analysis in section 8-2.1 to the *quadratic* functions $-$BP and $\pm$LP leads to the single-element passive sensitivities shown in Table 8-1 and to the worstcase-sensitivity estimates in equations (8.22) and (8.23) and the Schoeffler-sensitivity estimates in equation (8.27A). The passive element worstcase-sensitivity terms $WS_{R,C}^{G_b(\omega)}$ are included in Table 8-2 for comparison with other circuits, where it should be noted that they are strictly valid only for the quadratic $\pm$LP and BP functions $G_b(\omega)$. The *active-sensitivity performance* is provided by the analysis in section 8-3.1 and the results in equations (8.59) and (8.60). The Q_P self-enhancement feature applies to this circuit if TIL-3B is used and if the OP AMPs have accurate matching of gain bandwidth products, so that $B_1 \approx B_2 \approx B_3 \equiv B$. The fractional shift of ω_P is given by $-\omega_P/B$; thus, in Example 8.1 we expect that the use of a quad-OP AMP with $B = 10^6$ rad/sec will cause a downward shift in ω_P of 1% ($\omega_p/B = 0.01$), which may easily be compensated for by predistorting the capacitors C according to equation (8.54), which gives $C = 9.90$ nF for this particular numerical example. [Before proceeding to the description of other biquadratic circuits, it should be noted that an assumption has been implied concerning the effect(s) of the input summing resistors at nodes $N1$, $N2$, and/or $N3$ of the TIL-3A or B circuits. The high-frequency active-sensitivity analysis in sections 8-3.1 and 8-3.2 is based on Figs. 8.3(b) and (c), where *there is not any connection of summing resistors to N1, N2, and N3.* The inquiring reader may be concerned that the introduction of summing resistors at these nodes may detract from the accuracy of such equations as (8.42), (8.46), (8.47), (8.48), (8.49), (8.53), (8.54), (8.59), (8.60), and (8.61). Provided that these summing resistors are much larger than R (the TIL resistors), it may be shown that the active-sensitivity results given here are accurate. However, if these summing resistors are as low as say R, then inaccuracies occur; for example, $\Delta\omega_P/\omega_P$ deteriorates to $-3\omega_P/2B$ in this case.]

The MI/DF biquadratic circuit, Table 8-3: This circuit, shown in Table 8-3, has multiple-input summation at nodes $N1$, $N2$, and $N3$ of either TIL-3A or TIL-3B. The circuit is equivalent to the SFG shown in Fig. 7.9(c), and the reader may find it useful to review the discussion of this SFG in section 7-3.2. It is observed from Table 8-3 that the circuit uses only three OP AMPs, but it does require three capacitors (compared with four OP AMPs and two capacitors for the LBI/DF circuit). The input impedance is not constant over ω because of the capacitor C_4, which may cause a loading problem on the source circuit at high frequencies because $1/j\omega C_4 \to 0$ as $\omega \to \infty$.

TABLE 8-3

The MI/DF biquadratic circuit.

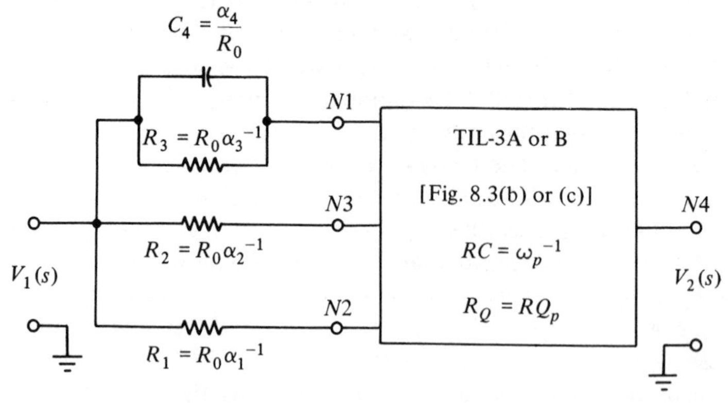

MI/DF circuit

Transfer function: $H_b(s) \equiv \dfrac{V_2(s)}{V_1(s)} = \left(-\dfrac{\alpha_4 \omega_p R}{R_0}\right) \dfrac{s^2 + \dfrac{1}{\alpha_4}[\alpha_3 - \alpha_2]s + \left[\dfrac{\alpha_1}{\alpha_4}\right]\omega_p}{s^2 + \left[\dfrac{\omega_p}{Q_p}\right]s + \omega_p^2}$

$$H_0 = -\alpha_4 \omega_p \left(\frac{R}{R_0}\right); \quad \omega_z^2 = \frac{\alpha_1 \omega_p}{\alpha_4}$$

$$\frac{\omega_z}{Q_z} = \frac{\alpha_3 - \alpha_2}{\alpha_4}$$

Q_p-Enhancement condition: Use TIL-3B so that

$$\hat{Q}_p \approx Q_p$$

if $B_1 \approx B_2 \approx B_3$

and $R_{1,2,3} \gg R$

$$\frac{\Delta\omega_p}{\omega_p} \approx -\frac{\omega_p}{B}$$

The following analysis of the MI/DF circuit reveals that it is highly flexible and useful. Direct calculation of $H_b(s)$ provides the transfer function shown in Table 8-3. The parameters α_1, α_2, α_3, and α_4 are used to obtain the desired H_0, Q_Z, and ω_Z; if Q_Z is negative, α_3 is set to zero and, thereby, R_3 is omitted from the circuit; conversely, if Q_Z is positive, α_2 is set to zero and, thereby, R_2 is omitted from the circuit. Note that the gain H_0 is negative.

The equations for H_0, ω_Z^2, and ω_Z/Q_Z in Table 8-3 may be rearranged to give the following design equations:

$$RC = \omega_P^{-1}, \text{ select C or R} \tag{8.71}$$

$$R_Q = Q_P R \tag{8.72}$$

$$C_4 = -H_0 C \tag{8.73}$$

$$R_1 \equiv R_0\alpha_1^{-1} = \omega_P/\omega_Z^2 C_4 \tag{8.74}$$

∴ *If* $Q_z < 0$, select $\alpha_3 = 0$.	∴ *If* $Q_z > 0$, select $\alpha_2 = 0$.
∴ $R_3 = \infty$, $R_2 \equiv R_0\alpha_2^{-1}$	∴ $R_2 = \infty$, $R_3 \equiv R_0\alpha_3^{-1}$
∴ $R_2 = -Q_Z/\omega_Z C_4$	∴ $R_3 = Q_Z/\omega_Z C_4$

If $Q_Z = \infty$, select $\alpha_2 = \alpha_3 = 0$, $R_2 = R_3 = \infty$. (8.75)

EXAMPLE 8.2: Repeat the design of Example 8.1 but for the MI/DF circuit.

SOLUTION: Select a TIL resistive level such that

$$\boxed{R = 10\ \text{k}\Omega}$$

so that, as before

$$\boxed{C = 10\ \text{nF}, \qquad R_Q = 100\ \text{k}\Omega}$$

Then, equation (8.73) gives

$$\boxed{C = 1\ \text{nF}}$$

and equation (8.74) gives

$$R_1 = \frac{\omega_P}{\omega_Z^2 C_4} = \frac{10^4}{(2 \times 10^4)^2 \times 10^{-9}} \Omega = 25 \text{ k}\Omega$$

and from equations (8.75) it is noted that, for $Q_Z = \infty$,

$$R_2 = R_3 = \infty$$

Two implementations of this design are shown in Fig. 8.6: one circuit uses TIL-3A and the other uses TIL-3B. The active-sensitivity performance of the latter is superior because of the self-enhancement property of TIL-3B.

Perhaps the most significant advantage of the MI/DF circuit is that it is easy to align (or tune); the parameters Q_P, Q_Z, ω_P, and ω_Z may be adjusted independently with four of the circuit resistors as follows:

(i) Adjust ω_P via a resistor R of the TIL.

(ii) Adjust Q_P via the damping resistor R_Q of the TIL.

(iii) Adjust ω_Z via the resistor R_1 of value $R_0\alpha_1^{-1}$.

(iv) Adjust Q_Z via the resistor R_2 of value $R_0\alpha_2^{-1}$ for right-half plane zeros ($\alpha_3 = 0$) and via the resistor R_3 of value $R_0\alpha_3^{-1}$ for left-half plane zeros ($\alpha_2 = 0$).

The passive-sensitivity performance of the MI/DF circuit is similar to that of the LBI/DF circuit because the analysis in section 8-2.1 is valid. The active-sensitivity performance at high frequencies is, of course, dependent on whether TIL-3A or TIL-3B is employed. Provided that the three input summation resistors R_1, R_2, and R_3 are chosen so that $R_{1,2,3} \gg R$, then the previous analysis in sections 8-3.1 and 8-3.2 remains valid and so, using TIL-3B for example, we find that

$$\hat{Q}_P \approx Q_P, \qquad \hat{\omega}_P \approx \omega_P\left(1 - \frac{\omega_P}{B}\right) \tag{8.76}$$

The MI/DF circuit, therefore, has similar active-sensitivity performance to that of the LBI/DF circuit.

The BHI/DF biquadratic circuit, Table 8-4: The basic SFG for the BHI/DF structure is shown in Fig. 7.9(b) and has the biquadratic transfer function provided by equation (7.28). The *feedforward* transmittance $-b_{1b}$

$H_0 \equiv -0.1$,	$\omega_p \equiv 10^4$ rad/sec, $Q_p \equiv 10$	$\omega_z = 2 \times 10^4$ rad/sec $Q_z \equiv \infty$

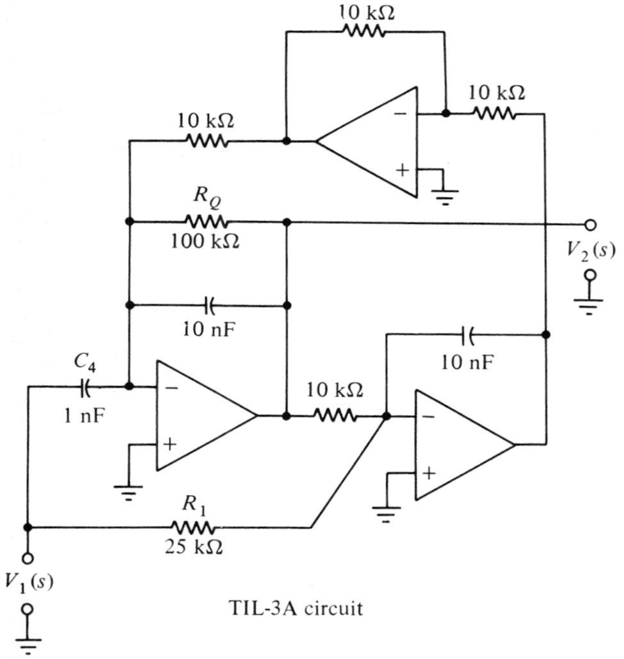

TIL-3A circuit

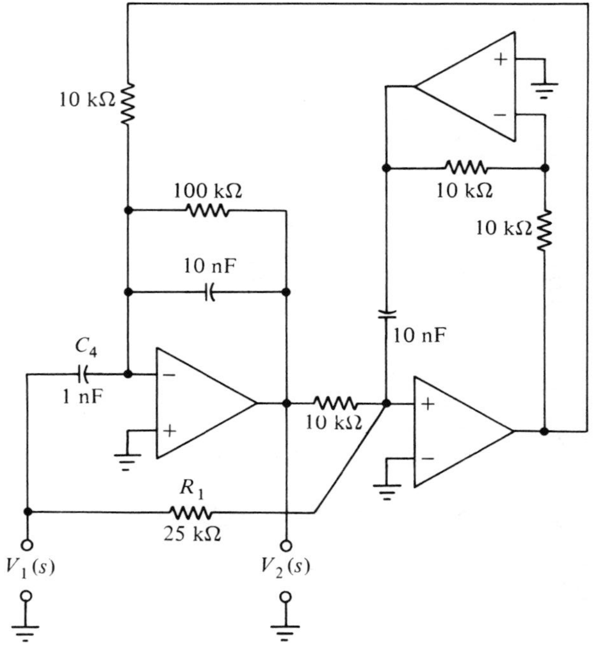

TIL-3B circuit

FIGURE 8.6 *Design examples, MI/DF circuits, example 8.2.*

TABLE 8-4

BHI/DF biquadratic circuits.

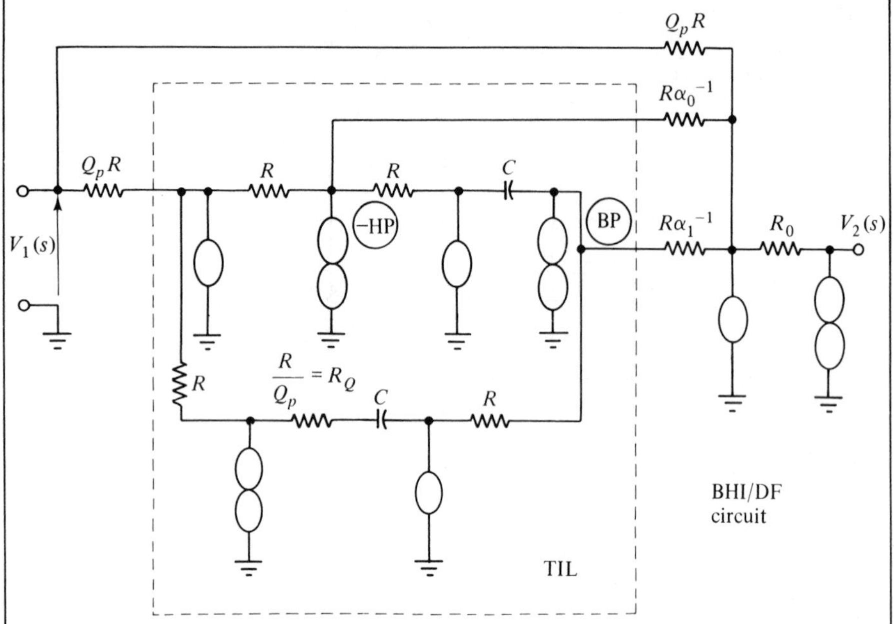

$$H_b(s) = H_0\left[\frac{s^2 + \left(\dfrac{\omega_z}{Q_z}\right)s + \omega_z^2}{s^2 + \left(\dfrac{\omega_p}{Q_p}\right)s + \omega_p^2}\right] \quad \text{where} \quad \omega_p^2 = \frac{1}{\text{CR}}$$

and $$\omega_z^2 = (1 - \alpha_0)^{-1}\,\omega_p^2; \qquad Q_z = \left(\frac{1}{Q_p} + \alpha_1\right)^{-1}\frac{\omega_p}{\omega_z}$$

and $$H_0 = \frac{R_0(\alpha_0 - 1)}{Q_p R}$$

Expressions for $\Delta\omega_p/\omega_p$, $WS_R^{G_b(\omega)}$, and $WS_C^{G_b(\omega)}$ are the same as LBI/DF in Table 8.2 if TIL-3B is used.

shown in Fig. 7.9(b) is realized by means of a resistance b_{1b} in series with the integrating capacitor, as shown in Fig. 8.4(b). See Problem 8-12.

A practical scheme for implementing the BHI/DF structure is shown in Table 8-4, where either type TIL-3A or TIL-3B may be used. Note that the $-$HP and BP outputs of the TIL are used with the input signal to provide the output $V_2(s)$ by summation. The transfer function of the

BHI/DF circuit in Table 8-4 is given by

$$H_b(s) = -\frac{R_0}{Q_P R}\left[\frac{(1-\alpha_0)s^2 + \left(\alpha_1 + \dfrac{1}{Q_P}\right)\omega_P s + \omega_P^2}{s^2 + \left(\dfrac{1}{Q_P}\right)\omega_P s + \omega_P^2}\right] \tag{8.77}$$

where $\omega_P^{-2} \equiv CR$. Comparing the numerator coefficients with equation (8.1) gives

$$Q_Z = \left(\frac{1}{Q_P} + \alpha_1\right)^{-1}\frac{\omega_P}{\omega_Z} \quad \text{and} \quad \omega_Z^2 = (1-\alpha_0)^{-1}\omega_P^2 \tag{8.78}$$

and

$$H_0 = \frac{R_0(\alpha_0 - 1)}{Q_P R} \tag{8.79}$$

The design procedure for the BHI/DF circuit is straightforward. From equations (8.78) and (8.79) we have

$$\left.\boxed{\begin{aligned} RC &= \omega_P^{-2}, \qquad R_Q = \frac{R}{Q_P} \\ R\alpha_1^{-1} &= R\left[\frac{\omega_P}{Q_Z\omega_Z} - \frac{1}{Q_P}\right]^{-1} \\ R\alpha_0^{-1} &= R\left[1 - \left(\frac{\omega_P}{\omega_Z}\right)^2\right]^{-1} \\ R_0 &= -R\left[\frac{H_0 Q_P \omega_Z^2}{\omega_P^2}\right] \end{aligned}} \quad \begin{aligned} &\text{BHI/DF} \\ &\text{DESIGN} \\ &\text{EQUATIONS} \end{aligned}\right\} \tag{8.80}$$

Resistor $R\alpha_1^{-1}$ is negative if $Q_z\omega_z > Q_p\omega_p$, resistor $R\alpha_0^{-1}$ is negative if $\omega_P > \omega_Z$, and resistor R_0 is negative if $H_0 > 0$. Inverting-amplifiers are required in the feedforward paths if these conditions occur.

EXAMPLE 8.3: Repeat the design of Example 8.1 but using the BHI/DF circuit of Table 8-4 and with $H_0 = -0.1$

SOLUTION: Substituting the given values

$$R = 10\ \text{k}\Omega, \qquad \omega = 10^4\ \text{rad/sec}$$
$$Q_P = 10, \qquad Q_Z = \infty, \qquad \omega_Z = 2 \times 10^4\ \text{rad/sec}$$
$$H_0 = -0.1$$

into equations (8.80), we obtain

$$\boxed{\begin{aligned} &R = 10\text{k}\Omega, \qquad C = 10\ \text{nF}, \qquad R = 1\ \text{k}\Omega \\ &R\alpha_1^{-1} = -100\ \text{k}\Omega, \qquad R\alpha_0^{-1} = 13{,}333\tfrac{1}{3}\ \Omega \\ &R_0 = 40\ \text{k}\Omega \end{aligned}} \tag{8.81}$$

Note that $R\alpha_1^{-1}$ is negative, so the designer should include an extra inverting-amplifier stage in the feedforward path from the BP output, as shown in the circuit realization of Fig. 8.5(b).

The *active-sensitivity* performance of the BHI/DF circuits may be shown to be similar to that of the LBI/DF circuit; that is, the phase error $M_1(j\omega)$ associated with the damped integrator is approximately $-\omega/B$ and, therefore, the use of TIL-3A (as in Fig. 8.5(a)) leads to $\Delta Q_P/Q_P$ of approximately $4\omega Q_P/B$, whereas use of TIL-3B leads to $\Delta Q_P/Q_P$ which is negligible ($\hat{Q}_P \approx Q_P$) compared with the effect of the downward shift of ω_P by the fractional amount ω_P/B.

The *passive-sensitivity* performance of the BHI/DF circuits may be shown to be similar to that of the LBI/DF circuits; that is, for quadratic functions $G_b(\omega)$,

$$WS_R^{G_b(\omega)} \approx 2 \times LWS_R^{G_b(\omega)} \tag{8.82}$$

and

$$WS_C^{G_b(\omega)} \approx LWS_C^{G_b(\omega)} \tag{8.83}$$

Remarks. Of the variety of three- and four-amplifier circuits that have been discussed in this subsection, the MI/DF circuit that uses TIL-3B (see Fig. 8.6) is highly recommended. It allows arbitrary placement of poles and zeros with just four resistive adjustments and without the need for extra feedforward inverting amplifiers. Furthermore, self-enhancement of Q-factor Q_P results in excellent insensitivity to variations of gain bandwidth products B if matched OP AMPs are employed.

Other types of TIL biquadratic circuits have been proposed and used but are not described here because of space limitations.

8-4 GENERALIZED IMMITTANCE CONVERTER (GIC) BIQUADRATIC CIRCUITS

The GIC biquadratic (or quadratic) circuits are especially useful because they may be designed to exhibit many of the passive-sensitivity and active-sensitivity advantages of the TIL circuits that have been described in section 8-3 but with the advantage that only *two* OP AMPs are required.

A general second-order GIC circuit is shown in Fig. 8.7(a); this circuit consists of the coupled-GIC that has been analyzed extensively in sections 5-5.4, 5-6, and 5-7. The GIC is terminated at ports 1 and 2 by the network containing Y_{1A}, Y_{1B}, Y_{6A}, Y_{6B}, and the voltage source $V_1(s)$. Depending on the application and the choice of *RC* elements for the admittances Y_{1A}, Y_{1B}, Y_2, Y_3, Y_4, Y_5, Y_{6A}, and Y_{6B}, the output voltage is chosen as either $V_2(s)$, $V_3(s)$, or $V_4(s)$, as shown in Fig. 8.7(a). There is, in fact, a large number of possible GIC circuits that are useful for realizing lowpass (LP), bandpass (BP), highpass (HP), notch (N), and allpass (AP) transfer functions, and no attempt will be made here to describe all of them. Some particularly useful LP, BP, HP, N, and AP circuits are described, and the nonideal performance of these circuits is derived by making use of the simulated-inductance and FDNR modeling equations that are derived in chapter 5 and summarized in Tables 5-6, 5-7 and 5-9.

Defining

$$Y_1 = Y_{1A} + Y_{1B}, \qquad Y_6 = Y_{6A} + Y_{6B}$$

we may analyze Fig. 8.7(a) by the *RC*-nullor nodal analysis technique described in chapter 3 (Problem 3.19) to obtain the transform voltage transfer functions:

$$\frac{V_2(s)}{V_1(s)} = \frac{Y_{6B}(Y_2Y_4 + Y_{1A}Y_4) + Y_{1B}(Y_3Y_5 - Y_{6A}Y_4)}{Y_1Y_3Y_5 + Y_2Y_4Y_6} \tag{8.84}$$

$$\frac{V_3(s)}{V_1(s)} = \frac{Y_{1B}Y_3Y_5 + Y_{6B}Y_2Y_4}{Y_1Y_3Y_5 + Y_2Y_4Y_6} \tag{8.85}$$

$$\frac{V_4(s)}{V_1(s)} = \frac{Y_{1B}(Y_3Y_5 + Y_{6A}Y_3) + Y_{6B}(Y_2Y_4 - Y_{1A}Y_3)}{Y_1Y_3Y_5 + Y_2Y_4Y_6} \tag{8.86}$$

GIC BIQUAD DESIGN EQUATIONS

There are many ways in which *RC* elements may be allocated to Y_1, Y_2, Y_3, Y_4, Y_5, and Y_6 such that the denominators of the foregoing design equations are quadratic functions. In this section, the particular coupled-GIC configurations of Tables 5-6, 5-7, and 5-9 are emphasized because the following Q_P-enhancement conditions have been derived in chapter 5 and shown to allow virtual independence of Q_P from B over a wide frequency

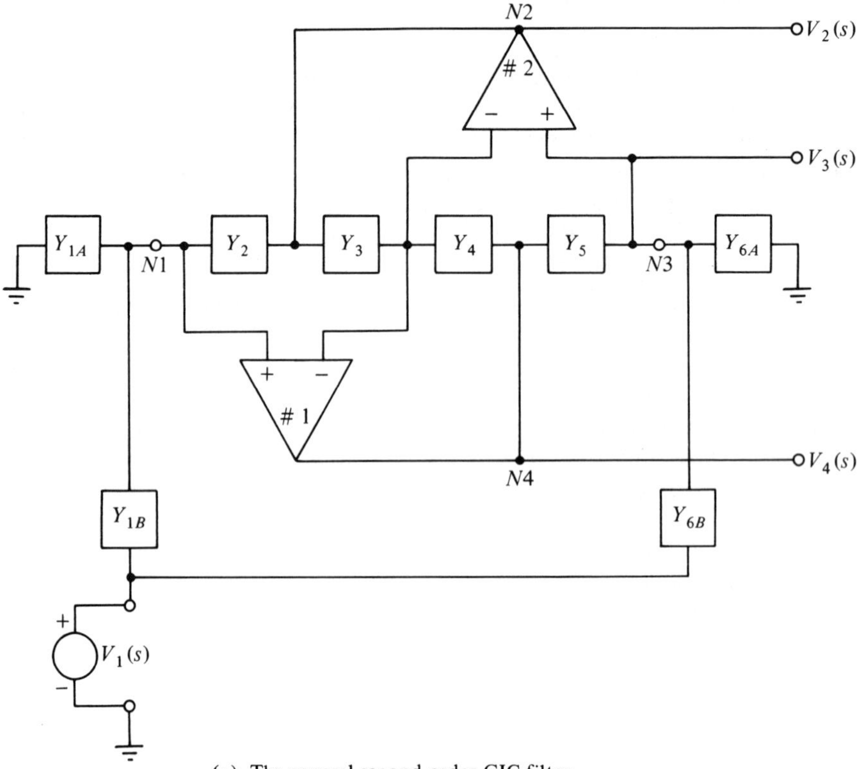

(a) The general second-order GIC filter.

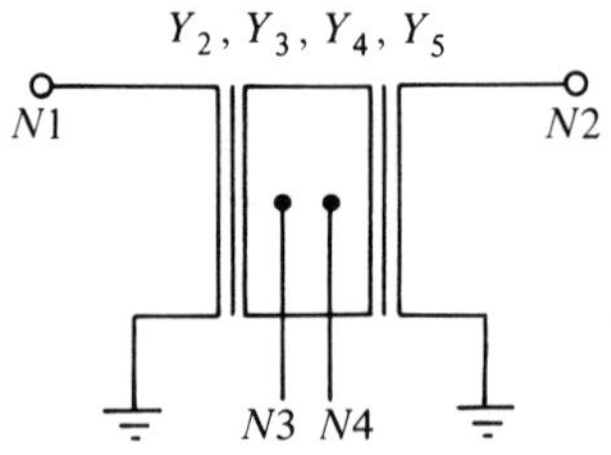

(b) Notation for describing the coupled–GIC

FIGURE 8.7

range:

$$Y_2 \equiv G_2, \quad Y_3 \equiv sC_3, \quad Y_4 \equiv G_4, \quad Y_5 \equiv G_5, \quad Y_6 \equiv G_6$$
$$\textit{enhancement conditions}: \quad B_1 G_6 = B_2 G_5 \tag{8.87}$$

TYPE 1 INDUCTANCE SIMULATION CIRCUIT

$$Y_2 \equiv G_2, \quad Y_3 \equiv G_3, \quad Y_4 \equiv G_4, \quad Y_5 \equiv sC_5, \quad Y_6 \equiv G_6$$
$$\textit{enhancement condition}: \quad G_3 = G_4 \tag{8.88}$$

TYPE 2 INDUCTANCE SIMULATION CIRCUIT

$$Y_2 \equiv sC_2, \quad Y_3 = G_3, \quad Y_4 \equiv sC_4, \quad Y_5 \equiv G_5, \quad Y_6 \equiv G_6$$
$$\textit{enhancement condition}: \quad B_1 G_6 \equiv B_2 G_5 \tag{8.89}$$

TYPE 1 *FDNR* CIRCUIT

The enhancement condition for the type 2 inductance-simulation circuit [equation (8.88)] requires only that $G_3 = G_4$ and does not require that $B_1 = B_2$. Since conductances may be made to match far more closely than the gain bandwidth products of OP AMPs, the quality of Q-enhancement achievable with a type 2 inductance-simulation circuit is particularly good (see section 5-6.2).

In the following subsections, the circuit diagrams for GIC biquadratic and quadratic circuits will be simplified by drawing the coupled-GIC shown in Fig. 8.7(a) according to the symbolic notation of Fig. 8.7(b). The nodes $N1$, $N2$ are external GIC nodes, and $N3$, $N4$ are internal GIC nodes corresponding to the nodes $N1$, $N2$, $N3$, and $N4$ in Fig. 8.7(a). The highpass circuit that is described in the following subsection is analyzed in considerable detail, whereas the remaining GIC circuits, which may be analyzed in a similar way, are treated less thoroughly. Major design equations are tabulated.

8-4.1 Highpass (HP) and Bandpass (BP) GIC circuits

The *highpass* function (HP) is realized using the *RLC* series circuit in Table 8-5, where the output voltage is across the inductance element. The corresponding simulated-inductance circuit is classified in Table 8-5 as GIC-2HP-1 (this mnemonic implies that we have a *GIC* circuit using *two* amplifiers to realize the *HP* function with a *type 1* simulated-inductance circuit). It follows from Table 5-6 that the simulated inductance is pro-

TABLE 8-5

GIC Highpass Circuits GIC-2HP-1

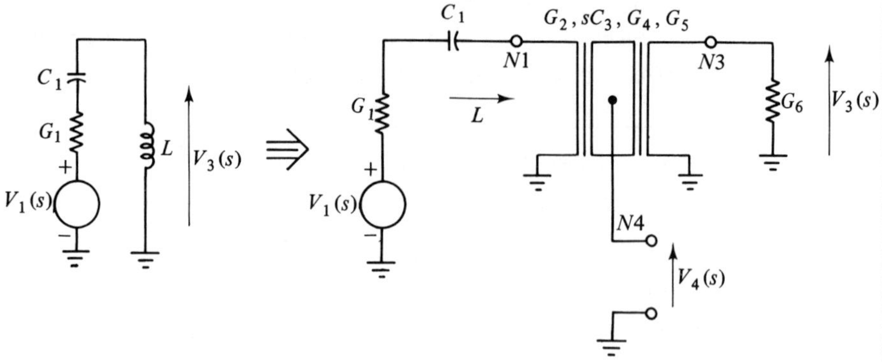

Highpass GIC filter type GIC-2HP-1

$$H_b(s) \equiv \frac{V_4(s)}{V_1(s)} = H_0 \left[\frac{s^2}{s^2 + \left(\frac{\omega_p}{Q_p}\right) s + \omega_p{}^2} \right]$$

where

$$\omega_p{}^{-2} = LC_1 = \frac{C_1 C_3 G_5}{G_2 G_4 G_6}; \qquad Q_p = \omega_p L G_1$$

$$L = \frac{C_3 G_5}{G_2 G_4 G_5}; \qquad H_0 = \frac{G_5 + G_6}{G_5}$$

Enhancement condition: $B_1 G_6 = B_2 G_5$

Using dual-OP AMPs,
$B_1 \approx B_2$, and then
$G_5 = G_6$ and $H_0 = 2$

Active-sensitivity: $\hat{Q}_p \approx Q_p$

$$\frac{\Delta\omega_p}{\omega_p} \approx -\frac{2\omega_p}{B}, \text{ uniform } C/\text{uniform } G$$

Passive-sensitivity: $WS_R^{G_b(\omega)} \approx 2\, LWS_R^{G_b(\omega)}$

$$WS_C^{G_b(\omega)} \approx LWS_C^{G_b(\omega)}$$

vided ideally by

$$L = \frac{C_3 G_5}{G_2 G_4 G_6} \tag{8.90}$$

from which it may be shown by analysis of the *LCR* circuit that the voltage $V_3(s)$ that appears across L is the HP function:

$$\boxed{\frac{V_3(s)}{V_1(s)} = \frac{s^2}{s^2 + \left(\frac{1}{LG_1}\right)s + \frac{1}{LC_1}}} \qquad \text{HP TRANSFER FUNCTION } (H_0 = 1) \tag{8.91}$$

so that

$$\omega_P^{-2} = LC_1 \tag{8.92}$$

and

$$Q_P = \omega_P L G_1 \tag{8.93}$$

It is also possible to obtain equation (8.91) directly by substituting $Y_2 = G_2$, $Y_3 = sC_3$, $Y_4 = G_4$, $Y_5 = G_5$, $Y_{6A} = G_6$, and $Y_{1A} = Y_{6B} = 0$ and $Y_{1B} = sC_1G_1/(1 + sC_1G_1)$ into equation (8.85) giving, after rearrangement of terms,

$$\frac{V_3(s)}{V_1(s)} = \left[\frac{s^2}{s^2 + \left(\frac{G_2G_4G_6}{G_1C_3G_5}\right)s + \left(\frac{G_2G_4G_6}{C_1C_3G_5}\right)}\right] \tag{8.94}$$

which is identical to equation (8.91) after substitution for L. The loading of the output voltage $V_3(s)$ at node 3 by a subsequent terminating network often presents a practical difficulty because the terminating impedance generally causes a deterioration in Q-factor and shift in natural frequency ω_p; fortunately, $V_4(s)$ is directly proportional to $V_3(s)$ whenever $Y_{6B}(s)$ is zero; substituting $Y_{6B}(s) = 0$ into equations (8.85) and (8.86), we obtain

$$V_4(s) = \left(\frac{Y_5 + Y_6}{Y_5}\right)V_3(s), \qquad Y_6 \equiv Y_{6A},\ Y_{6B} = 0 \tag{8.95}$$

so that for HP-GIC2-1 we combine equations (8.91) and (8.95) with

$Y_5 = G_5$ and $Y_6 = G_6$, to give

$$\frac{V_4(s)}{V_1(s)} = \frac{G_5 + G_6}{G_5}\left[\frac{s^2}{s^2 + \left(\frac{1}{LG_1}\right)s + \frac{1}{LC_1}}\right] \qquad \text{HP TRANSFER FUNCTION } (H_0 > 1) \tag{8.96}$$

corresponding to $H_0 = 1 + G_6/G_5$ and, more importantly, we now have a HP output voltage $V_4(s)$ *at the output of OP AMP #1*, which ideally provides the circuit with zero output impedance, thereby allowing similar GIC-2HP-1 sections to be directly cascaded without a requirement for interstage buffer amplifiers.

The *active-sensitivity performance* of this circuit has been derived in section 5-6.1, where it is shown that the fractional inductance error, equation (5.113), is

$$\epsilon_{Hl}(\omega) \approx (G_5 + G_6)\left(\frac{\omega^2 C_3}{B_2 G_4 G_6} + \frac{G_4}{B_1 C_3 G_5}\right) \tag{8.97}$$

and the nonideal Q-factor, equation (5.105), is given by

$$Q_l^{-1}(\omega) \approx (G_5 + G_6)\left[\frac{G_4}{A_{01} C_3 G_5 \omega} + \omega\left(\frac{1}{B_1 G_6} - \frac{1}{B_2 G_5} - \frac{\omega^2 C_3}{B_1 B_2 G_4 G_6}\right)\right] \tag{8.98}$$

where a term involving A_{02} has been omitted here because it is usually negligible. Now, to compare this circuit with the TIL circuits of the previous section it is useful to derive the effective Q-factor $\hat{Q}_P$ and effective natural frequency $\hat{\omega}_P$. The details of the analysis are requested in Problem 8.22, where it may be shown that if $|\epsilon_{Hl}(\omega)| \ll 1$ and $|Q_l(\omega)| \gg 1$ then

$$\hat{Q}_P^{-1} \approx Q_P^{-1} + Q_l^{-1}(\omega_P) \tag{8.99}$$

and

$$\hat{\omega}_P \approx \omega_P\left[1 - \frac{\epsilon_{Hl}(\omega_P)}{2}\right] \tag{8.100}$$

EFFECTIVE Q-FACTOR AND EFFECTIVE NATURAL FREQUENCY

For the widely used *uniform-C/uniform-G* case, we have $C \equiv C_1 = C_3$, $G \equiv G_2 = G_4 = G_5 = G_6$, so that, from equations (8.97), (8.98), (8.99), and

(8.100), we have

$$\hat{Q}_P^{-1} \approx Q_P^{-1} + \frac{2\omega_P}{B_1}\left(1 - m - \frac{\omega_P m}{B_1}\right) \qquad \text{UNIFORM } C \qquad (8.101)$$

UNIFORM G CASE

and

$$\hat{\omega}_P \approx \omega_P\left[1 - \frac{\omega_P}{B_1}(m+1)\right] \qquad \text{GIC-2HP-1} \qquad (8.102)$$

where $m \equiv B_1/B_2$. Thus, the deviation ΔQ_p of $\hat{Q}_p$ from the ideal Q_p is usually minimized by employing dual OP AMPs, so that $B_1 \approx B_2 = B$ or $m \approx 1$; then equations (8.101) and (8.102) indicate that

$$\frac{\Delta Q_P}{Q_P} \approx \frac{2Q_P\omega_P^2}{B^2} \qquad (8.103)$$

and

$$\frac{\Delta\omega_P}{\omega_P} \approx -\frac{2\omega_P}{B} \qquad (8.104)$$

APPROXIMATE EXPRESSIONS FOR GIC-2HP-1 MATCHED B's

The *magnitude of* $\Delta Q_P/Q_P$ in equation (8.103) is usually *insignificantly small over the frequency range of interest* ($\omega_P \ll B$) so that $\Delta Q_P/Q_p$ is virtually independent of B. The previously described TIL-3B circuit possesses a similar dependence of $\Delta Q_P/Q_P$ on ω_P, Q_P, and B, and it is concluded that the self-enhancement of Q-factor for GIC-2HP-1 is similar to that obtained for the circuits that use TIL-3B. However, a disadvantage of the GIC biquadratic circuits is typified by equation (8.104), where $|\Delta\omega_P/\omega_P|$ *is twice that obtained for TIL-3A/B circuits* (see equations (8.49) and (8.60). Thus, it is concluded that the uniform-G/uniform-C design of *GIC-2HP-1 is inferior to the LBI/DF and MI/DF circuits (that employ TIL-3B) insofar as active-sensitivity performance is concerned.* A similar conclusion follows for each of the GIC two-amplifier quadratic circuits that are to be described.

The gain H_0 of GIC-2HP-1 is equal to $1 + G_6/G_5$ for the transfer function in equation (8.96). Therefore, if the designer wishes to meet the Q_P-enhancement condition of equation (8.87) in order to obtain good high-frequency performance then it is necessary that $G_5 = G_6$ and, therefore, that $H_0 = 2$.

The *passive-sensitivity* performance of GIC-2HP-1 is derived from equation (8.96). Using the methods that have been described in section

8-2.1, it may be shown that this circuit is similar to the TIL circuit; that is,

$$\boxed{\begin{aligned} WS_R^{G_b(\omega)} &\approx 2LWS_R^{G_b(\omega)} \\ WS_C^{G_b(\omega)} &\approx LWS_C^{G_b(\omega)} \end{aligned}} \qquad \begin{matrix}\text{Approximate Worstcase}\\ \text{Sensitivity Performance}\\ \text{GIC-2HP-1}\end{matrix} \qquad (8.105)$$

If $Q_P \gg 1$. The *passive-sensitivity performance is, therefore, near-optimum in the worstcase sense.*

EXAMPLE 8.4: Use the GIC-2HP-1 circuit to implement the normalized transfer function

$$H(s) = \frac{Hs^4}{(s^2 + 0.1s + 1)(s^2 + 0.2s + 4)} \qquad (8.106)$$

where H should be chosen to achieve Q_P-enhancement.

SOLUTION: We realize $H(s)$ as a cascade connection of $H_1(s)$ and $H_2(s)$, where each section has a transfer function of the form of $V_4(s)/V_1(s)$ shown in Table 8-5 (GIC-2HP-1). Then,

$$H_1(s) = \frac{H_{01}s^2}{(s^2 + 0.1s + 1)} \qquad H_2(s) = \frac{H_{02}s^2}{(s^2 + 0.2s + 4)} \qquad (8.107)$$

where $H_{01}H_{02} = H$. From equations (8.107) we have

$$\omega_{p1} = 1 \text{ rad/sec}, \qquad Q_{p1} = 10 \qquad (8.108)$$

for the first stage, and

$$\omega_{p2} = 2 \text{ rad/sec}, \qquad Q_{p2} = 20$$

for the second stage. Since Q_P-enhancement is required, $H_1 = H_2 = 2$ and $H = 4$. Therefore, a suitable design for stage one requires that the expressions for Q_P and ω_P in Table 8-5 be given by

$$\omega_P^{-2} = 1 = L_1C_1, \qquad Q_{P1} = 10 = \omega_p L_1 G_1, \qquad G_5 = G_6 \qquad (8.109)$$

where $L_1 = C_3G_5/G_2G_4G_6$. We therefore *select*

$$C_1 = C_3 = 1 \text{ F}, \quad G_2 = G_4 = G_5 = G_6 = 1 \text{ mho}, \quad G_1 = 10 \text{ mho}$$

so that equation (8.109) is satisfied. Similarly, for the second stage we *select*

$$C_1 = C_3 = 2 \text{ F}, \quad G_2 = G_4 = G_5 = G_6 = 1 \text{ mho}, \quad G_1 = 20 \text{ mho}$$

to satisfy $\omega_P^{-2} = \frac{1}{4}$, $Q_P = 20$, and $G_5 = G_6$. The complete normalized circuit implementation is illustrated in Fig. 8.8.

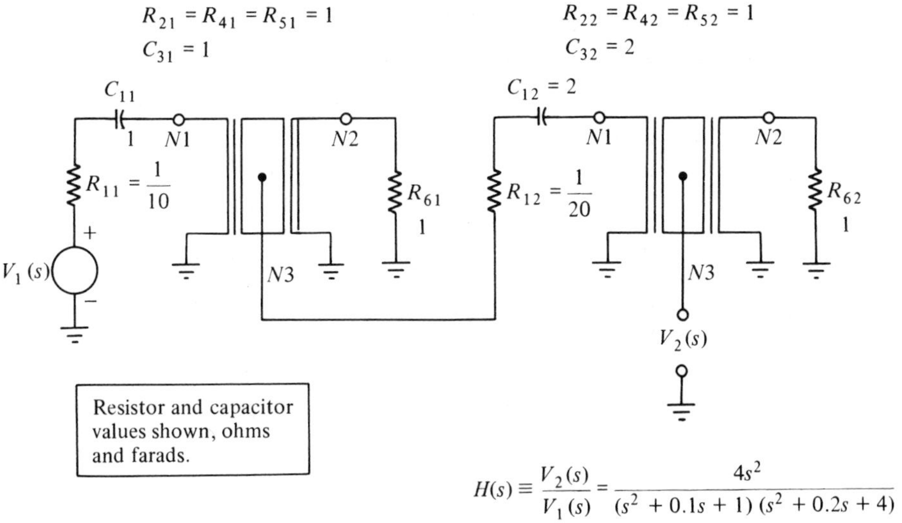

FIGURE 8.8 *Design example 8.4; fourth-order highpass circuit using GIC-2HP-1.*

The reader should note that implementation of the HP transfer function using TIL circuits always involves at least three OP AMPs; for example, the LBI/DF circuits require four OP AMPs, and since the HP function is not available directly the BP and LP terms in the output must be cancelled exactly to zero by choosing $\alpha_1 = Q_P^{-1}$ and $\alpha_0 = \pm 1$; the MI/DF circuit use three OP AMPs and three capacitors to realize the HP function. The GIC circuit GIC-2HP-1 of Table 8.5 is, therefore, particularly useful due to its near-optimum passive-sensitivity performance, self-enhancement of Q-factor, and use of only two OP AMPs and two capacitors.

A second *GIC highpass circuit GIC-2HP-2* is obtained from Table 8-5 by replacing the type 1 inductance-simulation circuit by a type 2 inductance-simulation circuit; this simply involves interchanging the elements in Y_3 and Y_5 so that $Y_3 \equiv G_3$ and $Y_5 \equiv sC_5$. This circuit has the advantage that the Q_P self-enhancement condition ($G_3 = G_4$) is independent of B_1 and B_2 but, unfortunately, $V_2(s)$ or $V_4(s)$ is *not* proportional to $V_3(s)$, and we cannot now find a voltage source output.

The *GIC bandpass circuits, GIC-2BP-1 and GIC-2BP-2*, are identical to the highpass circuits except that the simulated inductance is connected in *parallel* with a capacitance C_1 and conductance G_1 as shown in Table 8-6. The analysis of these circuits is similar to that of the highpass circuits, and the summarized results in Table 8-6 are the subject of Problem 8.25. The GIC-2BP-1 circuit is especially useful, because it has near-optimum passive sensitivity, Q_p self-enhancement, and uses only two OP AMPs and two capacitors. Problem 8.27 provides the reader with some design experience with this circuit.

TABLE 8-6
GIC Bandpass Circuit GIC-2BP-1

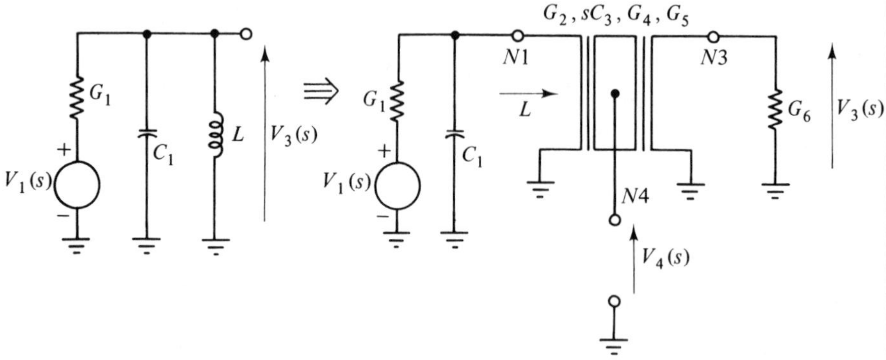

Bandpass GIC filter type GIC-2BP-1

$$H_b(s) = H_0 \left[\frac{s}{s^2 + \left(\frac{\omega_p}{Q_p}\right) s + \omega_p{}^2} \right] \equiv \frac{V_4(s)}{V_1(s)}$$

where

$$\omega_p{}^{-2} = LC_1 = \frac{C_1 C_3 G_5}{G_2 G_4 G_6}; \qquad Q_p = \frac{\omega_p C_1}{G_1}$$

$$L = \frac{C_3 G_5}{G_2 G_4 G_6}; \qquad H_0 = \frac{(G_5 + G_6)\omega_p}{G_5 Q_p}$$

Enhancement condition, active sensitivity, passive sensitivity as for GIC-2HP-1 in Table 8.5

8-4.2 Lowpass (LP) GIC circuits

The lowpass second-order function LP can, of course, be realized by means of the series *RLC* circuit shown in Fig. 8.9. This circuit is *not* recommended, because the output voltage $V_2(s)$ is across a capacitance element, and the usual deterioration of Q_P and ω_P due to any external loading is likely to be significant. A preferable configuration GIC-2LP-1 uses the concept of the FDNR D element (see sections 2.5 and Table 5.9); direct analysis of the parallel *RCD* circuit in Table 8-7 gives

$$\frac{V_3(s)}{V_1(s)} = \frac{1}{DR_1} \left[\frac{1}{s^2 + s\left(\frac{C_1}{D}\right) + \frac{1}{DR_1}} \right] \tag{8.110}$$

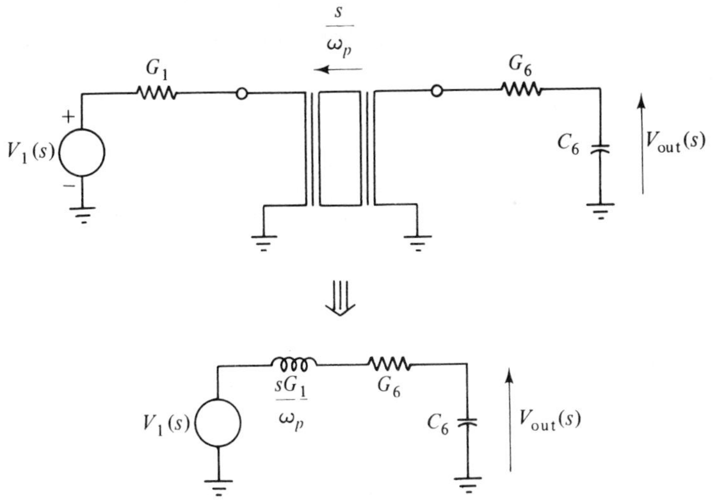

FIGURE 8.9 *A possible GIC lowpass circuit.*

implying that we have a resonant circuit at port 1 of the GIC in which the FDNR resonates with the resistance R_1 and where the damping is provided by a capacitance C_1. It follows from equation (8.110) that

$$\omega_P^{-2} = DR_1 \quad \text{and} \quad Q_P = \frac{\omega_P D}{C_1} \tag{8.111}$$

The value of the D element that is given in Table 5.9 for the type 1 FDNR circuit is

$$D = \frac{C_2 C_4 G_6}{G_3 G_5} \tag{8.112}$$

Since equation (8.95) implies that $V_4(s)$ and $V_3(s)$ are proportional, via $V_4(s) = V_3(s)[1 + G_6/G_5]$, then it follows from equation (8.110) that $V_4(s)$ may serve as a voltage source output and that

$$H_b(s) \equiv \frac{V_4(s)}{V_1(s)} = \frac{(G_5 + G_6)G_1}{G_5 D}\left[\frac{1}{s^2 + s\left(\frac{C_1}{D}\right) + \left(\frac{G_1}{D}\right)}\right] \tag{8.113}$$

The nonideal performance with respect to passive sensitivity, active sensitivity, etc. is similar to that of GIC-2HP-1 and is summarized in Table 8-7. The circuit possesses near-optimum passive sensitivity, Q_P self-enhancement if dual-OP AMPs are used *and* if $G_5 = G_6$, and it uses only two OP AMPs. The GIC-2LP-1 circuit of Table 8-7 is useful for the cascade

TABLE 8-7

GIC Lowpass Circuit GIC-2LP-1

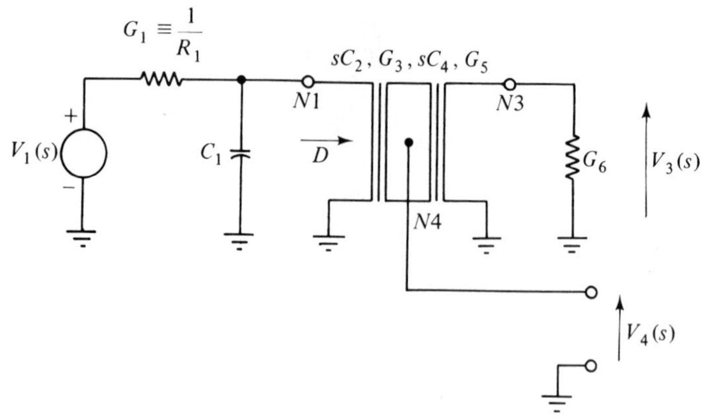

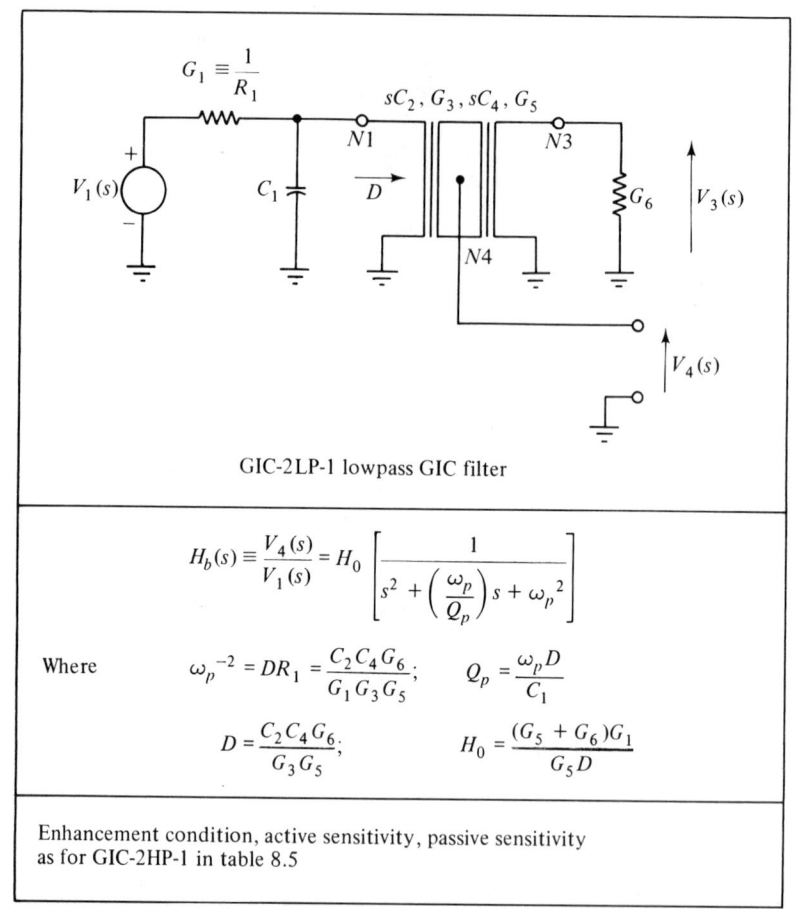

GIC-2LP-1 lowpass GIC filter

$$H_b(s) \equiv \frac{V_4(s)}{V_1(s)} = H_0 \left[\frac{1}{s^2 + \left(\frac{\omega_p}{Q_p}\right)s + \omega_p^{\,2}} \right]$$

Where

$$\omega_p^{-2} = DR_1 = \frac{C_2 C_4 G_6}{G_1 G_3 G_5}; \qquad Q_p = \frac{\omega_p D}{C_1}$$

$$D = \frac{C_2 C_4 G_6}{G_3 G_5}; \qquad H_0 = \frac{(G_5 + G_6)G_1}{G_5 D}$$

Enhancement condition, active sensitivity, passive sensitivity as for GIC-2HP-1 in table 8.5

realization of the *all-pole* lowpass transfer functions of the type considered in equation (7.39); that is, Chebyshev, Butterworth, and Bessel-type lowpass functions.

If zeros of $H(s)$ are required at some point other than the s-plane origin, then the cascade of these LP, BP, or HP circuits is not useful. It then becomes useful to consider GIC circuits that realize imaginary, real, or complex zeros.

8-4.3 Notch (N) and Allpass (AP) Circuits

The notch (N) function is characterized by $Q_Z \equiv \infty$ and, therefore, has a pair of imaginary zeros $\pm j\omega_Z$; the allpass (AP) function is characterized by $Q_Z \equiv -\,Q_P$ and $\omega_Z = \omega_P$ so that $|G_b(\omega)|$ is constant for all ω.

TABLE 8-8 *(a)*

GIC notch filters, $\omega_z > \omega_p$ and $\omega_z < \omega_p$; GIC-2N-2A and -2B.

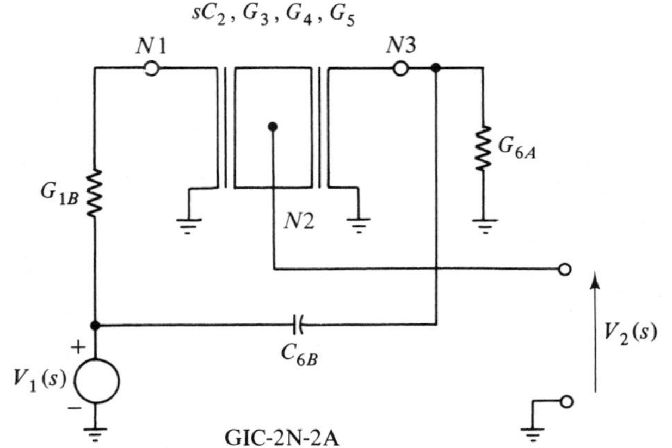

GIC-2N-2A
notch function
$\omega_z < \omega_p$

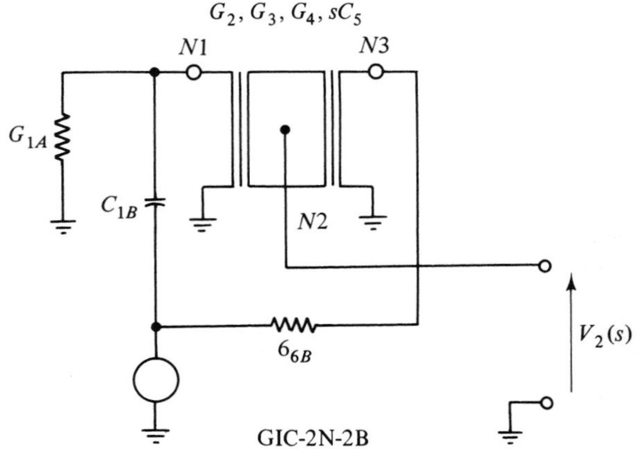

GIC-2N-2B
notch function
$\omega_z > \omega_p$

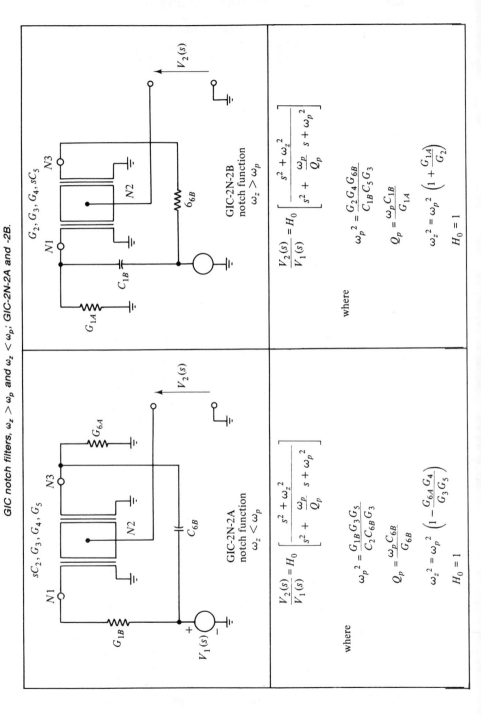

GIC-2N-2A	GIC-2N-2B
$\dfrac{V_2(s)}{V_1(s)} = H_0 \left[\dfrac{s^2 + \omega_z^2}{s^2 + \dfrac{\omega_p}{Q_p} s + \omega_p^2} \right]$	$\dfrac{V_2(s)}{V_1(s)} = H_0 \left[\dfrac{s^2 + \omega_z^2}{s^2 + \dfrac{\omega_p}{Q_p} s + \omega_p^2} \right]$
where	where
$\omega_p^2 = \dfrac{G_{1B} G_3 G_5}{C_2 C_{6B} G_3}$	$\omega_p^2 = \dfrac{G_2 G_4 G_{6B}}{C_{1B} C_5 G_3}$
$Q_p = \dfrac{\omega_p C_{6B}}{G_{6B}}$	$Q_p = \dfrac{\omega_p C_{1B}}{G_{1A}}$
$\omega_z^2 = \omega_p^2 \left(1 - \dfrac{G_{6A} G_4}{G_3 G_5}\right)$	$\omega_z^2 = \omega_p^2 \left(1 + \dfrac{G_{1A}}{G_2}\right)$
$H_0 = 1$	$H_0 = 1$

TABLE 8-8 (b)

GIC Allpass or Notch Circuit GIC-2N-1.

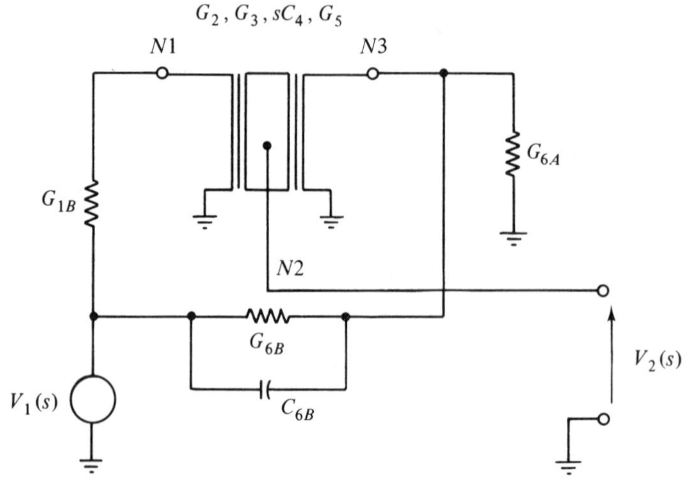

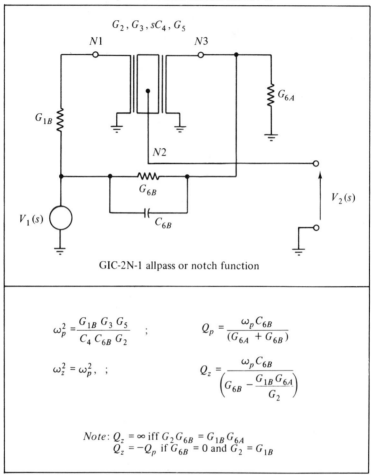

GIC-2N-1 allpass or notch function

$$\omega_p^2 = \frac{G_{1B}\,G_3\,G_5}{C_4\,C_{6B}\,G_2} \quad ; \qquad Q_p = \frac{\omega_p C_{6B}}{(G_{6A} + G_{6B})}$$

$$\omega_z^2 = \omega_p^2, \quad ; \qquad Q_z = \frac{\omega_p C_{6B}}{\left(G_{6B} - \dfrac{G_{1B}G_{6A}}{G_2}\right)}$$

Note: $Q_z = \infty$ iff $G_2 G_{6B} = G_{1B} G_{6A}$
$Q_z = -Q_p$ if $G_{6B} = 0$ and $G_2 = G_{1B}$

In this section, four circuits are described that use the coupled-GIC to realize the N and AP functions. All of these circuits are special cases of Fig. 8.7(a), and they are shown in Tables 8-8. By setting $V_1(s)$ to zero in each circuit in Table 8-8(a), the reader may verify that these two circuits are exactly equivalent to a type 2 inductance-simulation circuit that is terminated by the parallel admittance $(G_{1B,\,6A} + sC_{1B,\,6B})$. Similarly, setting $V_1(s)$ to zero in the circuit of Table 8-8(b) results in a type 1 inductance-simulation circuit that is terminated in the admittance $(G_{6A} + G_{6B} + sC_{6B})$. We therefore expect that the variations of ω_P and Q_P due to perturbations of $B_{1,\,2}$ and of the RC elements will be exactly the same for these circuits as for the corresponding type 1 and type 2 inductance-simulation circuits that are analyzed in chapter 5.

Notch function, $\omega_Z < \omega_P$, GIC-2N-2A: The circuit GIC-2N-2A is shown in Table 8-8(b), and the pertinent *ideal* design equations are also included in the table. The reader is asked in Problem 8.29 to verify these equations directly from equation (8.84).

The principal advantages of this circuit are that only two OP AMPs are required, and such circuits may be cascaded directly to realize higher-order functions having imaginary zeros. Also, the zeros lie on the imaginary axis for *all* values of the *RC* elements.

The *active-sensitivity* performance is not derived in detail here because it depends very much on the location of the zeros and the particular choice of *RC*-element values. However, if we *select* $G_3 = G_4$ *and use a dual OP AMP*, then it follows that the Q_P-enhancement condition for a type 2 inductance-simulation circuit is satisfied (equation (8.88)). With $V_1(s) = 0$, GIC-2N-2A is a circuit of this type and $\hat{Q}_P \approx Q_P$ and $\hat{\omega}_P \approx \omega_P(1 - \frac{2\omega_P}{B})$. The circuit has the property of Q_P self-enhancement.

A practical disadvantage of GIC-2N-2A is that ω_Z cannot be aligned without also altering ω_P and Q_P. Furthermore, it follows from the expression for ω_Z^2 in Table 8-8(a) that the conductance ratios involving G_{6A}, G_4, G_3, and G_5 must become large as $\omega_Z \to \omega_P$ and that, for this reason, the circuit cannot really be used if $\omega_Z = \omega_P$ or $\omega_Z \approx \omega_P$.

Notch function, $\omega_Z > \omega_P$, GIC-2N-2B: This circuit is used for notch functions only if $\omega_Z > \omega_P$. Similar comments apply to this circuit as apply to GIC-2N-2A. It is particularly useful as a building-block for realizing lowpass elliptic functions by cascading second-order sections.

Notch function ($\omega_Z = \omega_P, Q_Z = \infty$) and allpass function ($\omega_Z = \omega_P, Q_Z = -Q_P$): This circuit is given in Table 8-8(b); the mnemonic GIC-2N-1 indicates that a *two* amplifier *notch* function is realized by using a *type* 1 inductance-simulation circuit. The circuit is used as a band-reject notch filter or as an allpass filter. The transfer function given in Table 8-8(b) may be derived by direct substitution into equation (8.84) as requested in Problem 8.32.

By selecting $G_2G_{6B} = G_{1B}G_{6A}$ we obtain $\omega_Z = \omega_P$ and $Q_Z = \infty$ corresponding to the band-reject notch filter; by selecting $G_{6B} = 0$ and $G_2 = G_{1B}$ we obtain $\omega_Z = \omega_P$ and $Q_Z = -Q_P$ corresponding to the allpass function

$$\frac{V_2(s)}{V_1(s)} = \left[\frac{s^2 - \left(\frac{\omega_P}{Q_P}\right)s + \omega_P^2}{s^2 + \left(\frac{\omega_P}{Q_P}\right)s + \omega_P^2}\right] \qquad \text{ALLPASS GIC FUNCTION } G_{6B} = 0,\ G_2 = G_{1B} \tag{8.114}$$

The Q_P self-enhancement condition for this circuit requires that G_2 within the type 1 simulated-inductance circuit be chosen equal to G_{1B} (because the simulated inductance is seen between terminal 3 and ground when $V_1(s)$ is zero and *not* terminal 1 and ground as in Table 8-8). It may then be shown from the expressions for Q_P and Q_Z in Table 8-8(b) that with $G_2 = G_{1B}$ it is possible to realize $Q_Z = \infty$ for the N function or $Q_Z = -Q_P$ for the AP function; thus, Q_P self-enhancement is possible with GIC-2N-1. The proof is requested in Problem 8.35.

8-5 SUMMARY OF TIL AND GIC SECOND-ORDER CIRCUIT PERFORMANCE

The basic characteristics of the two-, three-, and four-amplifier biquadratic and quadratic circuits are summarized in Table 8-9. The passive-sensitivity performance is not included in the table because it is similar for all of these circuits. The $\Delta Q_P/Q_P$ term due to the finite GBPs B of the OP AMPs is much less than $Q_P\omega_P/B$ and, therefore, may be neglected by comparison with the perturbations $\Delta\omega_P/\omega_P$. Note that the use of TIL-3B is assumed for all the TIL circuits in Table 8-9; otherwise, using TIL-3A leads to $\Delta Q_P/Q_P$ of approximately $-4Q_P\omega_P/B$, which is certainly not negligible. The active-sensitivity performance of the GIC circuits is limited by $\Delta\omega_P/\omega_P \simeq -2\omega_P/B$, whereas that of the TIL circuits by $\Delta\omega_P/\omega_P \simeq -\omega_P/B$.

The GIC circuits are economical because they use only two OP AMPs and two capacitors. The LBI/DF and BHI/DF circuits require at least four OP AMPs and are not well suited for realizing the HP and LP functions, respectively. The circuit MI/DF is flexible and easily tuned; it uses three OP AMPs and has the disadvantages that it uses three capacitors and has a capacitive input immittance. At present, a capacitor element is usually more expensive than an OP AMP and therefore the MI/DF may not be economical by comparison with the other circuits shown in Table 8-9.

8-6 SINGLE-AMPLIFIER QUADRATIC CIRCUITS

The theoretical basis for the following introduction to single-amplifier circuit design is provided in section 7-4. To obtain a thorough understanding of the limitations of these circuits, it is recommended that the reader review the material in section 7-4 and take special note of the fundamental limitations imposed by equations (7.88) for type I structures, equations (7.103) for type IIA structures, and equations (7.118) for type IIB structures.

TABLE 8-9

Comparison of TIL and GIC Biquad Circuits

Circuit	Functions	No. of OP AMP's	No. of C's	$\frac{\Delta\omega_P}{\omega_P}$ Approx.	$\frac{\Delta Q_P}{Q_P}$ Approx.	Flexibility	8 General comments
Two-integrator Loop LBI/DF, Table 8-2	General biquad, LP, BP, HP, N, AP	4	2	$-\frac{\omega_P}{B}$	$\ll \frac{Q_P\omega_P}{B}$ with TIL-3B	Extra inverter needed if $Q_P\omega_Z > Q_Z\omega_P$.	Good active-sensitivity using TIL-3B; poor circuit for realizing HP function.
Two-integrator Loop BHI/DF, Table 8-4	General biquad, LP, BP, HP, N, AP	4	2	$-\frac{\omega_P}{B}$	$\ll \frac{Q_P\omega_P}{B}$ with TIL-3B	Extra interter needed if $Q_P\omega_P < Q_Z\omega_Z$ or $\omega_Z < \omega_P$.	Good active sensitivity using TIL-3B; poor circuit for realizing LP function.
Two-integrator Loop MI/DF, Table 8-3	General biquad, LP, BP, HP, N, AP	3	3	$-\frac{\omega_P}{B}$	$\ll \frac{Q_P\omega_P}{B}$ with TIL-3B	Excellent; no constraints; extra OP AMP is not required.	Good active sensitivity using TIL-3B; easily aligned; non-constant input impedance.
Generalized immittance converter, Fig. 8-7(a)						Basic circuit in Fig. 8.7(a) realizes all required functions, HP, BP, LP, N, AP.	Good active sensitivity; Q_P enhancement does *not* require matching of B_1 and B_2 for GIC-2N-2A/B. Excellent active sensitivity achievable by locating capacitor at Y_5 instead of Y_3 for GIC-2HP and GIC-2BP.
—GIC-2HP-1, Table 8-5	HP	2	2	$-\frac{2\omega_P}{B}$	$\ll \frac{Q_P\omega_P}{B}$		
—GIC-2BP-1, Table 8-6	BP	2	2	$-\frac{2\omega_P}{B}$	$\ll \frac{Q_P\omega_P}{B}$		
—GIC-2N-1, Table 8-7	N	2	2	$-\frac{2\omega_P}{B}$	$\ll \frac{Q_P\omega_P}{B}$		
—GIC-2N-A/B,	N, AP	2	2	$-\frac{2\omega_P}{B}$	$\ll \frac{Q_P\omega_P}{B}$		

The most significant conclusion from the active-sensitivity analysis in section 7-4 is that type I (that is, noninverting amplifier) circuits possess superior active-sensitivity performance to that of the type II (that is, inverting amplifier) circuits. The lower bound on $|S_B^{G_b(\omega)}|$ for the type I circuit is $\omega_P Q_P / B$ whereas, for the type II circuit this lower bound is $4\omega_P Q_P^3 / B$. Thus, for $Q_P \gg 1$, the type I circuits are preferable to type II circuits. The reader may verify in Problem 8.36 that the active-sensitivity performance of type II circuits severely limits their usefulness. In the remainder of this chapter, attention is therefore focused entirely on type I single-amplifier second-order filter circuits. The Sallen–Key *lowpass* circuit is probably the most widely used type I circuit, and it is analyzed in detail in section 8-6.1. Other Sallen–Key and other single-amplifier circuits are described, but in many cases the nonideal performance is mentioned briefly and a more detailed treatment is left to the reader.

8-6.1 Sallen-Key Lowpass Type I Circuit

The Sallen-Key lowpass circuit is shown in Table 8-10 with the equations for $H_b(s)$, H_0, ω_P and Q_P. In this subsection, it is shown that the sensitivity performance is highly dependent on the choice of the gain K of the non inverting amplifier (VCVS). It will be found that the designer is faced with a compromise between good sensitivity performance and large ratios between the two capacitor values and two resistor values. It is convenient for the purposes of the analysis to let

$$T \equiv \frac{C_1}{G_1} \tag{8.115}$$

$$\frac{G_2}{G_1} \equiv \alpha^2 \tag{8.116}$$

and

$$\frac{C_1}{C_2} \equiv \beta^2 \tag{8.117}$$

where α^2 and β^2 are direct measures of the *element spreads* in the network N_{RC}. Substituting equations (8.115), (8.116), and (8.117) into the expressions for H_0, ω_P, Q_P, and $H_b(s)$ of Table 8.10 we obtain

$$\boxed{H_b(s) = \frac{K(\alpha\beta)^2}{T^2}\left[\frac{1}{s^2 + \left[1 + \alpha^2 + \alpha^2\beta^2(1 - K)\right]\frac{s}{T} + \left(\frac{\alpha\beta}{T}\right)^2}\right]} \tag{8.118}$$

TABLE 8-10

Sallen-Key lowpass

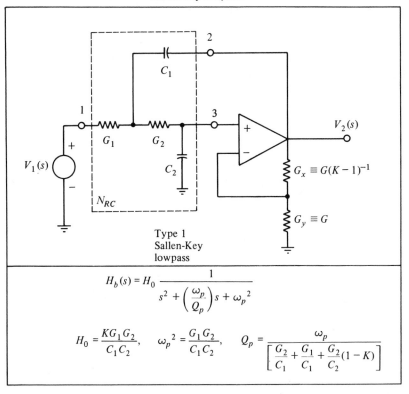

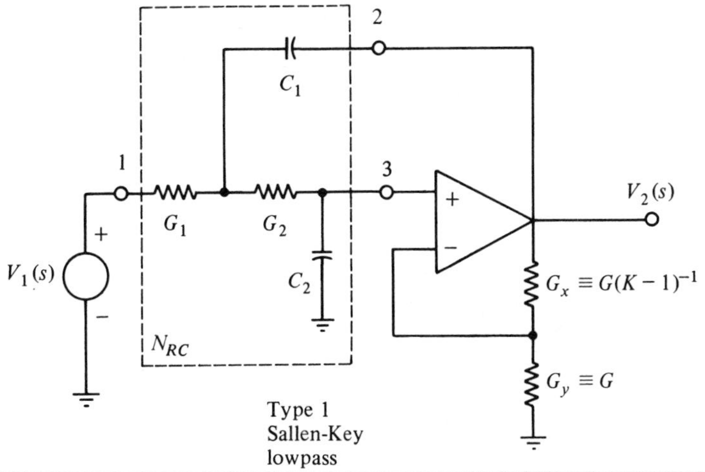

$$H_b(s) = H_0 \frac{1}{s^2 + \left(\dfrac{\omega_p}{Q_p}\right)s + \omega_p{}^2}$$

$$H_0 = \frac{KG_1G_2}{C_1C_2}, \quad \omega_p{}^2 = \frac{G_1G_2}{C_1C_2}, \quad Q_p = \frac{\omega_p}{\left[\dfrac{G_2}{C_1} + \dfrac{G_1}{C_1} + \dfrac{G_2}{C_2}(1-K)\right]}$$

The corresponding expressions for K, ω_P, and Q_P appear in Table 8-11, in the first row, labeled *general circuit equations*.

It is recalled from the sensitivity analysis of type I circuits in section 7-4.2 (see Table 7-4) that the active-sensitivity performance is written in terms of K, α_1, and ω_P as in equation (7.85). Now, the term α_1 is associated with the numerator of the *passive* transfer function $T_{32}(s)$ as defined in equation (7.52) and Fig. 7.13(b). We now derive α_1 for this particular circuit by first deriving $T_{32}(s)$ from the corresponding passive part of the circuit shown in Fig. 8.10. Direct analysis of Fig. 8.10 is requested in Problem 8.38 and gives

$$\frac{V_3(s)}{V_2(s)} \equiv T_{32}(s) = \frac{\dfrac{(\alpha\beta)^2}{T}s}{\left(s^2 + \dfrac{\omega_N}{Q_N}s + \omega_N^2\right)} \qquad \text{BANDPASS } T_{32}(s) \tag{8.119}$$

TABLE 8-11

Summary of Sallen-Key lowpass design equations

Type	*GC elements*	*Gain K*	Q_P^{-1}	ω_P	α_1	*Active-sensitivity term* $\frac{K^2\alpha_1}{\omega_P} \equiv \Phi^2(\eta),\ \eta \equiv \alpha\beta$
General circuit equations	$G_2 \equiv \alpha^2 G_1$ $C_1 \equiv \beta^2 C_2$	$1 + \beta^{-2} + \frac{1}{\alpha\beta}\left(\frac{1}{\alpha\beta} - \frac{1}{Q_P}\right)$	$\alpha\beta(1 + \beta^{-2} - K) + \frac{1}{\alpha\beta}$	$\frac{\alpha\beta}{T}$ where $T \equiv \frac{G_1}{G_1}$	$\frac{(\alpha\beta)^2}{T}$	$\frac{K^2\alpha_1}{\omega_P} = \left[\frac{1}{\eta^2} + (1 + \beta^{-2}) - \frac{Q_P^{-1}}{\eta}\right]^2 \eta$ where $\eta \equiv \alpha\beta$
Design No. 1: Uniform G/ uniform C	$G_2 = G_1$ $C_1 = C_2$ $\alpha = 1$ $\beta = 1$	$3 - Q_P^{-1}$	$3 - K$	$\frac{1}{T}$	$\frac{1}{T}$	$\frac{K^2\alpha_1}{\omega_P} = (3 - Q_P^{-1})^2$ $\approx 9 \quad \text{if } Q_P \gg \frac{1}{3}$
Design No. 2: Minimum active sensitivity	$G_2 \equiv \alpha^2 G_1$ $G \equiv \beta^2 C_2$ $\alpha\beta = \eta_{min}$	$1 + \beta^{-2} + \frac{1}{\eta_{min}}\left(\frac{1}{\eta_{min}} - \frac{1}{Q_P}\right)$	$\eta_{min}(1 + \beta^{-2} - K) + \frac{1}{\eta_{min}}$	$\frac{\eta_{min}}{T}$	$\frac{\eta_{min}^2}{T}$	$\frac{K^2\alpha_1}{\omega_P} = \Phi^2(\eta_{min})$ $\approx \frac{16(1 + \beta^{-2})}{3\sqrt{3}} \quad \text{if } Q_P \gg \frac{1}{3}$
Design No. 3: Minimum passive sensitivity ($K = 1$)	$G_2 = \alpha^2 G_1$ $C_1 = \beta^2 C_2$ $\alpha\beta \gg 1$ $\alpha/\beta \ll 1$	1	$\frac{1}{\alpha\beta} + \frac{\alpha}{\beta}$	$\frac{\alpha\beta}{T}$ $\omega_P \gg \frac{1}{T}$	$\frac{(\alpha\beta)^2}{T}$	$\frac{K^2\alpha_1}{\omega_P} > Q_P \quad \text{if } K = 1$

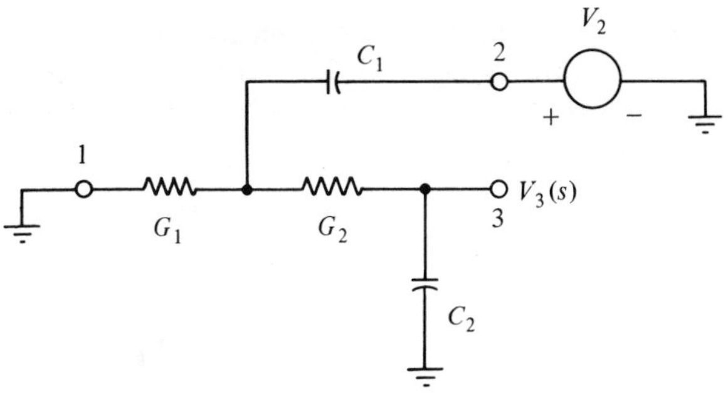

Circuit for $T_{32}(s) \equiv \frac{V_3(s)}{V_2(s)}$

FIGURE 8.10

where

$$\omega_N = \omega_P = \frac{\alpha\beta}{T} \tag{8.120}$$

and

$$Q_N^{-1} = \alpha\beta\left[1 + \beta^{-2}\right] + \frac{1}{\alpha\beta} \tag{8.121}$$

Comparing equations (8.119) and (7.52) supplies the required result that the numerator coefficient of $T_{32}(s)$ is

$$\boxed{\alpha_1 = \frac{(\alpha\beta)^2}{T}} \tag{8.122}$$

Of course, $T_{32}(s)$ is a *bandpass* function, so that the circuit is classified as type I according to equation (7.81). The expression above for α_1 is included in Table 8-11; the required expression for the active-sensitivity term $K^2\alpha_1/\omega_P$ is obtained from equations (8.120), (8.122), and the expression for K in Table 8-11 giving

$$\boxed{\frac{K^2\alpha_1}{\omega_P} = \left[\frac{1}{(\alpha\beta)^2} + 1 + \frac{1}{\beta^2} - \frac{Q_P^{-1}}{(\alpha\beta)}\right]^2(\alpha\beta)} \tag{8.123}$$

This expression may be used directly with equations (7.85) to determine the active-sensitivity performance.

The *passive single-element sensitivity* terms may be evaluated directly from the expressions for $H_b(s)$ in Table 8-10. For example, substituting H_0, ω_P, and Q_P into $H_b(s)$ in Table 8-10 and calculating the sensitivity with respect to G_1, we obtain:

$$S_{G_1}^{H_b(s)} = 1 + \frac{\left(\frac{G_1}{C_1}s + \frac{G_1G_2}{C_1C_2}\right)}{\left(s^2 + \frac{\omega_P s}{Q_P} + \omega_P^2\right)} \tag{8.124}$$

so that, by using equations (8.116), (8.117) and (8.124), we have

$$S_{G_1}^{H_b(s)} = 1 + \frac{1}{\omega_P \alpha\beta} t_{\mathrm{BP}}(s) + t_{\mathrm{HP}}(s) \tag{8.125}$$

Then, letting $s = j\omega$ and taking the real part provides the gain-sensitivity

$$\boxed{S_{G_1}^{G_b(\omega)} = 1 + \frac{1}{(\alpha\beta)} \operatorname{Re}[t_{\mathrm{BP}}(j\omega)] + \operatorname{Re}[t_{\mathrm{HP}}(j\omega)]} \tag{8.126}$$

where $\operatorname{Re}[t_{\mathrm{BP}}(j\omega)]$ and $\operatorname{Re}[t_{\mathrm{HP}}(j\omega)]$ are shown in Fig. 7.10 and have peak values of approximately $\pm Q_P/2$ if $Q_P \gg 1$. In a similar way, the gain sensitivities to the remaining passive elements G_2, C_1, C_2, G_X, and G_Y are derived and the resultant expressions are provided in Table 8-12.

Some standard design procedures are now described and the practical advantages of each are emphasized.

Design 1: uniform *G*/ uniform *C*: In this case, the element spread is minimized; thus,

$$G_1 \equiv G_2; \qquad C_1 \equiv C_2 \tag{8.127}$$

so that

$$\alpha \equiv \beta \equiv 1 \tag{8.128}$$

It then follows directly from the general circuit equations in the first row of Table 8-11 that the designer must choose $T = \omega_P^{-1}$ and

$$K = 3 - Q_P^{-1} \tag{8.129}$$

Clearly, if $Q_P \gg 1$, then

$$K \rightarrow 3 \tag{8.130}$$

The *passive-sensitivities* are derived by simply substituting $\alpha = \beta = 1$ and $K = 3 - Q_P^{-1}$ into Table 8-12; so that we obtain

$$\boxed{\begin{aligned}
S_{G_1}^{G_b(\omega)} &\simeq 1 + \text{Re}\left[t_{\text{BP}}(j\omega)\right] + \text{Re}\left[t_{\text{HP}}(j\omega)\right] \\
S_{G_2}^{G_b(\omega)} &\simeq 1 - \text{Re}\left[t_{\text{BP}}(j\omega)\right] + \text{Re}\left[t_{\text{HP}}(j\omega)\right] \\
S_{C_1}^{G_b(\omega)} &\simeq -1 - 2\,\text{Re}\left[t_{\text{BP}}(j\omega)\right] - \text{Re}\left[t_{\text{HP}}(j\omega)\right] \\
S_{C_2}^{G_b(\omega)} &\simeq -1 + 2\,\text{Re}\left[t_{\text{BP}}(j\omega)\right] - \text{Re}\left[t_{\text{HP}}(j\omega)\right] \\
S_{G_X}^{G_b(\omega)} &\simeq 0 + 2\,\text{Re}\left[t_{\text{BP}}(j\omega)\right] + 0 \\
S_{G_Y}^{G_b(\omega)} &\simeq 0 - 2\,\text{Re}\left[t_{\text{BP}}(j\omega)\right] + 0
\end{aligned}} \tag{8.131}$$

where we have assumed $Q_P \gg \frac{1}{3}$. The magnitudes of the first four equations may be estimated by consideration of Fig. 7.10. For example, the summation required by the expression above for $S_{G_1}^{G_b(\omega)}$ leads to a function with a peak value of approximately

$$Q_P\left(\frac{1}{2} + \frac{1}{\sqrt{2}}\right)$$

The worstcase sensitivity over all the passive elements is obtained by summing the moduli of all the equations in (8.131).

TABLE 8-12

$S_{x_i}^{G_b(\omega)}$	q_1	q_2	q_3
$S_{G_1}^{G_b(\omega)}$	1	$\dfrac{1}{(\alpha\beta)}$	1
$S_{G_2}^{G_b(\omega)}$	1	$\left[\dfrac{\alpha}{\beta} + (1 - K)\alpha\beta\right]$	1
$S_{C_1}^{G_b(\omega)}$	-1	$-\left[\dfrac{1}{(\alpha\beta)} + \dfrac{\alpha}{\beta}\right]$	-1
$S_{C_2}^{G_b(\omega)}$	-1	$-[\alpha\beta(1 - K)]$	-1
$S_{G_X}^{G_b(\omega)}$	0	$(K - 1)\alpha\beta$	0
$S_{G_Y}^{G_b(\omega)}$	0	$(1 - K)\alpha\beta$	0

All single-element sensitivities have the general form

$$S_{x_i}^{G_b(\omega)} = q_1 + q_2\,\text{Re}\,[t_{\text{BP}}(j\omega)] + q_3\,\text{Re}\,[t_{\text{HP}}(j\omega)]$$

where q_1, q_2, q_3 are shown in this Table.

The *active-sensitivity performance* is obtained by substituting $\alpha = \beta = 1$ into equation (8.123) and then equation (8.123) into equation (7.85); thus

$$S_B^{G_b(\omega)} = \left[-(3 - Q_P^{-1})^2 \frac{\omega_P}{B} \right] \text{Re}[\, t_{\text{HP}}(j\omega)] \tag{8.132}$$

and, for $Q_P \gg \frac{1}{3}$,

$S_B^{G_b(\omega)} \simeq -\left(\frac{9\omega_P}{B}\right) \text{Re}[t_{\text{HP}}(j\omega)]$	ACTIVE-SENSITIVITY SALLEN-KEY LOWPASS UNIFORM G/UNIFORM C	(8.133)

It may be shown (see Problem 8.42) that this result implies that

$$\frac{\Delta\omega_P}{\omega_P} \simeq -\frac{9\omega_P}{2B} \quad \text{and} \quad \frac{\Delta Q_P}{Q_P} \approx \frac{9\omega_P}{2B}, \qquad Q_P \gg 1 \tag{8.134}$$

Similarly, equation (7.85) gives, for $Q_P \gg \frac{1}{3}$,

$$S_{A_0}^{G_b(\omega)} \simeq -\left(\frac{9}{A_0}\right) \text{Re}[t_{\text{BP}}(j\omega)] \tag{8.135}$$

This last expression is usually negligible with modern OP AMPs, because $9/A_0$ is typically less than 10^{-3}.

In practical circuit design, the active-sensitivity result in equation (8.133) is usually the limiting factor in achieving a stable high-Q transfer function $H_b(s)$. The fractional error in $G_b(\omega)$ may be shown from equations (6.150) and (8.133) to be

$$\frac{\Delta G_b(\omega)}{G_b(\omega)} \approx \frac{9\omega_P}{B} \text{Re}[t_{\text{HP}}(j\omega)] \tag{8.136}$$

A justifiable question is whether it is possible to choose another design which would significantly reduce the magnitude of equation (8.133). We already know from equation (7.89) that the minimum achievable $|S_B^{G_b(\omega)}|$ occurs if $T_{32}(j\omega_P) \to 1$ and $Q_N \to \frac{1}{2}$, so that

$$|S_B^{G_b(\omega)}|_{\min} = \frac{2\omega_P}{B} |\, \text{Re}[\, t_{\text{HP}}(j\omega)]| \tag{8.137}$$

implying that equation (8.136) corresponds to a particular design that is a factor of $4\frac{1}{2}$ in excess of this theoretical "best" performance. We now develop a design procedure for minimizing $|S_B^{G_b(\omega)}|$ for this Sallen-Key lowpass circuit.

Design 2: minimum active-sensitivity $|S_B^{G_b(\omega)}|$: In this second design procedure, it is assumed that the designer is particularly concerned that the high-frequency performance be as good as possible; that is, the objective is to reduce $S_{B,A_0}^{G_b(\omega)}$ so that they are significantly smaller in magnitude than is obtained in design No. 1. We know from equation (7.85) that $|S_B^{G_b(\omega)}|$ and $|S_{A_0}^{G_b(\omega)}|$ are proportional to $K^2\alpha_1$. For convenience, we define a parameter η, where

$$\eta \equiv \alpha\beta \tag{8.138}$$

Furthermore, it is convenient to write $K^2\alpha_1/\omega_P$ as $\Phi^2(\eta)$ so that equation (8.123) becomes

$$\Phi^2(\eta) \equiv \frac{K^2\alpha_1}{\omega_P} = \left[\eta^{-2} + (1 + \beta^{-2}) - Q_P^{-1}\eta^{-1}\right]^2\eta \tag{8.139}$$

It is assumed in the following that the capacitance ratio β^2 is constant. The task of minimizing the active sensitivity is then equivalent to minimizing $\Phi(\eta)$, where we note that $\Phi^2(\eta) > 0$ because the right-hand side of equation (8.139) is necessarily positive. Differentiating $\Phi(\eta)$ with respect to η and equating to zero,

$$\frac{\partial\Phi(\eta)}{\partial\eta} = \tfrac{1}{2}\left[(1 + \beta^{-2})\eta^2 + Q_P^{-1}\eta - 3\right]\eta^{-5/2} = 0 \tag{8.140}$$

the solutions of which are provided by

$$\eta = 0 \quad \text{and} \quad \eta = \frac{1}{1 + \beta^{-2}}\left\{-\frac{Q_P^{-1}}{2} \pm \left[3(1 + \beta^{-2}) + \frac{Q_P^{-2}}{4}\right]^{1/2}\right\}$$

The reader may verify, by differentiating $\partial\Phi(\eta)/\partial\eta$ once more with respect to η, that the particular nontrivial positive solution of the equations above that corresponds to a *minimum* of $\Phi(\eta)$ is

$$\eta_{\min} = \frac{1}{1 + \beta^{-2}}\left\{-\frac{Q_P^{-1}}{2} \pm \left[3(1 + \beta^{-2}) + \frac{Q_P^{-2}}{4}\right]^{1/2}\right\} \tag{8.141}$$

Then equations (8.138), (8.123), (8.139), and (7.85) provide the corresponding α, K, $\Phi^2(\eta_{\min})$, and $S_B^{G_b(\omega)}|_{\min}$ as

$$\begin{aligned}
\alpha &= \frac{\eta_{\min}}{\beta} \\
K &= \frac{1}{\eta_{\min}^2} + (1+\beta^{-2}) - \frac{Q_P^{-1}}{\eta_{\min}} \\
\Phi^2_{\min} &= \left[\frac{1}{\eta_{\min}^2} + (1+\beta^{-2}) - \frac{Q_P^{-1}}{\eta_{\min}}\right]^2 \eta_{\min} \\
S_B^{G_b(\omega)}\Big|_{\min} &= -\frac{\omega_P \Phi^2_{\min}}{B}\,\mathrm{Re}[t_{HP}(j\omega)]
\end{aligned} \tag{8.142}$$

EXACT DESIGN EQUATIONS FOR MINIMUM ACTIVE-SENSITIVITY CONSTANT β^2

Near-minimum active sensitivity, $Q_P > 2$: In a practical situation, it is often possible to save design effort by using an approximation $\hat{\eta}_{\min}$ for the expression $\eta_{\min}$ in equation (8.141). If $Q_P > 2$, then to a good approximation we have

$$\eta_{\min} \simeq \left[\frac{3}{1+\beta^{-2}}\right]^{1/2} \equiv \hat{\eta}_{\min} \tag{8.143}$$

Thus, given a particular capacitance ratio $\hat{\beta}^2$ we derive the near-optimum design values for $\hat{\alpha}$, $\hat{K}$, $\hat{\Phi}_{\min}$, and $S_B^{G_b(\omega)}$ by substituting $\hat{\eta}_{\min}$ for $\eta_{\min}$ in equations (8.142). This leads immediately to

$$S_B^{G_b(\omega)}\Big|_{\min} \simeq -\frac{16(1+\hat{\beta}^{-2})^{3/2}}{3\sqrt{3}}\,\frac{\omega_P}{B}\,\mathrm{Re}[t_{HP}(j\omega)] \tag{8.144}$$

Thus, the active sensitivity depends on the capacitance ratio $\hat{\beta}^2$, and this should obviously be chosen to be as large as possible. Note that

$$\text{if} \quad \hat{\beta} \gg 1 \quad \text{then} \quad S_B^{G_b(\omega)}\Big|_{\min} \to -\left(\frac{3.079\omega_P}{B}\right)\mathrm{Re}[t_{HP}(j\omega)] \tag{8.145}$$

which is a factor of 3.079/2, or 1.539 in excess of the optimum active sensitivity that is possible using a type I circuit (inspect equation (7.89) with $Q_N \to \frac{1}{2}$ and $T_{32} \to 1$).

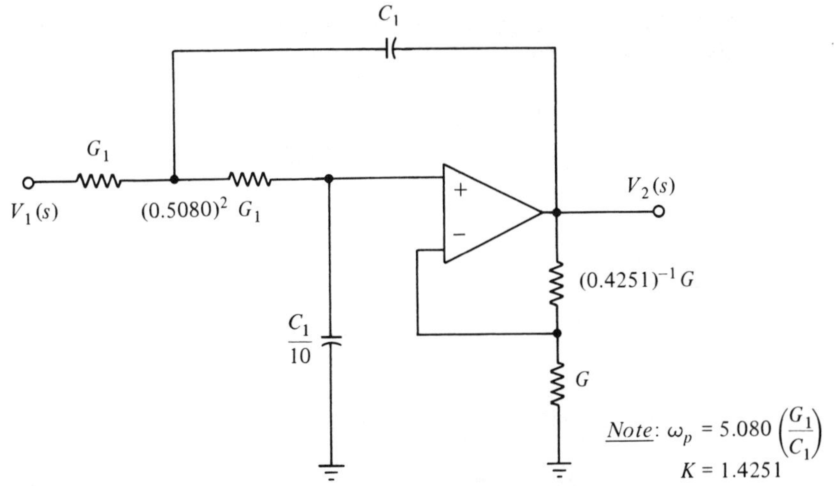

FIGURE 8.11 *Example 8.5 minimum active-sensitivity lowpass design; $Q_p = 10$, $\beta^2 = 10$.*

EXAMPLE 8.5: Design the *exact* minimum active-sensitivity lowpass Sallen-Key filter given that

$$\beta \equiv \sqrt{10} \quad \text{and} \quad Q_P \equiv 10 \tag{8.146}$$

SOLUTION: Substituting (8.146) into (8.141) we have

$$\eta_{\min} = 1.606616532$$

and substituting this value of $\eta_{\min}$ and equations (8.146) into equations (8.142), gives

$$\alpha = 0.50805675, \qquad K = 1.42517159, \qquad \Phi^2_{\min} = 3.26322$$

$$\boxed{S_B^{G_b(\omega)}\Big|_{\min} = -3.26322\frac{\omega_P}{B}\ \mathrm{Re}[t_{\mathrm{HP}}(j\omega)]} \tag{8.147}$$

The final circuit is shown in Fig. 8.11, where G_1 and C_1 must, of course, be selected so that equation (8.120) results in the desired value of ω_P. Note that this design exceeds the optimum $S_B^{G_b(\omega)}$ for a type I circuit by the factor 1.6316.

EXAMPLE 8.6: Repeat the previous example but employ the *near-minimum* active-sensitivity design method.

SOLUTION: We note that $Q_P > 2$, so that the expression $\hat{\eta}_{\min}$ in equation (8.143) may be used, to obtain

$$\hat{\eta}_{\min} = 1.651445648$$
$$\hat{\alpha} = \frac{1.651445}{\sqrt{10}} = 0.522232$$
$$\hat{K} = 1.406113660$$

and

$$S_B^{G_b(\omega)} = -\frac{3.265163\omega_P}{B} \operatorname{Re}[t_{\mathrm{HP}}(j\omega)] \tag{8.148}$$

which is, as expected, very close to the exact minimum active-sensitivity design [compare equations (8.147) and (8.148)].

Near minimum active-sensitivity, $Q_P > 2, \beta^2 \gg 1$: A further and most useful simplification is possible if the designer employs a capacitance ratio β^2 that is much greater than unity to realize $Q_P > 2$. Then, equations (8.141) and (8.142) simplify drastically. Equation (8.141) gives, for $Q_P > 2$ and $\beta \gg 1$,

$$\boxed{\eta_{\min} \simeq \sqrt{3} \equiv \hat{\hat{\eta}}_{\min}} \tag{8.149}$$

and equations (8.142) then give

$$\hat{\hat{\alpha}} = \frac{\sqrt{3}}{\beta}, \qquad \hat{\hat{K}} = \frac{4}{3} + \beta^{-2} - \frac{1}{\sqrt{3}\, Q_P} \simeq \frac{4}{3}$$

and

$$\boxed{S_B^{G_b(\omega)} \simeq -\left(\frac{16}{3\sqrt{3}}\right)\left(\frac{\omega_P}{B}\right) \operatorname{Re}[t_{\mathrm{HP}}(j\omega)]} \tag{8.150}$$

Usually, it is sufficient simply to *select* $\hat{\hat{K}} = \frac{4}{3}$ and then ensure that $\eta_{\min} \simeq \sqrt{3}$ and $\beta^2 \gg 1$. In Problem 8.47, the reader is asked to confirm this by deriving $S_B^{G_b(\omega)}$ for a design where $K \equiv \frac{4}{3}$, $\beta \equiv \sqrt{10}$, and where η is only *approximately* equal to $\sqrt{3}$.

It is concluded that *design no. 2 does allow the active sensitivity to be reduced significantly compared with design no. 1*. The minimum active-sensitivy design is summarized in Table 8-11.

The enquiring reader may have wondered whether the passive sensitivities are worsened [relative to equations (8.131)] by the minimization of the active sensitivity. The answer is that the passive sensitivities are, in fact, reduced. For example, we have

$$\alpha = 0.5080 \quad \text{and} \quad \beta = \sqrt{10} \quad \text{and} \quad K = 1.4251 \tag{8.151}$$

in the exact design of Example 8.5, which implies that

$$\alpha\beta \equiv \eta_{\min} = 1.6066 \quad \text{and} \quad \frac{\alpha}{\beta} = 0.1606 \tag{8.152}$$

The *passive sensitivities* are then obtained *exactly* by substituting equations (8.151) and (8.152) into Table 8-12 to provide the following six terms q_2 from row 1 to row 6:

$$\boxed{q_2 = [0.622,\ -0.522,\ -0.783,\ 0.682,\ 0.682,\ -0.682]} \quad \text{DESIGN No. 2} \tag{8.153}$$

Note that all of these terms q_2 are smaller in magnitude than the corresponding terms for design No. 1 which, from equation (8.131), are

$$\boxed{q_2 = [1,\ -1,\ -2,\ 2,\ 2,\ -2]} \quad \text{DESIGN No.1}$$

The passive sensitivities are particularly simple and straightforward to obtain if $Q_P \gg \frac{1}{3}$ and $\beta^2 \gg 1$ because we then have $\alpha\beta \equiv \eta \simeq \sqrt{3}$ and $\alpha/\beta \ll 1$ leading, via equation (8.131), to

$$\boxed{q_2 \simeq \frac{1}{\sqrt{3}},\ -\frac{1}{\sqrt{3}},\ -\frac{1}{\sqrt{3}},\ \frac{1}{\sqrt{3}},\ \frac{1}{\sqrt{3}},\ -\frac{1}{\sqrt{3}}} \quad \begin{array}{l}\text{DESIGN No. 2}\\ Q_P \gg \frac{1}{3}\\ \beta^2 \gg 1\end{array} \tag{8.154}$$

which is an interesting result implying that all the six single-element sensitivities in Table 8-12 have $|q_2| \simeq 1/\sqrt{3}$ in the case of this minimum active-sensitivity design.

The reader may acquire experience with minimum active-sensitivity design by answering Problems 8.44 to 8.47.

Design 3: Minimum passive sensitivity design, $K=1$: The designer is sometimes not particularly concerned about the high-frequency limitation $S_B^{G_b(\omega)}$ but more with using *RC elements that have the loosest possible tolerances*. That is, the designer may wish to select two of the three parameters K, α, and β so as to minimize the magnitudes of the single-element sensitivities $S_{x_i}^{G_b(\omega)}$ given in Table 8-12. Since q_1 and q_3 are independent of K, α and β, it follows that the designer should restrict attention to minimizing $|q_2|$. Inspection of the q_2 terms in Table 8-12 reveals that all of the terms $(K-1)\alpha\beta$ may be conveniently suppressed to zero by selecting the amplifier gain K so that

$$\boxed{K = 1} \qquad \text{AMPLIFIER GAIN CONDITION FOR DESIGN NO. 3} \tag{8.155}$$

Further inspection of the q_2 terms in Table 8-12 reveals that $|q_2|$ is then suppressed toward zero if, with $K = 1$,

$$\boxed{\alpha\beta \gg 1 \quad \text{and} \quad \frac{\beta}{\alpha} \gg 1} \qquad \text{ELEMENT SPREAD CONDITIONS FOR MINIMIZING PASSIVE SENSITIVITIES} \tag{8.156}$$

Now, with $K = 1$, the expression for Q_P^{-1} in the first row of Table 8-11 becomes

$$Q_P^{-1} = \frac{1}{\alpha\beta} + \frac{\alpha}{\beta} \tag{8.157}$$

implying that

$$\boxed{\alpha\beta > Q_P, \quad \frac{\beta}{\alpha} > Q_P, \quad \text{and} \quad \beta > Q_P} \tag{8.158}$$

It follows from equations (8.158) that the conditions of equation (8.156) are always satisfied if $Q_P \gg 1$ and $K = 1$. It is not difficult to show that α and β may be selected in a variety of ways to satisfy the conditions of equations (8.156). For example, an *equal conductance realization* ($G_1 \equiv G_2$) implies that $\alpha \equiv 1$; substituting $\alpha = 1$ into equation (8.157) supplies the design values

$$\boxed{\beta = 2Q_P, \quad \alpha = 1, \quad K = 1} \qquad \text{EQUAL CONDUCTANCE MINIMUM PASSIVE-SENSITIVITY DESIGN} \tag{8.159}$$

An unfortunate disadvantage of design No. 3 is that the first two conditions of equations (8.158) imply that

$$\beta > Q_P$$

so that for high-Q_P designs the capacitance spread β^2 is greater than Q_P^2, and this is sometimes too large for practical microelectronic fabrication of the capacitance elements.

We now show that design No. 3, with α and β chosen such that (8.158) is satisfied, possesses passive sensitivities that are close to the lower bounds for RC-active networks. Equations (8.158) imply that

$$|q_2| \ll 1$$

for all six single-element sensitivities in Table 8-12, so that

$$S_{G_1}^{G_b(\omega)} \simeq S_{G_2}^{G_b(\omega)} \simeq -S_{C_1}^{G_b(\omega)} \simeq -S_{C_2}^{G_b(\omega)} \simeq 1 + \text{Re}[t_{\text{HP}}(j\omega)] \quad (8.160)$$

and

$$S_{G_X}^{G_b(\omega)} = S_{G_Y}^{G_b(\omega)} = 0 \quad (K = 1) \quad (8.161)$$

Equations (8.160) and (8.161) give the worstcase sensitivity to all six passive elements as

$$WS_{RC}^{G_b(\omega)} \simeq 4|1 + \text{Re}[t_{\text{HP}}(j\omega)]| \quad (8.162)$$

For $Q_P \gg 1$, it may be shown that

$$|\text{Re}[t_{\text{HP}}(j\omega)]| \simeq \frac{1}{2}\left|\frac{\partial \ln G_b(\omega)}{\partial \ln \omega}\right| \gg 1 \quad (8.163)$$

so that equation (8.162) becomes

$$WS_{RC}^{G_b(\omega)} \simeq 2\left|\frac{\partial \ln G_b(\omega)}{\partial \ln \omega}\right|, \qquad Q_P \gg 1 \quad (8.164)$$

WORSTCASE PASSIVE-SENSITIVITY DESIGN NO. 3

This result is important because, by comparison with equation (6.126), it follows that the *worstcase passive sensitivity* $WS_{RC}^{G_b(\omega)}$ *is approximately lower bound.* [The reader may recall that none of the previously considered two-, three-, or four-amplifier second-order lowpass networks are as good as this; they are typically a factor of 1.5 in excess of lower-bound.] Insofar as passive-element tolerances are concerned, design No. 3 is near-optimum in the worstcase sense.

The excellent passive-sensitivity performance of design No. 3 is achieved at the expense of very poor active-sensitivity performance, which is proven by noting that, with $K = 1$ and using equations (8.120) and (8.122), we have

$$\frac{K^2\alpha_1}{\omega_P} = (1)^2\frac{(\alpha\beta)^2}{T}\cdot\frac{T}{(\alpha\beta)} = \alpha\beta$$

It follows from this equation and the first condition of equations (8.158) that

$$\frac{K^2\alpha_1}{\omega_P} > Q_P \qquad \text{if } K = 1$$

which, from equation (7.85), implies that

$$\boxed{|S_B^{G_b(\omega)}| > \frac{Q_P\omega_P}{B}\mathrm{Re}[t_{\mathrm{HP}}(j\omega)]} \qquad K = 1 \tag{8.165}$$

Inspection of Fig. 7.10 for $\mathrm{Re}[t_{\mathrm{HP}}(j\omega)]$ reveals that this function has extrema of approximate magnitude $\frac{1}{2}Q_P$, so that equation (8.165) implies

$$|S_B^{G_b(\omega)}|_{\max} > \frac{Q_P^2\omega_P}{2B} \tag{8.166}$$

where this maximum sensitivity occurs at $\omega \simeq \omega_P(1 \pm 1/2Q_P)$. Then,

$$\boxed{\left|\frac{\Delta G_b(\omega)}{G_b(\omega)}\right|_{\max} > \frac{Q_P^2\omega_P}{2B}\left(\frac{\Delta B}{B}\right)} \qquad \text{DUE TO PERTURBATIONS OF } B \tag{8.167}$$

The reader can verify that with quite moderate variations in bandwidth (say $\Delta B/B = 0.1$) the result in equation (8.167) implies that ω_P/B must be discouragingly low-valued if the maximum value of $|\Delta G_b(\omega)/G_b(\omega)|$ is to be tolerable. A numerical example of this serious active-sensitivity limitation is provided in Problem 8-48.

8-6.2 Sallen-Key Bandpass Type I Circuit

This circuit is shown in Table 8-13, with the equations for $H_b(s)$, H_0, ω_P, and Q_P. Detailed analysis of the circuit is similar to that of the previously considered lowpass Sallen-Key circuit; therefore, we consider here only the main features of the circuit performance.

We define new variables T, α, β, γ as

$$T \equiv \frac{C_1}{G_1}, \quad \alpha^2 \equiv \frac{G_2}{G_1}, \quad \beta^2 \equiv \frac{C_1}{C_2}, \quad \gamma^2 \equiv \frac{G_3}{G_1} \tag{8.168}$$

where α, β and γ are measures of the elements spreads. Substituting

TABLE 8-13

Type I Sallen-Key bandpass

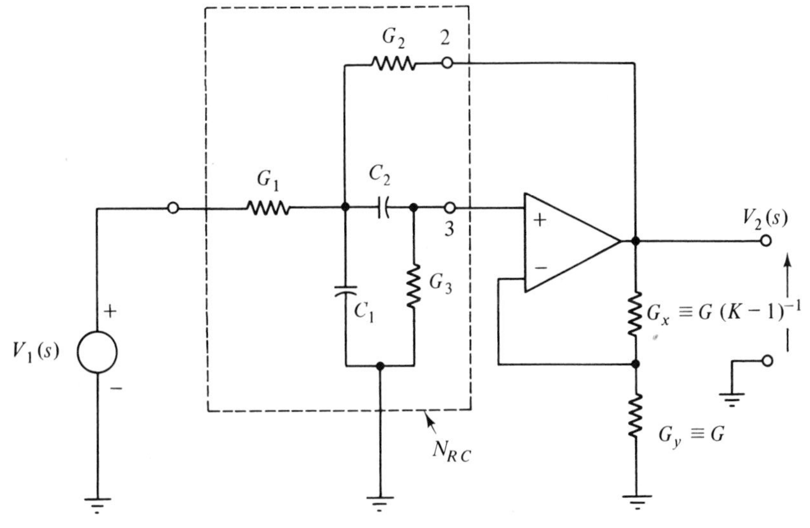

Type I Sallen-Key bandpass

$$H_b(s) = H_0 \frac{s}{s^2 + \left(\dfrac{\omega_p}{Q_p}\right) s + \omega_p{}^2}$$

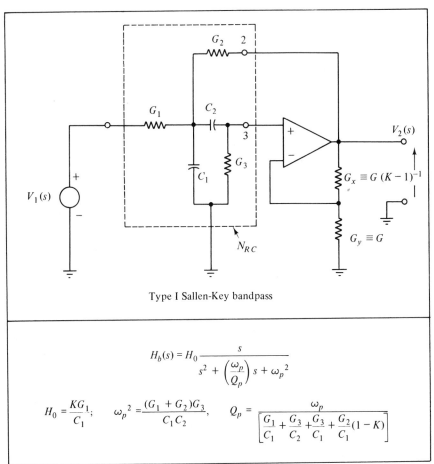

$$H_0 = \frac{KG_1}{C_1}; \qquad \omega_p{}^2 = \frac{(G_1 + G_2)G_3}{C_1 C_2}, \qquad Q_p = \frac{\omega_p}{\left[\dfrac{G_1}{C_1} + \dfrac{G_3}{C_2} + \dfrac{G_3}{C_1} + \dfrac{G_2}{C_1}(1 - K)\right]}$$

equations (8.168) into the expressions for H_0, ω_P, Q_P, and $H_b(s)$ of Table 8-13, we have

$$H_b(s) = \frac{K}{T}\left\{\frac{s}{s^2 + \left[1 + (\gamma\beta)^2 + \gamma^2 + \alpha^2(1 - K)\right]\frac{s}{T} + \frac{1}{T^2}(1 + \alpha^2)\gamma^2\beta^2}\right\} \tag{8.169}$$

$$\omega_P = \frac{(1 + \alpha^2)^{1/2}\gamma\beta}{T} \tag{8.170}$$

and

$$K = 1 + \frac{1}{\alpha^2} + \left(\frac{\gamma}{\alpha}\right)^2\left[\beta^2 + 1 - Q_P^{-1}\gamma^{-1}\beta(1 + \alpha^2)^{1/2}\right] \tag{8.171}$$

The *active-sensitivity* performance is derived by obtaining $K^2\alpha_1/\omega_P$ as in the previous section. Thus, we must first find α_1 by calculating $T_{32}(s)$. Direct analysis of the passive network N_{RC} in Table 8-13 for $T_{32}(s)$ reveals that

$$T_{32}(s) = \frac{\left(\frac{\alpha^2}{T}\right)s}{s^2 + \frac{\omega_P}{Q_N}s + \omega_P^2} \tag{8.172}$$

where Q_N is the Q-factor of the network N_{RC}. Thus, by comparison of equations (8.172) and (7.81) it follows that

$$\boxed{\alpha_1 = \frac{\alpha^2}{T}} \tag{8.173}$$

Then, equations (8.170), (8.171) and (8.172) provide the required quantity

$$\boxed{\frac{K^2\alpha_1}{\omega_P} = \left[1 + \frac{1}{\alpha^2} + \left(\frac{\gamma}{\alpha}\right)^2 \left[\beta^2 + 1 - Q_P^{-1}\gamma^{-1}\beta(1 + \alpha)^{1/2}\right]\right]^2\left[\frac{\alpha^2}{(1 + \alpha^2)^{1/2}\gamma\beta}\right]} \tag{8.174}$$

This term may be used directly in equation (7.85) to obtain the active-sensitivity performance. We consider one typical design example:

Design example: uniform G/uniform C: In this case, $G_1 \equiv G_2 \equiv G_3$ and $C_1 \equiv C_2$, so that

$$\boxed{\alpha = \beta = \gamma = 1}$$

Equation (8.171) gives

$$\boxed{K = 4 - Q_P^{-1}\sqrt{2}} \tag{8.175}$$

and, from equation (8.174), we have

$$\frac{K^2\alpha_1}{\omega_P} = \frac{1}{\sqrt{2}}\left[4 - Q_P^{-1}\sqrt{2}\,\right]^2 \tag{8.176}$$

From equations (8.176) and (7.85), we find that

$$S_B^{G_b(\omega)} = -\frac{1}{\sqrt{2}}(4 - Q_P^{-1}\sqrt{2}\,)^2\frac{\omega_P}{B}\mathrm{Re}[\,t_{\mathrm{HP}}(j\omega)\,]$$

and, for $Q_P \gg \sqrt{2}\,/4$,

$$\boxed{S_B^{G_b(\omega)} \simeq -\left(\frac{16\omega_P}{\sqrt{2}\,B}\right)\mathrm{Re}[t_{\mathrm{HP}}(j\omega)]} \tag{8.177}$$

ACTIVE-SENSITIVITY
SALLEN-KEY BANDPASS
UNIFORM C/UNIFORM G

implying that

$$\frac{\Delta\omega_P}{\omega_P} \simeq -4\sqrt{2}\,\frac{\omega_P}{B} \tag{8.178}$$

The active-sensitivity performance of this bandpass circuit [equation (8.177)] is, therefore, similar to that of the lowpass version [equation (8.133)].

8-6.3 Sallen-Key Highpass Type I Circuit

The circuit diagram is shown in Fig. 8.12. The analysis follows essentially the same procedure as that outlined in section 8-6.1. and is, therefore, left as an exercise for the reader in Problem 8.50. Nonideal performance is similar to that of the lowpass circuit insofar as active- and passive-sensitivity behavior is concerned.

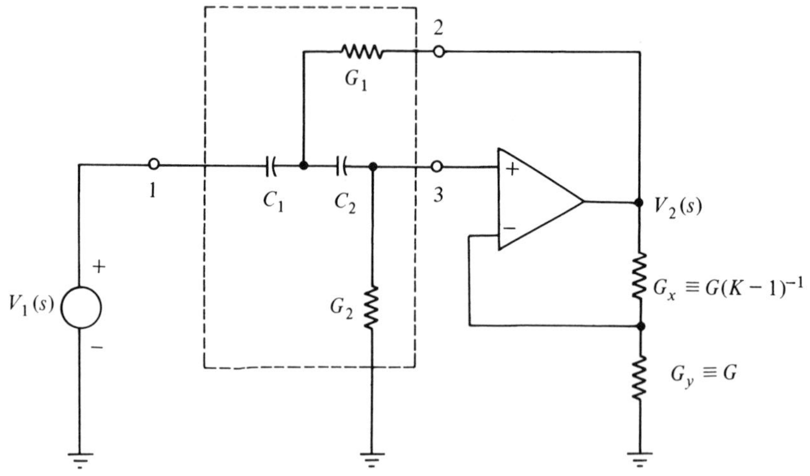

FIGURE 8.12 *Sallen-Key highpass type I circuit.*

8-6.4 Biquadratic Sallen-Key Type I Circuits

It is beyond the scope of this text to analyze a large variety of Sallen-Key and other types of single-amplifier structures. However, it should be noted that Sallen-Key type circuits exist for realizing numerators of the form $s^2 + (\omega_Z/Q_Z)s + \omega_Z^2$. In Table 8-14, for example, three circuits are shown that may be used to realize complex zeros in notch circuits (for which $\omega_Z = \omega_P$). General *biquadratic* type I Sallen-Key circuits may be realized by using somewhat more complicated networks N_{RC}. Problem 8-52 is concerned with a typical biquadratic circuit of this type.

8-7 SUMMARY

The topic of second-order *RC*-active filter design is extensive. It is not possible within a single chapter to mention more than a small selection of representative types of circuits. The three- and four-amplifier two-integrator loop (TIL) structures have been described in detail, being widely used as building-blocks in the realization of higher-order filters because they are easily tuned and designed and exhibit low harmonic distortion. It has been shown that a type 1B noninverting integrator may be employed in such a way that the Q-factor Q_P of the resultant TIL circuit is virtually independent of the finite OP AMP bandwidths; the principal factor limiting circuit performance is the effect of B on ω_P, which causes a fractional downward

TABLE 8-14 *Three Sallen-Key Notch circuits of the type 1 bandpass-feedback type.*

SALLEN-KEY NOTCH, $\omega_z = \omega_p$ (TYPE I CIRCUIT)

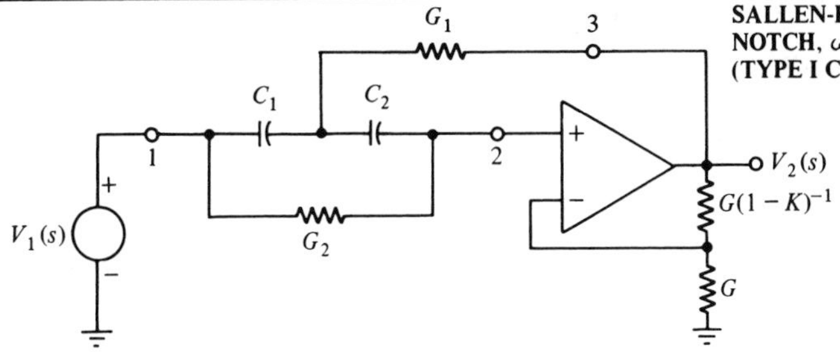

$$\frac{V_2(s)}{V_1(s)} = H_0 \frac{s^2 + \left(\frac{\omega_p}{Q_z}\right)s + \omega_p^2}{s^2 + \left(\frac{\omega_p}{Q_p}\right)s + \omega_p^2}$$

where $\omega_p = \frac{\alpha\beta}{T}$, $Q_z = \frac{\alpha\beta}{\alpha^2(1+\beta^2)}$, $H_0 = K$

$$Q_p = \frac{\alpha\beta}{\alpha^2(1+\beta^2) - (1-K)}$$

and $T \equiv \frac{C_1}{G_1}$, $G_2 \equiv \alpha^2 G_1$, $C_1 \equiv \beta^2 C_2$

SALLEN-KEY NOTCH, $\omega_z = \omega_p$ (TYPE I CIRCUIT)

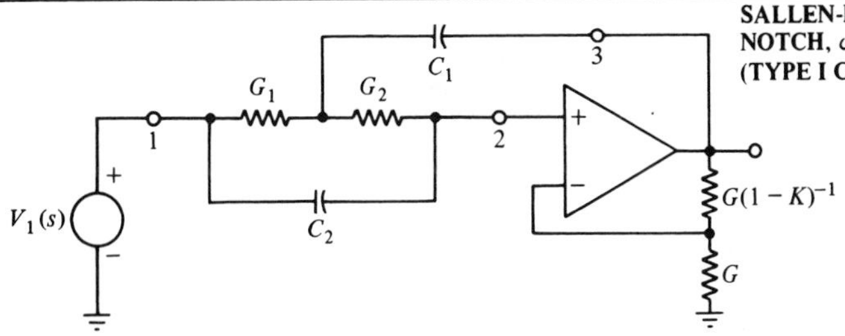

Expressions same as above except:

$$Q_z = \frac{\alpha\beta}{1+\alpha^2}, \qquad Q_p = \frac{\alpha\beta}{\beta^2(1+\alpha^2) + (1-K)\alpha^2\beta^2}$$

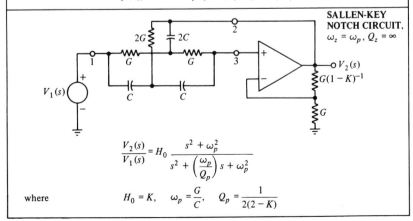

SALLEN-KEY NOTCH CIRCUIT, $\omega_z = \omega_p$, $Q_z = \infty$

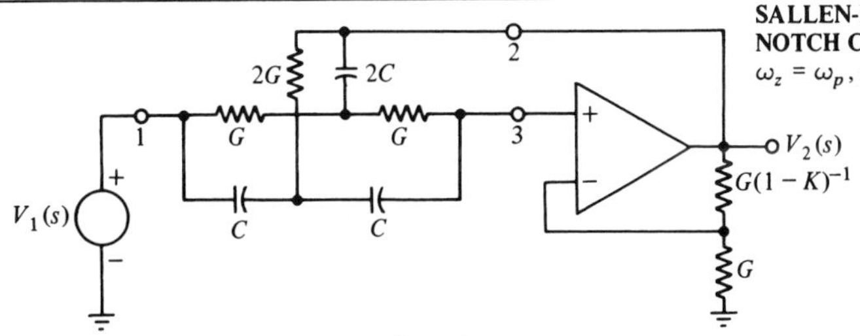

$$\frac{V_2(s)}{V_1(s)} = H_0 \frac{s^2 + \omega_p^2}{s^2 + \left(\frac{\omega_p}{Q_p}\right)s + \omega_p^2}$$

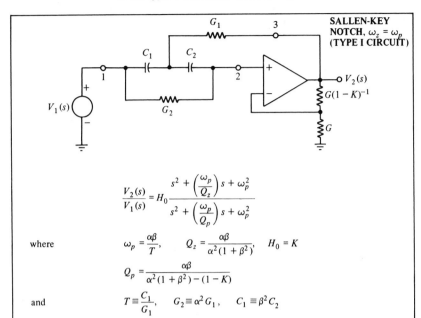

where $H_0 = K$, $\omega_p = \frac{G}{C}$, $Q_p = \frac{1}{2(2-K)}$

shift of ω_P by approximately ω_P/B. The generalized immittance converter (GIC) quadratic and biquadratic circuits have also been described in detail; they compare favourably with TIL circuits insofar as nonideal performance is concerned and most of them are realizable with two-OP AMP two-capacitor circuits. They are, therefore, likely to be more economical than TIL circuits.

The Sallen-Key circuits are described because they are typical type I (noninverting gain) structures that are especially useful for designs where Q_P is less than about 20. They are not as easily tuned as GIC or TIL circuits; however, they use just one OP AMP and by correct design it has been shown that the passive sensitivities are virtually lower bound in the worstcase sense. It is also possible, by careful design, to minimize the active sensitivity so that it is comparable to that of GIC/TIL circuits. The *lowpass* Sallen-Key circuit is treated in detail, and the principal circuit equations are given for bandpass, highpass and notch Sallen-Key circuits.

The type II single-amplifier structures that have been defined in chapter 6 lead to a large variety of circuits. They have not been pursued in this chapter because of the inherently poor active-sensitivity performance [equation (7.106)].

PROBLEMS

8-1. The two-integrator loop circuit of Fig. 8.1 employs resistors having tolerances given by

$$-0.01 \leqslant \epsilon_{iR} \leqslant 0.01$$

so that $\epsilon_{R_{\max}} = 0.01$. Use equation (8.23) to estimate the worstcase change in $G_b(\omega)$ that can occur due to these resistance tolerance errors, assuming $Q_P \gg 1$.

8-2. The two-integrator loop circuit of Fig. 8.1 is to be designed with $Q_P = 10$ so that the worstcase error in the gain $G_b(\omega)$ due to the tolerance of the two capacitors is less than 10%. Use equation (8.22) and Fig. 7.10 to estimate the maximum tolerance $\epsilon_{C_{\max}}$ of the two capacitors that can be permitted.

8-3. Assume that the input and output of the TIL-3A circuit [Fig. 8.3(b)] are chosen so that $N(j\omega)$ equals $I_1(j\omega)$ in equation (8.36), where $I_1(j\omega)$ and $I_2(j\omega)$ are given by equations (8.37) and (8.38). Show that it is justifiable to neglect the dependence of $N(j\omega)$ on gain bandwidth products B when deriving the high-frequency limitations of this circuit.

8-4. The basic TIL-3B loop of Fig. 8.3(c) has the following parameters:

$$R=20\text{ k}\Omega,\qquad R_Q=200\text{ k}\Omega$$
$$B\approx 2\times 10^6\text{ rad/sec}$$

where $B_1\approx B_2\approx B_3\approx B$. Select predistorted values for the capacitors C so that $\hat{\omega}_P$ is as close as possible to 10^5 rad/sec.

8-5. Design a type LBI/DF biquadratic circuit of the type shown in Table 8-2 using the two-integrator loop TIL-3B [Fig. 8.3(c)] as the basic building-block with

$$R=10\text{ k}\Omega,\quad R_1=100\text{ k}\Omega,\quad B=2\pi\times 10^6\text{ rad/sec}$$

having a general biquadratic response specified by $H_0=10$, $Q_P=15$, $\omega_P=10^4$ rad/sec, $Q_Z=\infty$, $\omega_Z=2\times 10^4$ rad/sec.
The design formulas are given in Table 8-2 and equations (8.63) to (8.67). The resistors R and R_1 are specified; therefore, equations (8.63) to (8.67) must be solved for C and for the resistors R_Q, $R_1\alpha_0^{-1}$, $R_1\alpha_1^{-1}$, and R_0. Finally, the designer should check equation (8.54) to determine whether C should be predistorted.
How would you tune ω_P, Q_P, and ω_Z?

8-6. Repeat Problem 8-5, but for $Q_Z=-15$ instead of infinity. Sketch the gain function $G_b(\omega)$ and phase function $\theta(\omega)$ for this allpass realization.

8-7. Repeat Problem 8-5, but for $Q_Z=25$. Note that equation (8.69) *is* satisfied, and you should therefore expect to employ an extra inverter at the $-BP$ output.

8-8. Repeat Problem 8-5, but for $\omega_P=10^5$ rad/sec and $\omega_Z=2\times 10^5$ rad/sec. This time it will be *important* that predistortion of capacitors C be considered.

8-9. Repeat Problem 8-5, but for $\omega_Z=\frac{1}{2}\times 10^4$ rad/sec.

8-10. Substitute the TIL-3B circuit from Fig. 8.3(c) into the MI/DF biquadratic circuit of Table 8-3 and prove that

$$H_b(s)=-\left(\frac{\alpha_4\omega_P R}{R_0}\right)\left[\frac{s^2+\dfrac{1}{\alpha_4}(\alpha_3-\alpha_2)s+\left(\dfrac{\alpha_1}{\alpha_4}\right)\omega_P}{s^2+\left(\dfrac{\omega_P}{Q_P}\right)s+\omega_P^2}\right]$$

where $\omega_P\equiv 1/RC$ and $Q_P\equiv R_Q/R$.

8-11. Design a type MI/DF biquadratic circuit of the kind shown in Table 8-3, using the two-integrator loop TIL-3B [Fig. 8.3(c)] as the basic building-block with

$$R = 100\ \text{k}\Omega, \quad R_0 = 100\ \text{k}\Omega, \quad B = 2\pi \times 10^6\ \text{rad/sec}$$

having the general biquadratic response specified as follows; $H_0 = -10$, $Q_P = 15$, $Q_Z = \infty$, $\omega_P = 10^4$ rad/sec, $\omega_Z = 2 \times 10^4$ rad/sec.

The design formulas are given in equations (8.71) to (8.75); you should find the values for C, C_4, $R_0\alpha_1^{-1}$, $R_0\alpha_2^{-1}$, and $R_0\alpha_3^{-1}$.

8-12. The BHI/DF signal-flow graph in Fig. 7.9(b) employs a negative *feedforward* transmittance $-b_{1b}$ to realize the term $b_{1b}s$ in the denominator of equation (7.28). Thus, the transmittance $-b_{1b}$ corresponds to the ω_P/Q_P term in the denominator of equation (8.77).

Prove that the inverting damped integrator circuit shown in Fig. P8.12(a) is equivalent to the transmittance in Fig. P8.12(b).

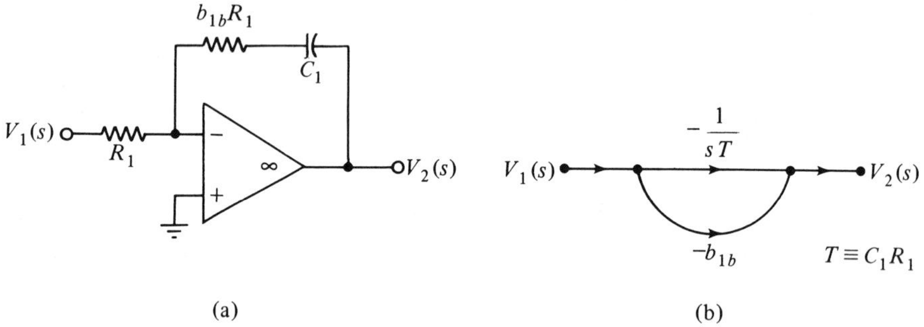

FIGURE P8.12

8-13. Analyze the type BHI/DF biquadratic circuit of the kind shown in Table 8-4, using the two-integrator loop TIL-3A (having the required *series* damping resistance $R_Q = R/Q_P$ associated with the output integrator) and thereby prove that equation (8.77) is valid.

8-14. Write a signal-flow graph for Fig. 8.5(b), and compare it with Fig. 7.9(b), identifying the terms b_{0b}, b_{1b}, α_0, α_1 with specific branch immittances of Fig. 8.5(b).

8-15. Design a type BHI/DF biquadratic circuit of the kind shown in Table 8-4, to have the following specifications: $\omega_P = 10^4$ rad/sec, $Q_P = 50$, $R = 10$ kΩ, $\omega_Z = \frac{1}{2} \times 10^4$ rad/sec, $Q_Z = 400$, $H_0 = -10$. Use an extra inverting amplifier if necessary.

8-16. The basic TIL structure shown in Table 8-4 uses a series resistance $R_Q = R/Q_P$ to realize the damped integrator. Assume that the output BP is denoted by the bandpass gain function $G_b(\omega)$ and then prove that equations (8.82) and (8.83) are valid for this TIL circuit.

8-17. Write signal-flow graphs for the circuits shown in Fig. 8.6 and compare with Fig. 7.9(c), identifying α_1, α_2, α_3, α_4, $-b_{0b}$ and b_{1b} with specific branches of Fig. 8.6.

8-18. Write a signal-flow graph for the circuit in Fig. 8.5(a), and compare with Fig. 7.9(a), identifying α_2, α_1, α_0, b_{0b}, and b_{1b} with specific branch immittances of Fig. 8.5(a).

8-19. A nullor equivalent circuit for the general second-order GIC filter of Fig. 8.7(a) is shown in Fig. P8-19. Transform the branches containing the voltage sources V_1 to their Norton equivalent circuits and then use the *RC*-nullor nodal analysis technique (see section 3.2) to solve for the node voltages V_2, V_3, and V_4. Check your answers with equations (8.84), (8.85), and (8.86).

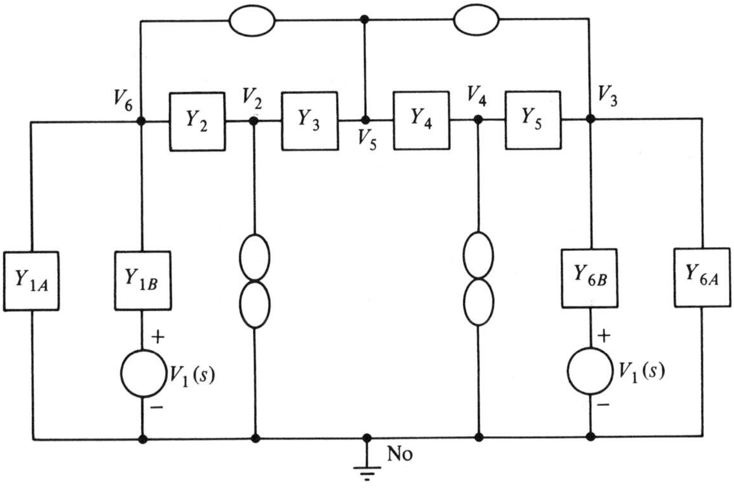

FIGURE P8.19

8-20. Prove that equation (8.94) is the ideal transfer function of the highpass second-order filter GIC-2HP-1 shown in Table 8-5.

8-21. A GIC-2HP-1 circuit has the following specifications: output voltage $V_4(s)$, $\omega_P = 10^4$ rad/sec, $Q_P = 25$, $H_0 = 2$, $G_2 = G_4 = 10^{-4}$ ohm^{-1}, $C_1 = C_3 = 10$ nF.

Find G_1, G_5, and G_6; then prove that the fractional inductance error $\epsilon_{Hl}(\omega)$ is provided by

$$\epsilon_{Hl}(\omega) \approx \frac{2\omega}{B}\left(\frac{\omega}{\omega_P} + \frac{\omega_P}{\omega}\right), \qquad B_1 \approx B_2 \equiv B$$

$$Q_l^{-1}(\omega) \approx 2\left[\frac{\omega_P}{\omega}\left(\frac{1}{A_{02}}\right) + \frac{\omega}{\omega_P}\left(\frac{\omega^2}{B^2}\right)\right]$$

Sketch the above functions $\epsilon_{Hl}(\omega)$ and $Q_l^{-1}(\omega)$ for $B = 10^7$ rad/sec and $A_{02} = 10^5$.

8-22. The GIC-2HP-1 circuit shown in Table 8.5 has a nonideal inductance characterized by $\epsilon_{Hl}(\omega)$ and $Q_l^{-1}(\omega)$. Prove that the *effective* Q-factor $\hat{Q}_P$ and the *effective* natural frequency $\hat{\omega}_P$ are given by equations (8.99) and (8.100). [*Note*: $\hat{\omega}_P$ is the *actual* value of ω_P; $\hat{Q}_P$ is the *actual* value of Q_P].

Using equations (8.101) and (8.102), derive the values of $\hat{\omega}_P$ and $\hat{Q}_P$ for the numerical example in Problem 8-21. Then, derive $\Delta Q_P/Q_P$ and $\Delta\omega_P/\omega_P$, using equations (8.103) and (8.104).

Suggest modified values for G_2, G_4, G_5, G_6 that will compensate for the perturbation $\Delta\omega_P/\omega_P$ due to finite gain bandwidth products.

8-23. Problem 8-21 results in a design that meets the enhancement condition, $G_5 = G_6$. Consider now a design specification that does not meet these conditions: $\omega_P = 10^4$ rad/sec, $Q_P = 25$, $H_0 = 5$, $G_2 = 10^{-4}$ ohm^{-1}, $G_4 = \frac{1}{4} \times 10^{-4}$ ohm^{-1}, $C_1 = C_3 = 10$ nF.

Find G_1, G_5, and G_6; then $\epsilon_{Hl}(\omega)$ and $Q_l^{-1}(\omega)$ as functions of ω, ω_P, B, and A_{02}. (Assume $B_1 \approx B_2 \equiv B$.) Sketch $\epsilon_{Hl}(\omega)$ and $Q_l^{-1}(\omega)$, and compare with the result in Problem 8-21.

8-24. Calculate $\hat{\omega}_P$ and $\hat{Q}_P$ for the design of Problem 8-23 and compare with that obtained in Problem 8-22. Comment on the significance of this comparison.

8-25. Prove that the second-order bandpass circuit GIC-2BP-1 has the transfer function $H_b(s)$ given in Table 8-6.

8-26. Prove that the second-order lowpass circuit GIC-2LP-1 has the transfer function $H_b(s)$ shown in Table 8-7.

8-27. Design a GIC-2BP-1 circuit, using 10 nF capacitors to have $G_2 = G_4 = G_5 = G_6$, $\omega_P = 2 \times 10^4$ rad/sec, and $Q_P = 100$.

Given that the mismatch factor m of the gain bandwidth products is anywhere in the range 0.8 to 1.2, use equation (8.101) to predict the corresponding range of $\hat{Q}_P$. Assume $B_1 = 2 \times 10^6$ rad/sec.

8-28. Design a GIC-2LP-1 circuit, of the type shown in Table 8-7, to have the following specifications: $C_1 = C_4 = 10$ nF, $\omega_P = 10^4$ rad/sec, $Q_P = 20$, and select a minimum bandwidth B such that $\hat{\omega}_P$ in equation (8.102) is within 0.1% of $\hat{\omega}_P$ for all values of m in the range 0.7 to 1.3. Find the corresponding range of values for $\hat{Q}_P$.

8-29. Prove that the notch circuits GIC-2N-2A and GIC-2N-2B have the transfer functions shown in Table 8.8(a). [*Hint*: Both circuits are special cases of Fig. 8.7(a); therefore, equations (8.84), (8.85) or (8.86) may be used.]

8-30. Design a GIC-2N-2A circuit having $C_2 = C_{6B} = 20$ nF, $\omega_P = 2 \times 10^4$ rad/sec, $Q_P = 5$, $\omega_Z = 10^4$ rad/sec, and $H_0 = 1$.

8-31. Design a GIC-2N-2B circuit having $G_2 = G_4 = G_3 = G_{6B} = 10^{-5}$ ohm^{-1}, $\omega_P = 10^3$ rad/sec, $Q_P = 4$, $\omega_Z = 10^2$ rad/sec, $H_0 = 1$.

8-32. Prove that the expressions in Table 8-8(b) for ω_P^2, Q_P, ω_Z^2, and Q_Z are valid for the GIC-2N-1 allpass-notch circuit.

8-33. Use GIC-2N-1 to design an allpass function having $\omega_P = 10^4$ rad/sec, $Q_P = 4.7$, according to equation (8.114). Use 10 nF capacitors.

8-34. Use GIC-2N-1 to design a notch function having $\omega_P = 10^4$ rad/sec, $Q_P = 4.7$, $Q_Z = \infty$, $\omega_Z = \omega_P$.

8-35. Prove that GIC-2N-1 can realize the allpass function ($\omega_P = \omega_Z$, $Q_P = -Q_Z$) or the notch ($\omega_P = \omega_Z$, $Q_Z = \infty$) and, simultaneously, satisfy the enhancement condition $G_2 = G_{1B}$.

8-36. The type IIA and type IIB single-amplifier RC-active filters employ inverting amplifiers of gain $-K$ with feedback functions $T_{32}(s)$ that are lowpass and highpass, respectively. For these structures, equation (7.106) states that

$$|S_B^{G_b(\omega)}|_{\max} \geqslant \frac{4\omega_P Q_P^3}{B}$$

Evaluate the lower bound on the right-hand side of this expression for $Q_P = 10$, $\omega_P = 10^4$ rad/sec, $B = 10^7$ rad/sec, and thereby determine the permissible $\Delta B/B$ above which $|\Delta G_b(\omega)/G_b(\omega)|$ exceeds 0.1 at some frequency ω. Comment on the practical significance of this result.

8-37. Prove that the expressions for H_0, ω_P, and Q_P in Table 8-10 are valid for the lowpass type I Sallen-Key circuit.

8-38. Prove that the voltage transfer function $T_{32}(s)$ of the type I Sallen-Key circuit in Table 8-10 is provided by equation (8.119).

8-39. Design a type I Sallen-Key lowpass circuit of the type shown in Table 8-10 with the specifications: $\omega_P = 10^4$ rad/sec, $Q_P = 10$, $C_1 = C_2 = 10$ nF, $G_1 = G_2$. That is, find G_1, G_2, and K. This is a uniform G/uniform C design.

8-40. Assuming the gain K in Problem 8-39 is realized by using two conductances G_x and G_y as shown in Table 8.10, calculate the six single-element sensitivities $S_{x_i}^{G_b(\omega)}$, where $x_i = C_1$, C_2, G_1, G_2, G_x, and G_y, for the numerical values given in the problem. Express these sensitivities in the form

$$S_{x_i}^{G_b(\omega)} = q_1 + q_2 \operatorname{Re}[t_{BP}(j\omega)] + q_3 \operatorname{Re}[t_{HP}(j\omega)]$$

and compare them with the approximate expressions in equation (8.131).

Use these results and Fig. 7.10 to sketch the worstcase passive-sensitivity function $WS_{RC}^{G_b(\omega)}$.

8-41. Sketch the *exact* active-sensitivity function $S_B^{G_b(\omega)}$, equation (8.132), for the design in Problem 8.39, assuming $B = 10^7$ rad/sec. Compare the active-sensitivity performance and passive-sensitivity performance of the circuit in

Problem 8-39, assuming

$$-0.1 \leqslant \frac{\Delta B}{B} \leqslant 0.1$$

and that the *passive* elements x_i are bounded by

$$-10^{-3} \leqslant x_i \leqslant 10^{-3}$$

8-42. Given the general lowpass function

$$H_b(s) = \frac{H_0}{s^2 + \dfrac{\omega_P s}{Q_P} + \omega_P^2}$$

and arbitrarily small perturbations $\Delta\omega_P/\omega_P$ and $\Delta Q_P/Q_P$, show that the resultant change in the gain function $G_b(\omega)$ is given by

$$\frac{\Delta G_b(\omega)}{G_b(\omega)} \approx \frac{2\Delta\omega_P}{\omega_P} \operatorname{Re}[t_{HP}(j\omega)] + \frac{1}{Q_P}\left(\frac{\Delta\omega_P}{\omega_P} + \frac{\Delta Q_P}{Q_P}\right) \operatorname{Re}[t_{BP}(j\omega)]$$

8-43. The result in equation (8.132) for the type I Sallen-Key lowpass circuit implies that the fractional change in $G_b(\omega)$ due to finite B is given by

$$\frac{\Delta G_b(\omega)}{G_b(\omega)} \approx -S_B^{G_b(\omega)} = -(3 - Q_P^{-1})^2\left(\frac{\omega_P}{B}\right) \operatorname{Re}[t_{HP}(j\omega)]$$

Compare this result with the result of Problem 8-42 to show that the effect of finite bandwidth B is equivalent to perturbations of ω_P and Q_P given by

$$\frac{\Delta\omega_P}{\omega_P} \approx -\frac{\Delta Q_P}{Q_P} \approx -(3 - Q_P^{-1})^2\left(\frac{\omega_P}{B}\right)$$

8-44. Repeat the Sallen-Key type I lowpass circuit design of Problem 8-39 but use the design No. 2 procedure of section 8.6.1 *Assume that the capacitance element spread is $\beta^2 = 20$;* the remaining specifications are $\omega_P = 10^4$ rad/sec, $Q_P = 10$, $C_2 = 1$ nF, ($C_1 = 20$ nF). It is recommended that equations (8.120), (8.141), and (8.142) be used to calculate the quantities $\eta_{\min}$, α, G_1, G_2, K, $\Phi^2_{\min}$, and $S_B^{G_b(\omega)}|_{\min}$. Assume that $B = 10^7$ rad/sec, and compare the minimum active-sensitivity performance of this design with that obtained in Problem 8-41 for the uniform G/uniform C design.

8-45. Repeat the Sallen-Key type I lowpass circuit design of Problem 8-39 ($\beta = 1$), using the near-minimum active-sensitivity design procedure of section 8.6.1. That is, use equation (8.143) to calculate $\hat{\eta}_{\min}$ to obtain near-optimum values $\hat{\alpha}$, $\hat{K}$, $\hat{\Phi}_{\min}$ and $S_B^{G_b(\omega)}$ in equations (8.142). Compare the active-sensitivity performance with that obtained in Problem 8-44.

8-46. Prove that the active-sensitivity expression in equation (8.144) implies that

$$\frac{\Delta\omega_P}{\omega_P} \approx -\frac{\Delta Q_P}{Q_P} \approx -\frac{8}{3\sqrt{3}}\left(\frac{\omega_P}{B}\right)$$

for $Q_P \gg 1$ and $\hat{\beta} \gg 1$. Comment on the active-sensitivity performance of a near-minimum active-sensitivity Sallen-Key type I circuit as compared with the corresponding TIL and GIC circuits.

8-47. An approximate minimum active-sensitivity design procedure is described in section 8.6.1 for the case where $Q_P > 2$ *and* $\beta^2 \gg 1$. This procedure involves *selecting* $K = 4/3$. In this case, equation (8.150) is an estimate of the active-sensitivity performance. For the Sallen-Key lowpass design considered in Problem 8-44, that is, $\omega_P = 10^4$ rad/sec, $Q_P = 10$, $C_2 = 1$ nF, and $C_1 = 20$ nF, select $K = \frac{4}{3}$ and calculate α, G_1, G_2, η, Φ^2, and $S_B^{G_b(\omega)}$. Compare this active-sensitivity term $S_B^{G_b(\omega)}$ with that obtained in Problems 8-41 and 8-44.

8-48. An equal-conductance ($G_1 = G_2$) lowpass Sallen-Key type I circuit is required with minimum passive-sensitivity performance [equations (8.159)]. The circuit specifications are $\omega_P = 10^4$ rad/sec, $G_1 = G_2 = 10^{-4}$ ohm^{-1}, $Q_P = 10$, $B = 10^7$ rad/sec. Calculate C_1, C_2, and K. Then calculate $S_B^{G_b(\omega)}$, and compare with the corresponding results in Problems 8-41, 8-44, and 8-47.

8-49. Design a Sallen-Key type I *bandpass* circuit of the kind shown in Table 8-13 having the specifications: $\omega_P = 10^4$ rad/sec, $G_1 = G_2 = G_3 = 10^{-4}$ ohm $^{-1}$, $C_1 = C_2$, $B = 10^7$ rad/sec. Calculate C_1, C_2, and K. Then calculate $S_B^{G_b(\omega)}$.

8-50. Derive the voltage transfer function of the Sallen-Key *highpass* type I circuit shown in Fig. 8.12 as a function of C_1, C_2, G_1, G_2, and K.

8-51. Verify the transfer functions of the three Sallen-Key *notch* circuits of Table 8-14.

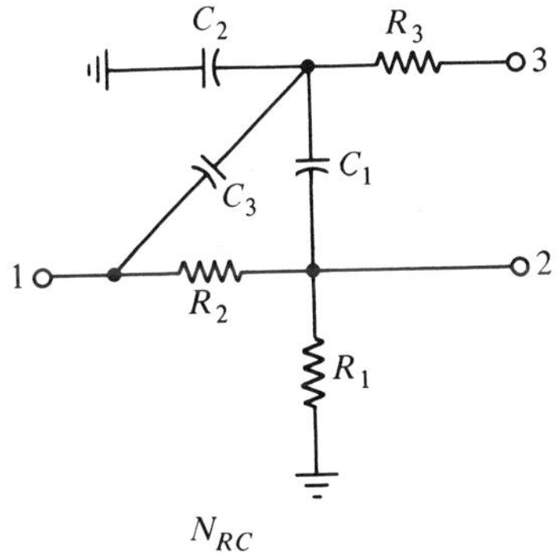

FIGURE P8.52

8-52. Consider the passive network N_{RC} shown in Fig. P8.52. Show that this network may be substituted for N_{RC} in Fig. 8.12 to realize a general *biquadratic* type I Sallen-Key circuit having a transfer function of the form given in equation (8.1).

BIBLIOGRAPHY

ACKERBERG, D., and K. MOSSBERG, "A Versatile Active RC Building Block with Inherent Compensation for the Finite Bandwidth of the Amplifiers," *IEEE Trans. Circuits Syst.*, vol. CAS-21, p. 75-78, Jan. 1974.

BRUTON, L.T., 'Biquadratic Sections Using Generalized Impedance Converters,' *IEEE Jour.*, vol. 41, No. 11, p. 510-512, Nov. 1971.

BRUTON, L.T., "Sensitivity Comparison of High-Q Second-Order Active Filter Synthesis Techniques," *IEEE Trans. Circuits Syst.*, p. 32-38, Jan. 1975.

GIRLING, F.E.J., and E. F. GOOD, "Active Filters 12 and 13: The Leapfrog or Active Ladder Synthesis; Applications of the Active Ladder Synthesis," *Wireless World*, vol. 76, p. 341-345, July 1970 and p. 445-450, Sept. 1970.

SALLEN, P.R., and E.L. KEY, "A Practical Method Of Designing RC-Active Filters," *IRE Trans. Circuit Theory*, vol. CT-2, p. 74-85, 1955.

SARAGA, W., "Sensitivity of 2nd-Order Sallen-Key Type Active RC Filters," *Electronics Letters*, vol. 3, No. 10, Oct. 1967.

TEMES, G.C., and J.W. LAPATRA, *Circuit Synthesis And Design*, McGraw-Hill, 1977, chapter 7, p. 303-309.

THOMAS, L.C., "The Biquad: Part I—Some Practical Design Considerations," *IEEE Trans. Circuit Theory*, vol. CT-18, p. 350-357, May 1971.

TOW, J., "A Step-By-Step Active-Filter Design," *IEEE Spectrum*, vol. 6, p. 64-68, Dec. 1969.

TOW, J., "Design Formulas for Active RC Filters Using Operational Amplifier Biquad.," *Electronics Letters*, p. 339-341, July 1969.

9

LADDER FILTER SYNTHESIS USING IMMITTANCE SIMULATION

9-1 INTRODUCTION

The material in this chapter is concerned with the synthesis and design of *RC*-active ladder filters based on the direct simulation of the doubly terminated *LC* filter. It is explained in chapter 7 that doubly terminated *LC* filters have the useful property that they can be designed in such a way that the magnitude steady-state frequency response, given by $M(\omega) \equiv |H(j\omega)|$, is highly insensitive to first-order perturbations of the *LC* elements and of the resistive terminations. By reviewing the sensitivity properties of *LC* filters that are described in section 7-5, the reader may verify that $M(\omega)$ is virtually insensitive to perturbations of the *LC* elements over the passband region of ω if the normalized transmission function $|t(j\omega)|$ is close to unity throughout the passband. It is this extraordinary insensitivity to element perturbations that to a large extent accounts for the successful practical applications of passive *LC* filters. It is, therefore, quite natural that *RC*-active filter designers should attempt to retain the good sensitivity behavior of doubly terminated *LC* filters by simulating their performance using active inductance-simulation circuits (of the types described in chapters 5 and 6). In this chapter, the applications of transconductance-type gyrators and GICs to the design of active "pseudo-lossless" ladder filters are considered in some detail. Such study is warranted because these methods have been used successfully in major telecommunication and instrumentation systems to replace high-quality "brick-wall"-type LC filters.

9-2 REVIEW OF *LC*-FILTER CIRCUITS

For the purposes of subsequent discussion, it is useful to consider some of the widely used LC-filter circuits; it should be noted that all of these filters may be designed in such a way that $|t(j\omega)|$ is close to unity throughout the passband, thereby resulting in the abovementioned insensitivity of the magnitude response $M(\omega)$ to the LC elements throughout the passband. The five LC topologies shown in Table 9.1 are commonly encountered and are readily designed by using LC-filter design tables or LC-filter computer-aided design algorithms.

The type A topology is that of a *lowpass* response $M(\omega)$ with zeros of transmission (that is, $M(\omega)$ equal to zero) wherever the shunt arms L_2C_2, L_4C_4, . . . are series resonant. A typical response is shown in Fig. 9.1(a). This type of filter is widely encountered, occurring in large volume in association with *each voice frequency (or baseband) channel* of frequency division multiplex and time division multiplex telecommunication systems.

The type B topology corresponds to a *highpass* response $M(\omega)$, as shown in Fig. 9.1(b). The transmission zeros occur where the shunt arms L_2C_2, L_4C_4, . . . , are series resonant.

The type C topology is that of a *bandpass* filter with the type of frequency response $M(\omega)$ shown in Fig. 9.1(c). This type of filter is often designed from a corresponding lowpass prototype filter of the type shown in Fig. 9.2(a), which is a lowpass structure having no finite zeros of $M(\omega)$. Standard LC design tables may be consulted to determine the LC-element values for various types of $M(\omega)$ functions such as Butterworth, Chebychev, Bessel, etc. To design a *bandpass* filter from the LC prototype shown in Fig. 9.2(a), we use the lowpass-to-bandpass transformation

$$s \rightarrow \frac{\omega_0\omega_{CL}}{\mathrm{BW}}\left(\frac{s}{\omega_0}+\frac{\omega_0}{s}\right) \qquad \text{LOWPASS-TO-BANDPASS TRANSFORMATION} \tag{9.1}$$

where ω_{CL} is the cut-off frequency of the lowpass filter and ω_0 and BW are the center frequency and bandwidth, respectively, of the bandpass filter (all in units of radians per second). Consider the transform impedance sL_i of a general series branch shown in Fig. 9.2(a), which is transformed

TABLE 9-1 *Some typical LC ladder filters.*

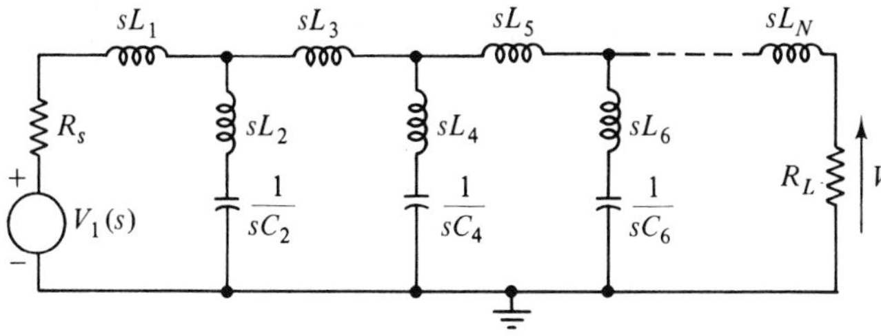

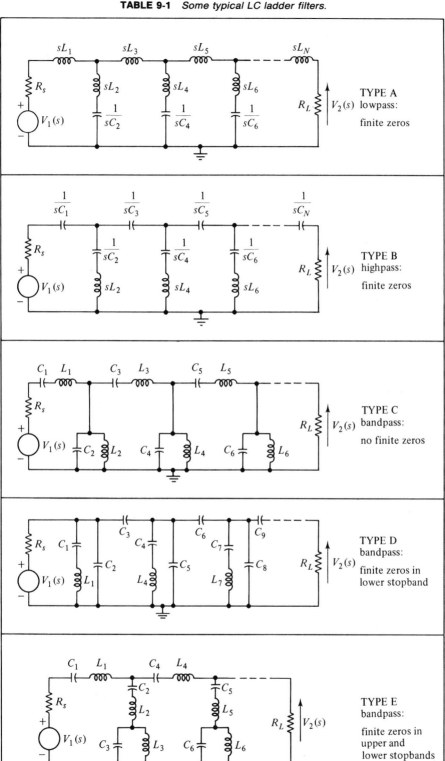

TYPE A
lowpass:
finite zeros

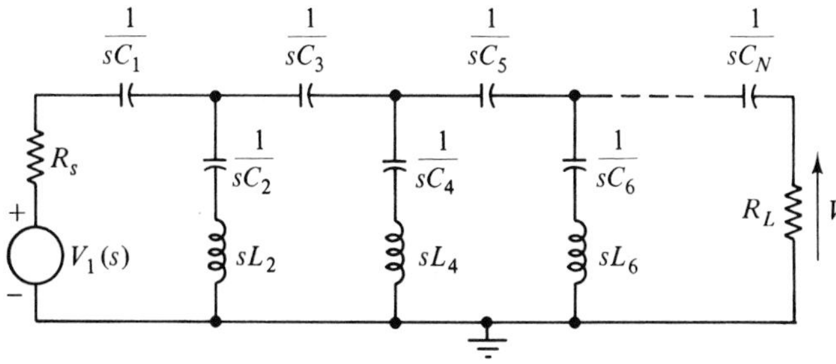

TYPE B
highpass:
finite zeros

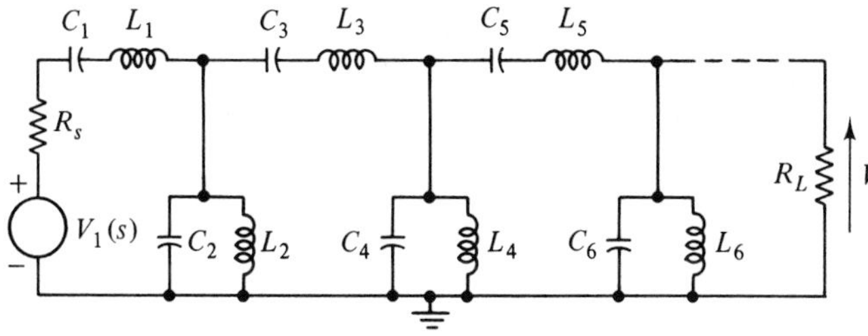

TYPE C
bandpass:
no finite zeros

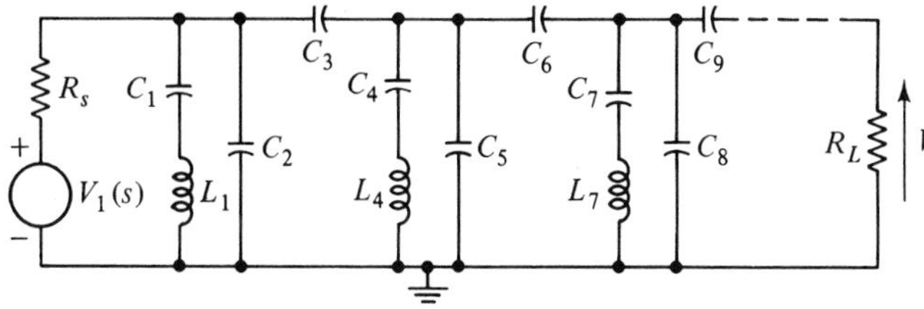

TYPE D
bandpass:
finite zeros in
lower stopband

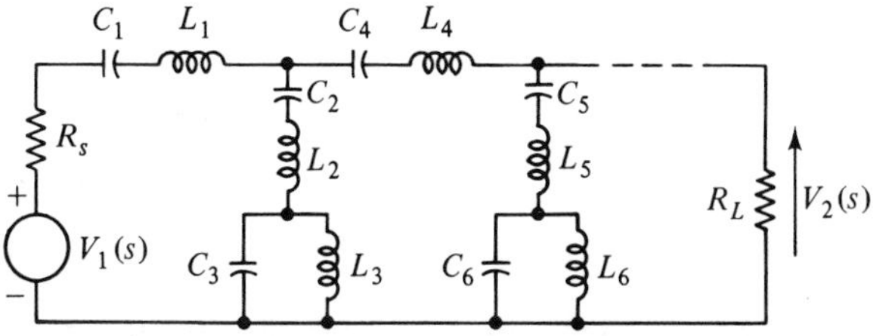

TYPE E
bandpass:
finite zeros in
upper and
lower stopbands

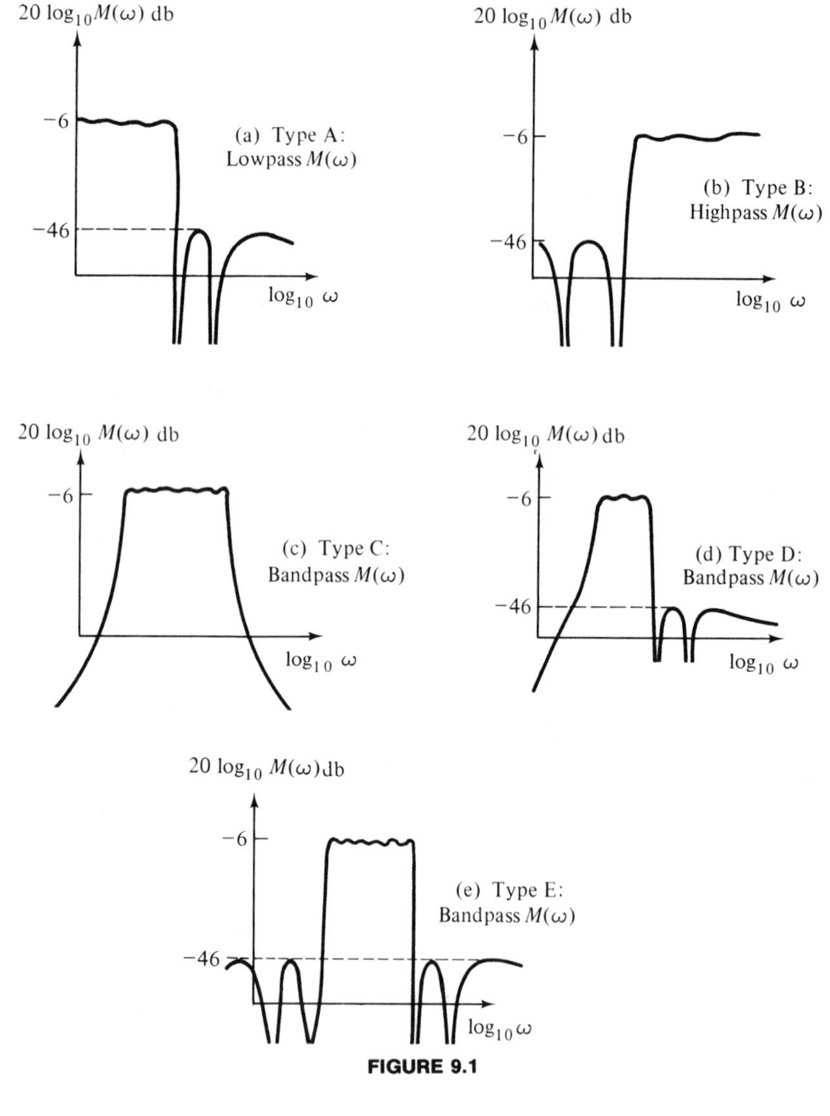

FIGURE 9.1

according to equation (9.1) as follows:

$$\underset{\substack{\text{lowpass series}\\ \text{branch impedance}}}{sL_i} \rightarrow \frac{\omega_0\omega_{CL}}{\text{BW}}\left(\frac{s}{\omega_0} + \frac{\omega_0}{s}\right)L_i = \underset{\substack{\text{bandpass series}\\ \text{branch impedance}}}{s\left(\frac{\omega_{CL}L_i}{\text{BW}}\right) + \frac{1}{s}\left(\frac{\omega_{CL}\omega_0^2 L_i}{\text{BW}}\right)} \tag{9.2}$$

We note that the right-hand side of equation (9.2) corresponds to the *series*

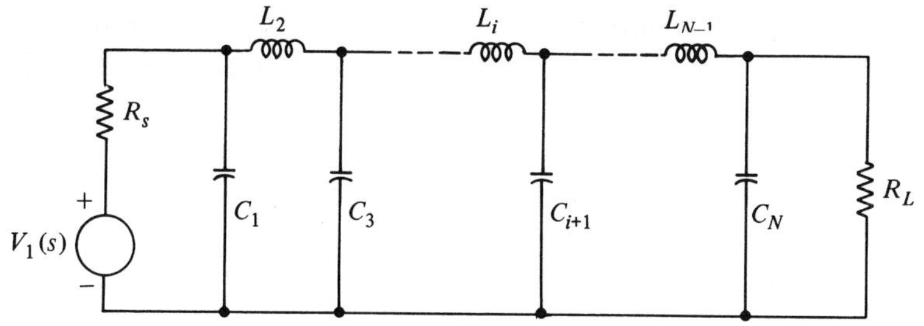

(a) An LC filter prototype: lowpass, cut-off frequency ω_{CL}.

Lowpass element	*Corresponding bandpass element*	*Corresponding highpass elements*
L_i	$C_i' = \dfrac{BW}{\omega_{CL}\,\omega_0^2 L_i}$ $L_i' = \dfrac{L_i \omega_{CL}}{BW}$	$C_i' = \dfrac{1}{\omega_{CL}\,\omega_{CH} L_i}$
C_i	$L_i' = \dfrac{BW}{\omega_{CL}\,\omega_0^2 C_i}$ $C_i' = \dfrac{C_i \omega_{CL}}{BW}$	$L_i' = \dfrac{1}{\omega_{CH}\,\omega_{CL} C_i}$

(b) Bandpass and highpass design values using transformations in equations (9.1) and (9.9), respectively.

FIGURE 9.2

connection of an inductance L_i' and capacitance C_i', where

$$L_i' = \frac{\omega_{CL} L_i}{BW}, \qquad C_i' = \frac{BW}{\omega_{CL}\omega_0^2 L_i} \qquad \text{SERIES ELEMENTS OF BANDPASS FILTER} \qquad (9.3)$$

The LC elements in the shunt branches of a type C filter are similarly found by applying the transformation in equation (9.1) to the transform admittance sC_i of the general shunt branch of the lowpass prototype filter shown in Fig. 9.2(a), to obtain

$$\underset{\text{lowpass shunt branch admittance}}{sC_i} \rightarrow \underset{\text{bandpass shunt branch admittance}}{s\left(\frac{\omega_{CL}C_i}{\text{BW}}\right) + \frac{1}{s}\left(\frac{\omega_{CL}\omega_0^2 C_i}{\text{BW}}\right)} \tag{9.4}$$

where the right-hand side of equation (9.4) corresponds to the *parallel* connection of an inductance L_i' and capacitance C_i', where

$$L_i' = \frac{\text{BW}}{\omega_{CL}\omega_0^2 C_i}, \qquad C_i' = \frac{\omega_{CL}C_i}{\text{BW}} \qquad \textsc{Shunt Elements of Bandpass Filter} \tag{9.5}$$

Equations (9.3) and (9.5) allow a bandpass LC filter to be designed directly from a lowpass prototype. The element values of the resultant bandpass filter are shown in the second column of Fig. 9.2(b).

EXAMPLE 9.1: Design of a Tenth-Order Chebychev Bandpass Filter from the Fifth-Order Lowpass Prototype. Given the lowpass fifth-order prototype Chebychev LC filter of Fig. 9.3(a), which has a bandpass ripple of 0.1 db and a cut-off frequency ω_{CL} of 1 rad/sec, we now design a corresponding bandpass filter with a center frequency ω_0 of 1 rad/sec and a bandwidth **BW** of 0.1 rad/sec.

SOLUTION: Consider the first shunt branch with capacitance

$$C_1 = 1.3013 \text{ ; F} \tag{9.6}$$

Then, using equations (9.5), $\omega_0 = 1$, and BW = 0.1, we have

$$L_1' = 0.076846 \text{ H}, \qquad C_1' = 13.013 \text{ F} \tag{9.7}$$

Thus, L_1' and C_1' form a parallel resonant circuit in the first shunt arm of the bandpass ladder circuit in Fig. 9.3(b). In a similar way, the $L_2(= 1.5559 \text{ H})$ element of the lowpass prototype is used in equations (9.3) to obtain

$$L_2' = 15.559 \text{ H}, \qquad C_2' = 0.064271 \text{ F} \tag{9.8}$$

where L_2' and C_2' form a series resonant circuit in the first series branch of the LC bandpass filter in Fig. 9.3(b). This process is continued until all series and shunt LC elements are obtained as shown in Fig. 9.3(b). The passband response $M(\omega)$ of this filter is shown in Fig. 9.3(c), from which it is noted that the passband extends from approximately 0.95 to 1.05 and that the ripple is 0.1 db, as expected.

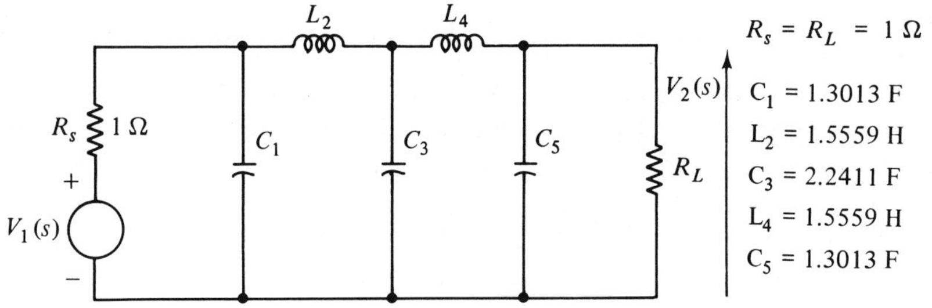

(a) Fifth-order Chebychev lowpass prototype: $\omega_0 = 1$ rad/sec, passband ripple = 0.1 dB.

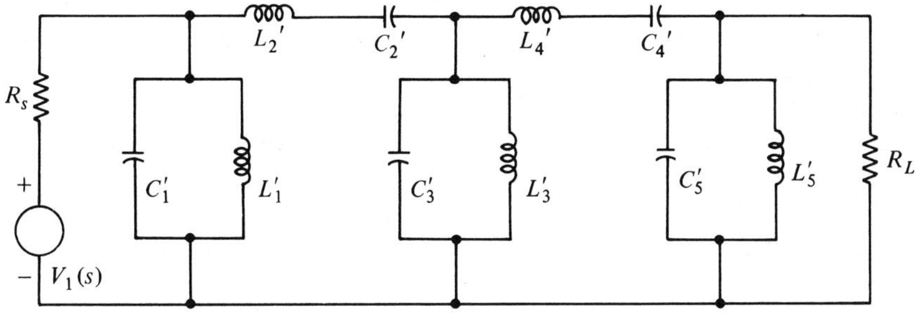

$C_1' = C_5' = 13.013$, $C_2' = C_4' = 0.0642714$, $C_3' = 22.411$

$L_1' = L_5' = 0.076846$, $L_2' = L_4' = 15.559$, $L_3' = 0.0446209$

(b) Tenth-order Chebychev bandpass design; center frequency $\omega_0 = 1$ rad/sec; bandwidth BW = 0.1 rad/sec.

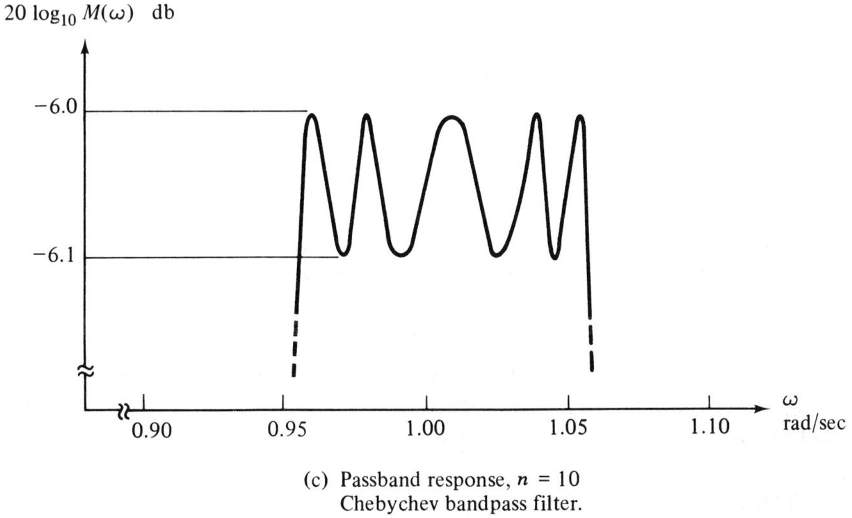

(c) Passband response, $n = 10$ Chebychev bandpass filter.

FIGURE 9.3

The reader may practice the design of bandpass type C filters by answering Problem 9.1. The type D bandpass filter shown in Table 9-1 is characterized by the fact that all inductance elements are grounded; a typical response is shown in Fig. 9.1(d).

Finally, the type E bandpass filter (Table 9-1) is obtained if one applies the transformation of equation (9.1) to the complete type A lowpass filter also shown in Table 9.-1. The typical response $M(\omega)$ of this bandpass filter is shown in Fig. 9.1(e) and is characterized by highly selective transition regions, which are due to the fact that each shunt branch ($L_2C_2L_3C_3$, for example) produces *two zeros of transmission*: one above and one below the passband. Unfortunately, this filter uses twice as many inductance elements as the type C or type D filters of equivalent order; furthermore, two-thirds of these inductance elements are *ungrounded*, which presents practical problems insofar as active inductance simulation is concerned.

A highpass LC filter may be designed from a lowpass LC filter by means of the lowpass-to-highpass transformation

$$\boxed{s \rightarrow \frac{\omega_{CL}\omega_{CH}}{s}} \quad \text{LOWPASS-TO-HIGHPASS TRANSFORMATION} \tag{9.9}$$

where ω_{CL} is the cut-off frequency of the lowpass filter and ω_{CH} that of the highpass filter. Therefore, a capacitance C_i transforms to an inductance L_i' and an inductance L_i to a capacitance C_i' as follows:

$$C_i' = \frac{1}{\omega_{CL}\omega_{CH}L_i} \tag{9.10}$$

$$L_i' = \frac{1}{\omega_{CL}\omega_{CH}C_i} \tag{9.11}$$

These results are summarized in Fig. 9.2(b).

9-2.1 The Effect of Finite *LC*-Quality Factors on the Magnitude Frequency Response *M*(ω)

A useful result from classical LC-filter theory which we present here without proof provides a good estimate of the deviation $\Delta M(\omega)$ of the magnitude frequency response $M(\omega)$ that is caused by the finite Q-factors, $Q_l(\omega)$ and $Q_c(\omega)$, of the LC elements. The result is that

$$\frac{\Delta M(\omega)}{M(\omega)} \approx -\frac{1}{2}\left(\frac{1}{Q_l(\omega)} + \frac{1}{Q_c(\omega)}\right)\omega T(\omega) + \frac{1}{2}\left(\frac{1}{Q_l(\omega)} - \frac{1}{Q_c(\omega)}\right)|\rho_1 + \rho_2| \tag{9.12}$$

where all inductors have the same Q-factor $Q_l(\omega)$, all capacitors have the same Q-factor $Q_c(\omega)$, $\tau(\omega)$ is the group delay (defined in Problem 6.32), $\rho_1(j\omega)$ is the reflection coefficient at the input port, and $\rho_2(j\omega)$ is the reflection coefficient at the output port. For high-quality "brick-wall"-type functions $M(\omega)$, the group delay $\tau(\omega)$ is a widely varying function of ω which may, of course, be uniquely determined from the transform transfer function $H(s)$. For almost all practical cases of interest, the term containing $\tau(\omega)$ in equation (9.12) is dominant almost everywhere in the passband, transition region and stopband because, over large regions of ω,

$$\omega|\tau(\omega)| \gg 1 \tag{9.13}$$

whereas the reflection coefficients are bounded by

$$0 \leqslant |\rho_{1,2}(j\omega)| \leqslant 1 \tag{9.14}$$

With the assumptions in equations (9.13) and (9.14), equation (9.12) is simplified to

$$\frac{\Delta M(\omega)}{M(\omega)} \approx -\frac{1}{2}\left[\frac{1}{Q_l(\omega)} + \frac{1}{Q_c(\omega)}\right]\omega\tau(\omega) \tag{9.15}$$

APPROXIMATE GAIN ERROR DUE TO FINITE Q-FACTORS

This result is most helpful for estimating passband errors of *active* ladder filters due to the finite gain bandwidth products of the OP AMPs.

Assuming a typical group delay function $\tau(\omega)$ as shown in Fig. 9.4(a) for a lowpass "flat-passband" LC filter with cut-off frequency ω_0, the group delay has a maximum of approximately 5 seconds near the band-edge; if all the inductors have Q-factors of 50 and all the capacitors have Q-factors of 200 then, by equation (9.15),

$$\frac{\Delta M(\omega)}{M(\omega)} \approx -\frac{1}{2}\left(\frac{1}{50} + \frac{1}{200}\right)\omega\tau(\omega) \tag{9.16}$$

leading to the resultant curve for $\Delta M(\omega)/M(\omega)$ shown in Fig. 9.4(b); the largest fractional deviation of $M(\omega)$ is $-5/80$ and occurs at the frequency where the group delay is a maximum; that is where $\omega \approx \omega_0 = 1$.

9-3 INDUCTANCE SIMULATION USING TRANSCONDUCTANCE-TYPE GYRATORS OR POSITIVE IMMITTANCE INVERTERS

A straightforward approach to the simulation of LC filters is to simply replace each inductance element L by an RC-active circuit that possesses a

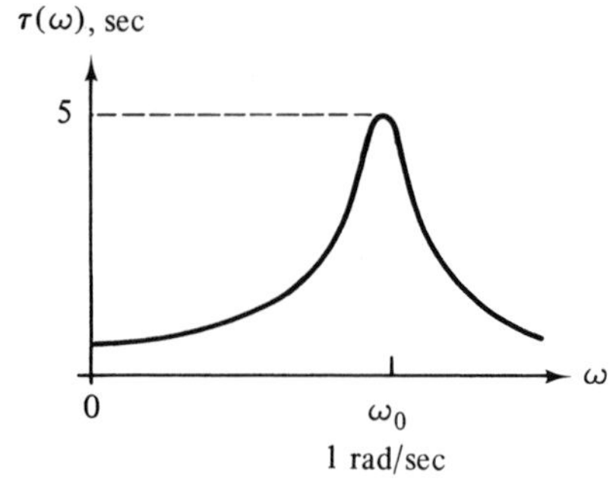

(a)

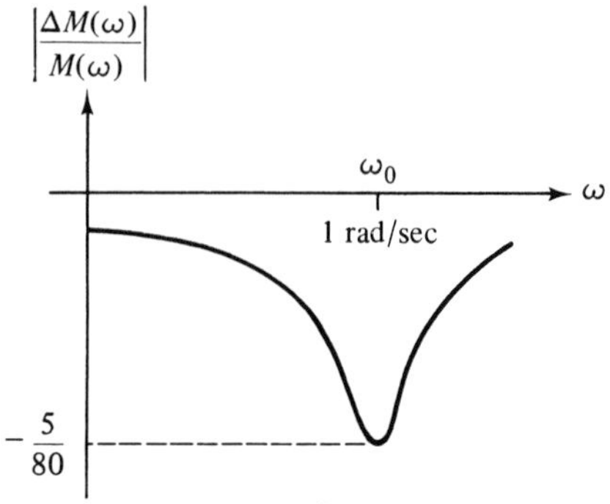

(b)

FIGURE 9.4 *Relationship between group delay $\tau(\omega)$ and fractional gain error $\frac{\Delta M(\omega)}{M(\omega)}$; $Q_c = 200$, $Q_l = 50$.*

driving-point immittance approximating that of an inductance over the frequency range of interest. In this section, we consider the use of capacitively terminated gyrators and PIIs for this purpose. (The reader may refer to chapters 4 and 5 for detailed descriptions of transistor and OP AMP circuits that realize RC-active gyrator and PII/GIC circuits.) Recall that a PII is characterized by the transmission matrix representation in Table 2-5; that is,

$$\begin{bmatrix} V_1 \\ I_1 \end{bmatrix} = \begin{bmatrix} 0 & \pm r_1 \\ \pm r_2^{-1} & 0 \end{bmatrix} \begin{bmatrix} V_2 \\ -I_2 \end{bmatrix} \quad \text{PII TRANSMISSION MATRIX} \quad (9.17)$$

which may be implemented by means of two voltage-controlled current sources (VCCS) with transconductances r_1^{-1} and $-r_2^{-1}$, as shown in Fig. 9.5(a). The PII circuit described in section 4.6 is an excellent example of a corresponding transistor circuit realization; the two VCCS circuits are

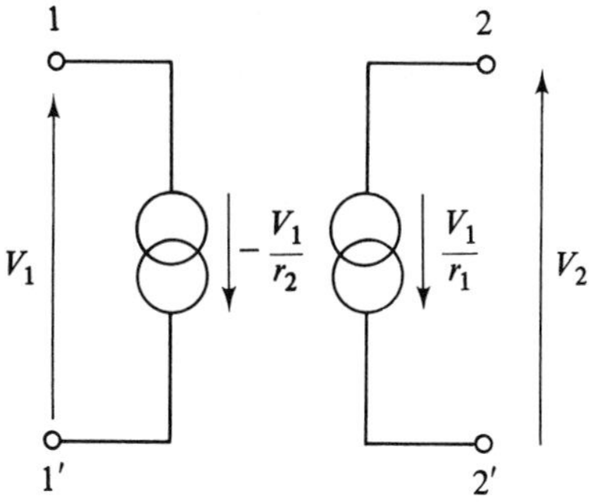

$$T = \begin{bmatrix} 0 & -r_1 \\ -r_2^{-1} & 0 \end{bmatrix}$$

NB. Gyrator iff $r_1 = r_2$.

(a) VCCS implementation of a PII-gyrator.

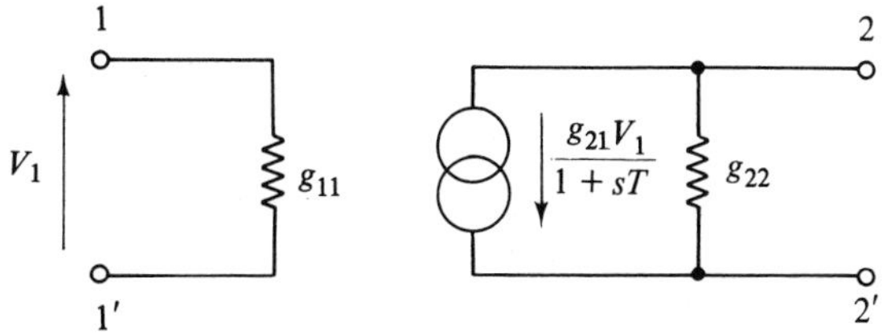

(b) Nonideal VCCS model.

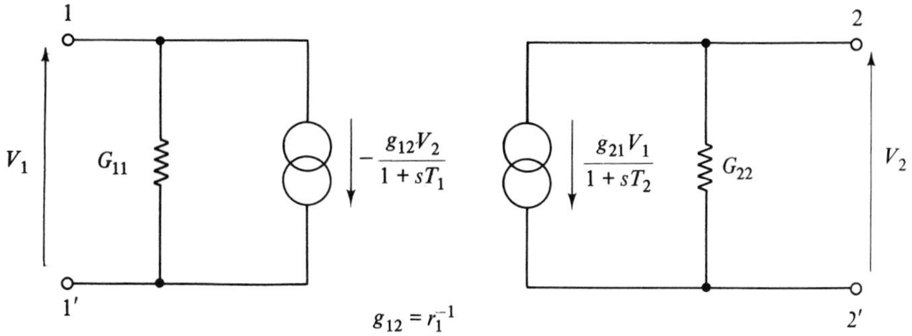

$g_{12} = r_1^{-1}$

$g_{21} = r_2^{-1}$

(c) Nonideal PII-gyrator model.

FIGURE 9.5

usually described as *transconductance amplifiers*. The PII circuit shown in Fig. 4.19 may be used to realize an ungrounded (that is, fully floating) inductance by means of a grounded capacitance termination at port 2. Equation (4.53) provides the user with an estimate of the accuracy with which the magnitude of the inductance may be realized at low frequencies. We now derive the general performance limitations of this type of PII-gyrator.

9-3.1 Performance Limitations of Transconductance-Type Gyrators-PIIs

A first-order analysis of the nonideal performance of this type of PII-gyrator requires that the nonideal nature of the transconductance (VCCS) amplifiers be taken into account. Ideally, these amplifiers have infinite input and output impedances and infinite bandwidth. In practice, the input and output impedances are finite and approximately real and the frequency response may be assumed to be "dominant pole." Thus, the transform model of Fig. 9.5(b) is used for the transconductance amplifier, where $-1/T$ is the location of the dominant pole. The parallel connection of two of these nonideal VCCS models, as required by Fig. 9.5(a), leads to the model shown in Fig. 9.5(c) for the PII-gyrator. The conductance G_{22} models the effect of the finite output impedance of the VCCS $g_{21}V_1/(1 + sT_2)$ in parallel with the finite impedance of VCCS $-g_{12}V_2/(1 + sT_1)$. The conductance G_{11} models a similar finite input impedance at port 2. For circuits of the type shown in Fig. 4.19, G_{11}^{-1} and G_{22}^{-1} are usually greater than 500 kΩ.

Let port 2 of Fig. 9.5(c) be terminated in a capacitance C_2; then straightforward analysis (see Problem 9-3) provides the following expression for the driving-point transform admittance $Y_{ld}(s)$ at port 1:

$$Y_{1d}(s) = \frac{G_{11}(sC_2 + G_{22})(1 + sT_1)(1 + sT_2) + g_{12}g_{21}}{(sC_2 + G_{22})(1 + sT_1)(1 + sT_2)} \tag{9.18}$$

Under ideal conditions, we have

$$G_{11} = G_{22} = T_1 = T_2 = 0$$

and then, from equation (9.18), we have

$$\underset{\text{IDEAL}}{Y_{1d}(s) = \frac{g_{12}g_{21}}{sC_2}} \qquad \text{IDEAL DRIVING-POINT ADMITTANCE} \tag{9.19}$$

corresponding to an ideal active inductance L_{eq} such that

$$\boxed{L_{eq} = \frac{C_2}{g_{12}g_{21}}} \quad \text{IDEAL INDUCTANCE} \tag{9.20}$$

Q-factor and stability: The impedance $Y_{1d}(j\omega)$ is obtained from equation (9.18) by letting $s = j\omega$; thus

$$Y_{1d}(j\omega) = G_{11} + \frac{g_{12}g_{21}}{(j\omega C_2 + G_{22})(1 + j\omega T_2)(1 + j\omega T_1)} \tag{9.21}$$

Now, it is assumed that G_{22}, T_1, and T_2 are sufficiently small that, over the frequency range of interest,

$$\omega \gg \frac{G_{22}}{C_2}, \qquad \omega \ll \frac{1}{T_2}, \qquad \omega \ll \frac{1}{T_1} \tag{9.22}$$

so that equation (9.21) leads, after some manipulation, to

$$Y_{1d}(j\omega) \approx \left[G_{11} + \frac{g_{12}g_{21}G_{22}}{\omega^2 C_2^2} - \frac{g_{12}g_{21}(T_1 + T_2)}{C_2} \right] + \left[\frac{g_{12}g_{21}}{j\omega C_2} \right] \tag{9.23}$$

This approximate expression for $Y_{1d}(j\omega)$ may be used to estimate the nonideal performance of the active inductance. The first square-bracketed term describes the (ideally-zero) real part of the driving-point admittance. The Q-factor $Q_l(\omega)$ of an inductance element is defined from equation (2.12) as

$$Q_l(\omega) \equiv \frac{\text{Im}\left[Z_{1d}(j\omega) \right]}{\text{Re}\left[Z_{1d}(j\omega) \right]} = \frac{\omega L}{r_s} \tag{9.24}$$

If we assume that

$$|Q_l(\omega)| \gg 1 \tag{9.25}$$

then it follows from equation (9.24) that

$$Q_l(\omega) \approx -\frac{\text{Im}\left[Y_{1d}(j\omega) \right]}{\text{Re}\left[Y_{1d}(j\omega) \right]} \tag{9.26}$$

Using equation (9.23) in equation (9.26), we obtain

$$\boxed{Q_l(\omega) \approx \frac{1}{\dfrac{G_{11}\omega C_2}{g_{12}g_{21}} + \dfrac{G_{22}}{\omega C_2} - \omega(T_1 + T_2)}} \qquad \text{Q-FACTOR} \qquad (9.27)$$

In terms of equation (9.20), we have

$$Q_l(\omega) \approx \left\{\left[G_{11}\omega L_{eq} + \frac{G_{22}}{\omega C_2}\right] - [\omega(T_1 + T_2)]\right\}^{-1} \qquad (9.28)$$

where L_{eq} is the ideal active inductance at port 1. It is convenient to describe the square-bracketed terms in equation (9.28) as follows:

$$Q_{Ll}(\omega) \equiv \left(G_{11}\omega L_{eq} + \frac{G_{22}}{\omega C_2}\right)^{-1} \qquad \text{LOW-FREQUENCY } Q\text{-FACTOR} \qquad (9.29)$$

and

$$Q_{Hl}(\omega) \equiv -[\omega(T_1 + T_2)]^{-1} \qquad \text{HIGH-FREQUENCY } Q\text{-FACTOR} \qquad (9.30)$$

so that, from equations (9.28), (9.29), and (9.30),

$$\boxed{Q_l^{-1}(\omega) \approx Q_{Ll}^{-1}(\omega) + Q_{Hl}^{-1}(\omega)} \qquad (9.31)$$

The Q-factor $Q_l(\omega)$ exhibits four distinct types of behavior, depending primarily on the magnitude of the inductance L_{eq} that is to be simulated:

(i) **Large Values of $\mathbf{L_{eq}}$:** $|Q_{Hl}| \gg Q_{Ll}(\omega)$ for all ω.

In this case, L_{eq} is sufficiently large-valued that

$$|Q_{Hl}| \gg Q_{Ll}(\omega) \qquad \text{for all } \omega \qquad (9.32)$$

implying that, from equation (9.28),

$$\boxed{L_{eq}G_{11} \gg T_1 + T_2} \qquad \text{CONDITION FOR INSIGNIFICANT ENHANCEMENT} \qquad (9.33)$$

In most practical designs, the value of L_{eq} required to satisfy this condition is so large that the condition only applies to LC filters with passbands below a few hundred hertz. The curves for $|Q_{Hl}(\omega)|$ and $Q_{Ll}(\omega)$ are shown in Fig. 9.6(a) for the case where inequality (9.33) is satisfied. The maxima of the $Q_l(\omega) \approx Q_{Ll}(\omega)$ curve is obtained by differentiating equation (9.29) with respect to ω and equating the derivative to zero in the usual way, so as to obtain

$$Q_{l\ \max} = \frac{1}{2}\left(\frac{C_2}{G_{11}G_{22}L_{eq}}\right)^{1/2} \tag{9.34}$$

and the frequency $\omega_{\max}$ at which $Q_l(\omega) = Q_{l\ \max}$ as

$$\omega_{\max} = \left(\frac{G_{22}}{G_{11}L_{eq}C_2}\right)^{1/2} \tag{9.35}$$

$Q_{L\ \max}$ and $\omega_{\max}$ are shown in Fig. 9.6(a). Since G_{11} and G_{22} are small-valued nonideal parameters that are beyond the precise control of the designer, it follows that the parameter C_2 is the only one which may be used to control $Q_{L\ \max}$ and $\omega_{\max}$; of course, L_{eq} is a fixed parameter of the passive LC filter that is being simulated. If the designer wishes to increase $\omega_{\max}$ by decreasing the capacitance C_2 at port 2, then a proportional decrease of $Q_{L\ \max}$ will occur. A design compromise must often be reached between the upper limit for $\omega_{\max}$ and the upper limit of $Q_{L\ \max}$. In fact, from equations (9.34) and (9.35), the *product* $Q_{L\ \max}\omega_{\max}$ is given by

$$Q_{L\ \max}\omega_{\max} = \frac{1}{2G_{11}L_{eq}} \qquad \text{FIGURE OF MERIT} \tag{9.36}$$

which is a figure-of-merit in the sense that it is fixed by the original LC filter element L_{eq} and the nonideal input conductance G_{11} of the PII-gyrator. The designer cannot influence this figure-of-merit by employing different values of g_{12}, g_{21}, and C_2 to realize L_{eq} in equation (9.20). On the contrary, it is necessary to decrease G_{11} by redesigning the PII-gyrator if one wishes to improve the $Q_{L\ \max}\omega_{\max}$ product.

In the case of lowpass and bandpass LC-filter realizations, the group delay $\tau(\omega)$ usually exhibits its peak magnitude at or very close to the (bandedge) cut-off frequency ω_0. Thus, by comparison of the curve for $Q_l(\omega)$ in Fig. 9.6(a) with equation (9.12) it follows that it is usually sensible to minimize the passband error $\Delta M(\omega)$ by simply ensuring that $Q_{L\ \max}$ and $\tau(\omega_0)$ occur at the same frequency; that is, C_2 is selected in equation (9.35) so that

$$\omega_{\max} \approx \omega_0 \tag{9.37}$$

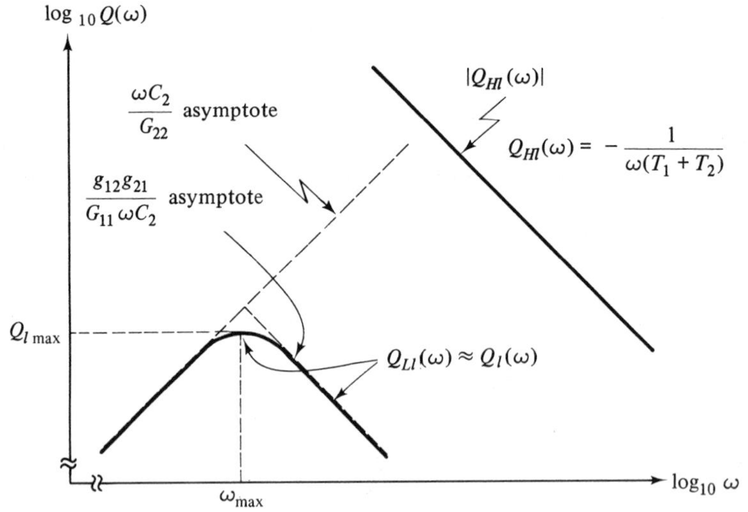

(a) Insignificant enhancement: $Q_I(\omega) \approx Q_{LI}(\omega)$.

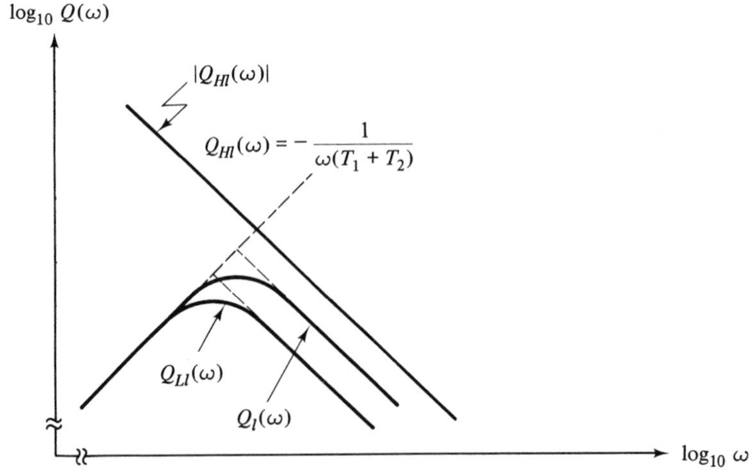

(b) Significant enhancement: $Q_I(\omega) \neq Q_{LI}(\omega)$.

FIGURE 9.6

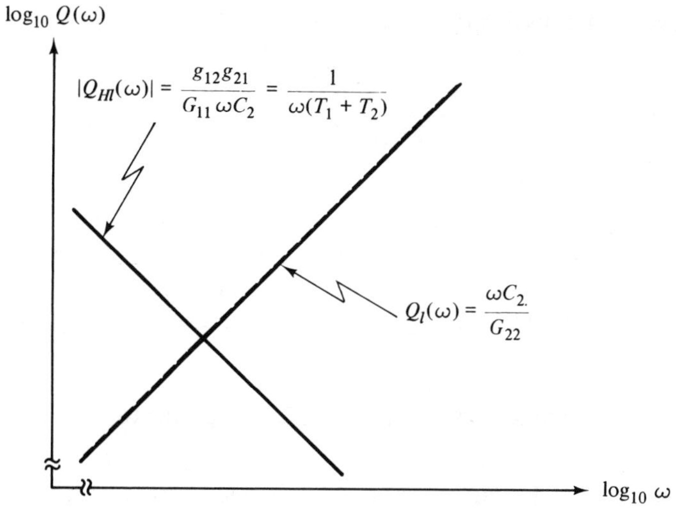

(c) Exact enhancement for $\omega \to \infty$: $Q_l(\omega) \approx \omega C_2 / G_{22}$.

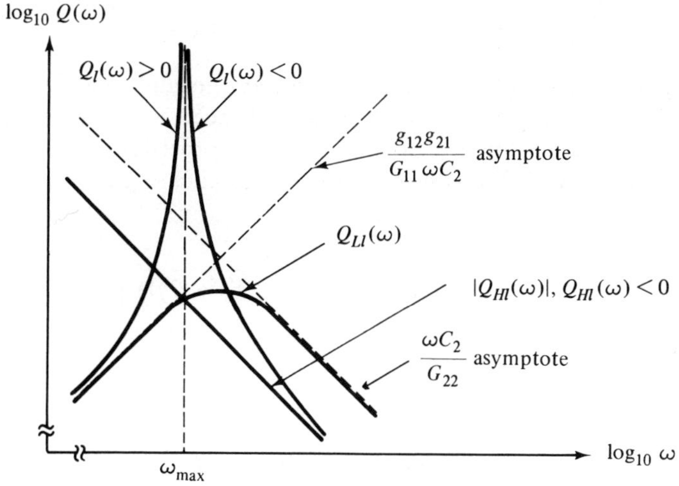

(d) Exact enhancement at a finite frequency $\omega_{\max}$.

FIGURE 9.6 *(Continued).*

implying, from equation (9.36), that

$$\boxed{Q_{l\ \max} \approx \frac{1}{2G_{11}\omega_0 L_{eq}}} \tag{9.38}$$

This is a useful guideline for the designer because it enables $Q_{l\ \max}$ to be determined rapidly in a particular LC simulation. For example, if G_{11}^{-1} for the PII-gyrator circuit is 500 kΩ and an inductor L_{eq} has a reactance of $+j\ 1000\ \Omega$ at the cut-off frequency of the filter then, from equation (9.38), it follows that $Q_{l\ \max} \approx 250$.

(ii) **Moderate Values of $\mathbf{L_{eq}}$:** $|Q_{Hl}(\omega)| > Q_{Ll}(\omega)$ for all ω.

When one considers active LC filters with passband frequencies somewhat above 1 kHz, the condition in equation (9.33) usually fails because L_{eq} is not sufficiently large. Thus, $Q_{Hl}(\omega)$ cannot be neglected, and we have the situation in Fig. 9.6(b), which is simply that

$$\boxed{|Q_{Hl}(\omega)| > Q_{Ll}(\omega) \quad \text{for all } \omega} \tag{9.39}$$

The high-frequency Q-factor $Q_{Hl}(\omega)$ is always negative and, consequently, causes the overall Q-factor $Q_l(\omega)$ to be enhanced to a higher value than $Q_{Ll}(\omega)$, as shown in Fig. 9.6(b). Following the procedure of differentiating $Q_l(\omega)$ in equation (9.27) with respect to ω and equating to zero, we have

$$\left.\begin{aligned} Q_{l\ \max} &= \frac{1}{2}\left[\frac{C_2}{G_{22}(G_{11}L_{eq} - (T_1 + T_2))}\right]^{1/2} \\ &\text{and} \\ \omega_{\max} &= \left[\frac{G_{22}}{C_2(G_{11}L_{eq} - (T_1 + T_2))}\right]^{1/2} \end{aligned}\right\} \tag{9.40}$$

so that

$$\boxed{Q_{l\ \max}\omega_{\max} = \frac{1}{2(G_{11}L_{eq} - (T_1 + T_2))}} \tag{9.41}$$

where proof is requested in Problem 9.7. Thus, realizing a smaller value of inductance L_{eq} will increase both $Q_{l\ max}$ and ω_{max}. The general "shape" of the $Q_l(\omega)$ curve is, however, the same as $Q_{Ll}(\omega)$; it simply is enhanced at the higher frequencies.

(iii) **Critical $\mathbf{L_{eq}}$ for Infinite ω_{max}.**

It follows from equation (9.41) that if L_{eq} is sufficiently low-valued then there exists some combination of L_{eq}, G_{11}, T_1, and T_2 such that

$$\boxed{G_{11}L_{eq} = T_1 + T_2} \quad \text{CONDITION FOR } \omega_{max} = \infty \tag{9.42}$$

then, $Q_{L\ max} = \omega_{max} = \infty$ and we have, from equation (9.28),

$$\boxed{Q_l(\omega) \approx \frac{\omega C_2}{G_{22}}} \tag{9.43}$$

which is clearly the condition to strive for if a high Q-factor is required over a *wide band at high frequencies*, as shown in Fig. 9.6(c). Of course, the condition in equation (9.43) cannot be chosen *accurately* because neither G_{11} or $(T_1 + T_2)$ are known with any real precision.

(iv) **Small Values of $\mathbf{L_{eq}}$:** $Q_l(\omega_{max}) = \infty$, $\omega_{max} < \infty$.

For sufficiently small L_{eq}, corresponding to an active-LC filter design at a sufficiently high frequency,

$$\boxed{G_{11}L_{eq} < T_1 + T_2} \quad \begin{array}{l}\text{CONDITION FOR}\\ \text{INFINITE } Q_l(\omega)\\ \text{AT A FINITE FREQUENCY}\end{array} \tag{9.44}$$

This condition is often satisfied for active-LC filters with passbands in the region above a few kilohertz. The situation is described in Fig. 9.6(d), from which it is observed that there must exist some frequency ω_{max} at which $Q_{Ll}(\omega) = |Q_{Hl}(\omega)|$, resulting in the $Q_l(\omega)$ curve shown in the diagram, where

$$Q_l(\omega_{max}) = \infty \tag{9.45}$$

Solving equation (9.28) for the value of ω corresponding to $Q_l(\omega_{max}) = \infty$, we have

$$\omega_{\max} = \left[\frac{G_{22}}{(T_1 + T_2 - G_{11}L_{eq})C_2} \right]^{1/2} \tag{9.46}$$

Note from Fig. 9.6(d) that $Q_l(\omega) > 0$ for $\omega < \omega_{\max}$, and $Q_l(\omega) < 0$ for $\omega > \omega_{\max}$. The designer may again select C_2 in equation (9.46) so that $\omega_{\max}$ corresponds to the frequency ($\approx \omega_0$) where the group delay $\tau(\omega)$ peaks, thereby minimizing $\Delta M(\omega)$ in equation (9.15). This causes $Q_l(\omega)$ to exhibit its "infinite-magnitude" peak at or close to the edge of the passband.

(v) **Absolute High-Frequency Limitation:** $Q_l(\omega) \approx Q_{Hl}(\omega)$.

The upper frequency at which the transconductance-type PII-gyrator may be used is determined by the high-frequency Q-factor $Q_{Hl}(\omega)$; that is, when L_{eq} is so small that

$$G_{11}L_{eq} \ll T_1 + T_2 \qquad \text{HIGH-FREQUENCY CONDITION} \tag{9.46A}$$

so that, at high frequencies, equation (9.28) indicates that

$$Q_l(\omega) \approx Q_{Hl}(\omega) = -\frac{1}{\omega(T_1 + T_2)} \tag{9.47}$$

over the region of ω corresponding to $\omega \gg \omega_{\max}$ in Fig. 9.6(d). Thus, in this region of ω, all inductances in the corresponding LC filter have approximately the same Q-factor [equation (9.47)] and, furthermore, the Q-factors of the capacitors are usually sufficiently high that

$$|Q_l(\omega)| \ll Q_c(\omega) \tag{9.48}$$

Then, equations (9.15), (9.48), and (9.30) lead to the simple result that

$$\left| \frac{\Delta M(\omega)}{M(\omega)} \right| \approx \frac{1}{2}\omega^2(T_1 + T_2)\tau(\omega) \qquad \text{PASSBAND ERROR DUE TO BANDWIDTH LIMITATIONS} \tag{9.49}$$

This result is useful for determining the bandwidth required of the transconductance amplifiers. Since $|\Delta M(\omega)/M(\omega)|$ is proportional to group delay $\tau(\omega)$, it follows that the designer must select sufficiently large bandwidths $1/T_{1,2}$ to satisfy the allowable deviation $\Delta M(\omega)$ in the pass-band. Of course, $\tau(\omega)$ is a function of the specified transfer function and is not at the designer's direct control.

EXAMPLE 9.2: An active inductance L_{eq} is implemented using a transconductance-type PII-gyrator, where L_{eq} is used in a particular lowpass *LC* filter for which the reactance $j\omega_0 L_{eq}$ is equal to $j10^4$ ohms at the cut-off frequency ω_0. The nonideal PII-gyrator parameters are

$$G_{11} \equiv G_{22} \equiv \frac{1}{2} \times 10^{-6}\,\text{ohm}^{-1}, \qquad T_1 \equiv T_2 \equiv \frac{1}{4\pi}\,\mu\text{sec} \qquad (9.50)$$

PROBLEM: The problem is to find $Q_{L\,\max}$ for $\omega_{\max} \equiv \omega_0 \equiv 10^3$ rad/sec. and to determine the corresponding values for the terminating capicitance C_2 and transconductance product $g_{12}g_{21}$. Repeat this problem for $\omega_{\max} \equiv \omega_0 \equiv 10^4$ rad/sec. and $\omega_{\max} \equiv \omega_0 \equiv 10^5$ rad/sec.

SOLUTION: (a) **Cut-off frequency $\omega_0 \equiv 10^3$ rad/sec; $L_{eq} = 10$H**

This is a low-frequency design. We have

$$G_{11}L_{eq} = \frac{1}{2} \times 10^{-5}\,\text{sec}, \qquad T_1 + T_2 = \frac{1}{2\pi} \times 10^{-6}\,\text{sec}$$

from which it follows that inequality (9.33) is valid. Therefore,

$$Q_l(\omega) \approx Q_{Ll}(\omega)$$

and, from equation (9.38),

$$Q_{L\,\max} \approx \frac{1}{2G_{11}\omega_0 L_{eq}}$$

Substituting for G_{11}, L_{eq}, and ω_0, we obtain

$$Q_{L\,\max} \approx 100 \quad \text{at } \omega_{\max},\ \omega_0 = 10^3\,\text{rad/sec}$$

and $Q_l(\omega)$ exhibits the behavior sketched in Fig. 9.6(a). It follows from equation (9.34) that

$$C_2 = 0.1\,\mu\text{F}$$

and from equation (9.20) that

$$g_{12}g_{21} = 10^{-8}\,\text{ohm}^{-2}$$

(b) Cut-Off Frequency $\omega_0 \equiv 10^4$ rad/sec; $L_{eq} = 1$ H.

This is a medium-frequency design where inequality (9.33) is no longer valid and this may be checked by substituting for G_{11}, L_{eq}, $T_{1,2}$ in (9.33). Thus, we use the result in equation (9.41) to obtain

$$Q_{L\ max} = 146.69 \quad \text{at} \quad \omega_{max}, \omega_0 = 10^4 \text{ rad/sec}$$

corresponding to the $Q_l(\omega)$ curve shown in Fig. 9.6(b). From equations (9.40) and (9.20) we obtain

$$C_2 = 14.668 \text{ nF}, \quad g_{12}g_{21} = 1.4668 \times 10^{-8} \text{ ohm}^{-2}$$

(c) Cut-Off Frequency $\omega_0 \equiv 10^5$ rad/sec; $L_{eq} = 0.1$ H.

This is a high-frequency design where by direct substitution of numerical values we obtain

$$L_{eq}G_{11} = \frac{1}{20} \times 10^{-6} \text{ sec}, \quad T_1 + T_2 = \frac{1}{2\pi} \times 10^{-6} \text{ sec} \tag{9.51}$$

so that inequality (9.44) is valid; therefore, it is possible to design so that $Q_l(\omega_0)$ is infinite. Substituting equations (9.51), $\omega_0 \equiv \omega_{max} \equiv 10^5$ rad/sec and $G_{22} \equiv \frac{1}{2} \times 10^{-6}$ ohm^{-1} into equation (9.46), we have

$$C_2 = 458 \text{ pF}$$

corresponding, in equation (9.20), to

$$g_{12}g_{21} = 4.58 \times 10^{-8} \text{ ohm}^{-2}$$

and

$$Q_l(\omega_0) = \infty \quad \text{at } \omega_0 \equiv 10^5 \text{ rad/sec}$$

which corresponds to the $Q_l(\omega)$ curve shown in Fig. 9.6(d).

9-3.2 A Lowpass Chebychev Filter Design Study

In this subsection, the seventh-order Chebychev lowpass LC filter shown in Fig. 7.17(a) is used to design an active-LC filter with a cut-off frequency of 10^5 rad/sec (15.9 kHz) kΩ and between resistive terminations of 10 kΩ. The corresponding denormalized LC filter is shown in Fig. 9.7(a); it is required that each of the ungrounded inductances L_3, L_5, L_7 be simulated by using a PII-gyrator circuit of the type shown in Fig. 4.19 and analyzed in section 9-3.1.

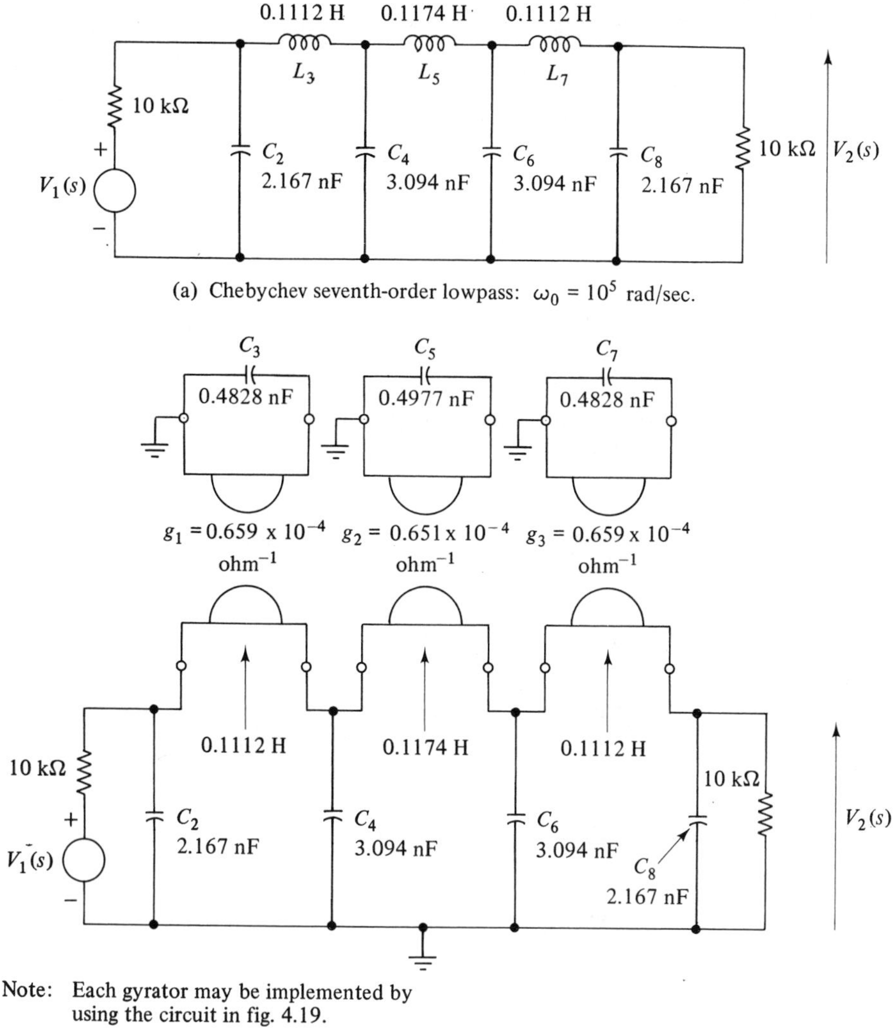

(a) Chebychev seventh-order lowpass: $\omega_0 = 10^5$ rad/sec.

(b) Corresponding active-LC filter; inductive Q-factor optimized to peak near the cut-off frequency.

FIGURE 9.7

It is assumed that equation (9.50) describes the approximate impedance and bandwidth limitations of each of the three PII-gyrator circuits. Further, for simplicity, the transconductances g_{21}, g_{12} are chosen equal so that we use gyrators rather than PIIs. The capacitances C_3, C_5, C_7 and gyrator transconductances g_1, g_2, g_3 are now derived so that the nonideal Q-factors $Q_{l1}(\omega)$, $Q_{l2}(\omega)$, $Q_{l3}(\omega)$ peak at approximately the cut-off frequency ω_0 so as to minimize the passband error.

Active inductances L_3 and L_7 give

$$G_{11}L_{3,7} = \left(\tfrac{1}{2} \times 10^{-6}\right)(0.1112)\ \text{sec} = 0.0556\ \mu\text{sec}$$

where

$$T_1 = T_2 = \frac{1}{4\pi}\ \mu\text{sec} = 0.0796\ \mu\text{sec}$$

so that inequality (9.44) is clearly valid and the design may, therefore, be performed so that $Q_{l3,7}$ peak at infinity at or close to the cut-off frequency. Thus, selecting $\omega_{\max} \equiv \omega_0 \equiv 10^5$ rad/sec and substituting for T_1, T_2, G_{22}, G_{11}, and $L_{3,7}$ in equation (9.46), we find that corresponding termination capacitances $C_{3,7}$ are

$$C_{3,7} = \frac{G_{22}}{[(T_1 + T_2) - G_{11}L_{3,7}]\omega_{\max}^2} = 0.4828\ \text{nF}$$

and, by equation (9.20), the transconductances $g_{1,3}$ of gyrators 1 and 3 are

$$g_{1,3} = \frac{C_{3,7}}{L_{3,7}} = \left[\frac{0.4828 \times 10^{-9}}{0.1112}\right]^{1/2} = 0.659 \times 10^{-4}\ \text{ohm}^{-1}$$

The Q-factors $Q_{l3,7}$ should then peak to infinity in the neighborhood of the cut-off frequency ω_0 as required. In a similar way, it may be shown that

$$C_5 = 0.4977\ \text{nF}, \qquad g_2 = 0.6511 \times 10^{-4}\ \text{ohm}^{-1}$$

for the second gyrator of Fig. 9.7(b). In this particular design, the transconductances $g_{1,2,3}$ are close to each other in value; this is *not* the case if the inductances in the original LC prototype are widely different in value. In such cases, it often becomes particularly important to select the transconductances carefully to maximize the Q-factor at the bandedge ω_0.

Practical circuits of the type described here have been implemented successfully in microelectronic form. They have the disadvantage that the basic gyrator circuit (Fig. 4.19) is not widely available as a low-cost active microelectronic device, so for many users it may prove economically prohibitive to custom-manufacture the microelectronic transconductance amplifiers of the required quality. In the remaining sections of this chapter, the low-cost widely available operational amplifier is the basic building-block.

9-4 GROUNDED-INDUCTANCE SIMULATION USING GICs

The coupled-GIC circuit may be used to implement a grounded active inductance by means of either the type 1 circuit shown in Table 5-6 or the type 2 circuit in Table 5-7. In this section, the use of the preferred type 2 GIC circuit in Table 5-7 is described as a means of implementing ultra-high-quality active ladder filters. The major restriction is, of course, that the active inductances must be grounded if the type 2 circuit of Table 5-7 is to be used; this restricts the *LC* topologies to the highpass type B and bandpass type D circuits shown in Table 9-1.

It is recalled from study of Table 5-7 that by maintenance of the enhancement condition $G_3 = G_4$ it is possible to ensure that the Q-factor $Q_l(\omega)$ of the active inductance is large-valued over a wide frequency range. For example, the expression in equation (5.119) for dissipation factor $d_l(\omega)$ may be simplified if the term $2\omega\tau/A_0$ is negligible (as is usually the case), so that

$$\boxed{Q_l^{-1}(\omega) \equiv d_l(\omega) \approx \frac{2}{A_0}\left(\frac{1}{\omega\tau}\right) - \omega^3\frac{\tau}{B_1B_2}} \qquad (9.52)$$

which may be estimated by sketching asymptotes as shown in Fig. 9.8(a). The term $\omega^3\tau/B_1B_2$ is usually the limiting factor in applications involving active-*LC* filters with passbands above a few kilohertz.

EXAMPLE 9.3: The type 2 inductance-simulation GIC circuit shown in Fig. 9.8(b) is used to realize an L_{eq} of 1 henry, where

$$G \equiv 10^{-4}\ \text{ohm}^{-1}, \quad C \equiv 10\ \text{nF}, \quad \tau = 10^{-4}\ \text{sec}$$

Assume

$$A_0 \approx 10^5, \quad B_1 \approx B_2 \approx 2\pi \times 10^6\ \text{rad/sec}$$

and use equation (9.52) to determine the frequency range over which $|Q_l(\omega)| > 100$.

SOLUTION: At high frequencies, equation (9.52) indicates that

$$Q_l^{-1}(\omega) \equiv d_l(\omega) \approx -\omega^3\frac{\tau}{B_1B_2}$$

Solving this equation for $\tau = 10^{-4}$ sec, $B_{1,2} = 2\pi \times 10^6$ rad/sec, and $Q_l = -100$, we obtain

$$\underline{\underline{\omega \approx 158047\ \text{rad/sec}\ (25.15\ \text{kHz})}}$$

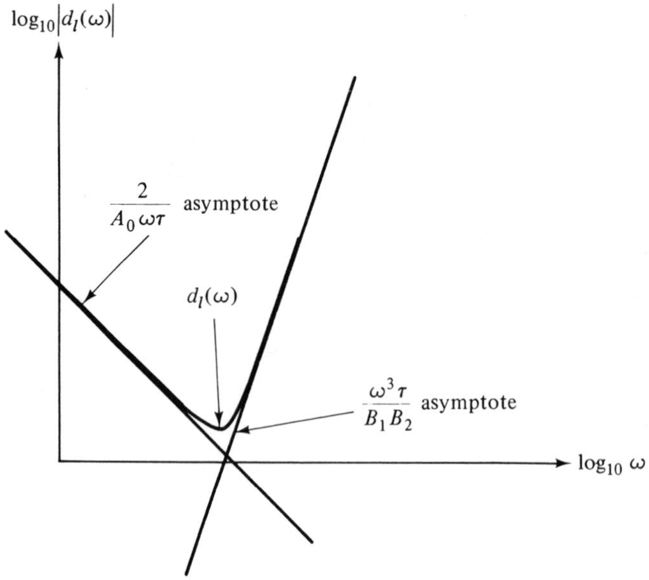

(a) Construction of asymptotes for $d_e(\omega)$; $G_3 = G_4$.

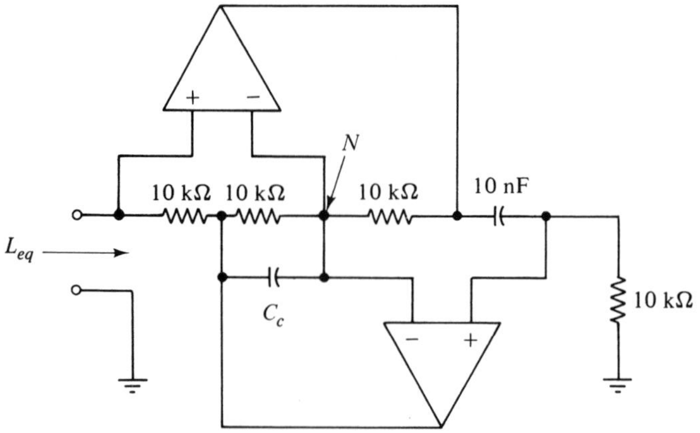

(b) A typical type 2 inductance-simulation GIC circuit.

FIGURE 9.8

as the upper frequency at which $|Q_l|$ is 100. Then, at low frequencies

$$Q_l^{-1}(\omega) \equiv d_l(\omega) \approx \frac{2}{A_0 \omega \tau}$$

Solving this equation for $Q_l(\omega) = 100$, $A_0 = 10^5$, and $\tau = 10^{-4}$ sec, we have

$$\underline{\underline{\omega \approx 20 \text{ rad/sec (3.18 Hz)}}}$$

as the lower frequency at which $Q_l = 100$. Thus, the circuit is capable of realizing $|Q_l|$ in excess of 100 over the approximate frequency range 3.18 Hz to 25.15 kHz. [The reader should note that this result only models one nonideal effect; that is, the nonideal frequency response of the operational amplifiers. Other parameters, such as finite input-output immittances of the operational amplifiers may further deteriorate $Q_l(\omega)$.]

9-4.1 A Narrowband High-frequency Simulated-Inductance Design Example Using GICs

In this subsection, a design example is described which illustrates the use of the type 2 simulated-inductance GIC circuit in a stringent application. The *LC* filter is of type D shown in Table 9.1 and is illustrated in Fig. 9.9 where each grounded inductance is realized with the type 2 GIC circuit. There are 6 grounded active inductors and 15 capacitors used to realize a highly selective bandpass response $M(\omega)$ with a passband from 20.3 kHz to 23.9 kHz, as shown in Fig. 9.10(a). The transfer function is of order 12 and the Q of the highest pole-pair is 178; there are four transmission zeros in the lower stopband and the passband ripple is ideally 0.03 db, as shown in Fig. 9.10(b). Passive *LC* filters of this type may be designed by using commerically available filter synthesis computer programs.

The measured passband response in Fig. 9.10(b) is obtained by using operational amplifiers for which $A_0 \simeq 10^5$ and $B \simeq 2\pi \times 10^6$ rad/sec. At the 20 kHz bandedge, the measured response "droops" below the theoretical response by about 0.5 db, and it may be shown [using equation (9.15)] and the group delay $\tau(\omega)$ in Fig. 9.10(c) that this amount of droop would be caused by effective-inductance Q-factors Q_l' of $+900$.

It is possible to almost exactly cancel out the abovementioned droop at the bandedge by simply modifying the type 2 GIC inductance-simulation circuit by means of a compensation capacitor C_c, as in Fig. 9.8(b), where

$$C_c = \frac{C}{Q_l'} \tag{9.53}$$

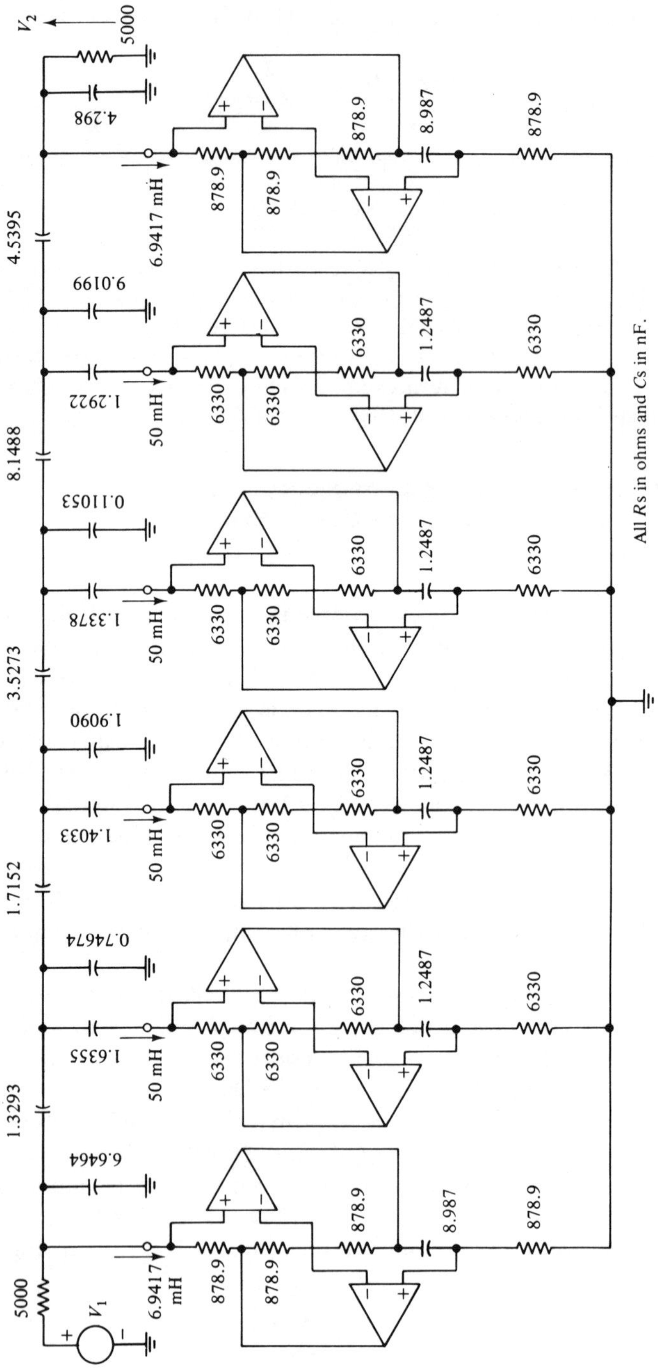

FIGURE 9.9 *20-24 kHz bandpass twelfth-order simulated inductance filter.*

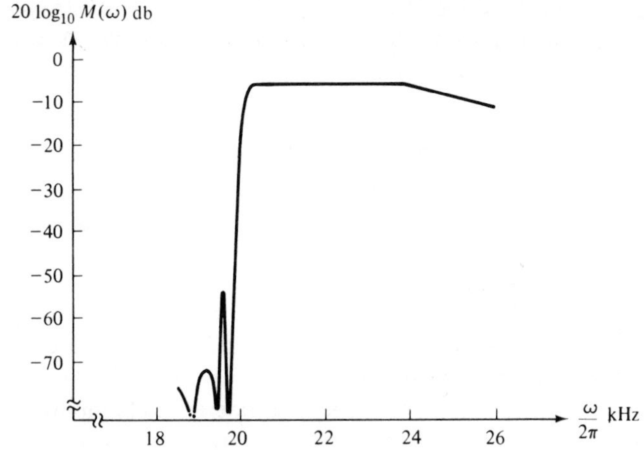

(a) Narrowband twelfth-order high-frequency active-filter response: passband 20–24 kHz.

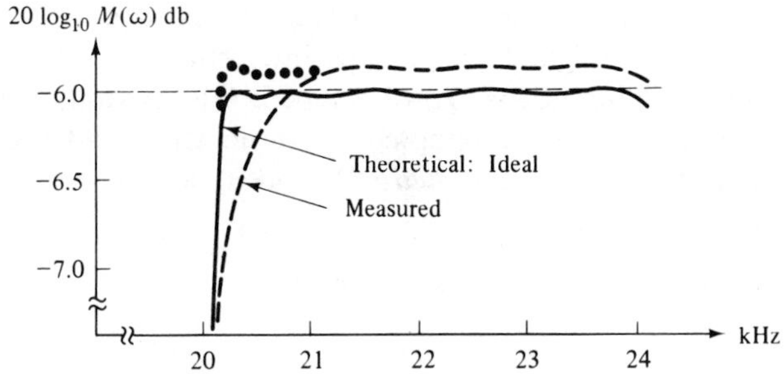

(b) Theoretical and measured passband responses.

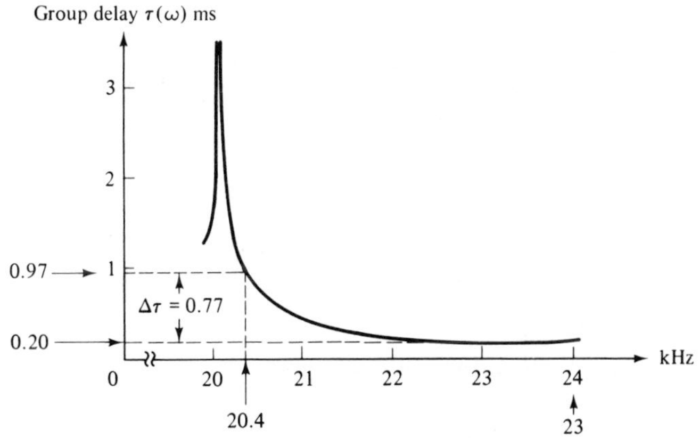

(c) Group delay in passband of filter in Fig. 9.9.

FIGURE 9.10

For this design, the value of C used to realize the 50 mH active inductance is 1248.7 pF so that, with $Q_l' = +900$ in equation (9.53), we have compensation capacitors given by

$$C_c = \frac{1248.7}{900} = 1.38 \text{ pF} \tag{9.54}$$

The measured effect of employing 1.5 pF compensating capacitors C_c is shown as black dots in Fig. 9.10(b) and confirms that the droop is indeed adequately corrected at the bandedge.

The practical version of Fig. 9.9 has been constructed by using R elements with temperature coefficients of ± 100 ppm/°C and C elements with temperature coefficients of -110 ± 30 ppm/°C. The GICs are realized by using Motorola 1458-type dual operational amplifiers. Temperature tests reveal that the response $M(\omega)$ varies by less than 0.1 db in the passband (and less than 2 db elsewhere) over the temperature range 25°C to 65°C.

This study illustrates the maximum quality that is currently available in RC-active filter design. It is perhaps interesting to note that the Q-enhancement condition ($G_3 = G_4$) of a type 2 GIC is vital in this design. Suppose a design had been attempted where no effort had been made to match B_1 and B_2 (by not using *dual* OP AMP's) so that, for example, we might have

$$B_2 = \tfrac{1}{2} B_1$$

and suppose that

$$G_3 = 2G_4 \tag{9.55}$$

Then, the terms containing $\left(\dfrac{G_3}{G_4} - 1\right)$ and $\left(1 - \dfrac{G_4}{G_3}\right)$ in equation (5.116) can no longer be assumed to be zero; they in fact *dominate* the overall expression, so that we have

$$d_l(\omega) \simeq \omega\left[\frac{1}{B_2}\left(\frac{G_3}{G_4} - 1\right) + \frac{1}{B_1}\left(1 - \frac{G_4}{G_3}\right)\right] \tag{9.56}$$

which combines with equations (9.55) to give

$$Q_l(\omega)^{-1} \equiv d_l(\omega) \approx \frac{5\omega}{2B_1}$$

so that, in our case, at 20.4 kHz we substitute $\omega = 2\pi \times 20.4 \times$

10^3 rad/sec and $B_1 \approx 2\pi \times 10^6$ rad/sec to obtain

$$Q_l(\omega) \approx 19.6 \tag{9.57}$$

which is far too low-valued to be of any use in realizing the filter. By using *dual* amplifiers and employing $G_4 = G_3$ the value of $|Q_l(\omega)|$ is enhanced in excess of $+1000$, thereby allowing the filter to be implemented successfully.

9-5 FREQUENCY-DEPENDENT NEGATIVE-RESISTANCE FILTERS

The frequency-dependent negative-resistance (FDNR) elements are defined in section 2-5.1, and GIC circuit implementations of FDNRs have been analyzed extensively in section 5-7. The reader may wish to review those sections before proceeding. In the following subsections, the basic concepts of FDNR filter implementation are explained.

9-5.1 Transfer Function Invariance Under Generalized Scaling of Branch Impedances

Let all the branches of a network be transform impedance-scaled by a multiplicative scaling function $K(s)$ as indicated diagrammatically in Fig. 9.11. The transform impedance $Z_i(s)$ in the diagram refers to the general ith branch. It may be obvious intuitively to the reader that *dimensionless* network functions, such as voltage and current transfer functions, are unaltered by this impedance-scaling operation, in which case the reader may wish to move on directly to statement (9.67). A formal proof is now given that *voltage* transfer functions are unaltered by the impedance-scaling function $K(s)$. Let the network shown in Fig. 9.11(a), prior to impedance-scaling, have L-loops with loop equations given by

$$\underset{L\times 1}{[\mathbf{V}]} = \underset{L\times L}{[\mathbf{Z}]}\ \underset{L\times 1}{[\mathbf{I}]} \tag{9.58}$$

The L-row column matrix [V] contains the voltage excitation generators and the L-row column matrix [I] contains the L-loop currents $I_1, I_2, \ldots, I_L$. Solving equation (9.58) by Cramer's rule, we obtain

$$I_j = \frac{\sum_{i=1}^{L} V_i \Delta_{ij}}{\Delta_z}, \qquad j = 1, 2, 3, \ldots, L \tag{9.59}$$

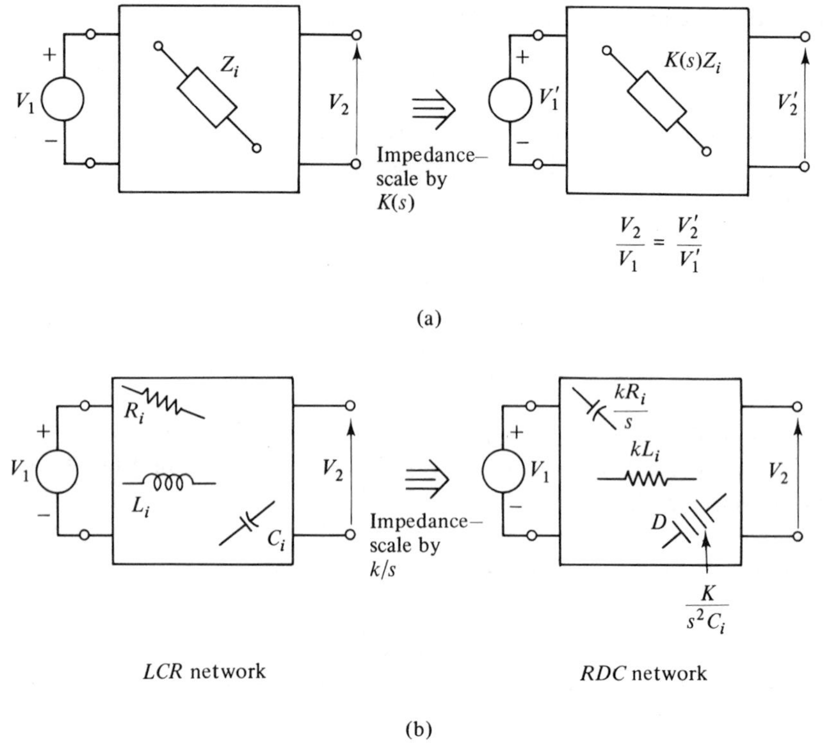

FIGURE 9.11 *Impedance-scaling.*

where I_j is the jth-loop current, Δ_z is the $(L \times L)$ determinant of the matrix $[\mathbf{Z}]$, Δ_{ij} is the cofactor of the ijth element of Δ_z, and V_i is the ith element of $[\mathbf{V}]$. The cofactor Δ_{ij} has dimension $(L - 1) \times (L - 1)$.

After impedance-scaling by $K(s)$, the resultant network is topologically unaltered; assuming the same excitation generators, it follows that $[\mathbf{V}]$ is unaltered. Therefore, the loop currents of the impedance-scaled network I_j' are

$$I_j' = \frac{\sum_{i=1}^{L} V_i \Delta_{ij}'}{\Delta_z'}, \qquad j = 1, 2, 3, \ldots, L \tag{9.60}$$

The next step is to obtain a simple relationship between Δ_z and Δ_z'. Each element z_{ij} of Δ_z is of the general form

$$z_{ij} = sL_{ij} + \frac{1}{sC_{ij}} + R_{ij} \tag{9.61}$$

so that each element of Δ_z' is of the general form

$$z_{ij}' = K(s)\left(sL_{ij} + \frac{1}{sC_{ij}} + R_{ij}\right) \tag{9.62}$$

or

$$z_{ij}' = K(s)z_{ij} \tag{9.63}$$

Since multiplying each element of a $L \times L$ determinant by $K(s)$ is equivalent to multiplying that determinant by $[K(s)]^L$, then it follows from equation (9.63) that

$$\Delta_z' = [K(s)]^L \Delta_z \tag{9.64}$$

and, by similar reasoning, that

$$\Delta_{ij}' = [K(s)]^{L-1} \Delta_{ij} \tag{9.65}$$

Equations (9.64), (9.65), (9.59), and (9.60) give

$$\boxed{I_j' = \frac{I_j(s)}{K(s)}}\quad , \qquad j = 1, 2, \ldots, L \tag{9.66}$$

That is, *impedance-scaling by $K(s)$ has the effect of dividing all transform loop currents by $K(s)$*. However, *since all branch impedances* are *multiplied by $K(s)$* as a result of impedance-scaling, it follows that *all branch voltages are unaltered by this impedance-scaling*. The *node* voltages are, therefore, also unaltered; thus,

IMPEDANCE-SCALING BY $K(s)$ DOES NOT ALTER ANY VOLTAGE TRANSFER FUNCTION $H(s)$.	(9.67)

A similar proof, using nodal equations and excitation current generators, may be given to show that *current transfer functions are also unaltered by impedance-scaling*. The result in equation (9.67) is illustrated in Fig. 9.11(a).

9-5.2 The $\frac{k}{s}$ Impedance Transformation

A useful impedance-scaling function $K(s)$ is

$$\boxed{K(s) \equiv \frac{k}{s}}\quad , \qquad k \text{ constant, rad/sec} \tag{9.68}$$

Inductive, resistive, and capacitive branches are transformed as follows:

Inductance,	sL_i	$\Rightarrow$	Resistance,	kL_i
Resistance,	R_i	$\Rightarrow$	Capacitive, branch	$\frac{kR_i}{s}$
Capacitive, branch	$\frac{1}{sC_i}$	$\Rightarrow$	FDNR branch,	$\frac{k}{s^2C_i}$

(9.69)

The network remains topologically identical, as shown in Fig. 9.11(b), and the voltage transfer function is unaltered. A passive *LCR* network is transformed to an *RC*-active interconnection of *R*, *D*, and *C* elements, where a *D* element has been defined in section 2-5.1 as an FDNR element. The FDNR branch described in equation (9.69) has transform impedance k/s^2C_i and, therefore, the value of D_i is simply C_i/k farad-seconds.

9-5.3 FDNR Lowpass Active Ladder Filters

An immediate application for the k/s impedance transformation is illustrated in Fig. 9.12(a) for the case of a lowpass *LCR* ladder filter. The inductance elements are all *ungrounded*, so it is not possible to use the type 1 or type 2 *grounded* inductance-simulation GIC circuits to simulate these elements. However, the k/s transformation conveniently transforms these ungrounded inductance branches sL_i to resistances kL_i; as a consequence, the grounded capacitance branches $1/sC_i$ must be realized as grounded *D* element branches $1/s^2D_i$ or k/s^2C_i.

Note in Fig. 9.12(a) that there are approximately twice as many ungrounded inductances in the lowpass *LCR* filter as there are grounded FDNR *D* elements in the corresponding FDNR filter. The number of active elements to be realized is, therefore, less by a factor of two as a result of the k/s impedance-scaling; furthermore, the active elements *are* realizable with a two-amplifier GIC circuit (see Table 5-9) to a high degree of accuracy.

The single-element sensitivity performance remains unaltered as a result of this k/s transformation. The dimensionless voltage transfer function $H(s)$ of the *LCR* filter is a function of the branch impedances and may be written as

$$H\left(sL_i, \frac{1}{sC_i}, R_i\right) \tag{9.70}$$

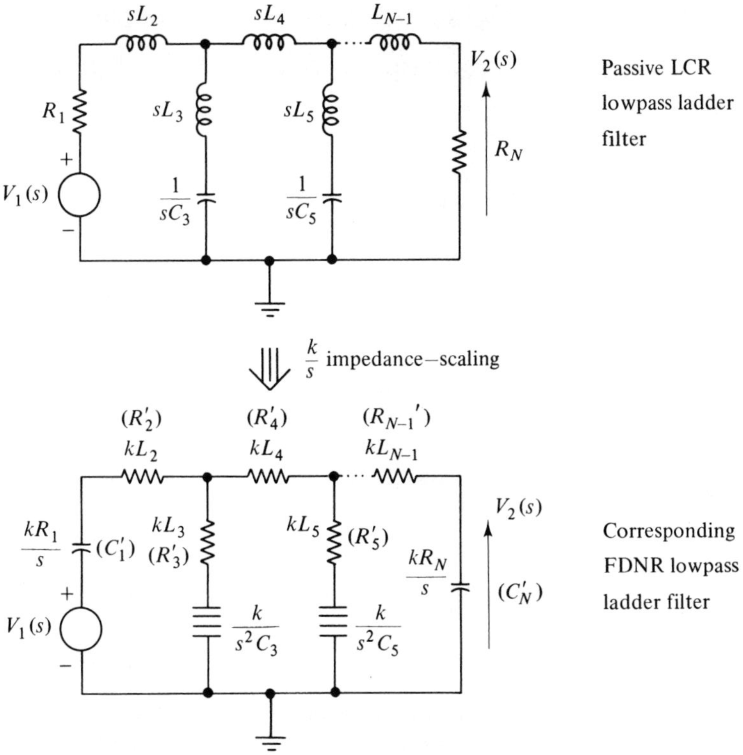

(a) FDNR lowpass filter realization from the LC prototype filter.

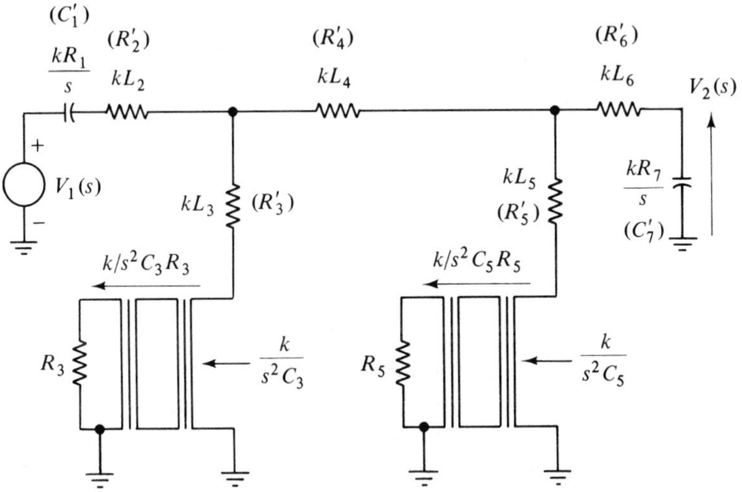

(b) Fifth-order FDNR lowpass filter using type 1 FDNR GIC circuits.

FIGURE 9.12

Since the k/s transformation does not alter the transfer function, it follows that

$$\underset{\text{LCR filter}}{H\left(sL_i, \frac{1}{sC_i}, R_i\right)} = \underset{\text{FDNR filter}}{H\left(kL_i, \frac{k}{s^2C_i}, \frac{kR_i}{s}\right)} \tag{9.71}$$

and, therefore,

$$S_{sL_i}^{H(s)} = S_{kL_i}^{H(s)}, \qquad S_{sC_i}^{H(s)} = S_{s^2C_i/k}^{H(s)}, \qquad S_{R_i}^{H(s)} = S_{kR_i/s}^{H(s)} \tag{9.72}$$

Writing the resistance elements, D elements, and capacitance elements of the FDNR filter as R_i', D_i', and C_i', we have

$$R_i' = kL_i, \qquad D_i' = \frac{C_i}{k}, \qquad C_i' = \frac{1}{kR_i} \tag{9.73}$$

corresponding to the transfer function $H(R_i', 1/s^2D_i', 1/sC_i')$ in terms of branch impedances. Equations (9.72) and (9.73) provide the single-element sensitivities of the FDNR filter in terms of the single-element sensitivities of the corresponding LCR filter as follows:

FDNR filter		LC filter	
$S_{R_i'}^{H(s)}$	=	$S_{L_i}^{H(s)}$	(9.74)
$S_{D_i'}^{H(s)}$	=	$S_{C_i}^{H(s)}$	(9.75)
$S_{C_i'}^{H(s)}$	=	$-S_{R_i}^{H(s)}$	(9.76)

This result is important because it implies that if the LC filter has been designed to possess the excellent insensitivity of $M(\omega)$ to the LC elements throughout the passband (as described in section 7-5) then, by equations (9.74) and (9.75), the FDNR circuit will have exactly the same excellent sensitivity behavior with respect to the R_i' and D_i'.

The type 2 FDNR GIC circuit shown in Table 5-9 is useful for realizing the high-quality grounded D elements. A fifth-order example is shown in Fig. 9.12(b), where the type 2 FDNR GIC circuits are used to realize the FDNR impedances k/s^2C_3 and k/s^2C_5. Major advantages of the FDNR lowpass circuit are that it retains the excellent sensitivity properties of the LC filter and it uses not more than N operational amplifiers to realize the Nth order transfer function.

If the FDNR GIC is terminated in a resistance R_i [that is, R_3 and R_5 in Fig. 9.12(b)] then the required conversion function $K_i(s)$ for this GIC is

$$K_i(s) = \frac{k}{s^2 C_i R_i}, \qquad i = 3, 5 \qquad \text{CONVERSION FUNCTIONS FOR FDNR GICs} \tag{9.77}$$

so that the required FDNR impedance k/s^2C_i is obtained for each FDNR element.

Simplified design procedure: A straightforward design procedure is to select k *to equal the cut-off frequency* ω_0 of the LC filter and the GIC terminating resistors R_i *to equal the magnitude of the reactance of* C_i *at the cut-off frequency*. That is,

SELECTION OF k AND R_i'S

$$k \equiv \omega_0 \tag{9.78}$$

$$R_i \equiv \frac{1}{\omega_0 C_i} \tag{9.79}$$

Equation (9.77) for the conversion function of the GICs is then simply

$$K_i(s) = \left(\frac{\omega_0}{s}\right)^2 \qquad \text{CONVERSION FUNCTIONS FOR FDNR GICs} \tag{9.80}$$

and the resistances R_i' that simulate inductances L_i in the LC prototype are then given by

$$R_i' = \omega_0 L_i \qquad \text{RESISTORS IN FDNR FILTER} \tag{9.81}$$

and the terminating capacitances C_1' and C_N' by

$$C_1' = \frac{1}{\omega_0 R_1}, \qquad C_N' = \frac{1}{\omega_0 R_N} \qquad \text{TERMINATING CAPACITORS IN FDNR FILTER} \tag{9.82}$$

The preceding five equations are the simplified ideal design equations for FDNR lowpass ladder filters. They are further simplified if the LC prototype filter has a cut-off frequency ω_0 of 1 rad/sec.

EXAMPLE 9.4: Design the FDNR version of the *normalized* seventh-order 1.0 db passband ripple Chebychev lowpass LC ladder filter shown in Fig. 9.13(a). Use the simplified design procedure described above.

SOLUTION: The cut-off frequency ω_0 is 1 rad/sec, and the LCR-element values are shown in Fig. 9.13(a). Therefore, applying equations (9.78) to (9.82), we have

$$R'_{3,7} = \frac{1}{C_{3,7}} = 0.8997\ \Omega, \qquad R'_5 = \frac{1}{C_5} = 0.8521\ \Omega$$

for the resistive terminations of the three GICs, and

$$R'_{2,8} = 2.166\ \Omega, \qquad R'_{4,6} = 3.0936\ \Omega$$

for the four resistance elements that replace the four inductance elements $L_{2,8}$, $L_{4,6}$, and

$$C'_{1,9} = 1\text{ F}$$

for the two terminating capacitance elements that replace the terminating resistance $R_{1,9}$. The final FDNR circuit is shown in Fig. 9.13(b). Equation (9.80) requires that

$$K_{3,5,7}(s) = \frac{1}{s^2} \tag{9.83}$$

It follows from equation (5.123) that, for a type 1 FDNR GIC circuit,

$$Z_{1d_i} = \left(\frac{1}{s^2 C_{2i} R_{3i} C_{4i} R_{5i}}\right) R_i, \qquad i = 3, 5, 7$$

corresponding to conversion functions $K_i(s)$ provided by

$$K_i(s) = \frac{1}{s^2 C_{2i} R_{3i} C_{4i} R_{5i}}, \qquad i = 3, 5, 7 \tag{9.84}$$

where C_{2i}, R_{3i}, C_{4i}, R_{5i} are the CR elements within the GICs. Equations (9.80) and (9.84) require that these CR elements are constrained such that

$$C_{2i} R_{3i} C_{4i} R_{5i} = 1, \qquad i = 3, 5, 7 \tag{9.85}$$

which is achieved by selecting

$$C_{2i} = C_{4i} = R_{3i} = R_{5i} = 1, \qquad i = 3, 5, 7 \tag{9.86}$$

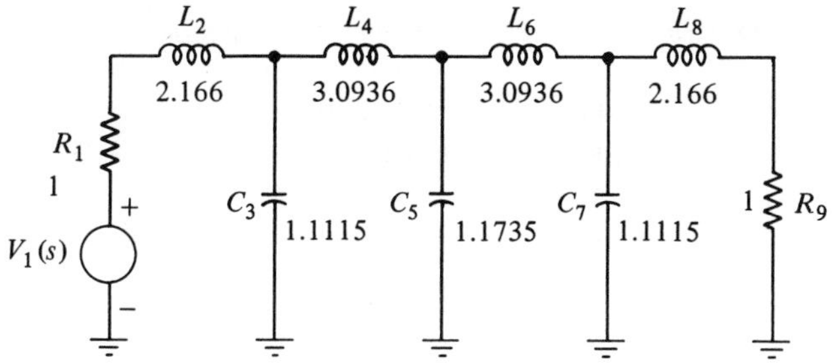

(a) Lowpass 1.0 dB passband seventh-order Chebychev prototype: $\omega_0 = 1$ rad/sec.

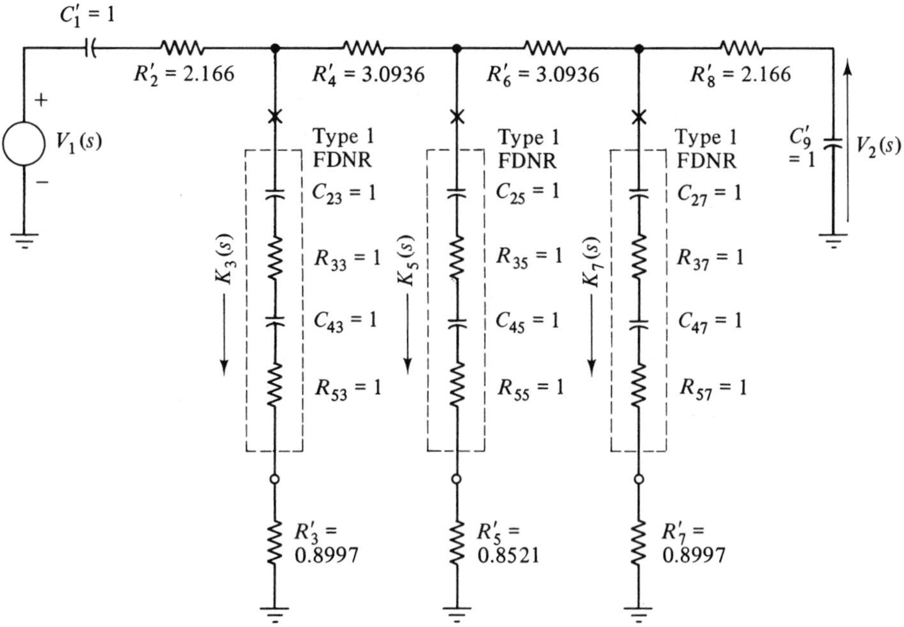

(b) Corresponding normalized FDNR filter: $\omega_0 = 1$ rad/sec. OP AMPs omitted for simplicity.

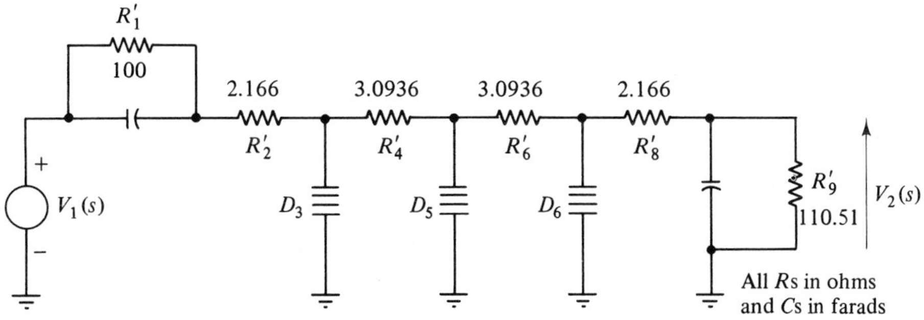

(c) Bias resistances R_1' and R_9'.

FIGURE 9.13

Note that $R_i' \neq R_{5i}$ and, therefore, the Q-enhancement condition in Table 5-9 is not satisfied. If this design is denormalized to cut-off frequencies above a few kilohertz, it usually is necessary to choose C_{2i}, C_{4i}, R_{3i}, and R_{5i} within the GICs so that the enhancement condition $R_{5i} = R_i$ is satisfied. This modification is illustrated in section 9-6.

9-5.4 Solution of the OP AMP Bias Problem in Lowpass FDNR Filters

Consider the noninverting input terminal of OP AMP A_2 of the type 1 FDNR circuit shown in Table 5-9. The practical nature of an OP AMP is that a small dc bias current must be drawn at each input terminal in order for the OP AMP to bias in the linear region; this implies that there must exist a resistive path to ground for all OP AMP input terminals. However, the noninverting input to OP AMP A_2 does not necessarily possess this resistive path to ground; this will depend on the nature of the network into which this FDNR circuit is embedded. In the case of the FDNR lowpass circuit of Fig. 9.13(b), no such resistive paths to ground exist from any one of the points marked $\times$ and, consequently, a practical version of this circuit will not bias because of lack of bias currents at the noninverting inputs of OP AMPs A_2 in each of the three FDNR circuits. One simple but straightforward solution to this problem is to place two large-valued resistance elements, R_1' and R_9', across the terminating capacitances C_1' and C_9', as shown in Fig. 9.13(c). Note that it is necessary to make R_1' and R_9' large enough that they do not significantly alter the passband response; for example, in this case R_1' is selected as

$$R_1' \equiv 100\ \Omega \tag{9.87}$$

for the normalized circuit shown in Fig. 9.13(b). Then, R_9' is selected so as to ensure that the dc gain $M(0)$ is equal to the ideal value of 0.5. This is achieved by noting from Fig. 9.13(c) that

$$M(0) = \frac{R_9'}{R_1' + R_2' + R_4' + R_6' + R_8' + R_9'} \tag{9.88}$$

Substituting $M(0) = 0.5$, $R_1' = 100\ \Omega$, and the numerical values for $R_{2,4,6,8}'$, we obtain

$$R_9' = 110.51\ \Omega \tag{9.89}$$

9-6 CASE STUDY: LOWPASS FDNR FILTER WITH CUT-OFF FREQUENCY OF 15.91 kHz

In this case study, a practical Chebychev lowpass FDNR filter with 1.0 db passband ripple is designed to have a cut-off frequency of 15.91 kHz, using economy OP AMPs. The impedance denormalization is chosen so that the terminating capacitances $C'_{1,9}$ and also the FDNR circuit capacitances have the practically convenient values of 10 nF. Thus,

$$A_0 \equiv 10^5, \qquad B \equiv 2\pi \times 10^5 \text{ rad/sec} \tag{9.90}$$

for the economy OP AMPs, and

$$C'_{1,9} \equiv 10 \text{ nF} \tag{9.91}$$

First, we determine the impedance denormalization factor that is implied by equation (9.91). This is found by noting that the reactances $X'_{1,9}$ of the terminations at the cut-off frequency ($\omega_0 \equiv 2\pi \times 15.91 \times 10^3$ rad/sec) are given from equation (9.91) by

$$\begin{aligned} X'_{1,9}(\omega_0) &= -j\frac{1}{2\pi \times 15.91 \times 10^3 \times 10 \times 10^{-9}} \\ &= -j1000\ \Omega \quad \text{DENORMALIZED} \end{aligned} \tag{9.92}$$

The corresponding reactances of the terminations of the normalized circuit in Fig. 9.13(b) are

$$X'_{1,9}(\omega_0) = -j1\ \Omega \quad \text{NORMALIZED} \tag{9.93}$$

The impedance denormalization factor is, therefore, 10^3; the frequency denormalization factor is obviously the ratio between the required cut-off frequency (10^5 rad/sec) and that of the normalized circuit (1 rad/sec). Thus, the frequency denormalization factor is 10^5.

This design procedure departs slightly from that outlined in Example 9.4 because it will be recalled that $R_i \neq R_{5i}$ ($i = 3, 5, 7$) in Fig. 9.13(b) and, thereby, Q-enhancement is not achieved. In this practical design, the cut-off frequency is high enough that the finite gain bandwidth products B of the OP AMPs must be taken into account. Therefore, the Q-enhancement condition (Table 5-9) is satisfied by ensuring that

$$\boxed{R_{5i} = R'_i = 1, \qquad i = 3, 5, 7} \quad \text{NORMALIZED DESIGN WITH } Q\text{-ENHANCEMENT} \tag{9.94}$$

in each of the FDNR circuits. The values of the D elements are retained exactly as in Fig. 9.13(b) by simply leaving the FDNR capacitances at 1 Farad but modifying R_{33}, R_{37}, and R_{35} as follows:

$$\left.\begin{aligned} R_{33} &= R_{37} = (0.8997)^{-1} = 1.1115\ \Omega \\ R_{35} &= (0.8521)^{-1} = 1.1735\ \Omega \end{aligned}\right\} \quad \text{NORMALIZED DESIGN WITH } Q\text{-ENHANCEMENT} \tag{9.95}$$

Using these values for R_{53}, R_{55}, R_{57}, R_{33}, R_{37}, and R_{35}, with the remaining element values as calculated in Example 9.2, the circuit is denormalized by the frequency-scaling factor 10^5 and impedance-scaling factor 10^3 to obtain

$$\left.\begin{aligned} &R_2' = R_8' = 2.166\ \text{k}\Omega: \quad R_4' = R_6' = 3.0936\ \text{k}\Omega \\ &R_3' = R_5' = R_7' = R_{53} = R_{55} = R_{57} = 1\ \text{k}\Omega \\ &R_{33} = R_{37} = 1.1115\ \text{k}\Omega: \quad R_{35} = 1.1735\ \text{k}\Omega \\ &R_1' = 100\ \text{k}\Omega: \quad R_9' = 110.51\ \text{k}\Omega \\ &C_1' = C_9' = C_{23} = C_{43} = C_{25} \\ &= C_{45} = C_{27} = C_{47} = 10\ \text{nF} \end{aligned}\right\} \quad \text{FINAL DESIGN VALUES} \tag{9.96}$$

Passive-sensitivity analysis: The basic reason for employing active-ladder filters is to obtain good sensitivity performance. Curve B in Fig. 9.14 is the theoretical lower-bound passive worstcase sensitivity function $LWS_{RC}^{M(\omega)}$ for this particular Chebychev transfer function (see section 6-9.2 to review the concept of $LWS_{RC}^{M(\omega)}$). An RC-active network cannot be found which has a lower worstcase passive sensitivity performance than curve B. For comparative purposes, curve C in Fig. 9.14 shows the worstcase-sensitivity performance $WS_{RC}^{M(\omega)}$ for a circuit implementation of this Chebychev filter using a cascade connection of three two-integrator loop (TIL) circuits of the type shown in Fig. 8.1. The transition region $WS_{RC}^{M(\omega)}$ result for this case is about a factor of only 1.5 in excess of the lower bound of curve B, but in the passband this TIL implementation is far inferior to curve B, with a peak value of about 50.

The FDNR realization that is defined by equations (9.96) possesses twenty-three RC elements and, therefore, the same number of single-element passive sensitivities $S_{x_i}^{M(\omega)}$. Adding the moduli of these twenty-three terms $S_{x_i}^{M(\omega)}$ provides the $WS_{RC}^{M(\omega)}$ result shown by curve A of Fig. 9.14. The transition region performance is similar to curve C, but the passband performance is far superior, with a peak $WS_{RC}^{M(\omega)}$ of about 10 in this region. Clearly, the superior insensitivity of the passband response $M(\omega)$ to the passive RC elements is a significant advantage of the FDNR lowpass filter compared with the corresponding realizations that employ cascaded quadratic sections.

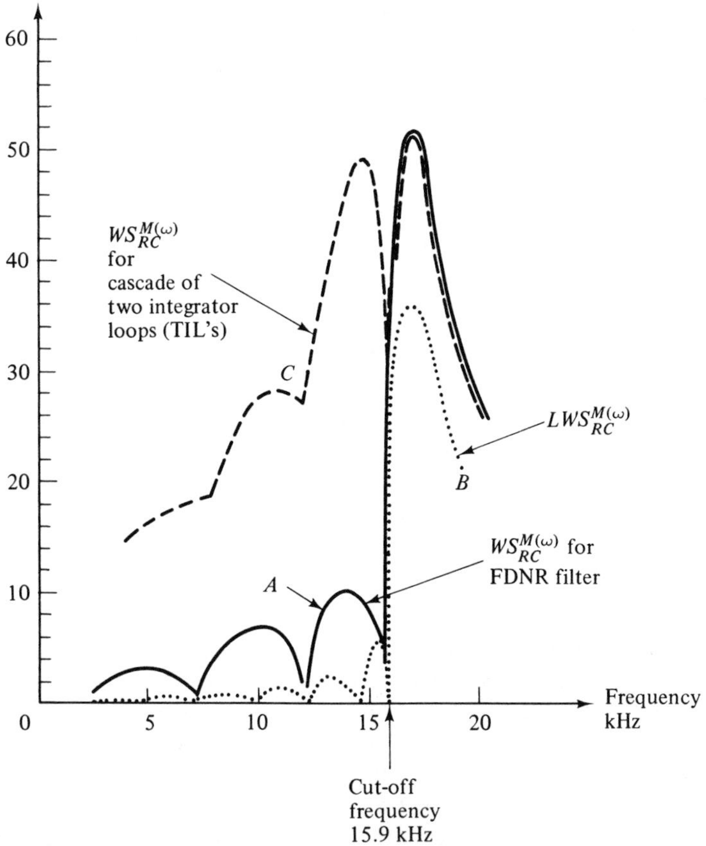

FIGURE 9.14 *Passive-sensitivity performance of FDNR versus TIL circuits: Chebychev lowpass 10 DB passband ripple.*

Active-sensitivity analysis: The Q-enhancement conditions have been met in this design [equation (9.94)]. However, recall from Table 5-9 that the *Q-enhancement condition also depends on the gain-bandwidth products $B_{1,2}$ of the FDNR OP AMPs being matched.* The minimal effect of $B_{1,2}$ in the matched case is evident from curve A of Fig. 9.15, which has been computed for $B_1 = B_2 = 10^7$ rad/sec in each FDNR circuit; the passband error is everywhere better than 2%. As expected, mismatching B_1 and B_2 by a significant factor (4 : 1) causes a significant departure from ideal behavior, as shown by curve B of Fig. 9.15. In this case, the geometric mean $\sqrt{B_1 B_2}$ of the gain bandwidth products is the same as for curve A, but the 4 : 1 mismatch causes an error of about -20% at the bandedge, thereby emphasizing the practical importance of using dual OP AMPs with the best possible matching between B_1 and B_2.

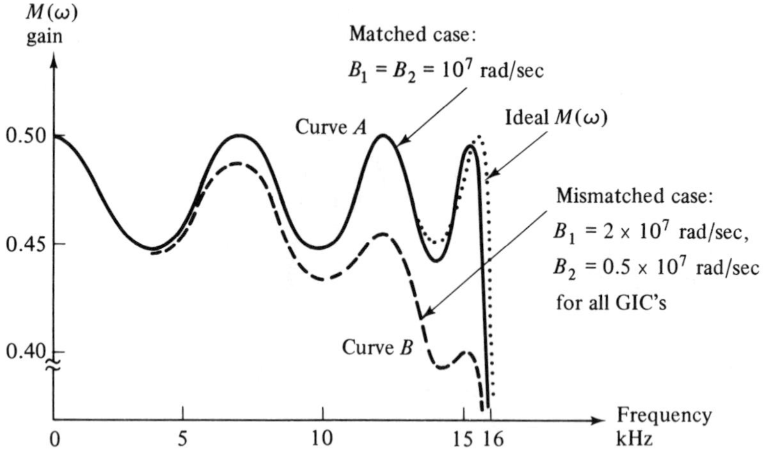

FIGURE 9.15 *The importance of matched gain bandwidth products on passbound response; circuit of Figure 9-13(b).*

Precorrection for errors in D_3, D_5, D_7: It is usually worthwhile to calculate the fractional errors $\epsilon_d(\omega)$ ($\approx \epsilon_{Hd}(\omega)$) in the magnitudes of the D elements by using the corresponding equation shown in Table 5-9. Substituting numerical values into this equation indicates that $\epsilon_{Hd}(\omega_0) \approx +4\%$ at the cut-off frequency of the filter for all three FDNR elements; the reader is asked to verify this in Problem 9.18. The +4% errors in D_3, D_5, and D_7 account for the fact that the cut-off frequency for curve A in Fig. 9.15 is approximately 2% lower than the ideal value of 15.9 kHz. Precorrection for this nonideal reduction of cut-off frequency is accomplished by simply decreasing the values of the resistances R_{33}, R_{35}, and R_{37} by 4%, thereby correcting the values of D_3, D_5, and D_7. In general, it is recommended that R_{3_i} in the ith branch of the ladder filter be precorrected to the value

$$\boxed{\frac{R_{3i}}{1+\epsilon_{Hd_i}(\omega_0)}} \qquad \text{PRECORRECTED VALUE OF } R_{3_i} \text{ IN FDNR GICs} \qquad (9.97)$$

where $\epsilon_{Hd_i}(\omega_0)$ is the fractional error in the D element D_i that is calculated from Table 5-9. Note that equation (5.131) may be used to determine $\epsilon_{Hd_i}(\omega_0)$ if each FDNR circuit uses equivalued C and equivalued R elements.]

9-7 IMPEDANCE-SCALING USING GIC-EMBEDDING TECHNIQUES

A basic limitation of the two-amplifier GIC circuits shown in Tables 5.6, 5.7, and 5.9 is that they may only be used to simulate *grounded* inductance and FDNR elements. However, in many applications it is necessary that *ungrounded* inductance and *ungrounded* FDNR elements be simulated.

We recall from Example 2.5 and Fig. 2.5 that an *ungrounded* inductance may be simulated by embedding an ungrounded resistance between two grounded GICs. In this section, this general concept of embedding a network between two GICs is considered in some detail.

9-7.1 The Basic GIC-Embedding Configuration

We are concerned with two three-terminal networks, N and N'. The network N is defined by its transmission matrix T_N, where

$$T_N \equiv \begin{bmatrix} a & b \\ c & d \end{bmatrix} \tag{9.98}$$

The network T'_N is defined to be an impedance-scaled version of T_N obtained by multiplying every branch transform impedance in T_N by the general impedance-scaling function $K_1(s)$, as is shown in Fig. 9.16(a). It is not difficult to prove that the transmission matrix of the impedance-scaled network N' is given by

$$T'_N = \begin{bmatrix} a & bK_1(s) \\ \dfrac{c}{K_1(s)} & d \end{bmatrix} \qquad \text{IMPEDANCE-SCALED NETWORK} \tag{9.99}$$

The general embedded structure under consideration is shown in Fig. 9.16(b); the network N with transmission matrix T_N is embedded between GIC1 and GIC2, where the GICs are defined as having transmission matrices

$$T_1 = \begin{bmatrix} 1 & 0 \\ 0 & K_1^{-1}(s) \end{bmatrix} \quad \text{and} \quad T_2 = \begin{bmatrix} 1 & 0 \\ 0 & K_2^{-1}(s) \end{bmatrix} \tag{9.100}$$

respectively. The resultant two-port is embedded between a source impedance Z_s and load impedance Z_L, as shown in Fig. 9.16(b), where the transmission matrices of the ungrounded Z_s and grounded Z_L are

$$\begin{bmatrix} 1 & Z_s \\ 0 & 1 \end{bmatrix} \quad \text{and} \quad \begin{bmatrix} 1 & 0 \\ \dfrac{1}{Z_L} & 1 \end{bmatrix} \tag{9.101}$$

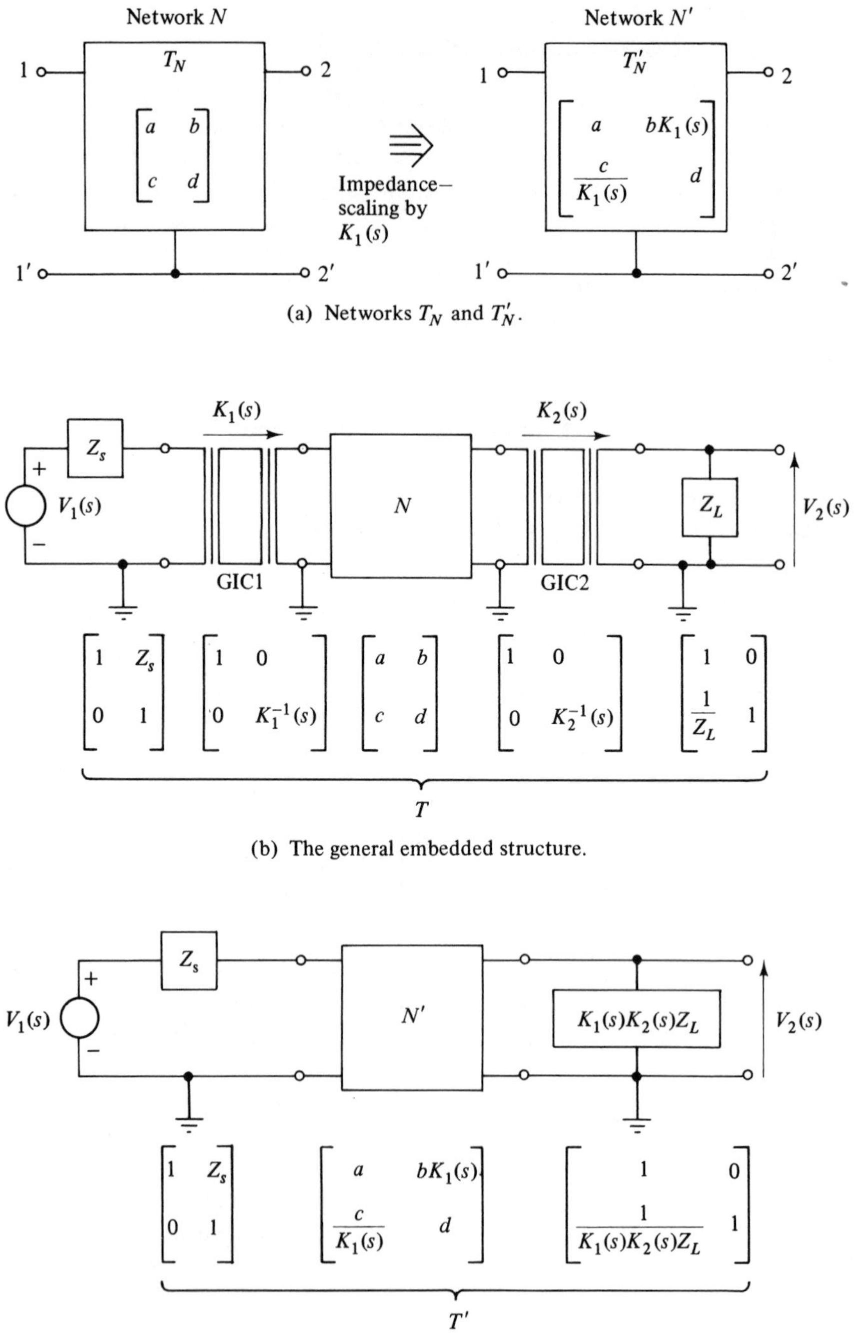

(a) Networks T_N and T'_N.

(b) The general embedded structure.

(c) The 'equivalent' nonembedded structure.

FIGURE 9.16

The five transmission matrices in equation (9.99) to (9.101) are shown in Fig. 9.16(b); their matrix product is defined as T. Thus,

$$T = \begin{bmatrix} 1 & Z_s \\ 0 & 1 \end{bmatrix} \begin{bmatrix} 1 & 0 \\ 0 & K_1^{-1}(s) \end{bmatrix} \begin{bmatrix} a & b \\ c & d \end{bmatrix} \begin{bmatrix} 1 & 0 \\ 0 & K_2^{-1}(s) \end{bmatrix} \begin{bmatrix} 1 & 0 \\ \dfrac{1}{Z_L} & 1 \end{bmatrix} \tag{9.102}$$

$$\equiv \begin{bmatrix} A & B \\ C & D \end{bmatrix}$$

The voltage ratio $V_1(s)/V_2(s)$ of Fig. 9.16(b) is given by the A element of matrix T above. Multiplying out the five matrices above, we have

$$A \equiv \frac{V_1(s)}{V_2(s)} = a + \frac{b}{K_2(s)Z_L} + Z_s\left[\frac{c}{K_1(s)} + \frac{d}{K_1(s)K_2(s)Z_L}\right] \tag{9.103}$$

The reciprocal of equation (9.103) is the voltage transfer function $V_2(s)/V_1(s)$ of Fig. 9.16(b).

Consider now a similar calculation of the function $V_1'(s)/V_2'(s)$ of Fig. 9.16(c). The overall transmission matrix T' of the two-port in Fig. 9.16(c) is

$$T' = \begin{bmatrix} 1 & Z_s \\ 0 & 1 \end{bmatrix} \begin{bmatrix} a & bK_1(s) \\ \dfrac{c}{K_1(s)} & d \end{bmatrix} \begin{bmatrix} 1 & 0 \\ \dfrac{1}{K_1(s)K_2(s)Z_L} & 1 \end{bmatrix} \equiv \begin{bmatrix} A' & B' \\ C' & D' \end{bmatrix} \tag{9.104}$$

Multiplying out the above matrices on the left to obtain A' gives

$$A' \equiv \frac{V_1'(s)}{V_2'(s)} = a + \frac{b}{K_2(s)Z_L} + Z_s\left[\frac{c}{K_1(s)} + \frac{d}{K_1(s)K_2(s)Z_L}\right] \tag{9.105}$$

which is identical to the result obtained in equation (9.103) for the network of Fig. 9.16(b). Thus, the implication is that the voltage transfer functions of Figs. 9.16(b) and Fig. 9.16(c) are identical; that is

EMBEDDING A NETWORK T_N BETWEEN GIC1 AND GIC2 HAS THE SAME EFFECT ON THE VOLTAGE TRANSFER FUNCTION AS IMPEDANCE-SCALING NETWORK T_N BY $K_1(s)$ AND ALTERING $Z_L(s)$ TO $K_1(s)K_2(s)Z_L(s)$.

(9.106)

In many applications, $K_2(s)$ is chosen equal to $K_1^{-1}(s)$; the effect of embedding is then directly equivalent to impedance-scaling T_N by $K_1(s)$ insofar as the voltage transfer function is concerned, and the voltage transfer function is, therefore, unaltered. This situation is described as *matched embedding*.

[A physical interpretation of the result above is that GIC1 behaves as an active-transformer (insofar as current is concerned) by impedance-scaling all impedances to the right of it by $K_1(s)$. Similarly, GIC2 causes all impedances to the right of itself to be scaled by $K_1(s)K_2(s)$. The voltage transfer function is unaltered.]

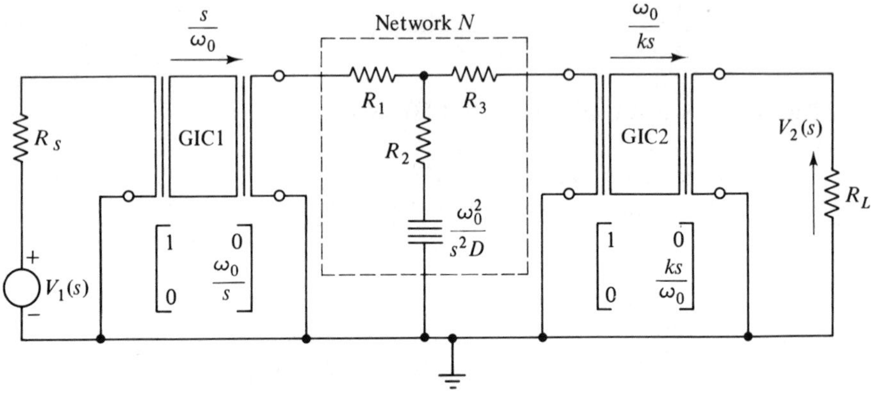

(a) Embedded resistor-FDNR network.

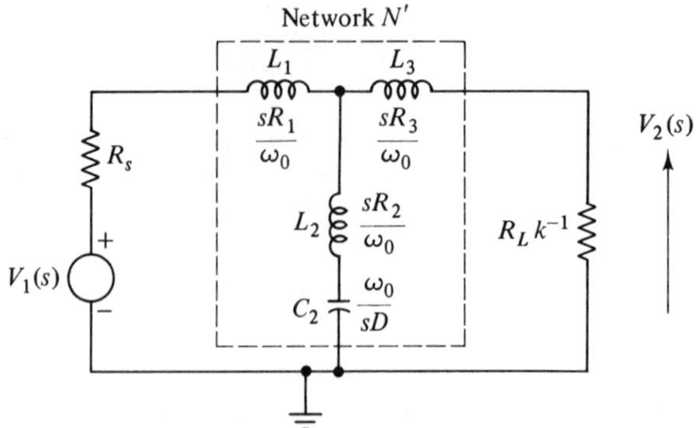

(b) 'Equivalent' resistively terminated LC network: same voltage transfer function as Fig. 9.17(a) if $k = 1$.

FIGURE 9.17

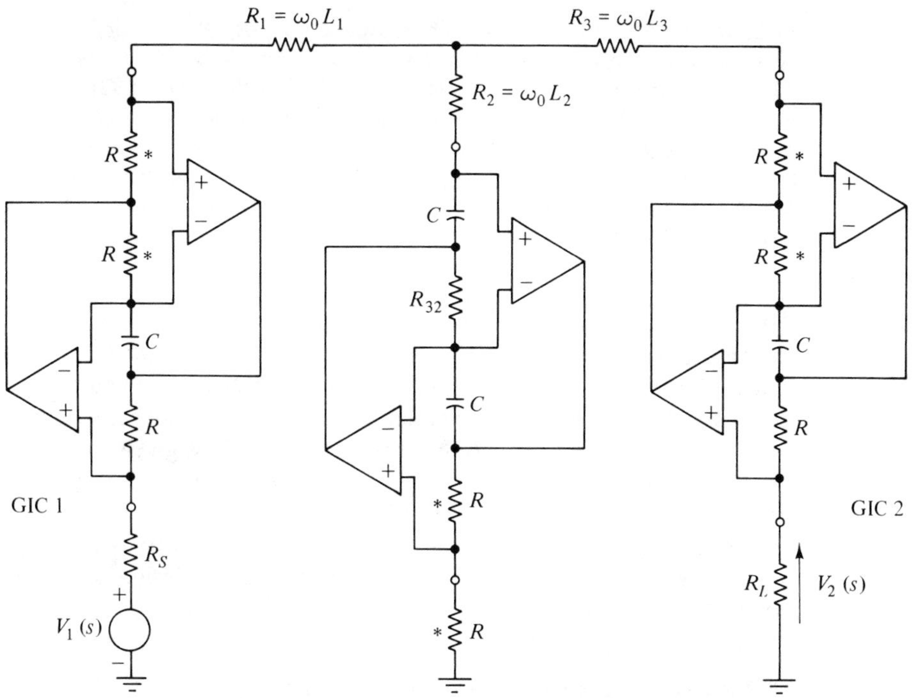

$$\frac{1}{CR} = \omega_0$$

$$R_{32} = \left[\frac{1}{\omega_0 C^2}\right] C_2$$

* indicates that resistors *must* be equal to maintain Q-enhancement condition in each GIC

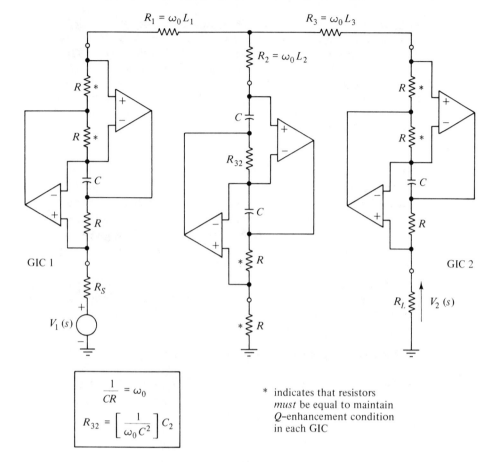

(c) Embedded-GIC implementation of Fig. 9.17(b) with Q-enhancement.

FIGURE 9.17 *(Continued).*

Clearly, it is possible to effectively impedance-scale a complete three-terminal network N by a particular function $K_1(s)$ by embedding it between two GICs.

EXAMPLE 9.5: A resistor-FDNR network N is embedded between two GICs, as shown in Fig. 9.17(a). The transmission matrices of GIC1 and GIC2 are also shown in the diagram; thus,

$$K_1(s) \equiv \frac{s}{\omega_0}, \qquad K_2(s) \equiv \frac{\omega_0}{ks}, \qquad k \text{ constant} \tag{9.107}$$

Find the equivalent network, corresponding to the impedance-scaled network N' shown in Fig. 9.16(c).

SOLUTION: Multiplying all branch impedances R_1, R_2, R_3, and ω_0^2/s^2D of network N by $K_1(s) = s/\omega_0$ provides the LC network N' shown in Fig. 9.17(b); multiplying the only branch impedance to the right of GIC2 by $K_1(s)K_2(s) = k^{-1}$ gives the equivalent terminating impedance $R_L k^{-1}$, as shown in Fig. 9.17(b). The voltage transfer functions of Figs. 9.17(a) and (b) are identical. Note that if $k = 1$, the two GICs are identical but connected in reverse directions and, further, the effective terminating resistances in Figs. 9.17(a) and (b) are then the same and so are the voltage functions.

It follows from the preceding example that lowpass LC filters of type A shown in Table 9-1 may be realized by using an embedded-GIC structure of the kind shown in Fig. 9.17(a). A circuit implementation is shown in Fig. 9.17(c), where the elements of the GICs are chosen to ensure Q-enhancement. The reader should prove that the values given for $1/CR$, R_1, R_2, R_{32}, R_3 in Fig. 9.17(c) imply that this circuit is a direct simulation of Fig. 9.17(b). Note that the asterisks identify resistances within each GIC that *must* be equivalued if the Q-enhancement conditions in Tables 5-6 and 5-9 are to be satisfied. Problems involving the design of FDNR and embedded-FDNR lowpass filters appear at the end of this chapter.

9-7.2 Series Connections of Embedded Networks

Consider now the series connection of embedded three-terminal networks $N_1, N_2, N_3, \ldots, N_m$ shown in Fig. 9.18(a), where the GICs have transmission matrices of the form

$$\begin{bmatrix} 1 & 0 \\ 0 & K_i^{-1}(s) \end{bmatrix}, \qquad \begin{matrix} i\text{th GIC in} \\ \text{Fig. 9.18} \end{matrix} \tag{9.108}$$

and where the impedance matrices of the networks $N_1, N_2, \ldots, N_m$ are written $\mathbf{Z}_1, \mathbf{Z}_2, \ldots, \mathbf{Z}_m$. Now, applying the result in (9.106) to the first embedded network N_1 of Fig. 9.18(a) leads directly to the "equivalent" network of Fig. 9.18(b); note that the first GIC is eliminated. Reapplying the result in (9.106) to Fig. 9.18(b) eliminates the second GIC. Thus, the process is continued until all GICs are eliminated, and the final equivalent network is that shown by Fig. 9.18(c). The networks of Fig. 9.18 have identical voltage transfer functions; the structure in Fig. 9.18(a) is described herein as a *series-connected embedded-GIC network*. Such networks are particularly useful for realizing high-performance *bandpass* active ladder filters. The essential feature of Fig. 9.18(c) is that the embedded network N_i of Fig. 9.18(a) has simply been replaced by a topologically similar network N_i' in which the impedance matrix is simply scaled by

$$K_1(s)K_2(s)K_3(s)\ldots K_i(s)$$

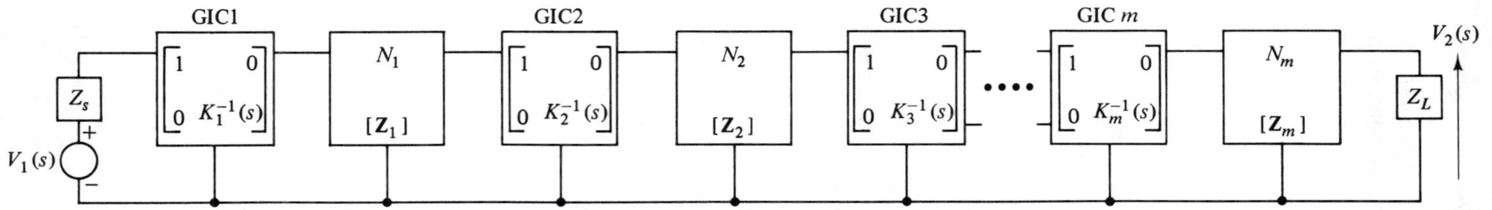

(a) General series-connected embedded network.

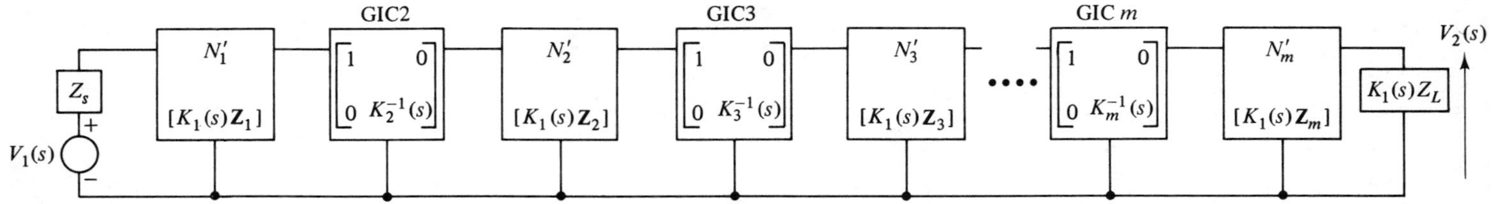

(b) Removal of GIC1 and resultant impedance-scaling of networks N_i' by $K_1(s)$.

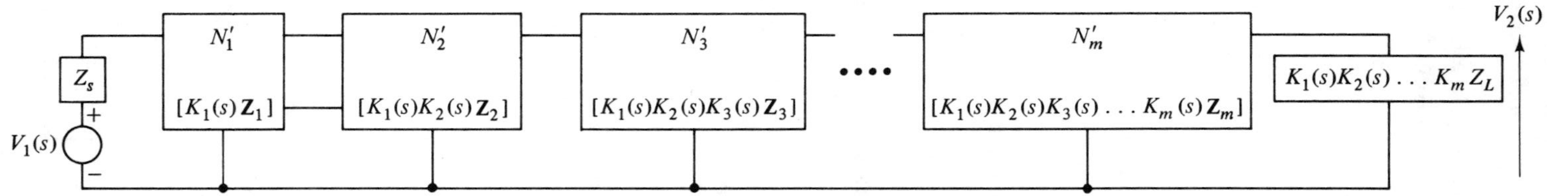

(c) Final equivalent circuit: all GIC's removed.

FIGURE 9.18

[corresponding to all GICs to the left of N_i in Fig. 9.18(a)]. It follows that N_i' is obtained from N_i by multiplying all *branch* transform impedances by

$$K_1(s)K_2(s)K_3(s)\ldots K_i(s)$$

9-7.3 Active *LC* Bandpass Ladder Filters Using GIC-Embedding Techniques

An immediate application for the series-connected embedded-GIC network of Fig. 9.18(a) is in the simulation of the type C *LC* bandpass filter structure of Table 9-1. (Other bandpass structures may also be implemented this way but are more complicated so discussion of them is postponed.)

Consider the fourteenth order doubly terminated *LC* prototype filter of Fig. 9.19; the networks N_1, N_2, and N_3 are shown in dashed lines to correspond to Fig. 9.18(a). Note that four embedding-type GICs are required to embed N_1, N_2, and N_3, as shown in Fig. 9.19. The *D* elements of the transmission matrices of GICs 1, 2, 3, and 4 are as follows:

$$\left.\begin{aligned} D_1(s) &= \frac{\omega_0}{k_1 s}, \qquad & D_2(s) &= \frac{s^2}{k_2\omega_0^2}, \\ D_3(s) &= \frac{\omega_0^2}{k_3 s^2}, \qquad & D_4(s) &= \frac{s}{k_4\omega_0} \end{aligned}\right\} \tag{9.109}$$

so that the N_1, N_2, N_3 in Fig. 9.19 are obtained from N_1', N_2', N_3' by the following impedance-scaling transformations:

$$\left.\begin{aligned} [\mathbf{Z}_1] &\to \left[\frac{\omega_0}{k_1 s}\mathbf{Z}_1\right] & &\text{NETWORK } N_1 \\ [\mathbf{Z}_2] &\to \left[\frac{s}{k_1 k_2 \omega_0}\mathbf{Z}_2\right] & &\text{NETWORK } N_2 \\ [\mathbf{Z}_3] &\to \left[\frac{\omega_0}{k_1 k_2 k_3 s}\mathbf{Z}_3\right] & &\text{NETWORK } N_3 \end{aligned}\right\} \tag{9.110}$$

Thus, the transform impedance sL_1 in the *LC* filter is impedance-scaled as follows:

$$\boxed{sL_1 \to \left(\frac{\omega_0}{k_1 s}\right) sL_1 = \frac{\omega_0 L_1}{k_1}} \qquad \text{RESISTANCE} \tag{9.111}$$

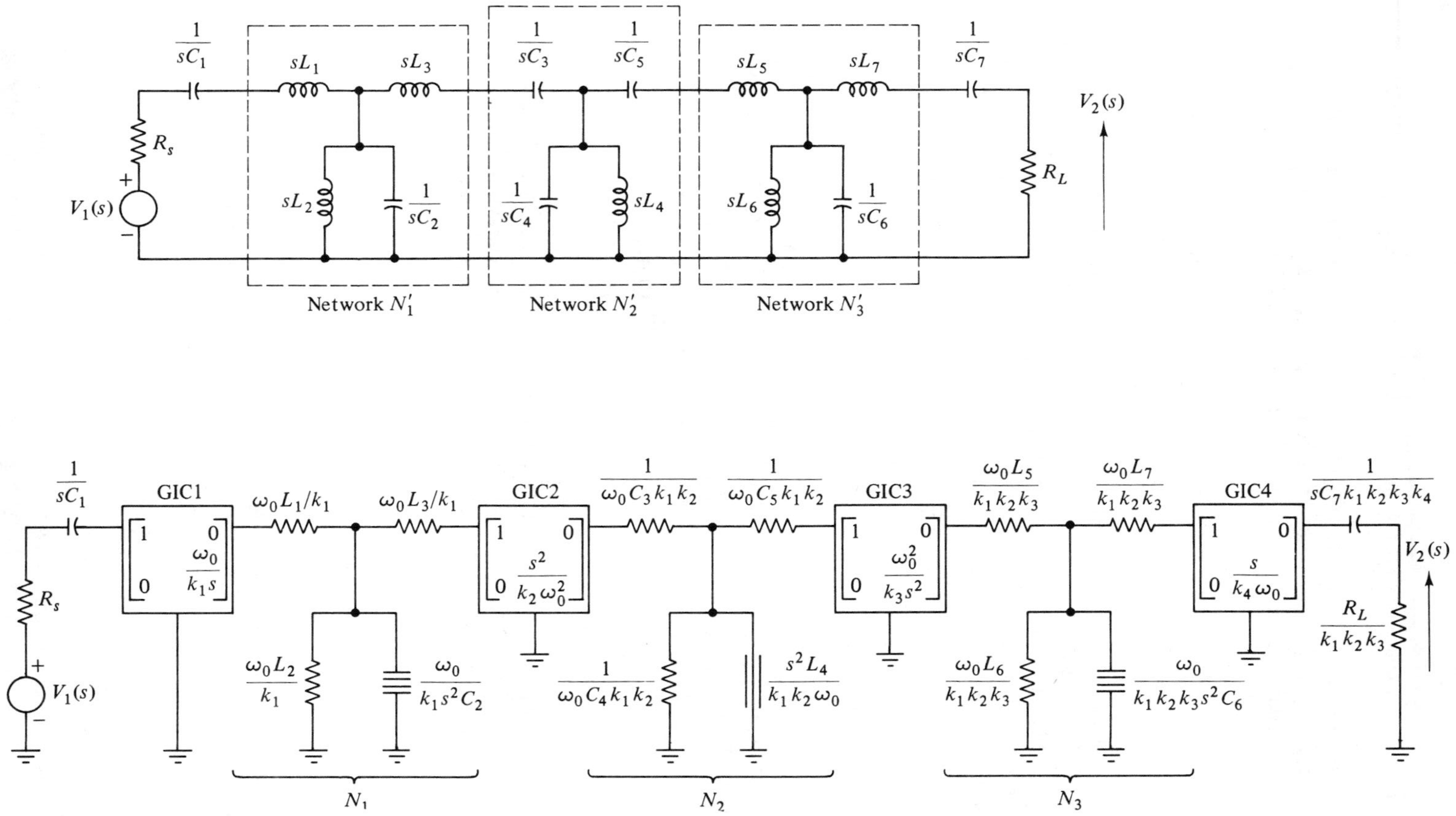

FIGURE 9.19 *Fourteenth-order bandpass active ladder realization using a series-connected embedded structure.*

and, therefore, appears in the embedded structure of Fig. 9.19 as a resistance $\omega_0 L_1/k_1$. A similar argument applies to sL_2 and sL_3. The transform impedance $1/sC_2$ is impedance-scaled as follows:

$$\boxed{\frac{1}{sC_2} \rightarrow \left(\frac{\omega_0}{k_1 s}\right)\frac{1}{sC_2} = \frac{\omega_0}{k_1 s^2 C_2}} \quad \text{FDNR } D \text{ ELEMENT} \tag{9.112}$$

which, of course, is an active FDNR D element that may be realized with the GIC circuit shown in Table 5-9. The transform impedance $1/sC_3$ of the LC filter is in network N_2, leading to an impedance-scaled network N_2' with both GIC1 and GIC2 to the left of itself. Thus, using the transformation in equation (9.110) for N_2, we have

$$\boxed{\frac{1}{sC_3} \rightarrow \left(\frac{s}{k_1 k_2 \omega_0}\right)\frac{1}{sC_3} = \frac{1}{k_1 k_2 \omega_0 C_3}} \quad \text{RESISTANCE} \tag{9.113}$$

where similar expressions apply to $1/sC_4$ and $1/sC_5$ of the LC prototype. The transform impedance sL_4 in the LC prototype is impedance-scaled as follows:

$$\boxed{sL_4 \rightarrow \left(\frac{s}{k_1 k_2 \omega_0}\right)sL_4 = \frac{s^2 L_4}{k_1 k_2 \omega_0}} \quad \text{FDNR } E \text{ ELEMENT} \tag{9.114}$$

and, therefore, is an active FDNR E element which is realized by using a single GIC circuit. For network N_3' of the LC prototype, equations (9.110) imply that the transform impedance sL_5 is impedance-scaled as follows:

$$\boxed{sL_5 \rightarrow \left(\frac{\omega_0}{k_1 k_2 k_3 s}\right)sL_5 = \frac{\omega_0 L_5}{k_1 k_2 k_3}} \quad \text{RESISTANCE} \tag{9.115}$$

where similar expressions apply to sL_6 and sL_7 of the LC prototype. The transform impedance $1/sC_6$ is impedance-scaled as follows:

$$\boxed{\frac{1}{sC_6} \rightarrow \left(\frac{\omega_0}{k_1 k_2 k_3 s}\right)\frac{1}{sC_6} = \frac{\omega_0}{k_1 k_2 k_3 s^2 C_6}} \quad \text{FDNR } D \text{ ELEMENT} \tag{9.116}$$

Finally, the terminating network of the LC filter is the series combination of $1/sC_7$ and R_L which, according to Fig. 9.18(c), is impedance-scaled as follows:

$$[Z_L] \rightarrow \left[\frac{1}{k_1 k_2 k_3 k_4} Z_L \right] \qquad (9.117)$$

so that, in this case,

$$\boxed{\frac{1}{sC_7} \rightarrow \left(\frac{1}{k_1 k_2 k_3 k_4} \right) \frac{1}{sC_7}} \quad \text{Capacitance} \qquad (9.118)$$

and

$$\boxed{R_L \rightarrow \frac{R_L}{k_1 k_2 k_3 k_4}} \quad \text{Resistance} \qquad (9.119)$$

A total of seven GIC's is required to implement this filter, four of which are required for embedding purposes and three for realizing the FDNR elements.

Choice of ω_0, k_1, k_2, k_3, k_4: The procedure described above allows the designer to select ω_0, k_1, k_2, k_3, and k_4. However, once this selection has been made, the values of the R and FDNR elements in networks N_1, N_2, and N_3 are then determined (see expressions shown in Fig. 9.19 for these elements). The simplest design procedure is to select ω_0 as the center frequency of the LC prototype filter (at which the series arms and shunt arms of the LC filter are series-resonant and parallel-resonant, respectively) and the parameters k_1, k_2, k_3, and k_4 as unity. Unfortunately, it will be shown that such a simple procedure does not allow the previously described Q-enhancement conditions to be maintained. We therefore describe *two* design procedures, the second of which maintains Q-enhancement conditions.

Design procedure 1: *matched-embedding, $k_i = 1$.* Select

$$\omega_0 = \text{center frequency} \qquad (9.120)$$

and

$$k_1 \equiv k_2 \equiv k_3 \equiv \ldots k_m \equiv 1 \qquad (9.121)$$

so that, in general, GIC1 at the source and GICm at the termination are matched (that is, identical but connected in *reverse* directions) and the remaining $(m - 2)$ GICs are similarly matched. The design procedure and resultant circuits are straightforward in this case because *the impedance levels of the branches of N_1, N_2, N_3, . . . and N_1', N_2', N_3' are identical at the center frequency* ω_0. For example, in the case of inductance L_1 and

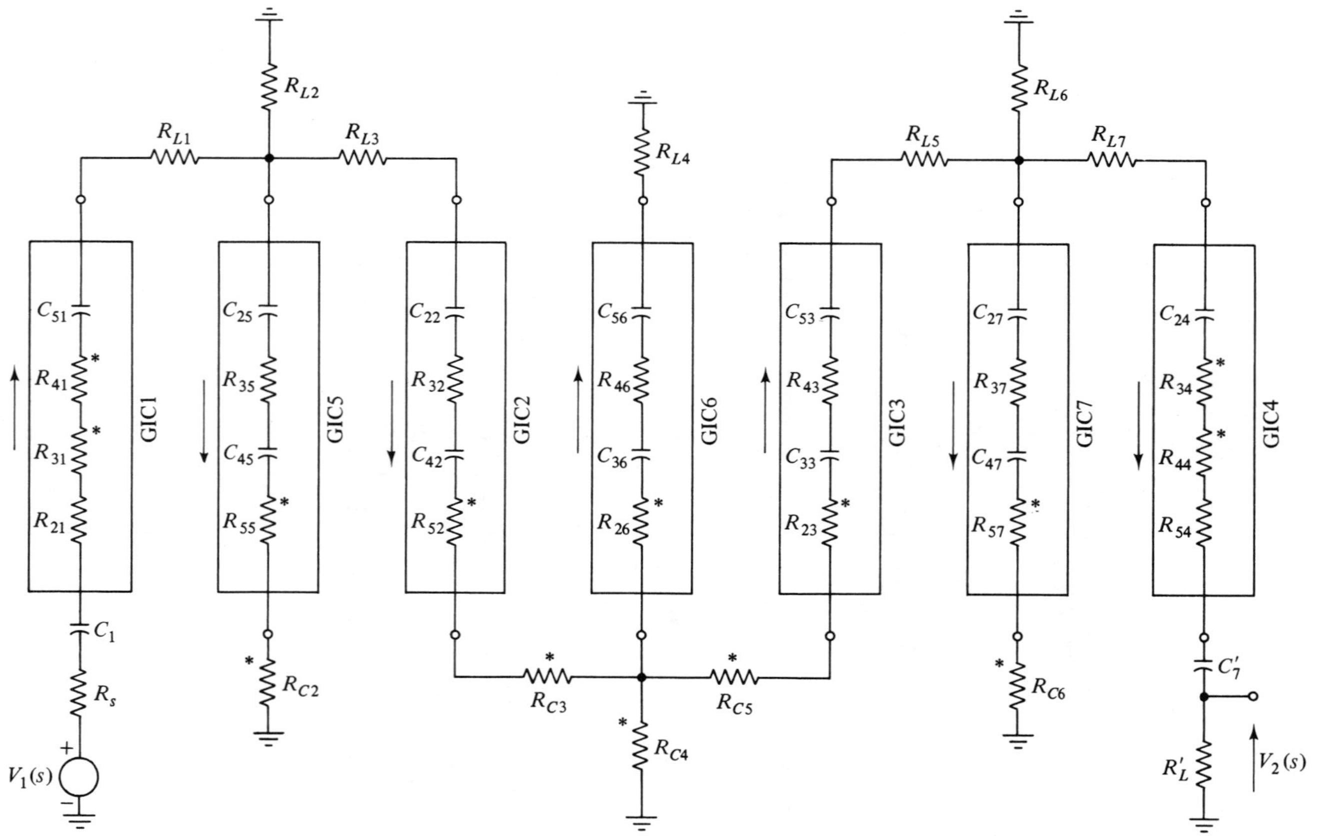

FIGURE 9.20 *Fourteenth-order bandpass circuit: OP AMPs omitted for simplicity.*

inductance L_4 the impedance transformations *at the frequency* ω_0 are

$$j\omega_0 L_1 \rightarrow \omega_0 L_1 \qquad \text{RESISTANCE} \qquad (9.122)$$

$$j\omega_0 L_4 \rightarrow -\omega_0 L_4 \qquad \text{FDNR IMPEDANCE AT } \omega_0 \qquad (9.123)$$

$\Big\}$ UNITY k_i

and for a frequency normalized prototype ($\omega_0 \equiv 1$) this design procedure is, therefore, straightforward. The circuit design of Fig. 9.20, which is a more detailed version of Fig. 9.19 with the GIC *RC* elements included, may be simplified drastically by selecting all internal GIC capacitors equal and all internal GIC resistors equal, so that

$$\begin{aligned} R_{21} &= R_{31} = R_{41} = R_{35} = R_{55} = R_{32} \\ &= R_{52} = R_{26} = R_{46} = R_{23} = R_{43} \\ &= R_{37} = R_{57} = R_{34} = R_{44} = R_{54} \equiv R \end{aligned} \qquad (9.124)$$

and

$$\begin{aligned} C_{51} &= C_{25} = C_{45} = C_{22} = C_{42} = C_{36} = C_{56} \\ &= C_{33} = C_{53} \\ &= C_{27} = C_{47} = C_{24} \equiv C \end{aligned} \qquad (9.125)$$

If we select

$$\boxed{CR = \frac{1}{\omega_0}} \qquad (9.126)$$

then $k_1 = k_2 = k_3 = k_4$, and the impedance conversion functions of the GICs are $(s/\omega_0)^2$ for GIC1, $(\omega_0/s)^2$ for GICs 5, 2, 7, $(s/\omega_0)^2$ for GICs 6, 3, and (ω_0/s) for GIC7. The resistors R_{Li} that correspond to inductances L_i in Fig. 9.20 are

$$\boxed{R_{Li} = \omega_0 L_i} \qquad i = 1, 2, 3, \ldots, 7 \qquad (9.127)$$

and the resistors R_{Ci} that correspond to capacitances C_i are

$$\boxed{R_{Ci} = \frac{1}{\omega_0 C_i}}, \qquad i = 2, 3, 4, 5, 6 \qquad (9.128)$$

Note that C_7 and R_L are then identical in both the *LC* prototype filter and the corresponding embedded-GIC filter. Equations (9.125) to (9.128) are the straightforward design equations for design procedure No. 1. (Example 9.1 illustrates this procedure.) Unfortunately, this procedure does not result in the *Q*-enhancement conditions being satisfied according to Tables 5-7 and 5-9. Thus, the active-sensitivity performance with respect to OP AMP gain bandwidth products is not optimal.

Design procedure 2: *optimum active-sensitivity*, $k_i = 1$. It is not difficult to select the RC elements of the GICs in such a way that the Q-enhancement conditions of Tables 5-7 and 5-9 *are* maintained. The *necessity* for obtaining optimum active-sensitivity performance in this way depends on the stringency of the filter specifications and the implied requirements for ultrahigh Q-factors. [A simple means of estimating the required element Q-factors is to employ equation (9.15) to estimate the necessary *uniform* Q-factors of the LC elements; then, the expressions for Q-factors in Tables 5-7 and 5-9 are evaluated and compared with the required Q-factors.]

GICs 1 and 4 in Fig. 9.20 are effectively type 2 inductance-simulation circuits of the kind shown in Table 5-7. Thus, the enhancement conditions are

$$\boxed{\begin{aligned} R_{31} &= R_{41} \\ R_{34} &= R_{44} \end{aligned}} \quad \begin{array}{l}\text{Q-ENHANCEMENT CONDITIONS} \\ \text{FOR TERMINATING GICs}\end{array} \tag{9.129}$$

The remaining GICs (GIC 2, 3, 5, 6, and 7) are effectively type 1 FDNR circuits of the kind shown in Table 5-9. (For example, GIC5 implements a FDNR D element that is parallel-resonant with resistance R_{L2} at the center frequency ω_0.) The Q-enhancement conditions are, according to Table 5-9, as follows:

$$\boxed{\begin{aligned} R_{C2} &= R_{55} \\ R_{C3} &= R_{52} \\ R_{C4} &= R_{26} \\ R_{C5} &= R_{23} \\ R_{C6} &= R_{57} \end{aligned}} \quad \begin{array}{l}\text{Q-ENHANCEMENT CONDITIONS} \\ \text{FOR INTERNAL GICs} \\ (B_1 = B_2 \text{ FOR ALL GICs})\end{array} \tag{9.130}$$

Conditions (9.129) are consistent with design procedure No. 1. However, conditions (9.130) require that the CR elements violate equation (9.124). A simple way of satisfying conditions (9.130) is to follow a similar procedure to that which led to Fig. 9.18; that is, we select R_{C2}, R_{C3}, R_{C4}, R_{C5}, R_{C6} according to equation (9.128), so that

$$\boxed{R_{Ci} = \frac{1}{\omega_0 C_i}}, \qquad i = 2, 3, 4, 5, 6 \tag{9.131}$$

Equations (9.130) and (9.131) constrain only one of the four RC elements in each GIC. Consider Fig. 9.21(a), which shows the RC elements for

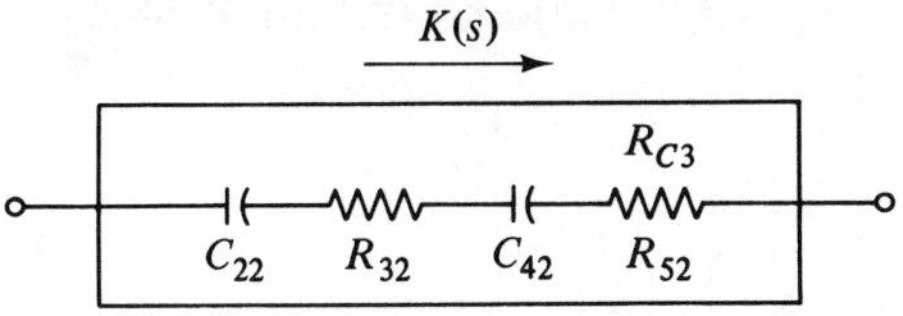

(a) RC elements of GIC2.

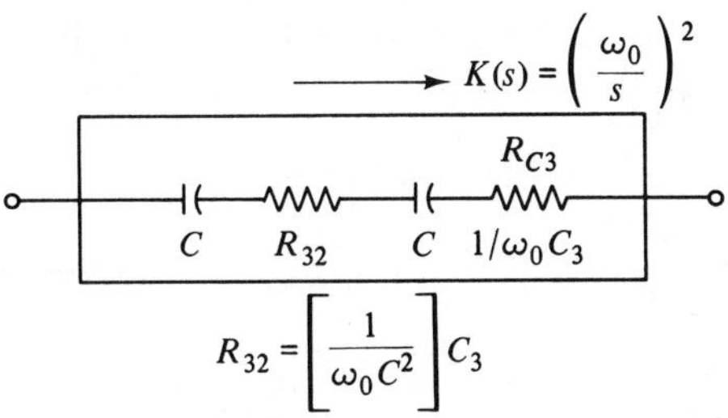

(b) Q-enhanced RC elements of GIC2.

FIGURE 9.21

GIC2. The impedance-conversion function $K(s)$ for this GIC is

$$K(s) = \frac{1}{s^2 C_{22} R_{32} C_{42} R_{52}} \quad \text{IMPEDANCE CONVERSION FUNCTION FOR GIC2} \tag{9.132}$$

and, since we have chosen all k_2 equal to unity, we must select C_{22}, R_{32}, C_{42}, R_{52} so that

$$K(s) = \frac{\omega_0^2}{s^2} \quad \text{REQUIRED IMPEDANCE CONVERSION FUNCTION FOR GIC2, } k_2 \equiv 1 \tag{9.133}$$

Equating (9.132) and (9.133),

$$\boxed{\omega_0^2 C_{22} R_{32} C_{42} R_{52} = 1} \tag{9.134}$$

For simplicity in implementation, we select

$$\boxed{C_{22} = C_{42} \equiv C} \tag{9.135}$$

and, therefore, equations (9.130) and (9.131) imply

$$\boxed{R_{52} = \frac{1}{\omega_0 C_3}} \tag{9.136}$$

We use equations (9.134), (9.135) and (9.136) to obtain

$$\boxed{R_{32} = \left(\frac{1}{\omega_0 C^2}\right) C_3} \tag{9.137}$$

The resultant Q-enhanced design for GIC2 is shown in Fig. 9.21(b). Similar calculations for GICs 3, 5, 6, and 7 lead to the following complete set of RC-element values for the Q-enhanced design:

$$\left.\begin{aligned}
&\text{Select } R \\
C &= \frac{1}{\omega_0 R} \\
C_{51} &= C_{25} = C_{45} = C_{22} = C_{42} = C_{36} = C_{56} \\
&= C_{33} = C_{53} = C_{27} = C_{47} = C_{24} \equiv C \\
C_7' &= C_7 \\
R_L' &= R_L \\
R_{21} &= R_{31} = R_{41} = R_{34} = R_{44} = R_{54} \equiv R, \text{ selected} \\
R_{55} &= R_{C2} = \frac{1}{\omega_0 C_2} \\
R_{52} &= R_{C3} = \frac{1}{\omega_0 C_3} \\
R_{26} &= R_{C4} = \frac{1}{\omega_0 C_4} \\
R_{23} &= R_{C5} = \frac{1}{\omega_0 C_5} \\
R_{57} &= R_{C6} = \frac{1}{\omega_0 C_6} \\
R_{L1} &= \omega_0 L_1, \quad R_{L2} = \omega_0 L_2, \quad R_{L3} = \omega_0 L_3 \\
R_{L4} &= \omega_0 L_4, \quad R_{L5} = \omega_0 L_5, \quad R_{L6} = \omega_0 L_6, \quad R_{L7} = \omega_0 L_7 \\
R_{35} &= \frac{1}{\omega_0 C^2} C_2, \quad R_{32} = \frac{1}{\omega_0 C^2} C_3 \\
R_{46} &= \frac{1}{\omega_0 C^2} C_4, \quad R_{43} = \frac{1}{\omega_0 C^2} C_5 \\
R_{37} &= \frac{1}{\omega_0 C^2} C_6
\end{aligned}\right\} \tag{9.138}$$

These RC-element values differ from those obtained in design procedure No. 1 in the values of R_{55}, R_{35}, R_{52}, R_{32}, R_{26}, R_{46}, R_{23}, R_{43}, R_{57}, R_{37}.

Design procedure No. 2 is highly recommended for applications involving stringent passband specifications. The computer-aided analysis of the previously considered fourteenth order bandpass filter reveals that, for ω_0/BW in excess of 10 with $\omega_0 > 10^3$, it is most important that the Q-enhancement feature of design procedure No. 2 be employed.

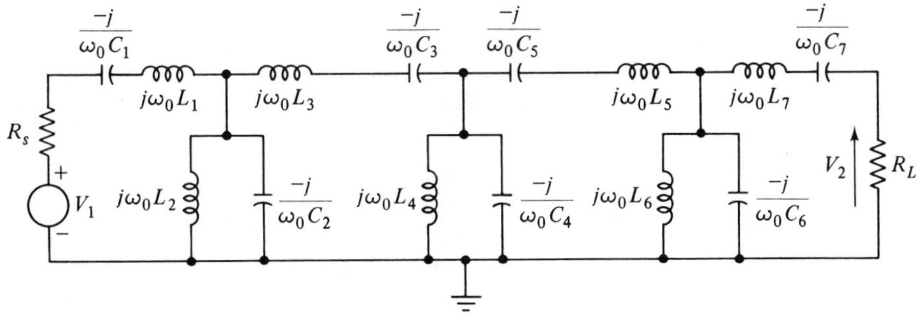

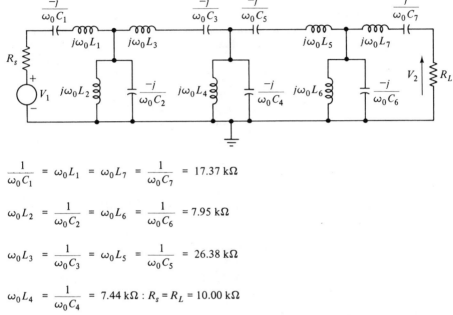

$$\frac{1}{\omega_0 C_1} = \omega_0 L_1 = \omega_0 L_7 = \frac{1}{\omega_0 C_7} = 17.37\ \text{k}\Omega$$

$$\omega_0 L_2 = \frac{1}{\omega_0 C_2} = \omega_0 L_6 = \frac{1}{\omega_0 C_6} = 7.95\ \text{k}\Omega$$

$$\omega_0 L_3 = \frac{1}{\omega_0 C_3} = \omega_0 L_5 = \frac{1}{\omega_0 C_5} = 26.38\ \text{k}\Omega$$

$$\omega_0 L_4 = \frac{1}{\omega_0 C_4} = 7.44\ \text{k}\Omega : R_s = R_L = 10.00\ \text{k}\Omega$$

FIGURE 9.22 *Fourteenth-order Chebychev bandpass LC prototype: passband ripple = 0.5 db; BW =* ω_0.

EXAMPLE 9.6: Use design procedure No. 1 to design an active ladder simulation of the LC filter shown in Fig. 9.22. This LC filter has been designed by lowpass-to-bandpass transformation of the corresponding seventh-order LC lowpass filter. The bandwidth BW is equal to the center frequency ω_0, and the passband ripple is 0.5 db. The active filter is required to have a center frequency of 1500 Hz.

SOLUTION: The fourteenth-order LC filter of Fig. 9.22 is implemented by means of the seven GIC-embedded active ladder structure of Fig. 9.20. GICs 1 and 4 are type 2 Inductance-simulation circuits (Table 5-7), and GICs 2, 3, 5, 6, and 7 are type 1 FDNR circuits (Table 5.9).

Equations (9.124), (9.125), and (9.126) are used to determine the RC elements of all seven GICs. Thus, the 12 GIC capacitance elements C are selected to have a convenient value:

$$C \equiv 10 \text{ nF} \tag{9.139}$$

and since

$$\omega_0 \equiv 2\pi \times 1500 \text{ rad/sec}$$

by equation (9.126) we find that

$$R = 10.610 \text{ k}\Omega \tag{9.140}$$

for the 16 GIC resistance elements. The resistance elements R_{Li} and R_{Ci} are obtained directly from the numerical values specified in Fig. 9.22 and equations (9.127) and (9.128), so that

$$\begin{aligned} R_{L1} &= R_{C1} = R_{L7} = R_{C7} = 17.37 \text{ k}\Omega \\ R_{L2} &= R_{C2} = R_{L6} = R_{C6} = 7.95 \text{ k}\Omega \\ R_{L3} &= R_{C3} = R_{L5} = R_{C5} = 26.38 \text{ k}\Omega \\ R_{L4} &= R_{C4} = 7.44 \text{ k}\Omega \\ R_s &= R_L = 10.00 \text{ k}\Omega \end{aligned} \tag{9.141}$$

C_1 and C_7' are the same as in the LC prototype, thus,

$$C_1 = C_7' = \frac{1}{2\pi \times 1500 \times 17{,}370} \text{F} = 6.108 \text{ nF} \tag{9.142}$$

Equations (9.139), (9.140), (9.141), and (9.142) provide the required RC element values for the final active ladder of Fig. 9.20.

EXAMPLE 9.7: Repeat Example 9.6, but use design procedure No. 2 to obtain optimum active-sensitivity performance.

SOLUTION: Equations (9.138) summarize this design procedure. The following RC elements *differ* in value from those used in Example 9.1:

$$\left.\begin{aligned} R_{55} &= R_{57} = 0.795 \text{ k}\Omega: && \text{GICs 5, 7} \\ R_{52} &= R_{23} = 26.38 \text{ k}\Omega: && \text{GICs 2, 3} \\ R_{26} &= 7.44 \text{ k}\Omega: && \text{GIC 6} \\ R_{35} &= R_{37} = 7.95 \text{ k}\Omega: && \text{GICs 5, 7} \\ R_{32} &= R_{43} = 14.16 \text{ k}\Omega: && \text{GICs 2, 3} \\ R_{46} &= 15.13 \text{ k}\Omega: && \text{GIC 6} \end{aligned}\right\} \tag{9.143}$$

A computer-aided analysis of a similar design to the above has been performed for OP AMPs with gain bandwidth products B equal to 1.5 MHz and for BW = 150 Hz. The resultant passband responses $M(\omega)$ are shown in Fig. 9.23 for design procedure No. 1 (curve A) and design procedure No. 2 (curve B). For this particular design, it is clear that curve A is unacceptable

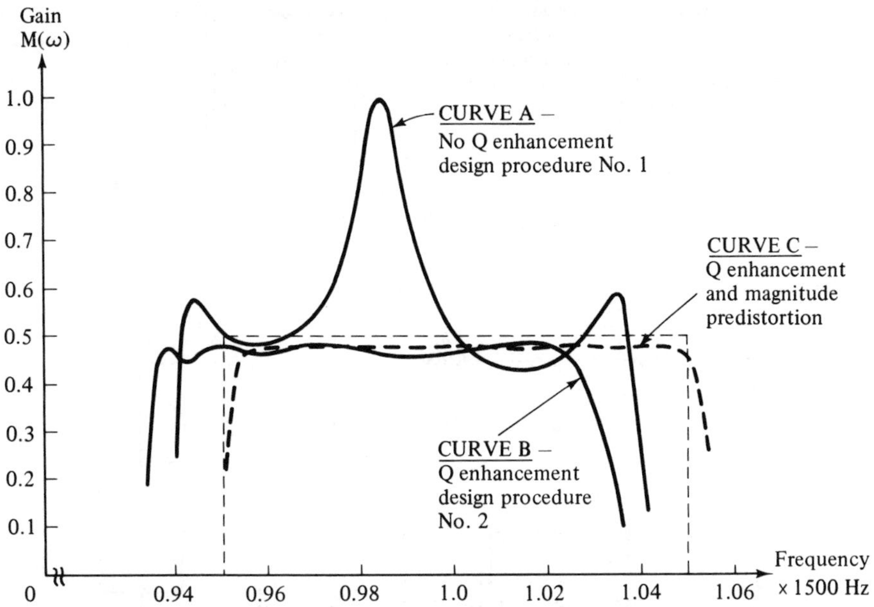

FIGURE 9.23 *Computed responses for fourteenth order Chebychev function; f_0 = 1.5 kHz; BW = 150 Hz; B = 1.5 MHz.*

whereas curve B, corresponding to a Q-enhanced design, gives a good approximation to the ideal response except that the bandedges are too low by about 2%. This is due to the magnitude error term ($\epsilon_{Hd}(\omega)$ in Table 5-9 for example); curve C shows the result of correcting for this magnitude distortion when selecting the capacitance elements.

9-7.4 Active Bandpass Ladder Filters Using FDNR *D* and *E* Elements

In this subsection, a method of realizing active ladder filters is described that is useful for realizing bandpass functions with zeros of transmission in the upper and lower stopbands. The method normally requires fewer OP AMPs than that described in subsection 9-7.3.

The conventional lowpass-to-bandpass transformation of equation (9.1) may be applied to the FDNR filter of Fig. 9.12(a) to realize the type of bandpass function in Fig. 9.1(c). It is shown here that this results in a FDNR bandpass structure containing grounded D and E type FDNR elements as shown in Fig. 9.24(b).

Consider the application of the lowpass-to-bandpass transformation of equation (9.1) to the admittance s^2C_i/k of the general FDNR shunt

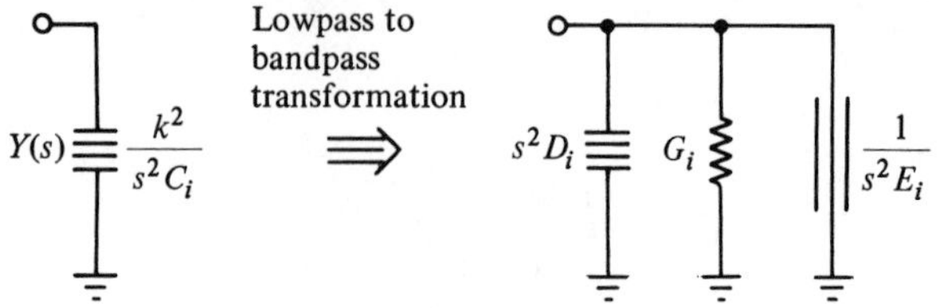

(a) Lowpass-to-bandpass transformation applied to the FDNR D element.

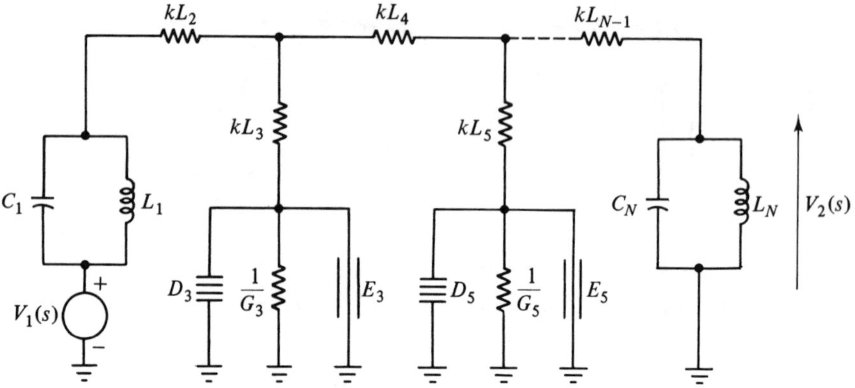

(b) Complete FDNR LC bandpass filter: obtained by bandpass transformation of FDNR lowpass filter in Fig. 9.12(a)

FIGURE 9.24

branch of the lowpass FDNR filter of Fig. 9.12(a); thus,

$$\frac{s^2 C_i}{k} \rightarrow \left[\frac{\omega_0 \omega_{CL}}{\text{BW}} \left(\frac{s}{\omega_0} + \frac{\omega_0}{s} \right) \right]^2 \frac{C_i}{k} \tag{9.144}$$

Expanding the right-hand side of the equation (9.144), we have

$$\boxed{\frac{s^2 C_i}{k} \rightarrow s^2 \text{D}_\text{i} + \text{G}_\text{i} + \frac{1}{s^2 E_i}} \quad \text{ADMITTANCE OF SHUNT BRANCH } i \tag{9.145}$$

where

$$\left.\begin{aligned} D_i &= \left(\frac{\omega_{CL}}{\text{BW}}\right)^2 \frac{C_i}{k} \\ E_i &= \left(\frac{\text{BW}}{\omega_0\omega_{CL}}\right)^2 \left(\frac{k}{\omega_0^2 C_i}\right) \\ \text{and} \qquad G_i &= 2\left(\frac{\omega_0\omega_{CL}}{\text{BW}}\right)^2 \left(\frac{C_i}{k}\right) \end{aligned}\right\} \tag{9.146}$$

Thus, the *FDNR element transforms to a parallel connection of a FDNR D element, a FDNR E element, and a conductance.* Since the FDNR element in the lowpass filter is grounded, both FDNR elements D_i and E_i are grounded, as shown in Fig. 9.24(a). Note that equations (9.146) are simplified considerably if k is chosen equal to ω_{CL}, the cut-off frequency of the prototype LC lowpass filter from which the FDNR lowpass filter in Fig. 9.12(a) is designed. Thus,

$$\boxed{\text{if} \quad k \equiv \omega_{CL}} \quad \text{rad/sec} \tag{9.147}$$

equations (9.146) simplify to

$$\boxed{D_i = \frac{C_i\omega_{CL}}{(\text{BW})^2}, \qquad E_i = \frac{(\text{BW})^2}{C_i\omega_0^4\omega_{CL}}, \qquad G_i = 2C_i\left(\frac{\omega_0}{\text{BW}}\right)^2 \omega_{CL}} \tag{9.148}$$

Consider now the terminating capacitances with admittances $s/kR_{1,N}$; applying the lowpass-to-bandpass transformation, we have

$$\frac{s}{kR_{1,N}} \to \frac{\omega_0\omega_{CL}}{\text{BW}}\left(\frac{s}{\omega_0} + \frac{\omega_0}{s}\right)\left(\frac{1}{kR_{1,N}}\right) \tag{9.149}$$

which may be written in the form

$$\boxed{\frac{s}{kR_{1,N}} \to sC_{1,N} + \frac{1}{sL_{1,N}}} \qquad \text{TERMINATING PARALLEL } LC \text{ CIRCUITS} \tag{9.150}$$

where

$$C_{1,N} = \frac{\omega_{CL}}{kR_{1,N}\text{BW}}, \qquad L_{1,N} = \frac{kR_{1,N}\text{BW}}{\omega_0^2\omega_{CL}} \tag{9.151}$$

If the original FDNR lowpass filter is designed such that $k = \omega_{CL}$, the cut-off frequency of the lowpass filter, then equations (9.151) simplify to

$$\boxed{C_{1,N} = \frac{1}{R_{1,N}\text{BW}}, \qquad L_{1,N} = \frac{R_{1,N}\text{BW}}{\omega_0^2}, \qquad k \equiv \omega_{CL}} \tag{9.152}$$

The transformation of the terminating capacitances to parallel LC circuits is illustrated in the final FDNR LC realization of Fig. 9.24(b). The really attractive feature of Fig. 9.24(b) is that both FDNR elements, D_i and E_i, in each shunt arm are *grounded.* Since the shunt arm requires two GICs (and therefore four OP AMPs), the series arm requires only the resistance kR_i. Thus, a fourteenth-order bandpass filter of this type would require just $3 \times 4 = 12$ OP AMPs to realize the FDNR elements in the three shunt arms, plus 4 OP AMPs for the source *floating* inductance L_1, and 2 OP AMPs for the load *grounded* inductance; this is a *total of 18 OP AMPs* to implement a transfer function with 14 poles and 6 imaginary zeros (three in each stopband). By comparison, the design of the Type E circuit in Table 9-1 using GIC-embedding techniques (see subsection 9-7.3) requires 6 OP AMPs for each of three shunt arms and 4 OP AMPs for each of four series arms corresponding to a *total requirement of 34 OP AMPs.* Thus, the FDNR LC structure of Fig. 9.2 results in a saving of 16 OP AMPs in the case of this particularly high-order filter. Unfortunately, the low-sensitivity properties of the LC prototype do *not* in any way transfer over to the FDNR LC structure of Fig. 9.24(b). The worstcase passive-sensitivity performance of the FDNR LC filter is quite troublesome and high-valued if ω_0/BW is much larger than unity. We shall not pursue this matter, and we state without proof that the FDNR LC structure is so highly sensitive to the RC-element values that it is *not useful for the narrowband case,* $\omega_0/\text{BW} \gg 1$.

Problem 9.26 provides the reader with an opportunity to design a FDNR LC bandpass filter with k equal to ω_{CL}.

9-7.5 Techniques for Simulating Important *LC* Bandpass Sections Having Finite Stopband Zeros

In this subsection, consideration is given to the realization of bandpass LC ladder filter sections that are widely used to realize finite zeros of

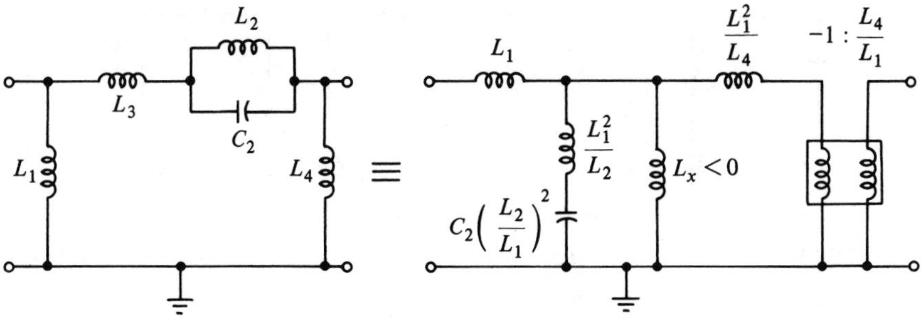

(a) Basic LC ladder section: bandpass with zero in lower stopband.

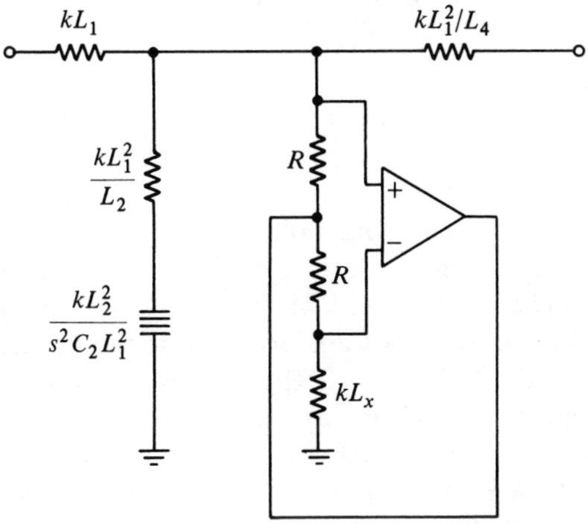

(b) FDNR/–R realization of Fig. 9.25(a): lower stopband zero.

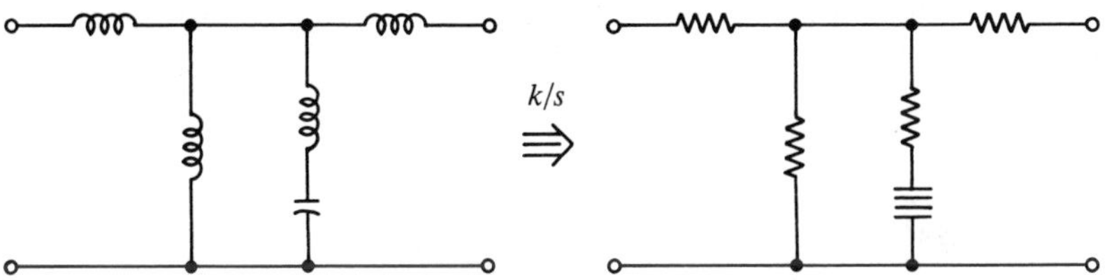

(c) FDNR realization of bandpass section with upper stopband zero.

FIGURE 9.25

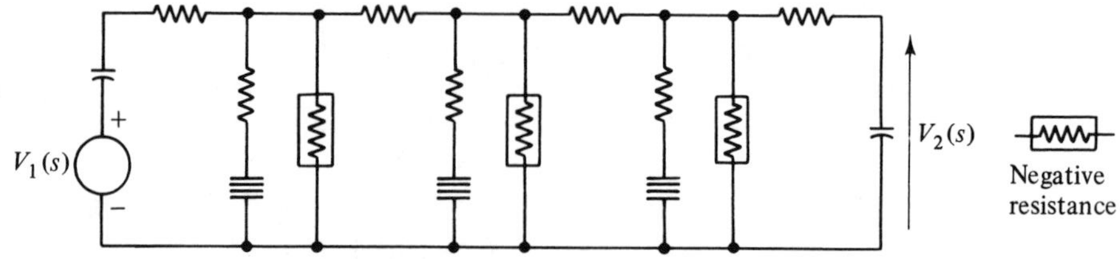

FIGURE 9.26 *Twelfth-order bandpass filter with upper stopband zeros; FDNR/-R.*

transmission in the upper or lower stopbands. The commonly used *LC* section of Fig. 9.25(a) realizes a lower stopband zero. The *LC* structure on the right side of Fig. 9.25(a) is equivalent; it contains a *grounded* capacitance $C_2(L_2/L_1)^2$ and a *ground negative inductance* L_x, where

$$L_x = -\left(\frac{L_1^2}{L_1 + L_2 + L_3 + L_4}\right) \tag{9.153}$$

and an ideal transformer with voltage ratio $-1 : (L_4/L_1)$. This transformer may be eliminated in a final *LC* realization by simply impedance-scaling all branches to the right of the transformer by the factor L_4/L_1. Applying the k/s impedance transformation to the *LC* section leads to the FDNR/$-R$ section of Fig. 9.25(b), where the transformer is omitted because it may be eliminated as described. The FDNR and $-R$ are *grounded* and have impedances $kL_2^2/s^2C_2L_1^2$ and $-kL_x$, respectively. Thus, three OP AMPs are required to realize this section.

The *LC* section that realizes an *upper* stopband zero is shown in Fig. 9.25(c). The FDNR realization uses just one grounded FDNR element, as shown in Fig. 9.25(c). A FDNR/$-R$ topology for implementing a bandpass filter with 12 poles and 6 upper stopband zeros is shown in Fig. 9.26. The resultant GIC circuit uses 12 OP AMPs.

9-8 SUMMARY

All of the active ladder synthesis and design techniques described in this chapter are based on the simulation of the branch immittances of resistively terminated *LC* filters or of their FDNR equivalents.

Inductance-simulation methods are described in sections 9-3 and 9-4. The transconductance-type PII (or gyrator) may be used to realize grounded or ungrounded active inductance elements and the resultant Q-factor $Q_l(\omega)$ may be determined as a function of frequency ω in terms of the nonideal bandwidths and nonideal input-output immittances of the transconductance amplifiers. Techniques are described for designing transconductance-type gyrators in such a way that $Q_l(\omega)$ has its maximum value at a critical frequency near the bandedge, where the group delay has a maximum value, allowing the nonideal passband deviation $\Delta M(\omega)$ to be minimized. High-quality filters may be implemented with transconductance amplifiers; however, transconductance amplifiers are not widely available at present as low-cost microelectronic devices. This fact tends to favor the GIC inductance-simulation methods described in section 9-4; GICs and OP AMPs *are* widely available as low-cost microelectronic devices. The excellent active-sensitivity performance of GIC inductance-simulation filters allows low cost OP AMPs ($B < 10^7$ rad/sec) to be used to design high-Q ($Q_P > 200$) filters to be realized at passband frequencies as high as 100 kHz. The magnitude error of L_{eq} may be taken into account in the original design. The principal disadvantage of GIC simulated inductors is that the GIC leads to a *grounded* inductance, thereby implying that the type A, C, and E filters of Table 9-1 cannot be realized by using simulated-inductance GIC methods because of the requirements for *ungrounded* inductance elements.

The concept of the k/s impedance transformation is used to realize type A lowpass FDNR filters. These filters are used widely in communication systems and are shown to have excellent passive-sensitivity and active-sensitivity performance. The use of *dual* OP AMPs for each FDNR element is important if the active-sensitivity performance is to be optimized; this allows each FDNR to be Q-enhanced.

Embedded-GIC techniques are useful for realizing ungrounded subnetworks of L, C, or FDNR elements. Bandpass active GIC ladder filters may be realized in this way, and active-sensitivity performance made optimum by taking particular care to maintain the Q-enhancement conditions for each of the GICs.

We do not pursue the nonideal performance of GIC active ladder filters beyond that provided in this chapter. However, the reader may be interested to know that researchers have shown that high-order ($N > 12$) simulations of LC filters may be implemented *in microelectronic form* over the frequency range 50 Hz to 100 kHz with Q pole-pairs in excess of 200 and with passband accuracies of better than 0.2 db over the temperature range 0°C to 70°C. The dynamic range is better than 90 db in such cases and, for a twelfth-order implementation, the typical power consumption is about 2 watts. Compared with the corresponding LC filter, the advantages of GIC microelectronic active ladders are:

(i) much smaller size and improved reliability,

(ii) may often be designed with fewer capacitors and with capacitors of equal and standard values,

(iii) may be functionally and permanently tuned by using automated (laser) trimming of microelectronic resistance elements.

The disadvantages are:

(i) the requirement for significant (≈ 2 watts, $N = 12$) dc power,

(ii) a slightly higher noise performance,

(iii) smaller signal-handling capability due to the limited dynamic range of the OP AMPs,

(iv) potential instability unless care taken during the design procedure.

At this time, GIC-type *RC*-active microelectronic filters are widely used in the telecommunications industry. They are particularly useful when ultraflat passbands are required at frequencies above a few kilohertz. The same type of high performance is possible without using GICs but by employing *RC*-active integrators to simulate the voltage-current relationship for each *L* and *C* element in the prototype filter. These methods are described in the next chapter; they usually require more OP AMPs than corresponding GIC methods.

PROBLEMS

9-1. Use the fifth-order lowpass *LC* prototype filter of Fig. 9.3(a) to design a tenth-order bandpass Chebychev *LC* filter having a center frequency of 10 kHz and a bandwidth $\mathrm{BW}/2\pi$ of 2 kHz.

9-2. Use the fifth-order lowpass *LC* prototype filter of Fig. 9.3(a) to design a fifth-order highpass Chebychev *LC* filter having a cut-off frequency $\omega_{CH}/2\pi$ of 5 kHz.

9-3. Terminate port 2 of the transconductance-type gyrator-PII model of Fig. 9.5(c) with a capacitance C_2, and then prove that the driving-point admittance at port 1 is given by equation (9.18).

9-4. A transconductance-type gyrator is used to simulate an inductance L_{eq}. The bandwidths of the transconductance amplifiers are sufficiently large that inequality (9.33) is valid. Prove that the Q-factor of the simulated inductance has a maximum value of $Q_{l\,max}$ given by equation (9.34), occurring at a frequency ω_{max} given by equation (9.35).

9-5. Inequality (9.33) is valid for a particular transconductance-type gyrator, and $L_{eq}=1$ H. The gyrator is designed to achieve a $Q_{l\,max}\omega_{max}$ product of greater than 10^6 rad/sec. Determine the maximum permissible value of the nonideal transconductance G_{11} at port 1 of the gyrator.

9-6. Assume that inequality (9.33) is valid and that, for a particular transconductance-type gyrator,

$$L_{eq}=1\text{ H}$$

$$G_{11}=G_{22}=0.5\times10^{-5}\text{ ohm}^{-1}$$

Find the terminating capacitance C_2 so that $Q_{l\,max}$ occurs at 10^4 rad/sec. What is the corresponding value of $Q_{l\,max}$?

If this *same* gyrator having a *different* terminating capacitance C_2 is used to realize the *same* inductance L_{eq} but with C_2 chosen so that ω_{max} is 10^3 rad/sec., what is the corresponding value of $Q_{l\,max}$?

9-7. If the gyrator transconductance time constants T_1 and T_2 in equation (9.28) cannot be neglected, prove that $Q_{l\,max}$, ω_{max}, and the product $Q_{l\,max}\omega_{max}$ are given by equations (9.40) and (9.41).

9-8. Consider a transconductance-type gyrator for which inequality (9.44) is valid; assume that $L_{eq}=1$ H, $G_{11}=G_{22}=0.5\times10^{-5}$ ohm^{-1}, $C_2=10$ nF, $T_1+T_2=2\times10^{-5}$ sec.

Find the frequency ω_{max} at which $Q_l(\omega_{max})$ is infinite, and compare this value of ω_{max} with that obtained in the first part of Problem 9-6.

Calculate $Q_l(\omega)$ at $\omega=5\times10^3$ rad/sec. and at $\omega=10^4$ rad/sec. Comment on the result, and identify these two points in Fig. 9.6(d).

9-9. Consider the application of a transconductance-type gyrator where inequality (9.46A) is valid. For example, assume that $L_{eq}=10$ mH, $G_{11}=G_{22}=\frac{1}{2}\times10^{-5}$ ohm^{-1}, $T_1+T_2=2\times10^{-5}$ sec., $C_2=100$ pF.

Check that inequality (9.44) is satisfied, and find ω_{max} by using equation (9.46). Then use the requirement that $|Q_l|\geqslant10$ to estimate the upper frequency at which this active inductance may be used. [*Hint*: Equation (9.47) may be employed.]

9-10. Assume that *identical* transconductance-type gyrators have been designed having transconductance amplifiers g_{21} and g_{12} with -3 db bandwidths, given by

$$\frac{1}{T_1}=\frac{1}{T_2}=0.5\times10^6\text{ rad/sec}$$

The gyrators are to be used to realize the type B and type D *LC* filters shown in Table 9.1 by direct simulation of the grounded inductances. Transfer function gain accuracy $|\Delta M(\omega)/M(\omega)|$ must be less than 0.05 (that is, $\approx\frac{1}{2}$ db) over the frequency range up to 10 kHz. Use equation (9.49) to estimate the extremum group delay $\tau(\omega)$ that can be permitted.

If Q_0 is the Q-factor of the pole-pair having the highest Q-factor, then it may be shown that

$$|\tau(\omega)|_{\max} \approx \frac{2Q_0}{\omega_{\max}}$$

for all the *LC* filters in Table 9.1, where $\omega_{\max}$ is the frequency at which the extremum group delay $|\tau(\omega)|_{\max}$ occurs. Use this result and the numerical values above to estimate the maximum Q_0 that is possible using this method of inductance simulation.

9-11. In a particular application, it is sometimes found that the transconductance amplifier bandwidths $1/T_1$ and $1/T_2$ are *too small* and thereby lead to high-frequency Q-factors $Q_{Hl}(\omega)$ in equation (9.47) that severely limit the high-frequency applications of the transconductance-type gyrator in inductance-simulation circuits.

Show that small capacitors in parallel with the resistors R_{E5}, R_{E6}, and R_{E7} in Fig. 4.19 introduce zeros into the expressions for g_{12} and g_{21} ($I_2(s)/V_1(s)$ and $I_1(s)/V_2(s)$ in equations (4.50) and (4.51)). Then, suggest a way of choosing the zeros to cancel exactly the poles at $-1/T_1$ and $-1/T_2$.

9-12. The design study in section 9-3.2 involves the use of three transconductance-type gyrators to realize L_3, L_5, and L_7 in Fig. 9.7(a). The gyrator transconductances $g_{1,3}$ and g_5 and the capacitances $C_{3,7}$ and C_5 are chosen so that the active inductance Q-factors are infinite at *approximately* the cut-off frequency ω_0 of 10^4 rad/sec. For the numerical values given in the design study, use the *exact* expression in equation (9.21) to calculate and sketch the Q-factor $Q_l(\omega)$ and inductances $L_{3,5}$ throughout the frequency range 0 to 10^5 rad/sec. Thereby, estimate the *accuracy* of equations (9.40) and (9.47) as design equations.

9-13. Equation (9.52) describes the Q-factor of a type 1 coupled-GIC simulation of the *grounded* inductance element, assuming *uniform* conductances and perfectly matched gain bandwidth products, $B_1 = B_2$. Sketch $Q_l(\omega)$ over the frequency range 0 to 10^5 rad/sec, assuming the typical numerical values $A_0 = 10^5$, $B = 2\pi \times 10^6$ rad/sec, $\tau = 10^{-4}$ sec.

Compare this $Q_l(\omega)$ function with that obtained in Problem 9.12.

9-14. The upper-frequency limitation of the type 2 GIC inductance-simulation circuit shown in Table 5-7 is provided by the last term in equation (5.116); that is, at sufficiently high frequencies,

$$Q_l(\omega) \approx -\frac{B_1 B_2 G_6}{\omega^3 C_5}$$

Writing C_5/G_6 as T seconds, show that the high-frequency performance of inductance-simulated filters, such as types B and D of Table 9-1, is limited by the relationship

$$\frac{\Delta M(\omega)}{M(\omega)} \approx \frac{\omega^4 T_T(\omega)}{2B^2}$$

9-15. The nonideal high-frequency performance of type 1 and type 2 GIC inductance-simulation circuits has been assumed to be primarily a function of the nonideal gain bandwidth products B_1 and B_2. However, the nonideal input capacitances C_{CM1} and C_{CM2} associated with Z_{CM1} and Z_{CM2} in the OP AMP model of Fig. 5.3(b) are sometimes significant. Assume that $C_{CM1} = C_{CM2} \equiv C_P$ for both OP AMPs in a uniform-G type 2 inductance-simulation GIC circuit, and show that the input admittance is as follows:

$$Y_{1d}(s) = \frac{G^2}{sC} + G\left(\frac{C_P}{C}\right) - sC_P$$

where G and C are the GIC branch elements. Derive the nonideal Q-factor $Q_l(\omega)$ and nonideal effective inductance L_{eq} as a function of G, C, C_P, and ω. Sketch $Q_l(\omega)$ for the typical numerical values $G = 10^{-4}$ ohm^{-1}, $C = 10$ nF, and $C_P = 2$ pF. Comment on the relative effects of C_P and B at high frequencies.

9-16. Prove that "impedance-scaling by $K(s)$ does not alter any *current* transfer function" of a network. The proof is similar to that in section 9-5.1 but uses the nodal equations instead of the loop equations, and Cramer's rule is used to solve for unknown loop currents as a function of excitation current generators.

9-17. Repeat Example 9.4, but with the cut-off frequency and impedance level denormalized to give a seventh-order FDNR lowpass Chebychev filter, with $\omega_0 = 10^4$ rad/sec. operating between terminating capacitances $C_1 = C_9 = 10$ nF, with all other capacitances equal to 10 nF, and with the Q-factor enhancement condition in Table 5-9 satisfied for all three FNDR GIC circuits.

Satisfy the bias conditions by selecting R_9' in Fig. 9.13(c) equal to 1 MΩ, and then select a suitable value for R_1' so that $M(0) = 0.5$.

9-18. Show that the D elements in the case study of section 9-6 are in error by approximately +4% at the cut-off frequency.

9-19. Prove that the expression (9.97) provides the correct predistorted value for the resistance R_{3_i} of the ith FDNR GIC circuit, where $\epsilon_{Hd_i}(\omega_0)$ is given in equation (5.131) for the uniform G/uniform C case and in equation (5.130) otherwise.

9-20. Given that $B = 2\pi \times 10^7$ rad/sec for every OP AMP, calculate the predistorted values for R_{33}, R_{35}, and R_{37} in Problem 9.17. Also, calculate the Q-factors $Q_d(\omega)$ for each of the three FDNR elements D_3, D_5, and D_7 at the cut-off frequency $\omega_0 = 10^4$ rad/sec.

9-21. Repeat Problem 9-20, using $B = 2\pi \times 10^6$ rad/sec.

9-22. Repeat Problem 9-17, but use GIC-embedding techniques, as described in section 9-7.1, to operate the filter between terminating resistances $R_s = R_L = 10$ kΩ. Choose $K = \omega_0$. Maintain GIC-enhancement conditions.

9-23. A third-order LC elliptic filter, as shown in Fig. 9.17(b), has the following element values: $L_1 = 0.1267$H, $L_2 = 0.0536$H, $L_3 = 0.1267$H, $R_s = R_L = 10$ kΩ, $C_2 = 7.480$ nF. Design a corresponding embedded-GIC realization, as shown in Fig. 9.17(a), with $k = 1$ and with *all* circuit capacitances equal to 10 nF.

9-24. Repeat Example 9.6 in the text to design a fourteenth-order symmetrical bandpass filter, using embedded GICs and design procedure No. 1 but with

$$\omega_0 \equiv \mathrm{BW} \equiv 10^3 \text{ rad/sec}$$

and using GIC capacitance elements such that

$$C = 100 \text{ nF}$$

9-25. Repeat Problem 9-24, but use design procedure No. 2 to achieve optimum active-sensitivity performance.

9-26. Use the lowpass seventh-order LC filter of Fig. 9.7(a) to design a corresponding fourteenth-order symmetrical bandpass active filter. The filter is to be designed using the FDNR method described in section 9-7.4. Why should the FDNR method not be recommended in this case?

9-27. The bandpass LC filter of Fig. P9.27 has two zeros in the lower stopband. Show that it may be realized with just six OP AMPs, and sketch a circuit configuration that could be used. Use FDNR/$-R$ methods.

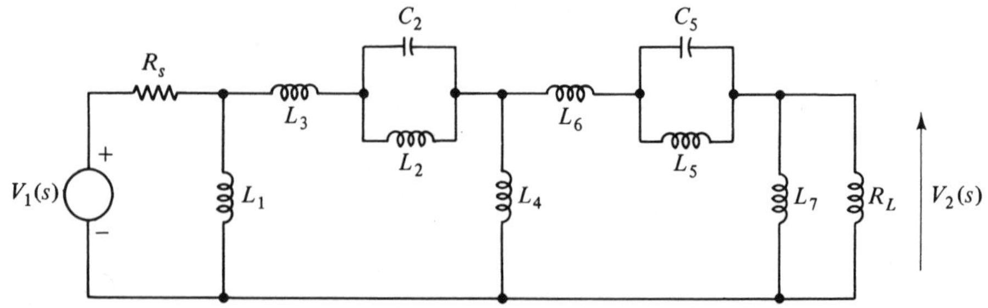

FIGURE P9.27

9-28. The bandpass *LC* filter of Fig. P9.28 has two zeros in the upper stopband. Show that it may be realized with just four OP AMPs, using FDNR methods.

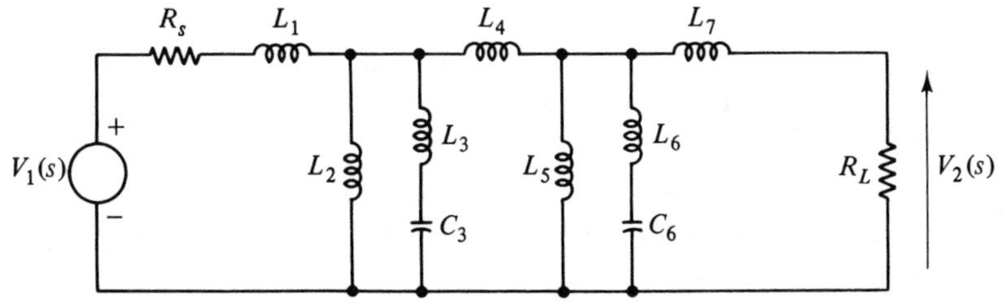

FIGURE P9.28

9-29. Use the *LC* prototype filter of Fig. P9.29 to design a four OP AMP FDNR version having a cut-off frequency of 25 kHz, using 5 nF capacitors and meeting the enhancement conditions for the FDNR elements.

Assume the OP AMPs have gain bandwidth products B equal to $2\pi \times 10^6$ rad/sec and predistort the values of D_2 and D_4 if necessary to take into account $\epsilon_{Hd}(\omega)$.

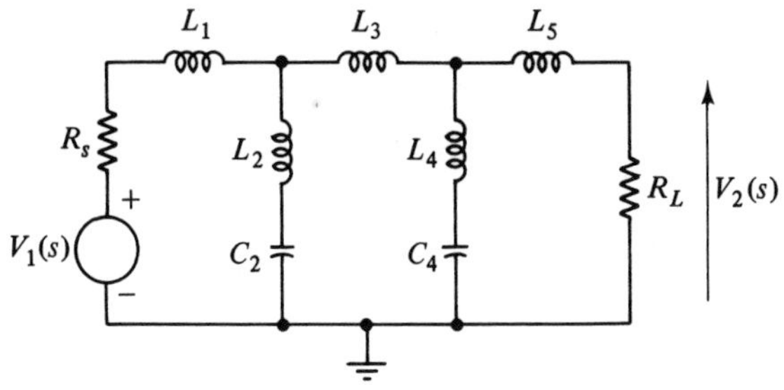

FIGURE P9.29

9-30. Prove that a coupled-GIC circuit may be used to realize a type 3 FDNR such that the *enhancement condition does not require matched B_1 and B_2* by selecting $Y_2 = sC$, $Y_3 = Y_4 = Y_5 = G$, and a termination $Y_6 = sC$ at port 2 in Table 5-10. [*Hint*: Use equation (5.67).] This is a lengthy proof.

BIBLIOGRAPHY

ANTONIOU, A., "Bandpass Transformation and Realization Using Frequency Dependent Negative-Resistance Elements," *IEEE Trans. Circuit Theory*, vol. CT-18, p. 297–299, March 1971.

BLOSTEIN, M. L., "Sensitivity Analysis of Parasitic Effects In Resistance Terminated LC Two-Ports," *IEEE Trans. Circuit Theory*, vol. CT-14, p. 21–25, March 1967.

BRUTON, L. T., "Network Transfer Functions Using The Concept Of Frequency Dependent Negative Resistance," *IEEE Trans. Circuit Theory*, vol. CT-16, p. 406–408, Aug. 1969.

BRUTON, L. T., J. T. LIM, and J. VALIHORA, "The Feasibility of Active Filtering in Frequency Division Multiplex Systems," *Proc. Int. Symp. on Circuit and System Theory*, San Francisco, April 1974.

BRUTON, L. T., and A. I. A. SALAMA, "Frequency Limitations of Coupled-Biquadratic Active Ladder Structures," *IEEE Jour. of Solid-State Circuits*, vol. SC-9, p. 69–71, April 1974.

BRUTON, L. T., and D. TRELEAVEN, "Active Filter Design Using Generalized Impedance Converters," *EDN*, p. 68–75, Feb. 1975.

GORSKI-POPIEL, J. 'RC-Active Systhesis Using PICs,' *Electronics Letters*, vol. 3, p. 381–382, 1967.

HOLMES, W. H., W. E. HEINLEIN, and S. GRUTZMANN, "Sharp Cut-off Low-Pass Filters Using Floating Gyrators," *IEEE Jour. Solid-State Circuits*, vol. SC-4, p. 38–50, Feb. 1969.

ZVEREV, A. I., *Handbook of Filter Synthesis*, Wiley, New York, 1967.

10

FILTER DESIGN USING SIGNAL-FLOW GRAPH SIMULATION

10-1 INTRODUCTION

In the previous chapter, techniques are described that allow high-quality high-order *RC*-active filters to be implemented by active simulation of the branch immittances of equiterminated *LC* filters. In this chapter, important *alternative* means of simulating *LC* filters are described. The methods to be explained here do *not* involve a simulation of each immittance in a direct way but, rather, they involve the use of electronic integrators, differentiators, summers, and inverters *to simulate the voltage-current equations* that describe the *LC* filter. For this reason, the reader may wish to review section 7.5 and particularly subsection 7-5.5, where equations (7.151) describe the general ladder topology corresponding to Fig. 7.19. A signal-flow graph (SFG) description of the ladder topology in Fig. 7.19 is shown in Fig. 7.20(b); note that *this SFG is characterized by the one-to-one correspondence between a branch immittance Z_i or Y_{i+1} in Fig. 7.19 and a unique transmittance in the SFG*. The corresponding SFG description of the *LC* filter of Fig. 7.18(a) is shown in Fig. 7.21 and is the starting point for the following discussions of active "leapfrog" ladder filters. The sensitivity performance and the design of active simulations of these SFG's are the major topics of interest in this chapter.

10-2 LOWPASS ("LEAPFROG") LADDER FILTERS USING INTEGRATORS TO SIMULATE THE SFG

From inspection of Fig. 7.21 it follows that the basic building-blocks required to realize this lowpass SFG are the inverting-integrator $(-1/sC_i)$ for each capacitance element C_i and the noninverting integrator $(1/sL_{i+1})$ for each inductance element L_{i+1}. Further, two-input summers are required at the inputs of each of the integrators. OP AMP circuit implementations of inverting and noninverting integrators and summers are described and analyzed in sections 5-3 and 5-4; the performance limitations due to the gain bandwidth product B of the OP AMP are summarized in Tables 5-1 and 5-2 and are used here to determine the high-frequency limitations of the resultant practical active ladder filters. The designer may implement the SFG of Fig. 7.21 by using either type 1A noninverting integrators, as shown in Fig. 10.1(a), or type 1B noninverting integrators, as shown in Fig. 10.1(b). The *ideal performance* of the circuits shown in Figs. 10.1(a) and 10.1(b) are, of course, identical because they correspond to the same RC-nullor network. However, it is shown in section 10-2.3 that the active-sensitivity performance of the type B lowpass "leapfrog" structure of Fig. 10.1(b) is far superior to that of the type A structure of Fig. 10.1(a).

Note that the structures shown in Figs. 10.1(a) and (b) contain approximately $3N/2$ OP AMPs ($N \gg 1$), where the order of the transfer function is N, whereas the GIC FDNR method of Fig. 9.12(a) requires only N OP AMPs. (Furthermore, the FDNR circuit is more general in that the approximately $N/2$ resistors R_3', R_5', etc. allow imaginary zeros of transmission to be easily implemented.)

10-2.1 Passive-Sensitivity Performance of Type A and Type B Lowpass "Leapfrog" Ladder Filters

The RC-nullor versions of Figs. 10.1(a) and 10.1(b) are identical, and this can be verified by replacing each OP AMP by a nullor in the circuits of Fig. 10.1. Thus, if the OP AMPs are ideal ($A_0 = \infty$, $B = \infty$) then the type A and type B lowpass "leapfrog" filters are entirely equivalent. Consequently, the passive-sensitivity performance is identical for type A and type B circuits.

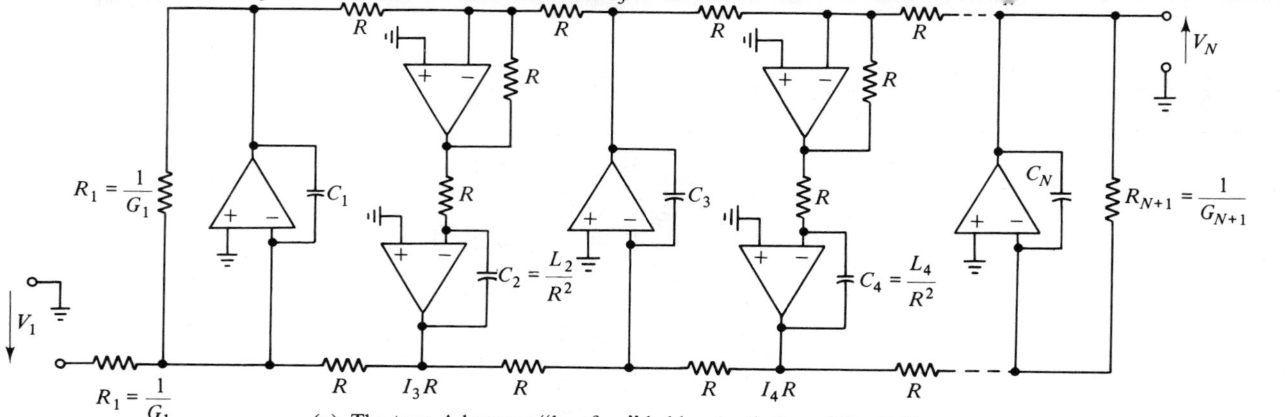

(a) The type A lowpass "leapfrog" ladder simulation of Fig. 7.21.

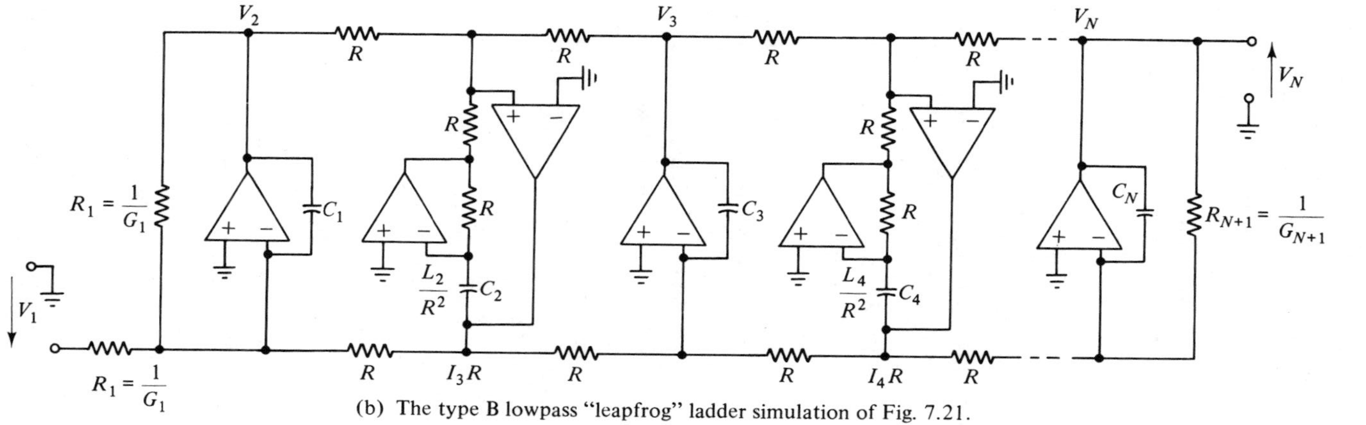

(b) The type B lowpass "leapfrog" ladder simulation of Fig. 7.21.

FIGURE 10.1

Consider now the dependence of the filter input-output magnitude response $M(\omega)$ on the RC elements of Figs. 10.1(a) and 10.1(b). The *vertically drawn RC* elements correspond, on a one-to-one basis, to the *gains* of the integrating transmittances in the SFG of Fig. 7.21 and, therefore, to a *particular LC* element or resistive termination element of the LC prototype filter. A perturbation of any one of these *vertically drawn RC* elements of Figs. 10.1(a) and 10.1(b) causes exactly the same perturbation of $H(s)$ and, therefore, of $M(\omega)$ as a similar perturbation of the corresponding L, C, or R element in the prototype LC filter. It therefore follows that the analysis and discussion in section 7-5.5 applies directly to the *vertically drawn RC* elements of Figs. 10.1(a) and 10.1(b). That is, if these active "leapfrog" ladder filters are designed from an equiterminated LC filter prototype, with $|t(j\omega)|$ equal approximately to unity throughout the passband, then the sensitivity of $M(\omega)$ to the *vertically drawn RC* elements is low-valued throughout the passband [see equations (7.148) and (7.150).] Wherever, $|t(j\omega)| = 1$, the sensitivity of $M(\omega)$ to the *vertically drawn RC* elements *is ideally zero.*

The preceding argument *cannot* be extended to the *horizontally drawn* resistance elements of Figs. 10.1(a) an 10.1(b) because these summing resistance elements do *not* correspond directly to a particular L, C, or R element of the prototype filter. However, if each *pair* of summing resistance elements connected to a particular inverter or inverting-integrator in Fig. 10.1(a) or 10.1(b) is *matched* (that is, equivalued) then a perturbation of a matched pair of summing resistances is equivalent to a perturbation of the corresponding L or C element in the LC prototype filter, and the results in equations (7.148) and (7.150) may then be extended to the *horizontally drawn* resistance elements in Figs. 10.1(a) and 10.1(b). It may be shown that $|t(j\omega)|$ remains insensitive to the horizontally drawn summing resistances even when they are not of equal value; the proof is, however, beyond the scope of this text.

To summarize, "leapfrog" structures that are designed from equiterminated LC prototypes such that $|t(j\omega)|$ is approximately unity in the passband possess excellent passband insensitivity of $M(\omega)$ to the vertically drawn RC elements and also to the remaining horizontally drawn R elements. It may be shown that the worstcase *transition region* passive-sensitivity performance of "leapfrog" ladder filters is as follows:

$$\boxed{WS_{RC}^{M(\omega)} \approx 3\left|\frac{\partial \ln M(\omega)}{\partial \ln \omega}\right|} \qquad \text{TRANSITION REGION} \qquad (10.1)$$

which is a factor of only 1.5 times the lower bound [see equations (6.115) and (6.116).]

10-2.2 Active-sensitivity Performance of Type 1A Lowpass "Leapfrog" Ladder Filters

Consider now the effect on $M(\omega)$ of the OP AMPs shown in Fig. 10.1(a), having nonideal voltage transfer functions of the dominant-pole type given in equation (5.1). The consequential nonideal voltage transfer functions of the inverting integrator and type 1A and 1B noninverting integrators are given in Tables 5-1 and 5-2 in terms of the nonideal multiplicative functions $M(j\omega)$. Thus, the inverting integrators and noninverting integrators of Fig. 10.1(a) have nonideal voltage transfer functions as follows:

$$\frac{Z_i(j\omega)}{R} = -\frac{1}{j\omega C_i R}[M_i(j\omega)], \qquad i = 2, 4, 6, \text{etc.} \tag{10.2}$$

NONIDEAL INVERTING-INTGEGRATOR GAINS

where the $Z_i(j\omega)$ terms correspond to the nonideal capacitive impedances of the shunt branches of the LC filter, and

$$\frac{R}{Z_i(j\omega)} = \frac{R}{j\omega L_i}[M_i(j\omega)], \qquad i = 1, 3, 5, \text{etc.} \tag{10.3}$$

NONIDEAL NONINVERTING INTEGRATOR GAINS

where the $Z_i(j\omega)$ terms correspond to the nonideal inductive impedances of the series branches of the LC filter. These expressions for the nonideal gains of the integrators are now used to determine the effective Q-factors, $Q_c(\omega)$ and $Q_l(\omega)$, of the C_i and L_i elements of the equivalent LC filter. The effective Q-factors of the nonideal impedances are as follows:

$$Q_{l,c}(\omega) \approx \frac{1}{\arg\left[M(j\omega)\right]}, \qquad \text{if } |Q_{l,c}(\omega)| \gg 1 \tag{10.4}$$

[See equation (5.85).] Thus, using equation (10.4) with the result from Table 5-1 that $\arg[M(j\omega)] \approx -\omega/B$, we find that

$$Q_c(\omega) \approx -\frac{B}{\omega} \tag{10.5}$$

EFFECTIVE-CAPACITOR Q-FACTORS, $|Q_c| \gg 1$

Similarly, the result shown in Table 5-2 for the type 1A noninverting integrator is that arg $[M(j\omega)] \approx -3\omega/B$ so that, from equation (10.4),

$$Q_l(\omega) \approx -\frac{B}{3\omega} \qquad \text{EFFECTIVE-INDUCTOR } Q\text{-FACTORS, } |Q_l| \gg 1 \tag{10.6}$$

It is interesting to note that, to a first approximation, $Q_l(\omega)$ and $Q_c(\omega)$ are independent of the values of L_i and C_i. The approximate effect of the phase errors arg $[M(j\omega)]$ on the overall filter magnitude function $M(\omega)$ are derived from equations (10.5), (10.6), and (9.15) as

$$\frac{\Delta M(\omega)}{M(\omega)} \approx \frac{2\omega^2\tau(\omega)}{B} \qquad \text{APPROXIMATE ERROR IN MAGNITUDE TRANSFER FUNCTION DUE TO INTEGRATOR PHASE ERRORS} \tag{10.7}$$

This simple result is sufficiently accurate for most applications to allow the passband error $\Delta M(\omega)/M(\omega)$ and the required OP AMP gain bandwidth products B to be determined reliably.

NUMERICAL EXAMPLE: Fifth-Order Lowpass Type A "Leapfrog" Filter; Chebychev Response: Passband Ripple = 1.0 *db; Cut-off Frequency* = 10 *kHz; OP AMP Bandwidths* = 3 *MHz.*

SOLUTION: A suitable *denormalized LC* prototype Chebychev filter is illustrated in Fig. 10.2(a). It follows by comparison of this *LC* filter with Fig. 10.1(a) that the corresponding element values for the type A lowpass "leapfrog" circuit of Fig. 10.1(a) are

$$\left.\begin{aligned} &R = R_1 = R_6 = 10\ k\Omega \\ &C_1 = C_5 = 3.39\ \text{nF}, \qquad C_3 = 4.77\ \text{nF} \\ &\text{and} \\ &C_2 = C_4 = \frac{L_{2,4}}{R^2} = \frac{0.173}{(10^4)^2} = 1.73\ \text{nF} \end{aligned}\right\} \tag{10.8}$$

The *ideal* passband gain $M(\omega)$ is shown in Fig. 10.3(b) as a *continuous* curve that is calculated on the assumption that the OP AMPs are equivalent to nullors ($A_0 = B = \infty$) and that all the *RC*-element values are precisely correct. The active-sensitivity performance is calculated from the group delay τ of the original *LC* filter [Fig. 10.2(a)]; the group delay is shown in Fig. 10.2(b), where it is observed to have a peak value of approximately 210 μs near the bandedge. Given that the OP AMPs possess gain bandwidth products of approximately 3 MHz ($B \approx 2\pi \times 3 \times 10^6$ rad/sec), it is possible

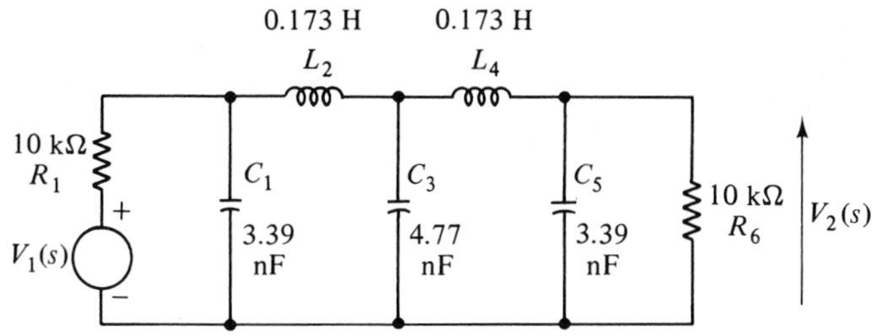

(a) Suitable LC prototype; Chebychev, 1 db passband ripple, cut-off frequency = 10 kHz.

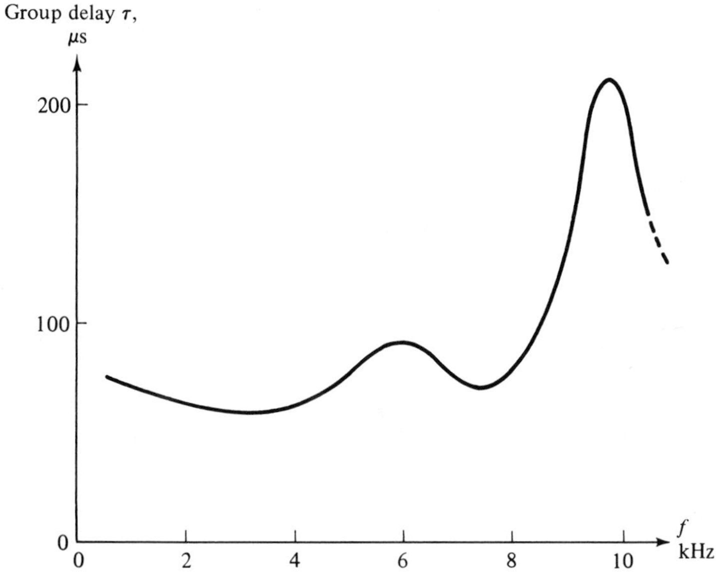

(b) Group delay of filter in Fig. 10.2(a).

FIGURE 10.2

to use equation (10.7) with Fig. 10.2(b) to obtain the corresponding fractional passband gain error $\Delta M(\omega)/M(\omega)$, as shown in Fig. 10.3(a). The *nonideal* passband response $M'(\omega)$ due to the effects of finite OP AMP bandwidths B is as follows:

$$M'(\omega) = M(\omega) + \Delta M(\omega) \approx M(\omega)\left[1 + \frac{2\omega^2\tau(\omega)}{B}\right] \qquad (10.9)$$

NONIDEAL PASSBAND GAIN: TYPE A "LEAPFROG" LADDER

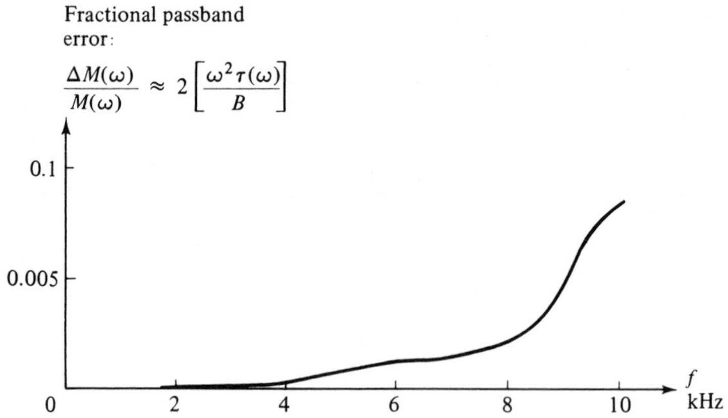

(a)

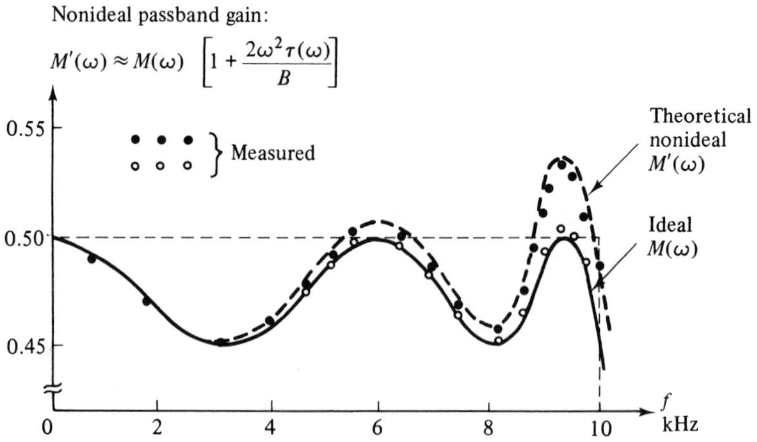

(b)

FIGURE 10.3 *(a) Approximate fractional passband error: type A lowpass Chebychev "leapfrog" filter; (b) nonideal passband response M(ω) for the type A lowpass Chebychev "leapfrog" filter; ideal, nonideal theoretical M′(ω), and measured.*

and is sketched in Fig. 10.3(b) as the *dashed* curve $M'(\omega)$. Verification of equation (10.9) may be obtained using a computer-aided analysis program with dominant-pole approximations for all the OP AMPs; in this case, it is found that equations (10.7) and (10.9) are highly accurate. These equations lose accuracy if the assumptions leading to equations (10.5) and (10.6) are inaccurate; that is, if $\omega \ll B$ and $\omega \ll 1/CR$ for all capacitors are not valid

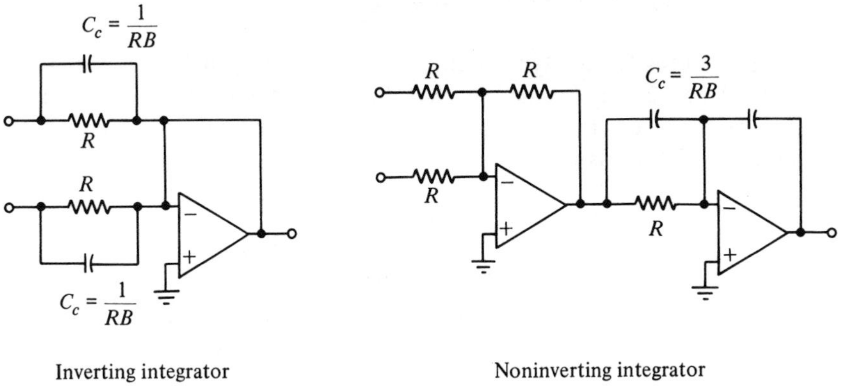

FIGURE 10.4 *Capacitive compensation scheme for type A "leapfrog" ladder filters.*

assumptions. In such cases, the *complete* expressions for $M(j\omega)$ in Tables 5-1 and 5-2 may be used to improve the accuracy of the calculation.

A capacitive compensation scheme: A straightforward technique for correcting the nonideal passband gain $M'(\omega)$ is to employ small compensating capacitors C_c in parallel with the input resistances R associated with each of the inverting-integrators and noninverting integrators of Fig. 10.1(a), as shown in Fig. 10.4. Thus,

$$C_c = \frac{3}{RB} \qquad \text{NONINVERTING INTEGRATORS} \tag{10.10}$$

and

$$C_c = \frac{1}{RB} \qquad \text{INVERTING INTEGRATORS} \tag{10.11}$$

from which it can be shown (Problem 10-3) that the equations for C_c given above provide the phase shift necessary to cancel the phase shifts of $-3\omega/B$ and $-\omega/B$ that are shown in Tables 5-1 and 5-2. In the previously considered numerical example, the compensation capacitors C_c for the integrators 1, 2, 3, 4, and 5 are supplied by equations (10.10) and (10.11) as 5.3 pF, 15.9 pF, 5.3 pF, 15.9 pF, and 5.3 pF, respectively.

Measured data: The above type 1A Chebychev lowpass ladder filter, constructed with OP AMPs for which $B \approx 3 \times 2\pi \times 10^6$, has been measured carefully in the laboratory. The black dots in Fig. 10.3(b) are measured without compensation capacitors C_c, and the white circles are measured after the introduction of compensation capacitors. These measurements substantially confirm the validity of the foregoing analysis of active-sensitivity performance.

10-2.3 Superior Active-Sensitivity Performance of Type B Lowpass "Leapfrog" Ladder Filters

The analysis for the type B circuit in Fig. 10.1(b) differs from that in subsection 10-2.2 only by the fact that the type 1B noninverting integrator (Table 5-2) is used, so that $M(j\omega)$ is approximately $+\omega/B$ (instead of $-3\omega/B$), which leads to

$$\boxed{Q_l(\omega) \approx +\frac{B}{\omega}} \qquad \begin{array}{l}\text{EFFECTIVE-INDUCTOR}\\ Q\text{-FACTOR}, |Q_l| \gg 1\\ \text{TYPE B "LEAPFROG" FILTER}\end{array} \tag{10.12}$$

instead of $-B/3\omega$ as in equation (10.6). Now, the implication of this is very important because, clearly, by comparison of equations (10.5) and (10.12),

$$Q_l(\omega) \approx -Q_c(\omega) \quad \text{TYPE 1B "LEAPFROG" CIRCUIT} \tag{10.13}$$

for all LC elements. Therefore, from equations (10.5), (10.12), and (9.15),

$$\boxed{\frac{\Delta M(\omega)}{M(\omega)} \ll \frac{\omega^2\tau(\omega)}{B}} \qquad \begin{array}{l}\text{TYPE 1B "LEAPFROG" CIRCUIT}\\ \text{PASSBAND DEVIATION}\end{array} \tag{10.14}$$

implying that *this circuit is self-compensating in the sense that the passband error produced by the inverting-integrators is approximately cancelled by that produced by the noninverting integrators*. This important advantage of type B "leapfrog" ladder filters makes them especially useful for high-frequency high-quality applications. Further qualitative evidence of this is given in the following section.

10-3 BANDPASS ("LEAPFROG") LADDER FILTERS USING INTEGRATORS

In this section, the design of bandpass "leapfrog" ladder filters is discussed; the LC prototype filter is shown in Fig. 7.22(a) and the corresponding SFG in Fig. 7.22(d). It follows by inspection of Fig. 7.22(d) that the SFG may be simulated with inverting-integrators $(-1/sC_i)$ and noninverting integrators $(+1/sL_i)$. Note that the basic structure in Fig. 7.22(d) is an interconnection of two integrator loops (TILs), where each TIL realizes a parallel LC branch or a series LC branch of the LC prototype filter.

A suitable implementation of Fig. 7.22(d) is obtained by replacing each inverting-integrator by the single OP AMP integrator of Fig. 5.5(b) and each noninverting integrator by *either* the type 1A *or* the type 1B circuit of Table 5.2. The resultant type A and type B bandpass "leapfrog" ladder circuits are shown in Figs. 10.6(a) and 10.6(b) for the sixth-order case.

Consider, for example, the sixth-order prototype LC filter of Fig. 10.5(a), where the numerical values of the LC elements correspond to the following normalized narrowband Chebychev magnitude transfer function specifications:

$$\left.\begin{array}{ll} \text{Center frequency:} & \omega_0 = 1 \text{ rad/sec} \\ \text{Bandwidth:} & \text{BW} = 0.1 \text{ rad/sec} \\ \text{Passband ripple:} & = 0.1 \text{ db} \\ \text{Order:} & N = 6 \end{array}\right\} \qquad (10.15)$$

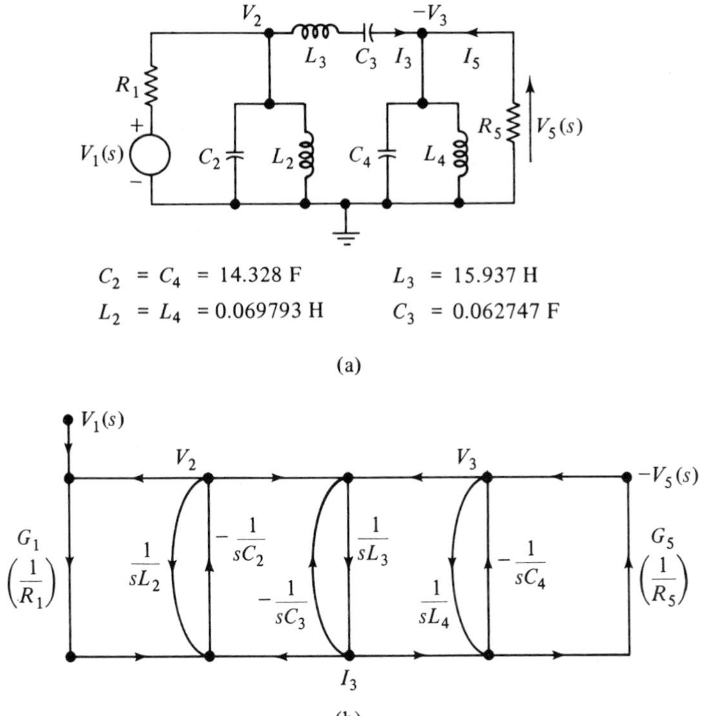

FIGURE 10.5 *(a) Normalized Chebychev bandpass LC prototype: passband ripple = 0.1 db, ω = 1 rad/sec; BW = 0.1 rad/sec; sixth order; (b) SFG for the sixth-order LC bandpass filter in Figure 10-5(a).*

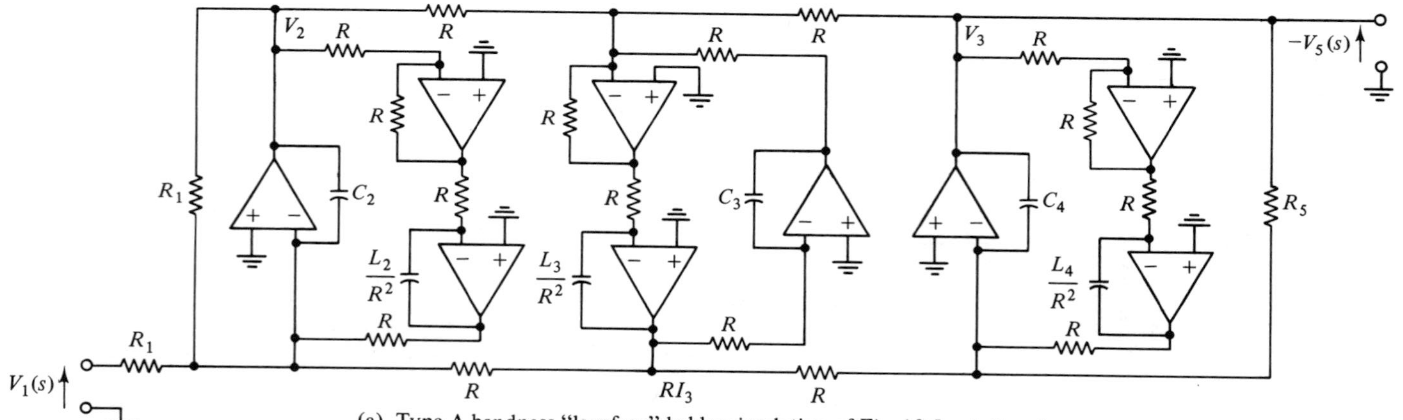

(a) Type A bandpass "leapfrog" ladder simulation of Fig. 10.5: sixth-order.

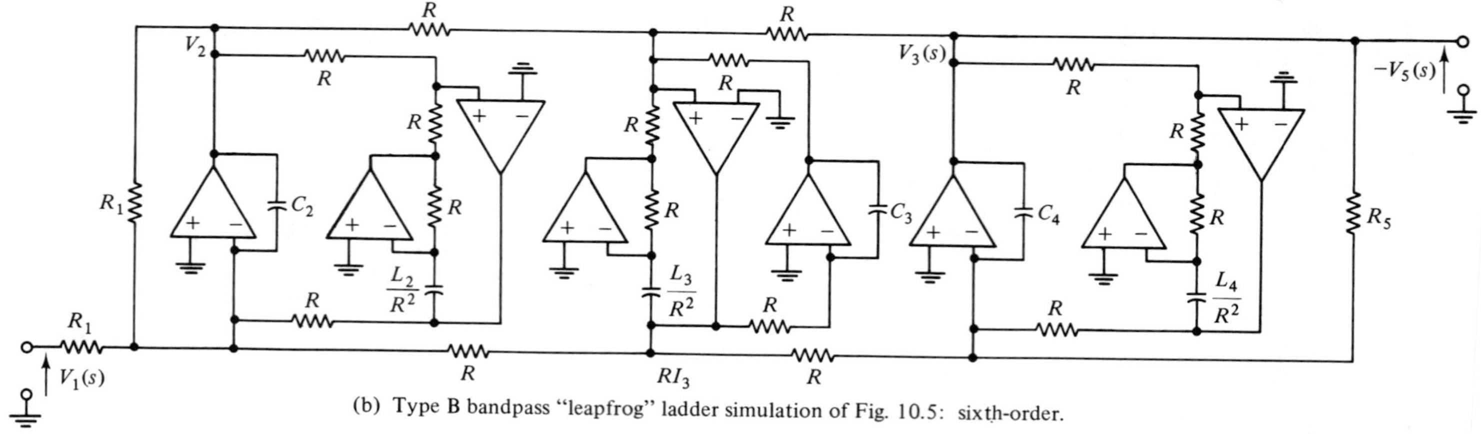

(b) Type B bandpass "leapfrog" ladder simulation of Fig. 10.5: sixth-order.

FIGURE 10.6

The corresponding SFG is illustrated in Fig. 10.5(b) and is a special case of Fig. 7.22(d). The node variables shown as V_2, I_3, V_3 correspond directly to the corresponding signals in Fig. 10.5(a). Note that the SFG in Fig. 10.5(b) realizes $-V_5(s)$, whereas the *conventional* output in Fig. 10.5(a) is $+V_5(s)$. The type A circuit implementation of Fig. 10.6(a) uses type 1A (phase-lag) noninverting integrators, and the designer must select the resistances R at a numerically convenient value; the six capacitance elements are then derived directly from the values of the LC elements in the corresponding LC filter. Thus, for this *normalized* design, we may select $R = 1$ so that C_2, C_3, and C_4 are obtained directly from Fig. 10.5(a). The remaining capacitances L_2/R^2, L_3/R^2, and L_4/R^2 are calculated by using $R = 1$ and the values of L_2, L_3, and L_4 in Fig. 10.5(a).

NUMERICAL EXAMPLE: Sixth-Order Bandpass Type A Leapfrog Filter; Chebychev Response: Passband Ripple = 0.1 db; Center Frequency = 1.5915 kHz, Bandwidth = 0.15915 kHz.

SOLUTION: The center frequency of the bandpass filter is specified as

$$\omega_0 = 2\pi \times 1.5915 \times 10^3 = 10^4 \text{ rad/sec} \tag{10.16}$$

and the bandwidth as

$$\text{BW} = 2\pi \times 0.15915 \times 10^3 = 10^3 \text{ rad/sec}$$

The first step in the design procedure is to use the relevant *normalized LC* filter ($\omega_0 = 1$, BW = 0.1) element values [see Fig. 10.5(a)] to implement a normalized version of Fig. 10.6(a). Thus,

$$\left.\begin{aligned} &R_1 = R_5 = R = 1\ \Omega \\ &C_2 = C_4 = 14.328\text{ F} \\ &C_3 = 0.062747\text{ F} \\ &\frac{L_2}{R^2} = \frac{L_4}{R^2} = 0.69793\text{ F} \\ &\frac{L_3}{R^2} = 15.937\text{ F} \end{aligned}\right\} \quad \begin{array}{l}\text{Types A and B}\\ \text{Bandpass}\\ \text{“Leapfrog”}\\ \text{Ladders}\\ \text{Normalized}\\ \text{Design}\end{array} \tag{10.17}$$

The required *denormalized* design depends on the choice of the resistances R. We choose

$$R = R_1 = R_5 = 10\text{ k}\Omega \tag{10.18}$$

so that all capacitances in equation (10.17) must be divided by 10^4 to achieve the required impedance denormalization and, further, by another factor of 10^4 to denormalize the center frequency from 1 rad/sec to 10^4 rad/sec.

Thus,

$$\left.\begin{aligned} C_2 &= C_4 = 143.28 \text{ nF} \\ C_3 &= 0.62747 \text{ nF} \\ \frac{L_2}{R^2} &= \frac{L_4}{R^2} = 6.9793 \text{ nF} \\ \frac{L_3}{R^2} &= 159.37 \text{ nF} \end{aligned}\right\} \text{DENORMALIZED } C \text{ ELEMENT VALUES} \qquad (10.19)$$

in Fig. 10.6(a).

The ideal ($A_0 = \infty$, $B = \infty$) passband magnitude response $M(\omega)$ is shown as the dashed curve in Fig. 10.7. It is, of course, identical to that obtained for the corresponding LC filter.

The preceding numerical example also applies to the type B bandpass "leapfrog" ladder circuit of Fig. 10.6(b), where type 1B noninverting integrators are used. Thus, the dashed curve in Fig. 10.7 also describes the ideal response of the type B bandpass "leapfrog" ladder circuit.

10-3.1 Passive-Sensitivity Performance of Type A and Type B Bandpass "Leapfrog" Ladder Filters

The comments in section 10-2.1 apply exactly to Figs. 10.6(a) and 10.6(b). Thus, the transfer function magnitude $M(\omega)$ has zero sensitivity to the six RC elements in each of the three TILs wherever $|t(j\omega)|$ is unity in the passband and the sensitivity of $M(\omega)$ to these RC elements is low-valued throughout the low-ripple passband. Equation (10.1) describes the transition region worstcase-sensitivity performance.

10-3.2 Active-Sensitivity Performance of Type A and Type B Bandpass "Leapfrog" Ladder Filters

The analysis in section 10-2.2 applies exactly to the *bandpass* case; thus, equations (10.7), (10.9), (10.10), (10.11), and (10.14) are valid. In particular, equations (10.7) and (10.14) imply that the active-sensitivity performance of the type B circuit is vastly superior to that of the type A circuit. *This superiority is particularly important in bandpass designs* because $|\tau(\omega)|$ is generally much larger than for corresponding lowpass designs.

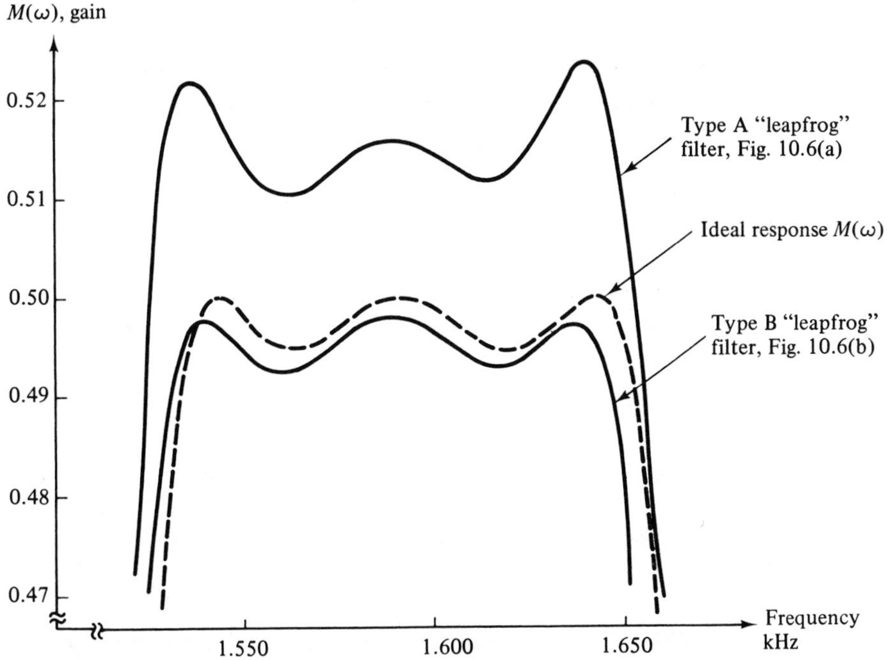

FIGURE 10.7 *Type A and B bandpass "leapfrog" filter responses: Chebychev sixth order: $\omega_{0_f}/BW = 10$, center frequency = 1.59 kHz, B = 4.8 MHz.*

NUMERICAL EXAMPLE: Center Frequency 1.5915 kHz; Gain Bandwidth Products 4.8 MHz.

SOLUTION: Consider the previous numerical example of a Chebychev bandpass "leapfrog" design with a center frequency of 1.5915 kHz and a bandwidth of 0.15915 kHz and with the ideal $M(\omega)$ passband response shown in Fig. 10.7 (dashed curve). Computer-aided analysis of the circuits shown in Figs. 10.6(a) and 10.6(b) for nonideal OP AMPs having

$$A_0 = 10^5 \quad \text{and} \quad B = 2\pi \times 4.8 \times 10^6 \text{ rad/sec} \tag{10.20}$$

provides the two solid curves of Fig. 10.7. Note the considerable superiority of the type B circuit.

It is interesting to note that it is not difficult to compensate the type B circuit to virtually remove the small transition region discrepancy between the solid curve and the ideal (dashed) curve; the *magnitudes* of the *gains* of the integrators (the $|M(j\omega)|$ terms in Tables 5-1 and 5-2) are taken into account, thereby resulting in a requirement for capacitance elements that are smaller by the approximate fractional quantity 1/BT (or 1/BCR) than is implied by the *ideal* equations (10.19). Predistorting the capacitance elements in this way causes the transition regions of the solid curve in Fig. 10.7 to correspond more closely to the ideal dashed curve. (The reader is asked to calculate the predistorted capacitance values in Problem 10-4.)

10-4 GENERALIZED *LC* FILTER SYNTHESIS USING INTEGRATORS OR DIFFERENTIATORS

The reader may have noticed that "leapfrog"-type SFGs have been derived for lowpass and bandpass filters having all their zeros of the transfer function $H(s)$ at zero and infinity. So far, no attempt has been made to explain how integrators might be used to realize *highpass* filters or how they might be used to realize finite (imaginary or complex) zeros of $H(s)$. In terms of Table 9-1, types A, B, D, and E realize imaginary zeros, but we have not so far explained how to write corresponding SFGs such that a one-to-one correspondence is retained between the gains of the integrators and the LC-element values. In this section, a general approach is described that allows the designer to write such a SFG for *any* resistively terminated LC filter structure.

10-4.1 The Four Basic Signal-Flow Graphs (SFGs)

In this subsection, the impedance-type and admittance-type SFG transmittances are defined, and four basic types of SFGs are described.

DEFINITIONS: *A transmittance in an SFG is an impedance-type (Z-type) if it has a current input variable and a voltage output variable.*

A transmittance in an SFG is an admittance-type (Y-type) if it has a voltage input variable and a current output variable.

For example, the transmittances $1/sC_i$ for Fig. 10.5(b) are Z-type transmittances, whereas the transmittances $1/sL_i$ are Y-type transmittances. Having defined these two types of transmittances, we now define four types of basic SFGs, two that simulate a parallel connection of two immittances and two that simulate a series connection of two immittances.

The parallel connection: Consider the parallel connection of the branches 1 and 2, shown in Fig. 10.8; this connection implies that

$$\left.\begin{array}{l} V = V_1 = V_2 \\ \text{and} \\ I = I_1 + I_2 \end{array}\right\} \text{PARALLEL-CONNECTION CONSTRAINTS} \qquad (10.21)$$

where the V and I variables are transform voltage and current variables. Further, in terms of the transform branch impedances $Z_{1,2}$ and $Y_{1,2}$ we

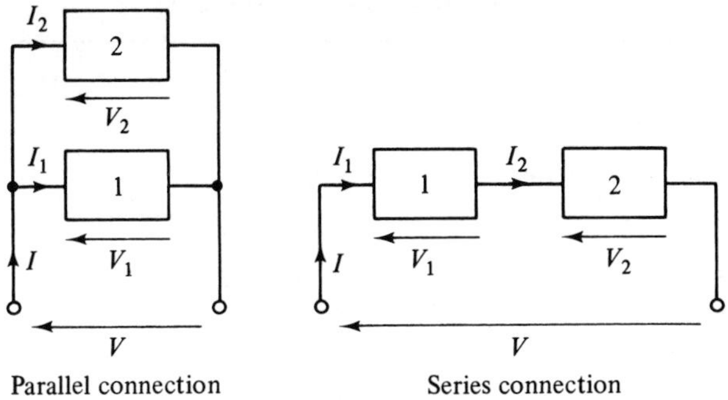

FIGURE 10.8

may write

$$V_1 = Z_1 I_1 \tag{10.22}$$

and

$$V_2 = Z_2 I_2 \tag{10.23}$$

BRANCH TRANSFORM-IMMITTANCE CONSTRAINTS

or

$$I_1 = Y_1 V_1 \tag{10.24}$$

and

$$I_2 = Y_2 V_2 \tag{10.25}$$

To write an SFG for the parallel connection of Fig. 10.8, we wish to eliminate V_1, I_1, V_2, and I_2 and simply obtain an SFG that relates the two terminal variables V and I via the branch immittances (Y_1, Y_2, Z_1, Z_2). One simple approach is to use equations (10.21) with equations (10.24) and (10.25) to obtain

$$\left.\begin{aligned} I_1 &= V_1 Y_1 = V Y_1 \\ I_2 &= V_2 Y_2 = V Y_2 \end{aligned}\right\} \tag{10.26}$$

so that

$$\boxed{I = Y_1 V + Y_2 V} \tag{10.27}$$

which has the SFG description shown as $Y_1 \| Y_2$ in Table 10-1. Note that the input node is a voltage variable, and the output node is a current variable, so that the transmittance existing between the input node and the output node is Y-type; further, the internal transmittances are Y_1, Y_2.

TABLE 10-1 *The three parallel-type SFGs.*

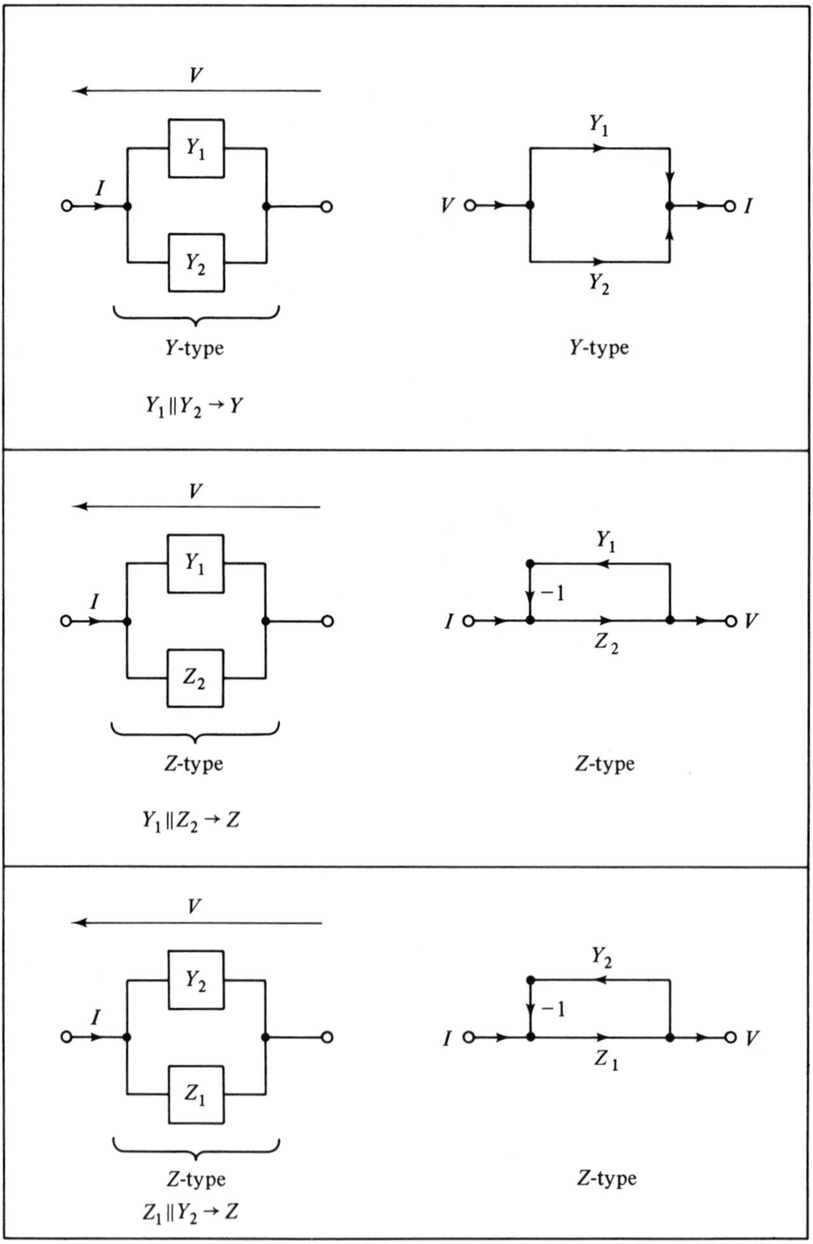

This situation is summarized by writing

$$Y_1 \| Y_2 \rightarrow Y \tag{10.28}$$

implying that a parallel connection may be simulated by using two Y-type transmittances, Y_1 and Y_2, to realize a Y-type SFG simulation.

Alternative SFG simulations of the parallel connection involve using equations (10.23) and (10.24) with (10.21) or (10.22) and (10.25) with (10.21). For example, using equations (10.21) with (10.23) and (10.24), we find that

$$\begin{aligned} V &= Z_2(I - I_1) \\ I_1 &= VY_1 \end{aligned} \tag{10.29}$$

so that

$$\boxed{V = Z_2(I - VY_1)} \tag{10.30}$$

which has the simple single-loop negative feedback SFG shown as $Y_1 \| Z_2$ in Table 10-1. Note that the transmittance from the input node to the output node is Z-type, so that we may write

$$Y_1 \| Z_2 \rightarrow Z \tag{10.31}$$

Similarly, it is readily shown that there exists an SFG

$$Z_1 \| Y_2 \rightarrow Z \tag{10.32}$$

corresponding to the third SFG simulation of the parallel connection shown in Table 10-1. *It is important to note that we do not have an SFG that allows a parallel connection to be simulated with Z-type transmittances Z_1 and Z_2.* This is a fundamental limitation of the SFG simulation methods.

Consider now the parallel connection of an *arbitrary number* of branches. Starting with a Y-type branch Y_1, as shown in Fig. 10.9(a), we may add further Y-type branches Y_2, Y_3, etc., in parallel and the connection remains Y-type, according to equation (10.28). However, if a *single* Z-type branch Z_i is introduced as a parallel branch, it then becomes a Z-type branch, according to equation (10.31), and as shown in Fig. 10.9(b). Any additional parallel branches *must* be Y-type, as shown in Fig. 10.9(c), because we do not have an SFG that allows a Z-type branch to be

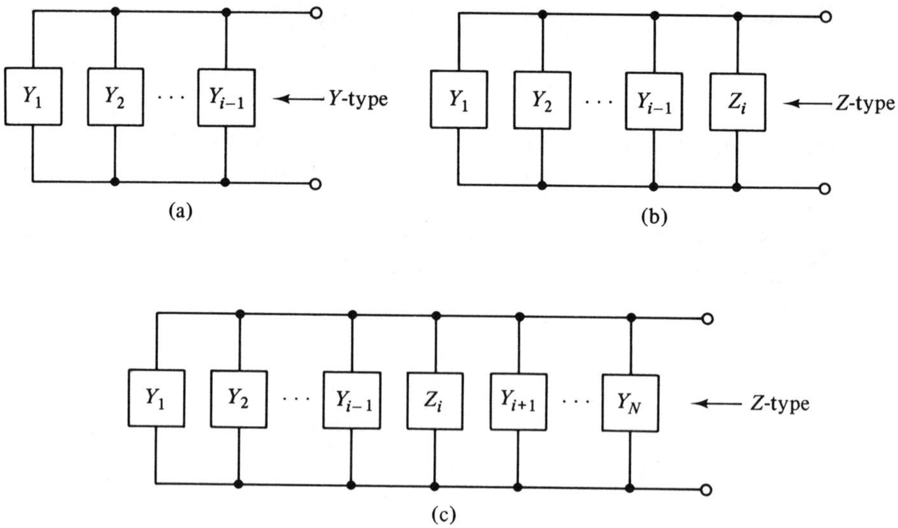

FIGURE 10.9 *The parallel connection, Z-type.*

connected in parallel with a Z-type branch. Consequently,

A PARALLEL CONNECTION OF BRANCHES CAN POSSESS NO MORE THAN ONE Z-TYPE TRANSMITTANCE IN ITS *SFG*.	(10.33)

We define a parallel Y-type SFG and parallel Z-type SFG as follows:

DEFINITIONS: *A parallel Y-type SFG is one obtained by realizing all branch transmittances of a parallel connection as Y-type thereby resulting in the Y-type SFG of Fig. 10.10(a).*

A parallel Z-type SFG is one obtained by realizing all branch transmittances except one as Y-type and the remaining one branch as Z-type thereby resulting in the Z-type SFG of Fig. 10.10(b).

The series connection: Consider the series connection of the branches 1 and 2 shown in Fig. 10.8; this connection implies that

$$\left.\begin{aligned} I &= I_1 = I_2 \\ V &= V_1 + V_2 \end{aligned}\right\} \quad \text{SERIES-CONNECTION CONSTRAINTS} \qquad (10.34)$$

Combining equations (10.34) with equations (10.22) to (10.25) allows us to write three *alternate* equations that relate V to I via Z_1, Z_2, Y_1, and Y_2: they are

$$\boxed{V = Z_1 I + Z_2 I} \tag{10.35}$$

or

$$\boxed{I = Y_2(V - IZ_1)} \tag{10.36}$$

or

$$\boxed{I = Y_1(V - IZ_2)} \tag{10.37}$$

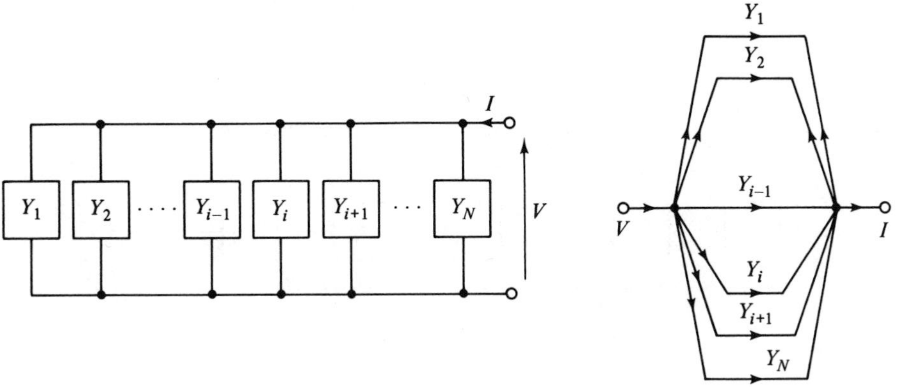

(a) The general parallel Y–type SFG.

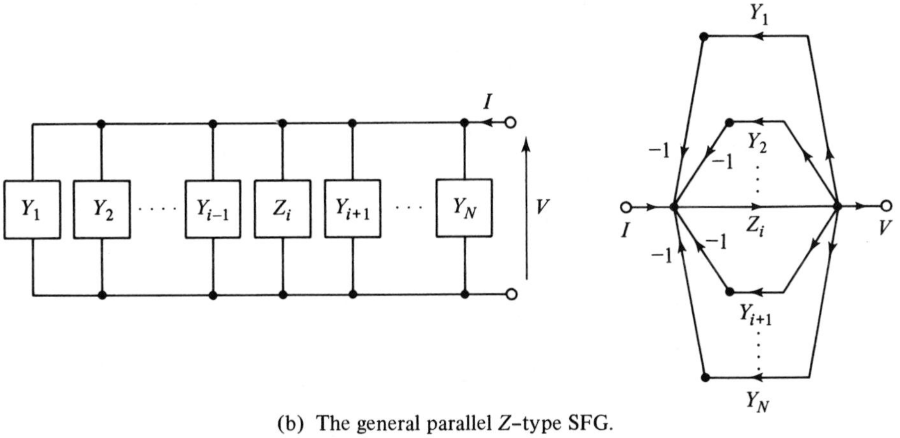

(b) The general parallel Z–type SFG.

FIGURE 10.10

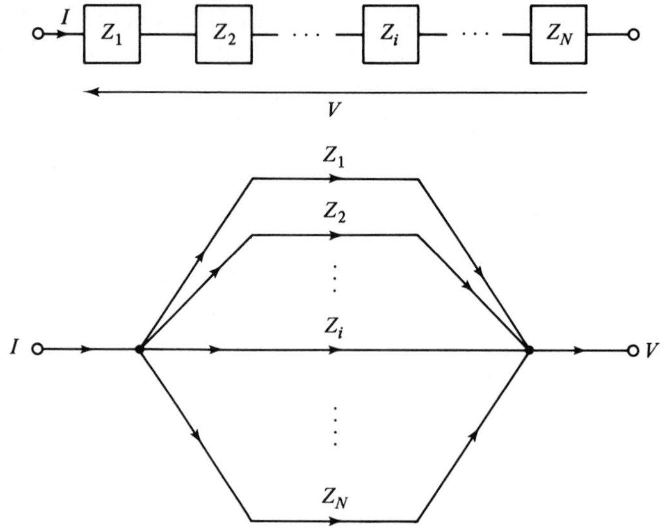

(c) The general series Z–type SFG.

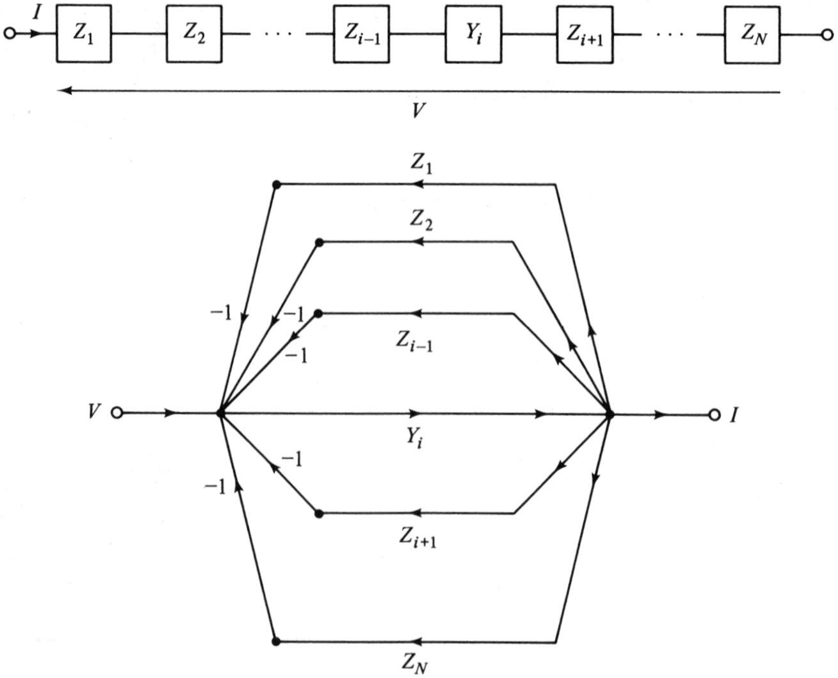

(d) The general series Y–type SFG.

FIGURE 10.10 *(Continued).*

TABLE 10-2 *The three series-type SFGs.*

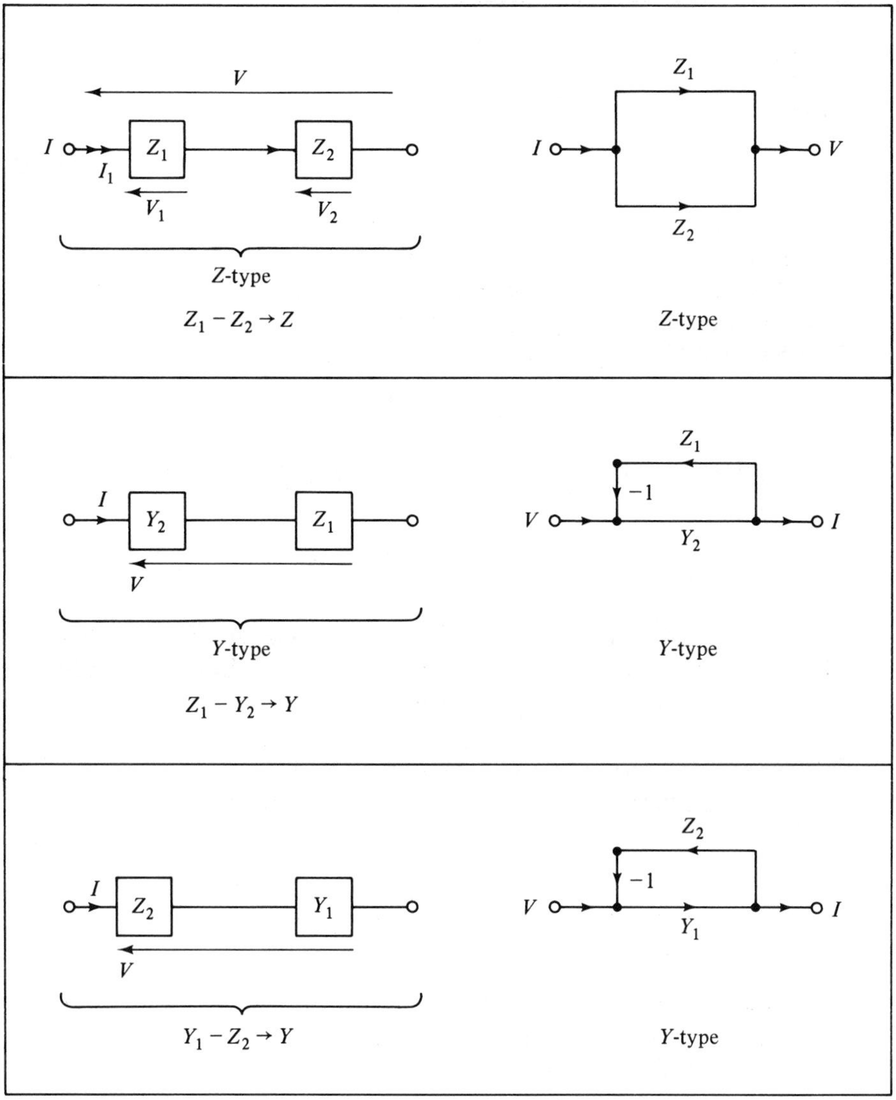

Equations (10.35), (10.36), and (10.37) lead, respectively, to the three SFG's shown in Table 10-2. Note that the series connection of two branches may be realized from two Z-type branches (Z_1 and Z_2), in which case the resultant SFG is Z-type. This statement is written symbolically as

$$Z_1 - Z_2 \rightarrow Z \tag{10.38}$$

corresponding to equation (10.35) and the first SFG of Table 10-2. Similarly, equations (10.36) and (10.37) and the second and third SFG's of Table 10-2 may be written as

$$Z_1 - Y_2 \rightarrow Y \tag{10.39}$$

and

$$Y_1 - Z_2 \rightarrow Y \tag{10.40}$$

Note that *an SFG for the series connection* $Y_1 - Y_2$ *does not exist*, being a limitation of the SFG simulation method.

The series connection of an *arbitrary number* of branches may therefore be realized by using Z-type transmittances to represent each branch; the resultant SFG is Z-type;

$$Z_1 - Z_2 - Z_3, \ldots, Z_N \rightarrow Z \tag{10.41}$$

One of the branches in a series connection may be simulated as a Y-type, in which case the SFG is Y-type:

$$Z_1 - Z_2 - Z_3, \ldots, Z_{i-1} - Y_i - Z_{i+1}, \ldots, Z_N \rightarrow Y \tag{10.42}$$

Equations (10.41) and (10.42) are illustrated in Fig. 10.10. More than one Y-type branch is not possible for a series connection because the SFG for the $Y_1 - Y_2$ connection does not exist: thus,

A SERIES CONNECTION OF BRANCHES CAN POSSESS NO MORE THAN ONE Y- TYPE TRANSMITTANCE IN ITS *SFG*.	(10.43)

DEFINITION: *A series Z-type SFG is one obtained by realizing all branch transmittances of a series connection as Z-type thereby resulting in the Z-type SFG of Fig. 10.10(c).*

A series Y-type SFG is one obtained by realizing all branch transmittances except one as Z-type and the remaining one branch as Y-type thereby resulting in the Y-type SFG of Fig. 10.10(d).

EXAMPLE 10.1: The Parallel LC Circuit. Derive the parallel Z-type and parallel Y-type SFGs for the LC connection shown in Fig. 10.11(a).

SOLUTION: The inductive branch Y_2 is realized as Y-type ($1/sL_2$) and the capacitive branch Y_1 as Y-type (sC_1), according to the $Y_1 \| Y_2 \rightarrow Y$ SFG shown in Table 10-1. This leads directly to the *parallel Y-type SFG* realiza-

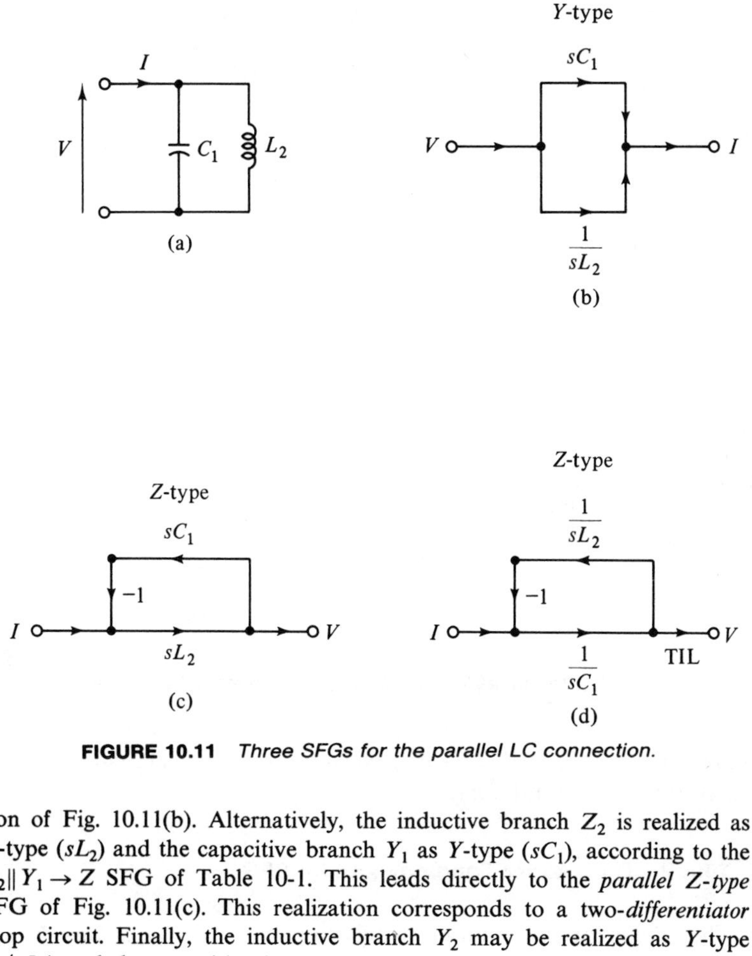

FIGURE 10.11 *Three SFGs for the parallel LC connection.*

tion of Fig. 10.11(b). Alternatively, the inductive branch Z_2 is realized as Z-type (sL_2) and the capacitive branch Y_1 as Y-type (sC_1), according to the $Z_2 \| Y_1 \rightarrow Z$ SFG of Table 10-1. This leads directly to the *parallel Z-type* SFG of Fig. 10.11(c). This realization corresponds to a two-*differentiator* loop circuit. Finally, the inductive branch Y_2 may be realized as Y-type ($1/sL_2$) and the capacitive branch Z_1 as Z-type ($1/sC_1$), according to the $Y_2 \| Z_1 \rightarrow Z$ SFG of Table 10-2, as shown in Fig. 10.11(d). This third possible SFG is the well-known two-*integrator* loop (TIL) circuit that is encountered in chapter 8 and in the earlier sections of this chapter. Clearly, *the TIL is only one of three possible ways to simulate a parallel LC circuit* such that there is a one-to-one correspondence between the LC element and a transmittance in the corresponding SFG.

In a similar way it may be shown that the *series LC* circuit may be simulated by any one of the three SFGs of Fig. 10.12. The TIL is just one of three possible SFGs.

The reader should by now be conscious of the fact that it is possible to make a variety of choices with regard to the Z-type or Y-type nature of the transmittances of the SFG. Creative use of these options allows a variety

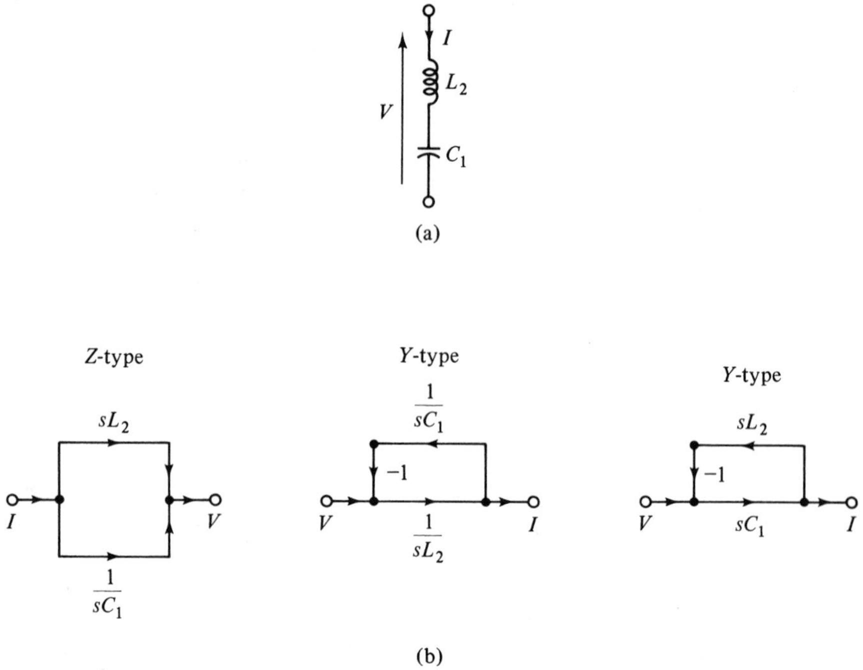

FIGURE 10.12 *Three SFGs for the series LC connection.*

of resistively terminated *LC*-filter structures to be simulated by using integrators (or differentiators).

10-4.2 The Ladder Topology and Its SFG

The ladder topology is used widely to realize *LC* filters; therefore, it is appropriate to consider the preceding rules and definitions as they apply to this structure. Consider first the ladder topology shown in Fig. 10.13(a), where we wish to *derive* an SFG in which the parallel branches of the ladder are implemented as *Z*-type transmittances and the series branches of the ladder as *Y*-type. Starting with branch 1, this is designated as *Y*-type and is to be implemented in the SFG as the transmittance Y_1. Then, introducing branch 2 as *Z*-type is permissible [equation (10.31)] but the resultant series connection of Y_1 and Z_2 is now *Y*-type, as shown in Figs. 10.13(b) and (c). Branch 3 is in parallel with the *Y*-type branch of Fig. 10.13(c); therefore, it may be *Z*-type or *Y*-type. *Selecting* branch 3 as *Z*-type creates the *Z*-type branch of Fig. 10.13(d). Branch 4 is to be in series with the *Z*-type branch of Fig. 10.13(d) and, therefore, must be *Y*-type, as shown in Fig. 10.13(e). Continuing in this way for a ladder

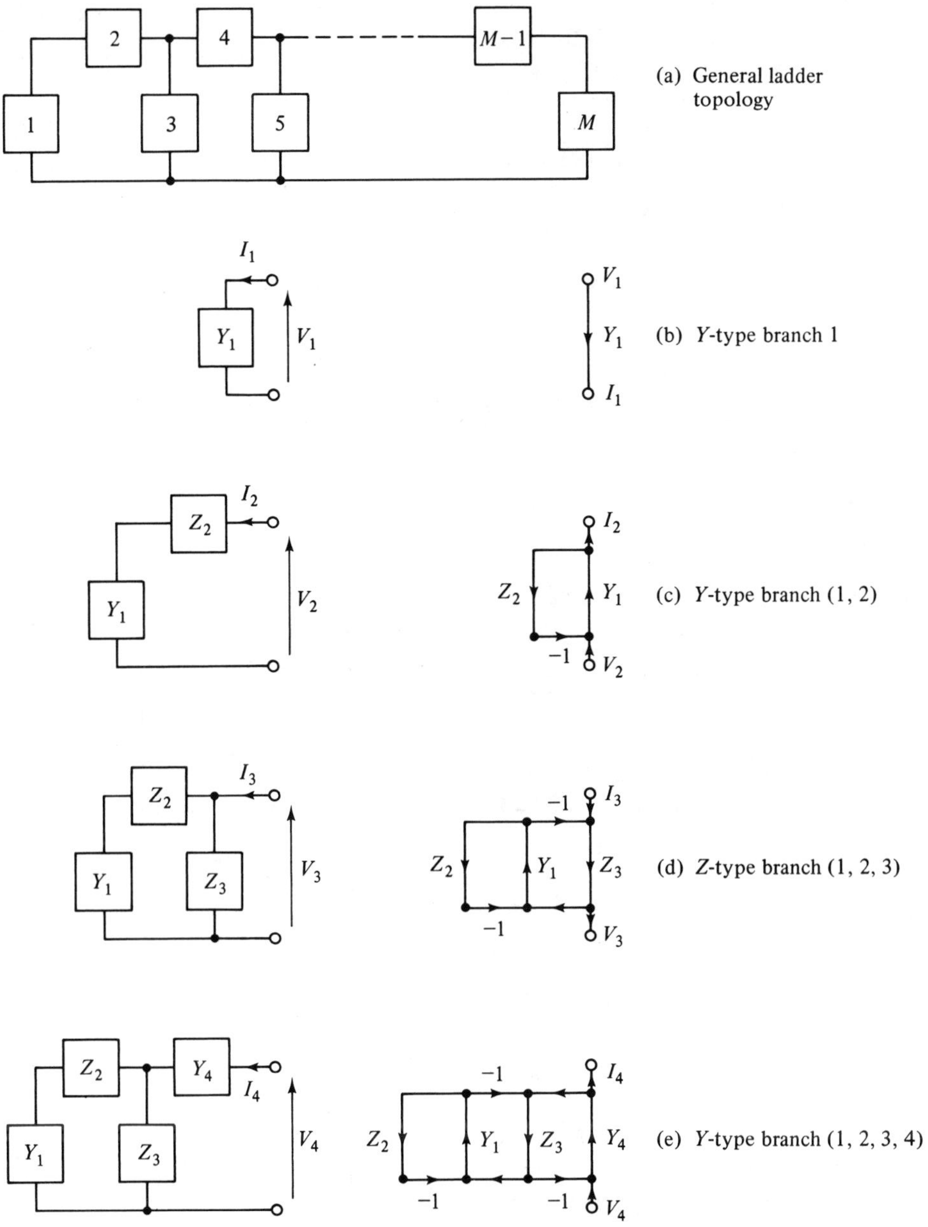

FIGURE 10.13 *Ladder SFGs.*

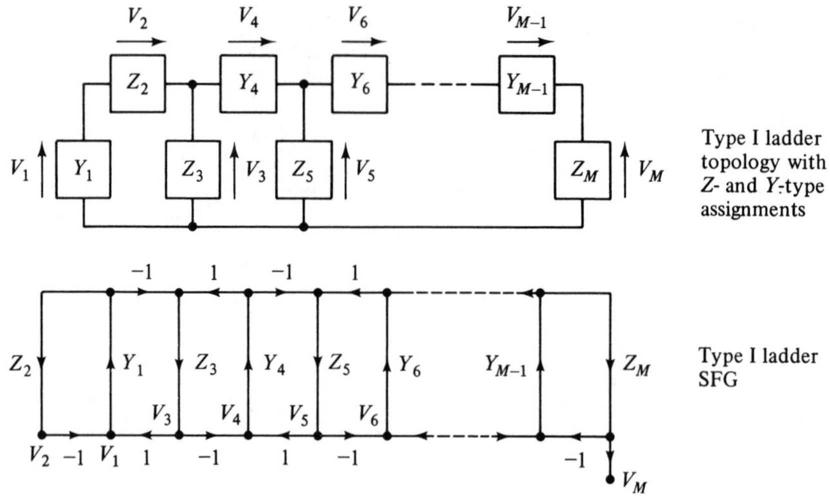

Type I ladder topology with Z- and Y-type assignments
Type I ladder SFG

(a) Type I ladder SFGs.

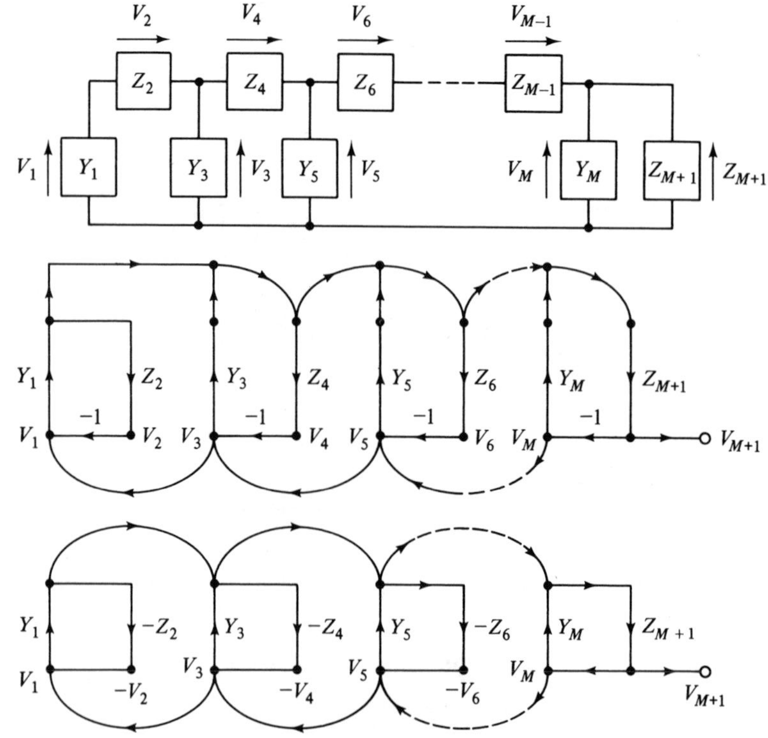

(b) Type II ladder SFGs.

FIGURE 10.14

structure of arbitrary length leads to the SFG of Fig. 10.14(a); this SFG is defined here as a type I ladder SFG. Note that our *selection* of branch 3 as Z-type implies that all subsequent shunt branches are Z-type and all series branches are Y-type.

This procedure may be repeated to arrive at a completely different SFG by simply selecting branch 3 as Y-type rather than Z-type. Thereafter, introducing branches 4, 5, 6, etc., leaves no option but for the shunt branches to be Y-type and the series branches to be Z-type. The corresponding SFG is shown in Fig. 10.14(b) along with a simplified version that is obtained by removing redundant unity transmittances. We define this SFG as a type II ladder SFG.

Note that all of the "leapfrog" structures mentioned in sections 10-1 and 10-2 are derived from type I ladder SFGs. In Problem 10-6, the reader should discover that OP AMP implementations of the type II ladder SFG usually involve the use of more resistance elements than the corresponding type I ladder SFG.

10-4.3 The Foster *LC* Immittances and Their SFGs

To implement resistively terminated *LC* ladder filters of *arbitrary topology*, it is necessary that the designer be able to obtain SFGs for the

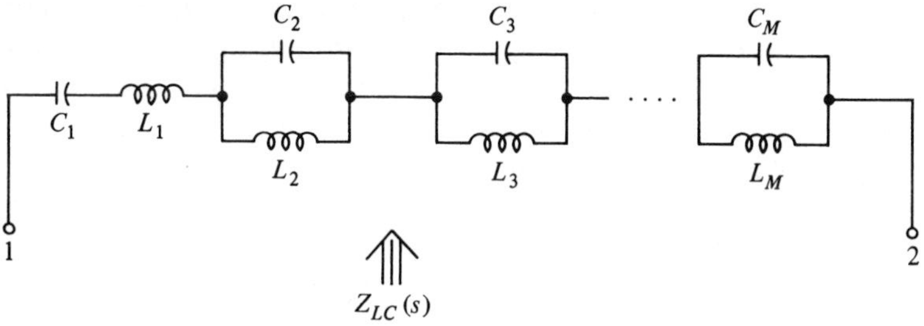

Foster I circuit for *LC* branch immittance

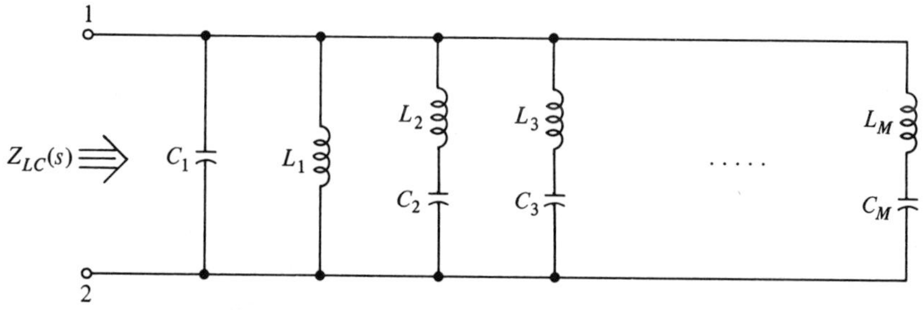

Foster II circuit for *LC* branch immittance

FIGURE 10.15 *Alternative Foster LC branch immitance circuits.*

LC branches of the ladder filter. *In general*, the *LC* branches have driving-point immittances that are realizable by means of *either* one of the two Foster *LC* one-port networks of Fig. 10.15. [The Foster I circuit is a series connection of branches realizing finite imaginary poles of $Z(s)$ at the natural resonant frequencies of parallel *LC* circuits, whereas the Foster II circuit is a parallel connection of branches realizing finite imaginary zeros of $Z(s)$ at the natural resonant frequencies of series *LC* circuits.] The important points are that it be understood that either of these two Foster forms may be used to realize *any LC* driving-point immittance and, further, that both forms are used widely as the branches of *LC* ladder filter networks.

Consider now the permissible assignments of Y-type and/or Z-type transmittances to the branches of the two Foster *LC* circuits. First, *it is assumed that we wish to realize the final* SFG*s with integrators* (rather than differentiators), so we should try to choose capacitive branches as Z-type (resulting in *integrating* transmittances $1/sC_i$) and inductive branches as Y-type (resulting in *integrating* transmittances $1/sL_i$). Making this choice where possible, there are two obvious ways of realizing the Foster I circuit, as shown in Table 10-3. The parallel *LC* circuits are realized by using a Z-type for each capacitance and Y-type for each inductance, resulting in a Z-type (TIL) SFG for each of them. The two versions shown in Table 10-3 differ only by the Z- or Y-type assignment to the final series inductance element L_1. If we require a Z-type branch between terminals 1 and 2, then L_1 *must* be assigned as Z-type, thereby introducing a *differentiating* transmittance sL_1 in the corresponding SFG. On the other hand, if we require a Y-type branch between terminals 1 and 2, then L_1 must be assigned as Y-type. In practical situations, the necessity for L_1 to be Z-type in a Z-type Foster I circuit and the necessity for a series Y-type element L_1 in the Y-type Foster I circuit can cause practical difficulties that may be overcome by using techniques to be described here.

The Z-type Foster II and Y-type Foster II branch assignments and corresponding SFGs appear in Table 10-4. Note that the element C_1 *must* be realized as a Y-type transmittance in the Y-type Foster II topology and that the element C_1 *must* be realized as a Z-type transmittance in the Z-type Foster II topology of Table 10-4. These constraints can also lead to practical difficulties that may be overcome by using the following techniques:

Replacement of differentiators by integrators: The elements marked with a single asterisk (*) in Tables 10-3 and 10-4 correspond to *differentiating* transmittances (sL_1 or sC_1) in the SFGs. For practical reasons, related to biasing problems and high-frequency performance, OP AMP integrator circuits are preferable to OP AMP differentiator circuits. Therefore, the

TABLE 10-3 *Foster I SFGs.*

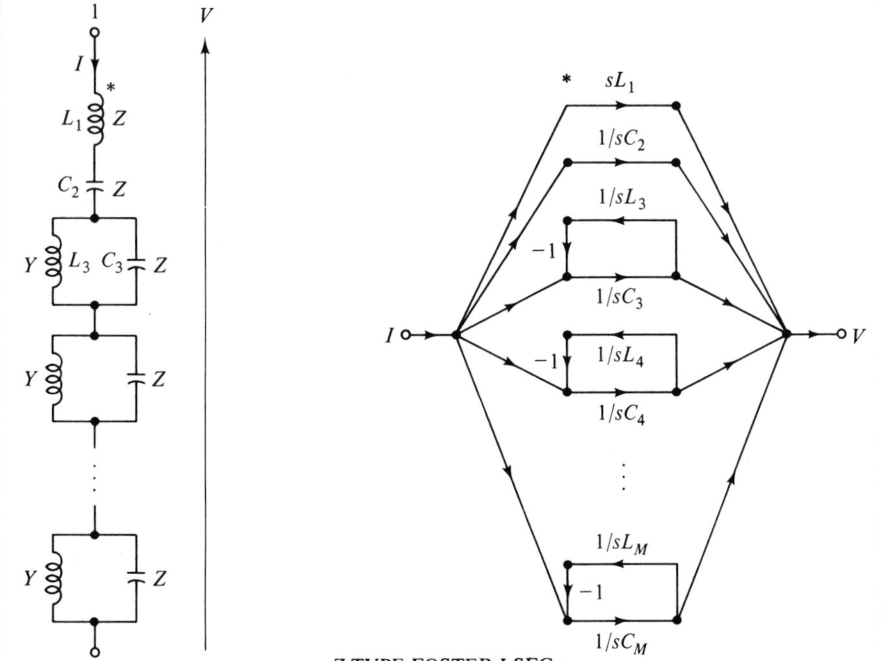

Z-TYPE FOSTER I SFG*s*.

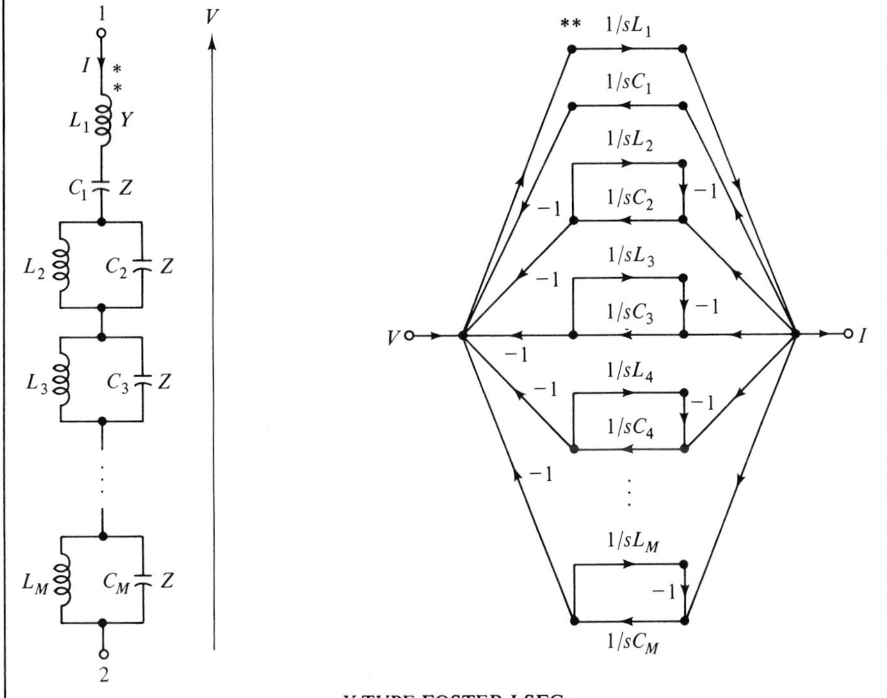

Y-TYPE FOSTER I SFG*s*.

TABLE 10-4 *Foster II SFGs.*

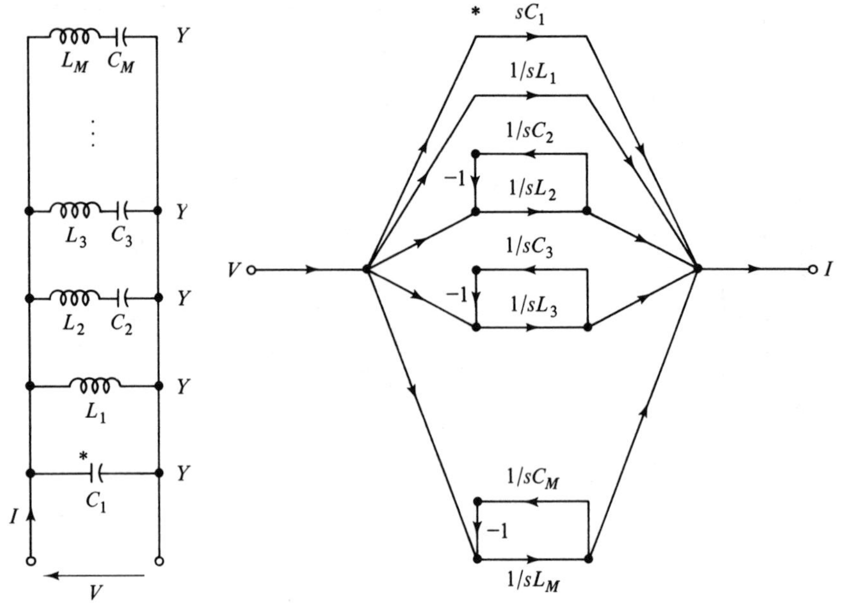

Y-TYPE FOSTER II SFG

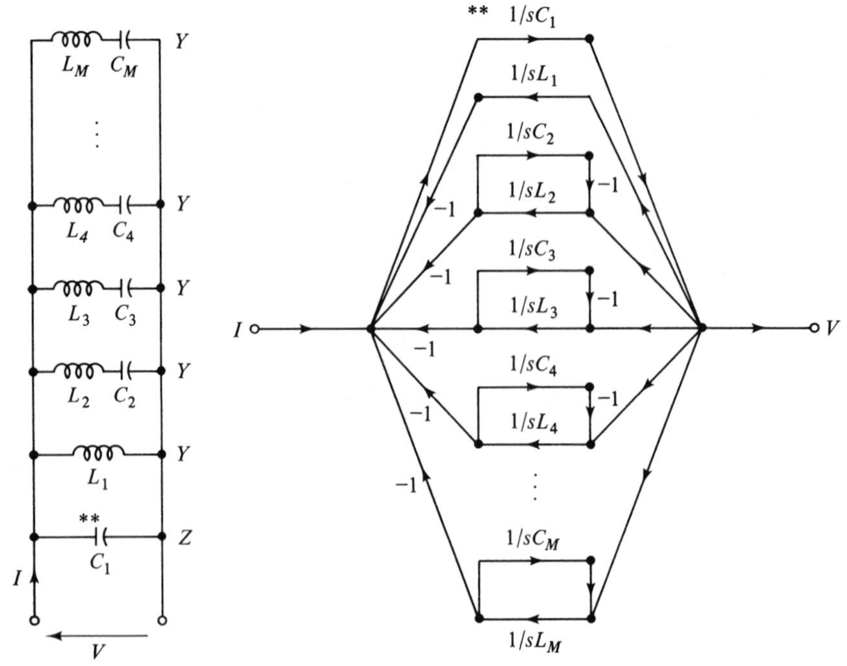

Z-TYPE FOSTER II SFG

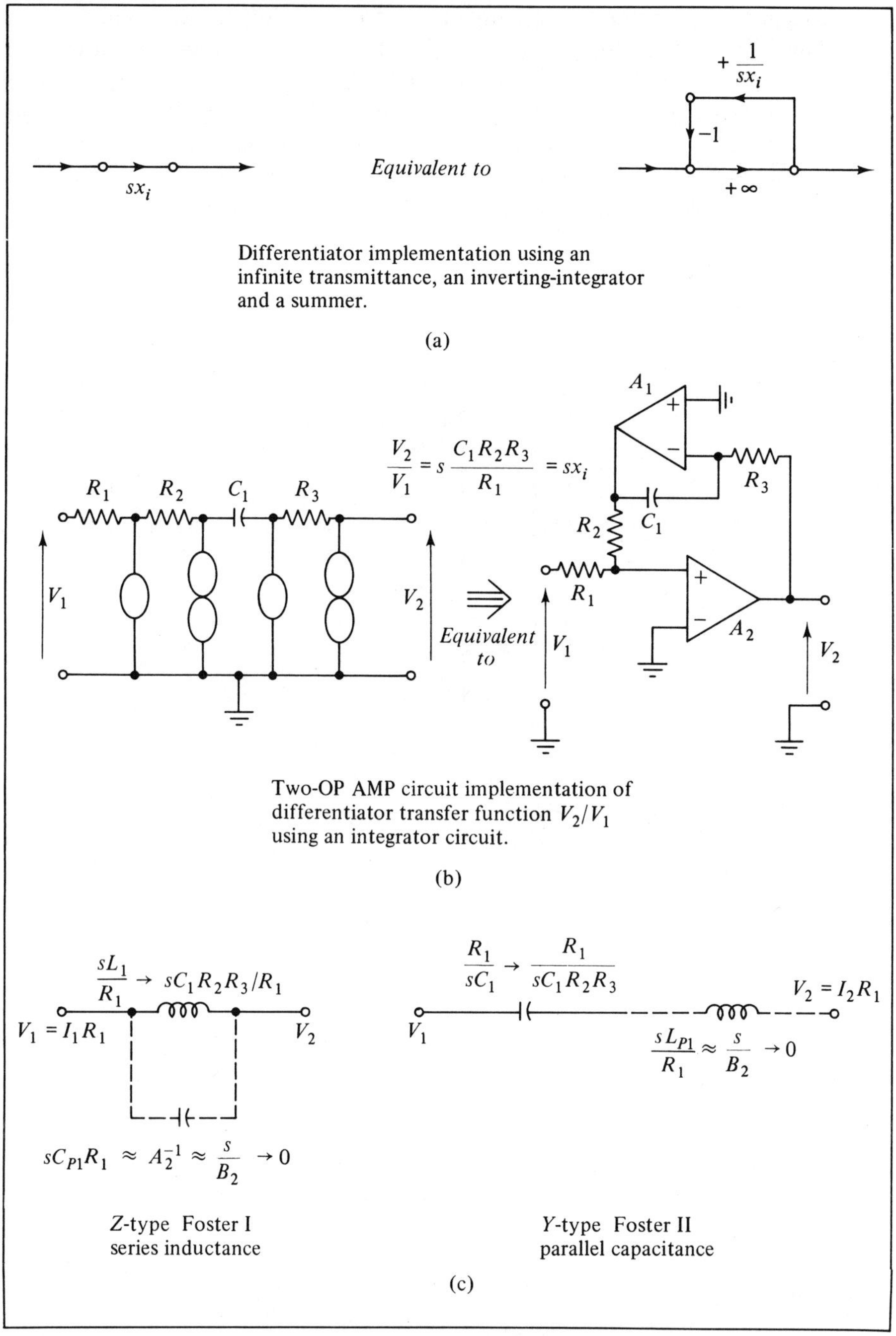

FIGURE 10.16 *Differentiator replacement with integrator.*

designer may prefer to implement a differentiating transmittance sx_i as shown in Figs. 10.16(a) and 10.16(b). The circuit involves the use of an OP AMP integrator circuit in the feedback path of an OP AMP inverter circuit. (Analysis of this circuit is requested in Problem 10-7). As shown in Fig. 10.16(c), this is equivalent to assuming that parasitic elements L_{p1} and C_{p1} of ideally-zero value are in series and parallel, respectively, with the single-asterisked elements C_1 and L_1 of Tables 10-3 and 10-4.

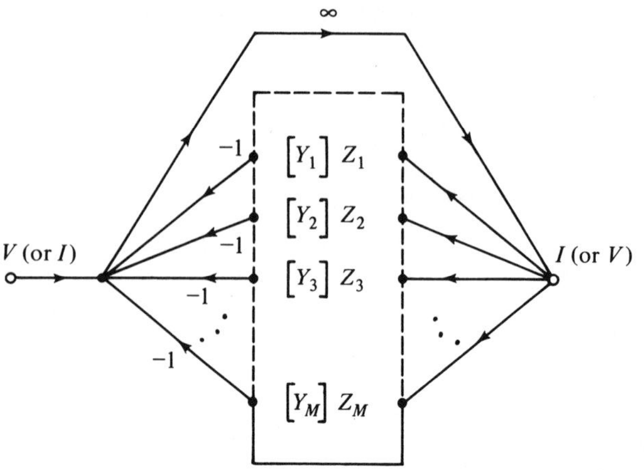

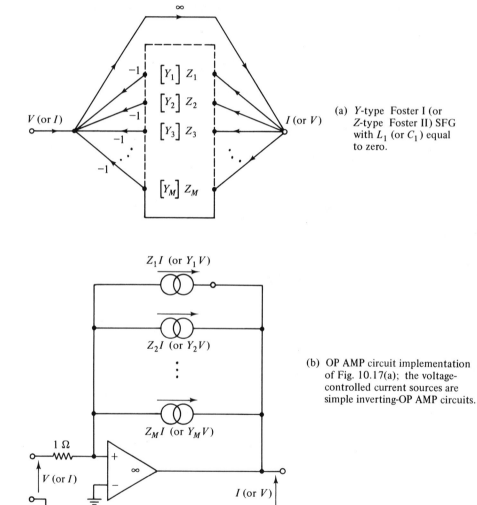

(a) Y-type Foster I (or Z-type Foster II) SFG with L_1 (or C_1) equal to zero.

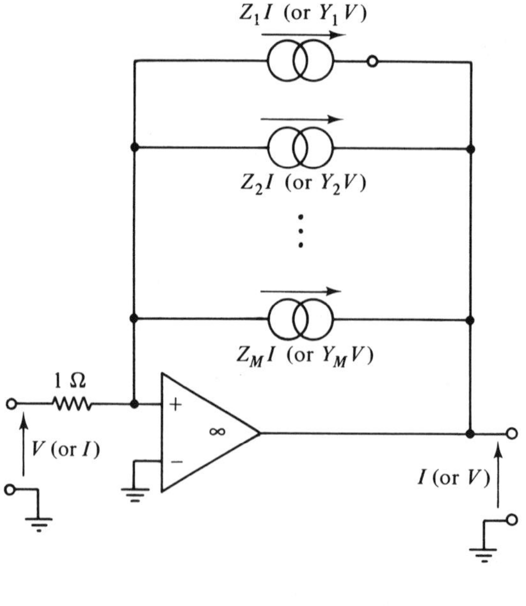

(b) OP AMP circuit implementation of Fig. 10.17(a); the voltage-controlled current sources are simple inverting-OP AMP circuits.

FIGURE 10.17

The use of dummy LC elements of zero value: The elements marked with a double asterisk (**) in Tables 10-3 and 10-4 *cannot* be omitted without changing the branch types from Z to Y or vice-versa. However, *in a practical situation their omission may be required.* This difficulty is overcome by allowing them to remain, but as dummy elements of zero value. Thus, their integrating transmittances $1/sC_1$ and $1/sL_1$ are simply replaced by *infinite gain transmittances*, as shown in Fig. 10.17.

TABLE 10-5 *SFG for independent sources in Z-type and Y-type transmittances.*

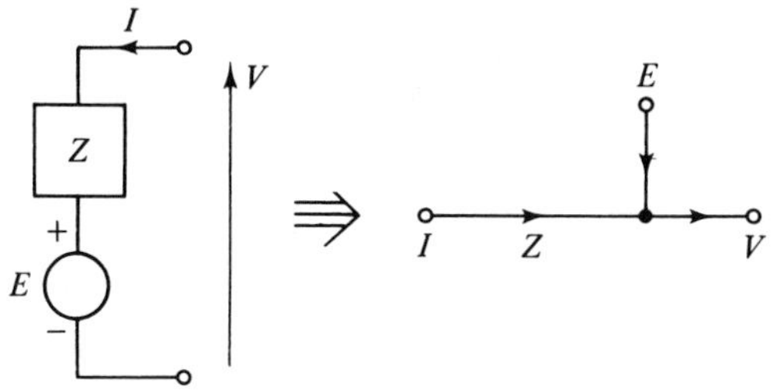	Voltage source Z-type branch
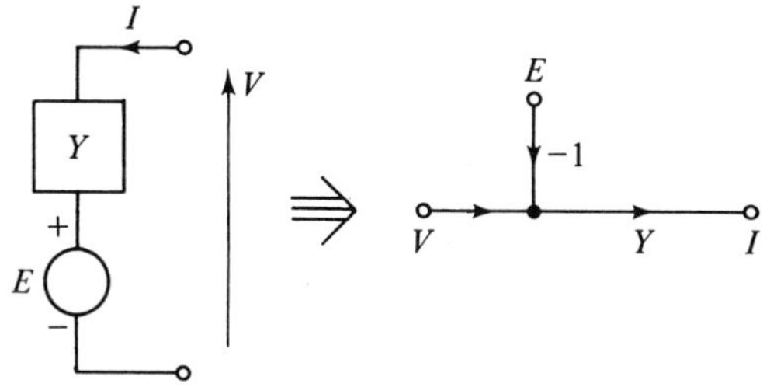	Voltage source Y-type branch
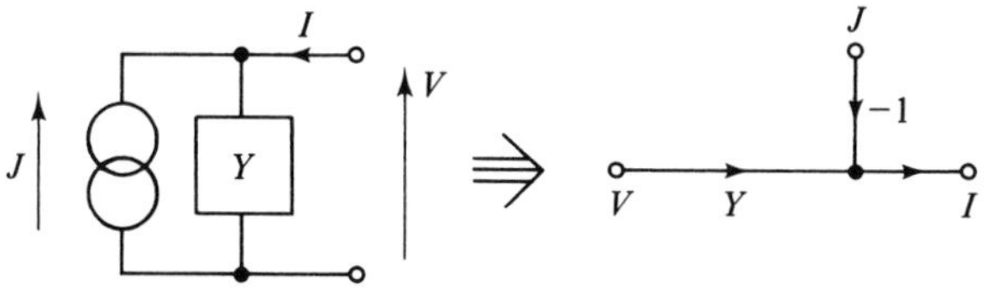	Current source Y-type branch
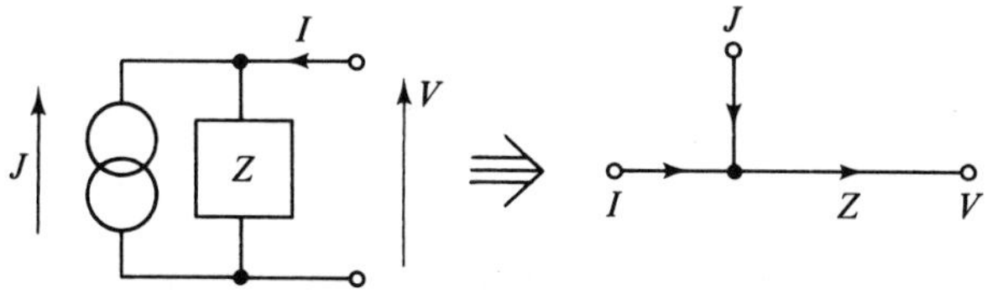	Current source Z-type branch

Signal-flow graph representations of input signals: The four possible ways of representing the independent voltage and current sources, E and J respectively, are shown in Table 10-5 for the Y- and Z-type transmittances.

Signal-flow graph representations of transformers and gyrators: It follows from the definitions of transformers and gyrators (see chapter 2) that the SFGs of Tables 10-6 and 10-7 describe the transformer and gyrator, respectively.

Note that *for the transformer* we can expect to actually *use* only the SFGs shown as (b) and (d) in Table 10.6 because the (a) and (c) types cannot be terminated in either Z-type or Y-type impedances since *both* the

TABLE 10-6 *Four SFGs for the transformer.*

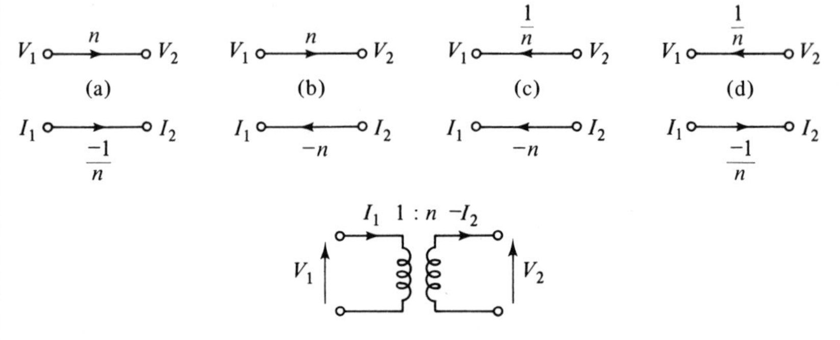

Four SFGs for the transformer

TABLE 10-7 *Four SFGs for the gyrator.*

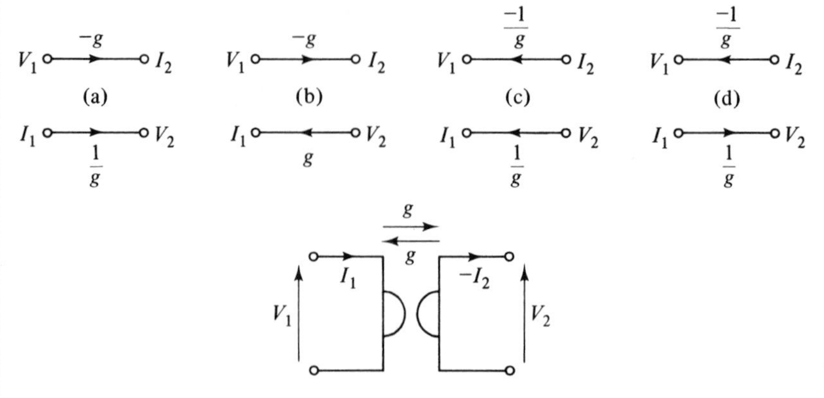

Four SFGs for the gyrator

$V_{1,2}$ and $I_{1,2}$ node variables have transmittances with arrows pointing in toward or away from them. On the other hand, type (b) SFGs may be terminated at port 1 with a Z-type branch and at port 2 with a Y-type branch. Similarly, the type (d) SFG may be terminated at port 1 with a Y-type branch and at port 2 with a Z-type branch. *For the gyrator,* only types (b) and (d) SFGs can be terminated in Z- or Y-type transmittances. The (b) type is terminated in Z-type transmittances at both ports and the (c) type in Y-type transmittances at both ports.

10-4.4 Highpass "Leapfrog" SFG Ladder Filters

We can implement, in a straightforward manner, the SFG for a resistively terminated *highpass LC* ladder filter of the type shown in Fig. 10.18(a). Two obvious approaches are possible: First, we may choose to

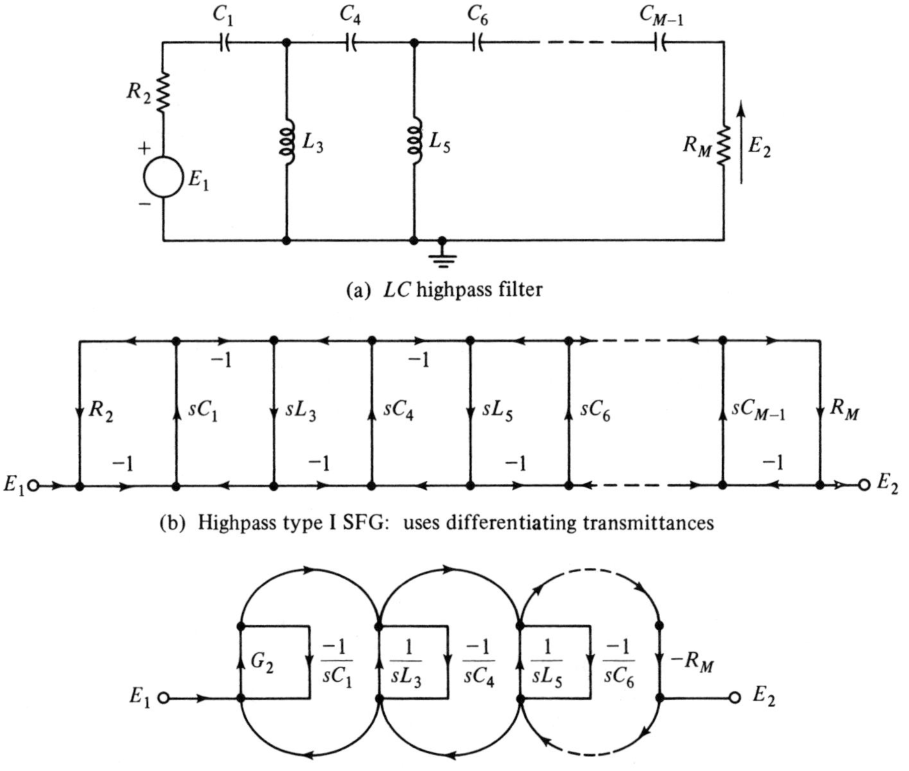

(a) LC highpass filter

(b) Highpass type I SFG: uses differentiating transmittances

(c) Highpass type II SFG: uses integrating transmittances

FIGURE 10.18 *Highpass leapfrog SFG filters.*

use the type I SFG of Fig. 10.14(a); if so, we must select

$$\left.\begin{aligned} &Y_i = sC_i, && i = 1, 4, 6, 8, \ldots, (M-1) \\ &Z_i = sL_i, && i = 3, 5, 7, 9, \ldots, (M-2) \\ &Z_2 = R_2 \\ &Z_M = R_M \end{aligned}\right\} \begin{array}{l}\text{HIGHPASS}\\ \text{TRANSMITTANCES}\\ \text{USING TYPE I}\\ \text{LADDER SFG}\end{array} \tag{10.44}$$

The resultant type I SFG is shown in Fig. 10.18(b), from which it is clear that *differentiating* transmittances sC_i and sL_i are required. This disadvantage may be overcome by eliminating the differentiators as shown in Fig. 10.16(a), but that results in a substantial increase in the required number of OP AMPs. A second approach to the simulation of the highpass filter of Fig. 10.18(a) employs the type II SFG of Fig. 10.14(b), which leads directly to

$$\left.\begin{aligned} &Z_i = \frac{1}{sC_i}, && i = 2, 4, 6, 8, \ldots, (M-1) \\ &Y_i = \frac{1}{sL_i}, && i = 3, 5, 7, 9, \ldots, (M-2). \\ &Y_1 = G_1 \\ &Z_M = R_M \end{aligned}\right\} \begin{array}{l}\text{HIGHPASS}\\ \text{TRANSMITTANCES}\\ \text{USING TYPE II}\\ \text{LADDER SFG}\end{array} \tag{10.45}$$

The resultant type II SFG is shown in Fig. 10.18(c), where it may be observed that *integrating-type* transmittances are used. Thus, the use of differentiating transmittances is avoided by using the type II ladder SFG. *For this reason, the type II ladder SFG is preferable to the type I ladder SFG for the purpose of realizing the highpass ladder filter.*

The reader will perhaps have noticed that almost all of the nodes in the type II ladder SFGs [Fig. 10.14(c)] have *two output-directed transmittances*. This property of the type II ladder SFG is a disadvantage because it does not allow the designer to use a simple *single current*-summing OP AMP circuit of the type shown in Fig. 5.7. Usually, the designer must create two distinct summing nodes, as shown in Fig. 10.19, to realize the two-output current node variables I_1 and I_2, which implies the use of an additional OP AMP to implement the equations at each node of Fig. 10.18(c) [as compared with Fig. 10.18(b)].

The highpass filter realizations of Fig. 10.18 have all the zeros of the transfer function $E_2(s)/E_1(s)$ at the origin of the s-plane. The realization of finite imaginary (stopband) zeros may be achieved by means of capacitance elements $C_3, C_5, C_7, \ldots$ *in series with* $L_3, L_5, L_7, \ldots,$ as shown in Fig. 10.20(a) or, alternatively, by means of inductance elements

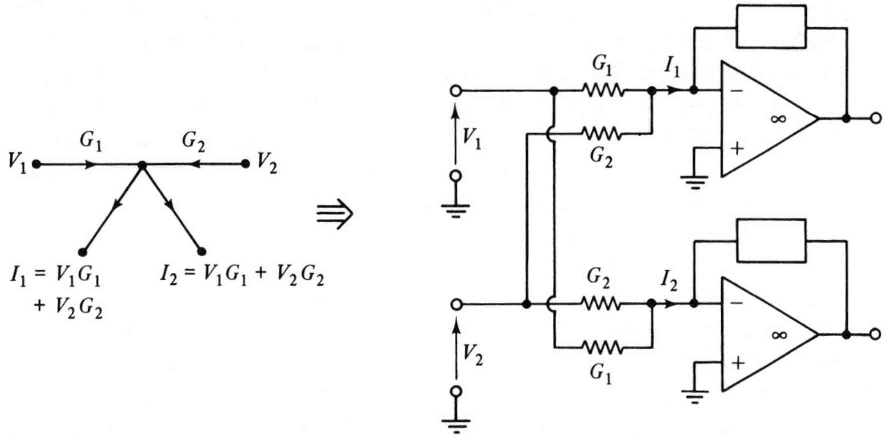

FIGURE 10.19 *An implementation of a two-input two-output SFG node, using current-summation at two nodes.*

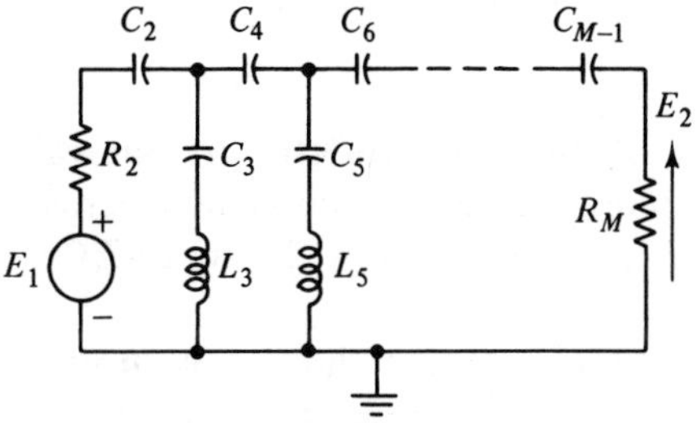

(a) Highpass with finite stopband zeros

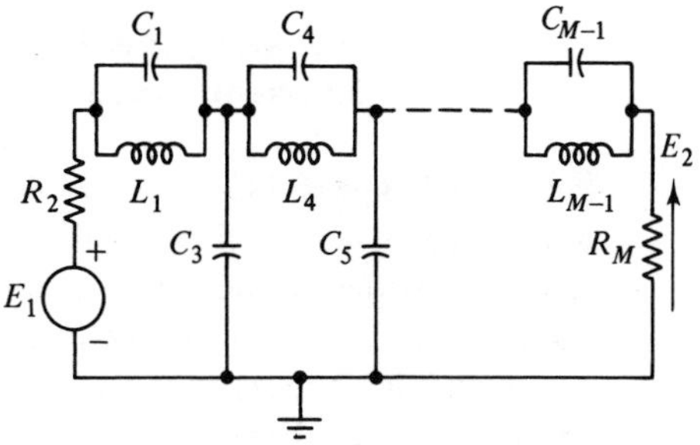

(b) Highpass with finite stopband zeros

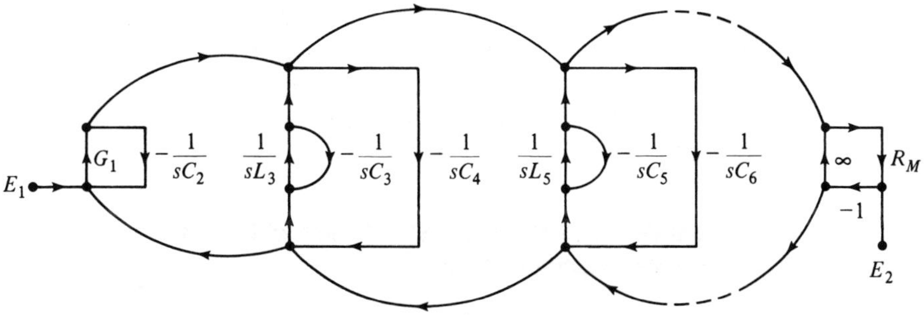

(c) Type II SFG implementation of the highpass filter in Fig. 10.20(a): finite stopband zeros

FIGURE 10.20 *Leapfrog highpass with stopband zeros.*

$L_1, L_4, L_6, \ldots,$ in parallel with $C_1, C_4, C_6, \ldots,$ as shown in Fig. 10.20(b). Structures of this kind are realizable by appropriate modifications of either of the two SFGs shown in Fig. 10.18. For example, let us modify the highpass type II SFG to realize the *LC* structure of Fig. 10.20(a). We must modify the *Y*-type transmittances $1/sL_3, 1/sL_5, 1/sL_7, \ldots,$ etc., so that they correspond to a *Y*-type series *LC* circuit. Inspection of Tables 10-3 and 10-4 reveals that a series *Y*-type *LC* circuit is a special case of *both* the *Y*-type Foster II SFG (L_2 and C_2) and the *Y*-type Foster I SFG (L_1 and C_1); in both cases, the series *LC* connections are realized as shown in Fig. 10.20(c). That is, the feedforward transmittances $1/sL_3, 1/sL_5, 1/sL_7, \ldots,$ etc., may be modified by the addition of feedback transmittances $-1/sC_3, -1/sC_5, -1/sC_7, \ldots,$ etc., to represent the series capacitances $C_3, C_5, C_7, \ldots,$ etc., of Fig. 10.20(a). [The reader is asked in Problem 10-9 to find a way of modifying Fig. 10.18(b) so that it realizes the alternate highpass structure of Fig. 10.20(b)].

10-4.5 Lowpass "Leapfrog" SFG Filters with Finite Stopband Zeros

The lowpass *LC* filters of Figs. 10.21(a) and 10.21(b) are equivalent and used widely; they realize finite imaginary (stopband) zeros at frequencies determined by the resonant frequencies of the series-*LC* and parallel-*LC* branches, respectively.

Consider the type I ladder SFG of Fig. 10.14(a) as a means of implementing Fig. 10.21(b). Clearly, the *Z*-type transmittances $Z_3, Z_5, Z_7, \ldots,$ etc., are simply integrators $1/sC_3, 1/sC_5, 1/sC_7, \ldots,$ etc. The *Y*-type transmittances $Y_1, Y_4, Y_6, Y_8, \ldots,$ etc., must describe *parallel-LC* circuits; consequently, we must use the *Y*-type Foster I SFG for each of $Y_1, Y_4, Y_6, Y_8, \ldots,$ etc., as shown in Fig. 10.21(c). Note that we *must use a dummy inductance* L_{pi}, where, ideally, $L_{pi} \to 0$, in order to maintain the SFG as *Y*-type. The complete SFG for the *LC* filter of Fig. 10.21(a) is shown in Fig. 10.21(d). [In Problem 10.10 the reader is asked to obtain an integrator-type SFG for the *LC* filter in Fig. 10.21(a).]

10-4.6 Bandpass "Leapfrog" SFG Filters with Finite Upper Stopband Zeros

An important bandpass *LC* filter section is shown in Fig. 10.22(a); it is used extensively to realize high-quality high-order bandpass functions with finite imaginary (stopband) zeros *above* the passband. Clearly, this case differs from that of Fig. 10.21(b) only by the addition of the series

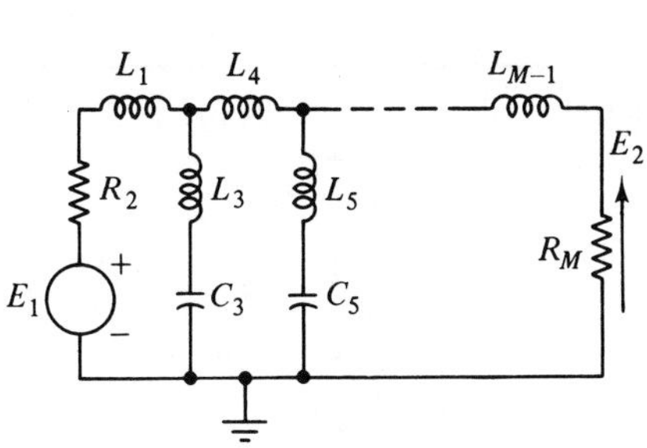

(a) Lowpass with finite stopband zeros

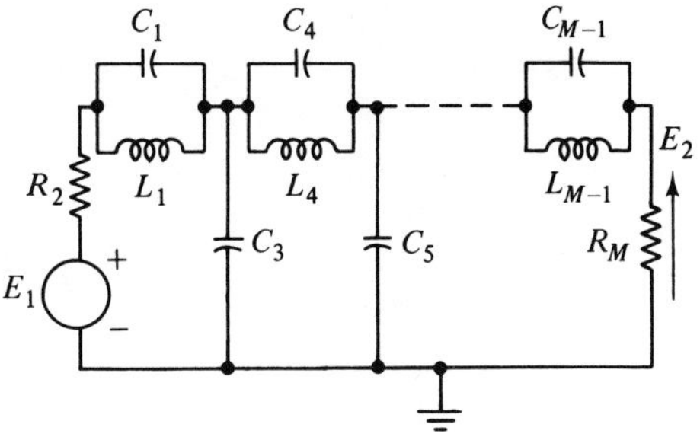

(b) Lowpass with finite stopband zeros

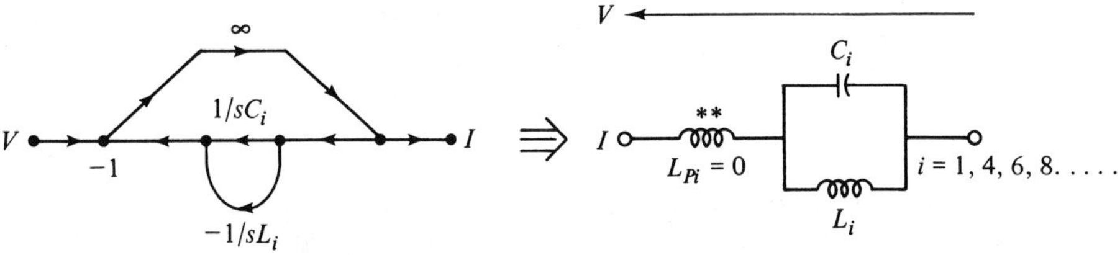

(c) Y-type implementation of series arms of the ladder in Fig. 10.21(b)

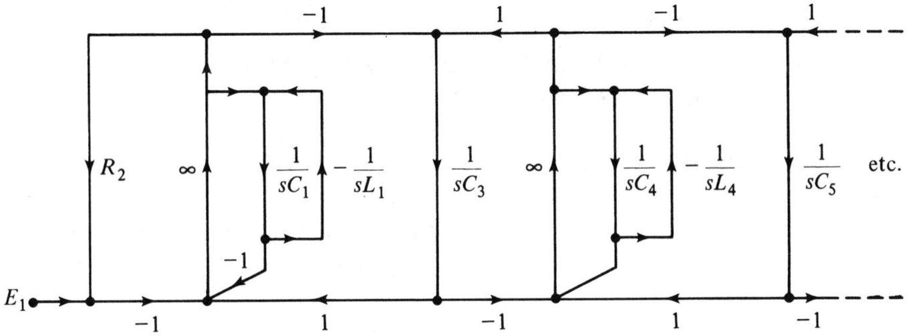

(d) Type I ladder SFG realization of the lowpass filter in Fig. 10.21(b)

FIGURE 10.21 *Leapfrog lowpass with stopband zeros.*

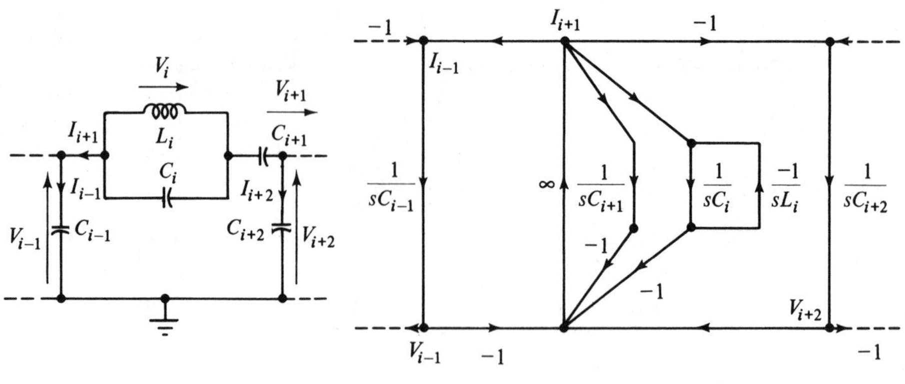

(a) Bandpass section with upper stopband zero

(b) Type I ladder SFG for Fig. 10.22(a)

FIGURE 10.22

capacitances in the series arms of the ladder. Again, we may use the Y-type Foster I SFG of Table 10-3 (with L_1 retained as a dummy inductance of ideally-zero value) so as to arrive at the SFG of Fig. 10.22(b). Other equally valid integrator-type SFGs may be derived for this LC filter structure (see Problem 10-11).

10-4.7 Bandpass "Leapfrog" SFG Filters with Lower Stopband Zeros

The LC bandpass section of Fig. 10.23(a) is used extensively to realize imaginary (stopband) zeros below the passband of the filter. This kind of filter is shown in Table 9-1 as type D and is straightforwardly implemented by using grounded-inductance simulation techniques as described in section 9.4.

An integrator-type SFG implementation may be achieved in a variety of ways. The shunt branch of Fig. 10.23(a) is a special case of the Foster II circuit and may be implemented by use of either the Z- *or* the Y-type Foster II SFG shown in Table 10-4. A somewhat simpler circuit is achieved by *selecting* the Z-type Foster II SFG to realize the shunt arms, so that the complete SFG is a type I ladder SFG of the type shown in Fig. 10.14(a); thus, the series branches of Fig. 10.23(a) must be realized as Z-type SFGs (sC_{i-1}, sC_{i+2}, . . . ,), which may be implemented with *integrating* transmittances ($-1/sC_{i-1}$, $-1/sC_{1+2}$, . . .), by using the infinite-transmittance scheme shown in Fig. 10.16(a). The resultant SFG section is illustrated in Fig. 10.23(b) for the bandpass section with a lower stopband zero.

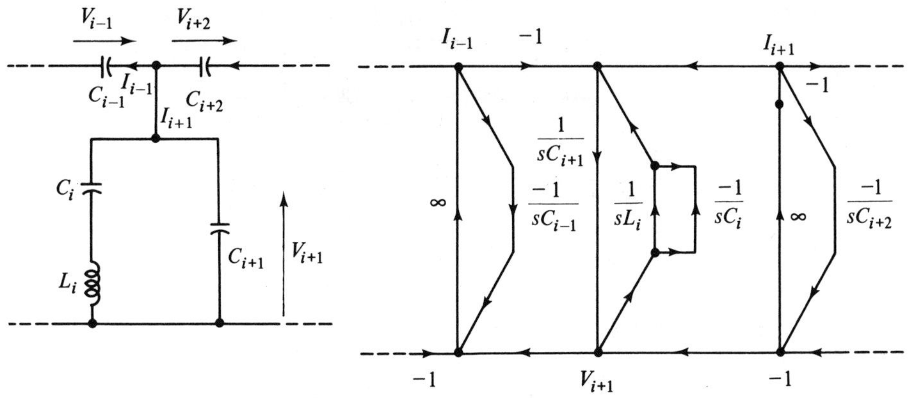

(a) Bandpass section with lower stopband zero

(b) Type I ladder SFG for Fig. 10.23(a)

FIGURE 10.23

10-5 ALLPASS CIRCUITS

The allpass transfer function is characterized by the fact that $|H(j\omega)|$ is constant over all ω and is used extensively to shape the phase characteristic of a signal without altering its magnitude characteristic. In particular, the allpass function may be used to reduce signal delay distortion.

10-5.1 Second-Order Allpass Functions

The two *impedance-normalized* circuits of Fig. 10.24 are used by designers of passive filters to implement the second-order allpass function

$$H(s) \equiv \frac{E_2(s)}{E_1(s)} = \left[\frac{s^2 - \left(\dfrac{\omega_P}{Q_P}\right)s + \omega_P^2}{s^2 + \left(\dfrac{\omega_P}{Q_P}\right)s + \omega_P^2}\right] \qquad \text{Normalized All-Pass Second Order} \qquad (10.46)$$

Analysis of these two circuits is requested in Problems 10-12 and 10-13. Circuits of this kind may be cascaded directly to realize higher-order allpass functions.

The use of GIC-embedding (see section 9-7) allows the circuit shown in Fig. 10.24(a) to be implemented with a total of six OP AMPs.

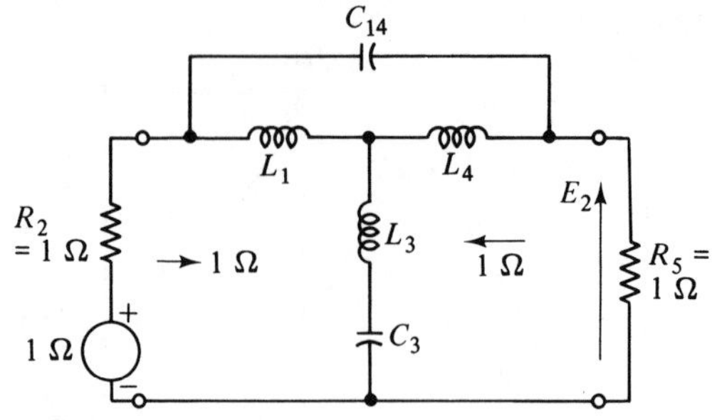

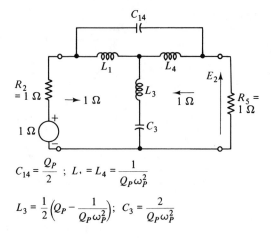

$$C_{14} = \frac{Q_P}{2}\ ;\ L_1 = L_4 = \frac{1}{Q_P \omega_P^2}$$

$$L_3 = \frac{1}{2}\left(Q_P - \frac{1}{Q_P \omega_P^2}\right);\ C_3 = \frac{2}{Q_P \omega_P^2}$$

(a) *LC* second-order allpass section

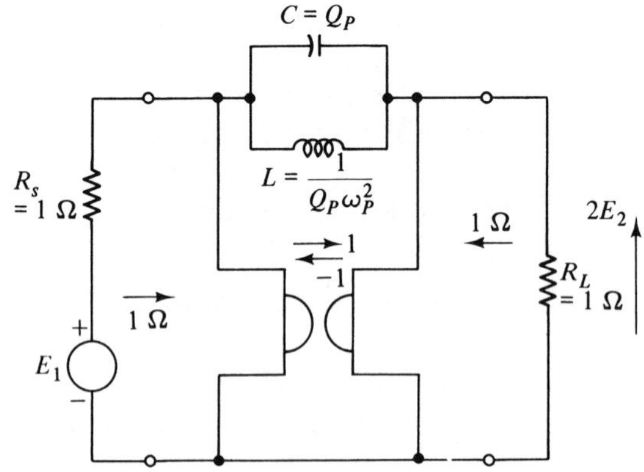

(b) *LC*–gyrator second-order allpass section

FIGURE 10.24 *Passive second order allpass circuits.*

A useful approach to the realization of the allpass function of equation (10.46) is obtained by writing it in the following form:

$$H(s) = \frac{1 - \left(\dfrac{s^2 + \omega_P^2}{\omega_P s / Q_P}\right)}{1 + \left(\dfrac{s^2 + \omega_P^2}{\omega_P s / Q_P}\right)} \tag{10.47}$$

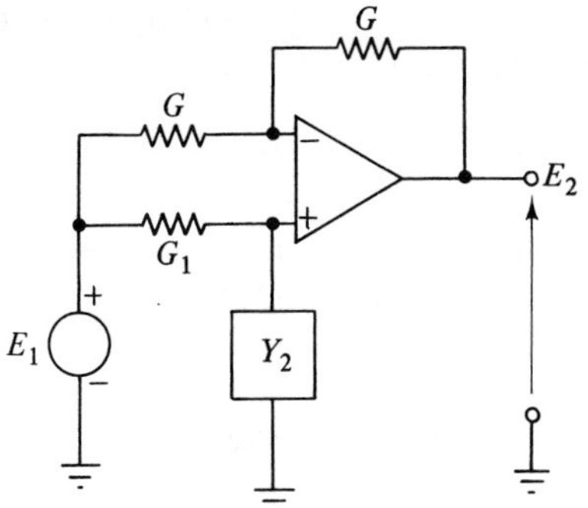

(a) Basic allpass circuit

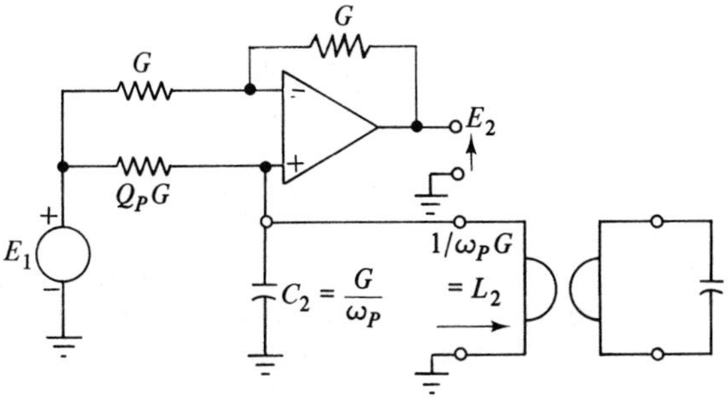

(b) Implementation of second-order allpass function: equation (10.46)

FIGURE 10.25

therefore

$$H(s) = \frac{Q_P - \left(\dfrac{s^2 + \omega_P^2}{\omega_P s}\right)}{Q_P + \left(\dfrac{s^2 + \omega_P^2}{\omega_P s}\right)} \tag{10.48}$$

Now, the circuit of Fig. 10.25(a) has a voltage transfer function given by

$$H(s) = \frac{G_1 - Y_2}{G_1 + Y_2} \tag{10.49}$$

so that equation (10.48) may be realized by selecting

$$G_1 = Q_P G \quad \text{and} \quad Y_2 = \left(\frac{s^2 + \omega_P^2}{\omega_P s}\right) G \tag{10.50}$$

This driving-point admittance Y_2 is realized by the parallel combination of an inductance L_2 and capacitance C_2, where

$$L_2 = \frac{1}{\omega_p G} \quad \text{and} \quad C_2 = \frac{G}{\omega_p} \tag{10.51}$$

The resultant second-order allpass circuit is shown in Fig. 10.25(b). The grounded-inductance may be realized by using a two-OP AMP GIC gyrator circuit of the type shown in Table 5-7.

10-5.2 High-Order Allpass Functions

Allpass functions of order greater than two may be realized directly by using the important result that *all stable* allpass functions may be written in the form

$$H(s) = \pm\left[\frac{1 - X(s)}{1 + X(s)}\right] \tag{10.52}$$

where $X(s)$ is a realizable LC driving-point immittance function. For example, in equation (10.47) the term $[Q_p(s^2 + \omega_p^2)/s\omega_p]$ is the realizable LC driving-point immittance function $X(s)$ for the second-order case. The properties of LC driving-point immittance functions $X(s)$ *of arbitrary order* are well known and are not explained in detail here. We state simply that:

(i) the poles and zeros of $X(s)$ alternate along the imaginary s-plane axis and are simple;

(ii) there must be a zero or a pole of $X(s)$ at the s-plane origin;

(iii) all functions $X(s)$ are realizable with the Foster I circuit and also with the Foster II circuit.

Consequently, the circuit of Fig. 10.25(a) may be used to realize *arbitrary-order* allpass functions $\pm H(s)$ simply by using a Foster I or Foster II LC circuit in place of the grounded admittance Y_2. If the Foster II circuit is used (Fig. 10.15) to realize Y_2, then *all* inductance elements are grounded and each one may, therefore, be realized by using the two-OP AMP GIC circuit shown in Table 5-7.

An alternative approach for the realization of *arbitrary-order* allpass functions uses the SFG representations of the Foster I and II circuits shown in Tables 10-3 and 10-4. Thus, $H(s)$ in equation (10.52) is realized as an SFG, as shown in Fig. 10.26(a). Then the transmittance $X(s)$ may be realized by using one of the SFGs from Tables 10-3 or 10-4. The general arrangement for using a Y-type Foster SFG (for any order) is shown in Fig. 10.26(b); a particular OP AMP circuit implementation is illustrated in Fig. 10.26(c) for the simple second-order case. Problems 10-15 to 10-17 provide the designer with some design experience with this technique.

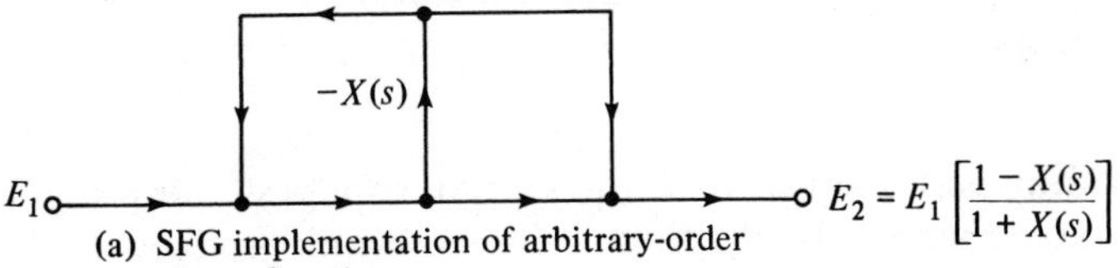

(a) SFG implementation of arbitrary-order allpass function.

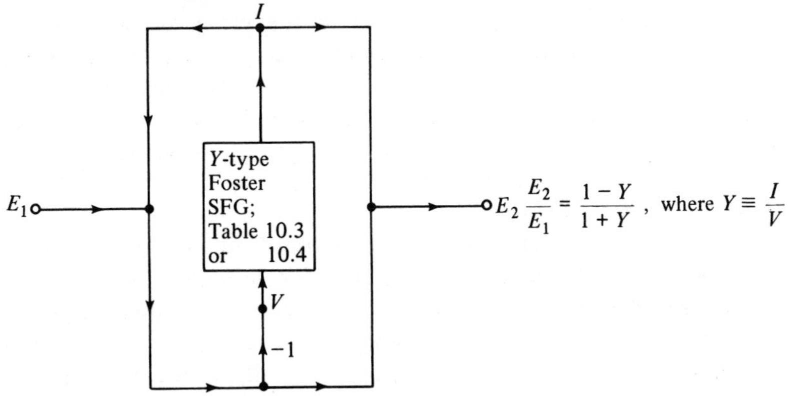

(b) SFG implementation of arbitrary-order allpass function using Y–type Foster SFGs.

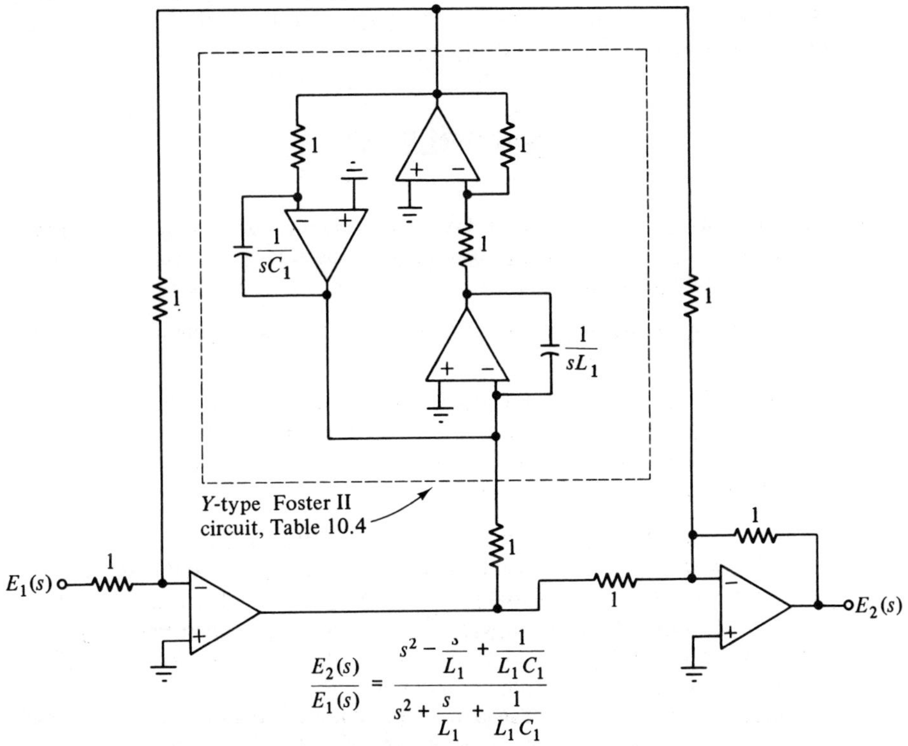

(c) A second-order allpass circuit using OP AMPs and the Foster II configuration from Table 10.4.

FIGURE 10.26

10-6 SUMMARY

The nonideal performance of lowpass and bandpass "leapfrog" ladder filters is caused usually by the finite gain bandwidth products B of the OP AMPs; it is shown in sections 10-2 and 10-3 that the deviation of the passband gain $M(\omega)$ is proportional to the group delay $\tau(\omega)$ and is a maximum at the edge of the passband. The type A realizations use type 1A noninverting integrators (phase-lag types) and lead to a passband fractional deviation given by $\Delta M(\omega)/M(\omega) \approx 2\omega^2\tau(\omega)/B$. The type B realizations employ type 1B noninverting integrators (phase-lag types) and are far superior to type A realizations insofar as the effects of finite gain bandwidths B are concerned.

A general approach to the synthesis of LC filters is described in section 10-4, where rules are established for writing integrator-type SFGs for any transformerless resistively terminated LC filter. In particular, SFGs are described for realizing imaginary s-plane zeros, because filters of this kind are of great practical importance. Allpass functions of arbitrary order may be implemented by using the Foster I and Foster II SFGs of Tables 10-3 and 10-4. The reader who wishes to design RC-active filter versions of the SFGs presented in this chapter is strongly advised to attempt the problems at the end of the chapter.

PROBLEMS

10-1. Write a signal-flow graph (SFG) for the seventh-order Chebychev lowpass filter of Fig. 9.7(a) and, thereby, design a type A lowpass "leapfrog" ladder simulation of the kind shown in Fig. 10.1(a). Let $R = 10$ kΩ.

10-2. Obtain a type B lowpass "leapfrog" ladder simulation of the filter shown in Fig. 9.7(a). Let $R = 10$ kΩ.

10-3. Prove that the compensation capacitors C_c, as defined by equations (10.10) and (10.11) and as shown in Fig. 10.4, provide the necessary phase lead compensation to compensate for the phase lags of $-B/\omega$ and $-B/3\omega$ given in equations (10.5) and (10.6).

10-4. In section 10-3.2 and Fig. 10.7, it is shown that a Chebychev bandpass "leapfrog" design with a center frequency of 1.5915 kHz and bandwidth of 0.15915 kHz may be realized by using the type B circuit. Inspection of Fig. 10.7 reveals that $M(\omega)$ for curve B is in good agreement with the ideal response; the transition regions are slightly too low in frequency. This small error is the result of the terms for $|M(j\omega)|$ in Table 5.1 (inverting-integrator) and in Table 5.2 (noninverting-integrator) being slightly less than unity. Derive the function $|M(j\omega)|$ for each integrator used in this design and, thereby, predistort the integrating capacitors C_2, L_2/R^2, L_3/R^2, C_3, C_4, L_4/R^2 to allow for these errors in $|M(j\omega)|$.

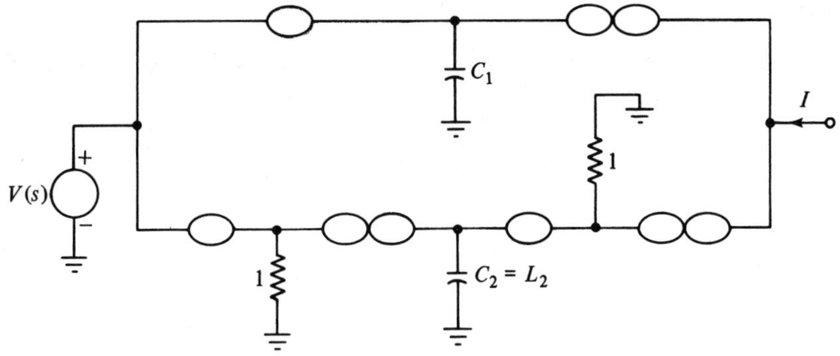

FIGURE P10.5

10-5. Prove that the three-nullor network of Fig. P10.5 is a realization of the SFG shown in Fig. 10.11(b). Suggest a small-signal transistor circuit version.

10-6. Use inverting-integrator and noninverting integrator OP AMP circuits to implement a type II ladder SFG of the type shown in Fig. 10.14(b) for the *highpass LC* filter prototype provided by

$$Y_1 = G_1,\quad Z_2 = \frac{1}{sC_2},\quad Y_3 = \frac{1}{sL_3},\quad Z_4 = \frac{1}{sC_4},\quad Y_5 = G_5$$

corresponding to a third-order transfer function. See Fig. 10.18(c).

10-7. Prove that the *two-integrator* OP AMP circuit of Fig. 10.16(b) may be used to realize the *differentiating* transmittance sx_i of Fig. 10.16(a).

10-8. Prove that the equivalent circuits of Fig. 10.16(c) are valid when the two-integrator OP AMP circuit of Fig. 10.16(b) is used to realize a Z-type Foster I series inductance and a Y-type Foster II parallel capacitance. The gain of OP AMP #2 is approximately B_2/s.

10-9. Find a way of modifying the SFG of Fig. 10.18(b) so that it corresponds to the *LC* highpass prototype of Fig. 10.20(b) having finite stopband zeros. Ensure that *integrating* transmittances are *not* used.

10-10. Obtain an SFG representation of the lowpass filter of Fig. 10.21(a) that uses integrating-type transmittances and does *not* use differentiating-type transmittances.

10-11. Obtain an SFG representation of the bandpass section in Fig. 10.22(a) by using the Z-type Foster I SFG (with $sL_1 = 0$) to realize the series branch and the Y-type Foster II SFG to realize the shunt capacitance branch. Eliminate differentiating-type transmittances from the Y-type Foster II SFGs.

10-12. Prove that Fig. 10.24(a) realizes the allpass function, and show that it may be simulated with six OP AMPs by using GIC-embedding techniques.

10-13. Prove that Fig. 10.24(b) realizes the allpass function.

10-14. With $G = 10^{-4}$ ohm^{-1}, use the circuit shown in Fig. 10.26(b) to realize the second-order allpass function, with $Q_p = 20$ and $\omega_p = 10^4$ rad/sec. Do not use more than three OP AMPs for the complete realization.

10-15. Use the structure shown in Fig. 10.26(a) to realize the third-order allpass function

$$H(s) = \frac{-s^3 + 6s^2 - 11s + 6}{s^3 + 6s^2 + 11s + 6}$$

Only three OP AMPs are required. [*Hints*: Divide numerator and denominator of $H(s)$ by $(6s^2 + 6)$, rearrange in the form of equation (10.52), and realize $X(s)$ by using a single GIC circuit.]

10-16. Prove that it is *always* possible to choose the decomposition of an allpass function so that $X(s)$ in equation (10.52) leads to a Foster II configuration for the admittance Y_2 shown in Fig. 10.26(a), thereby permitting all active inductance elements to be *grounded* in the active simulation of Y_2.

10-17. Implement the allpass function of Problem 10-15 by using the SFG scheme of Fig. 10.26(b). Choose a decomposition for $X(s)$ that minimizes the number of OP AMPs.

BIBLIOGRAPHY

Brackett, P. O., and A. S. Sedra, "Active Compensation for High-Frequency Effects in OP AMP Circuits with Applications to *RC*-Active Filters," *IRE Trans. Circuits Systems*, vol. CAS-23, p. 68–73, Feb. 1976.

Heinlein, W., and H. Holmes, *Active Filters for Integrated Circuits*, Prentice-Hall, 1974, chapter 8.

Sedra, A. S., and P. O. Brackett, *Filter Theory and Design: Active and Passive*, Matrix Publishers, 1978, chapter 12.

Wing, O., "Ladder Network Analysis by Signal Flow Graph—Application to Analog Computer Programming," *IRC Trans. Circuit Theory*, vol. CT-3, p. 289–294, Dec. 1956.

SUPPLEMENTARY BIBLIOGRAPHY

The purpose of this supplementary bibliography is to provide additional reading for the graduate level student or researcher who wishes to pursue the topic in depth.

GENERAL

BRUTON, L. T., "Multiple-Amplifier RC-Active Filter Design with Emphasis on GIC Realizations," *IEEE Trans. on Circuits and Systems*, Vol. CAS-25, No. 10, pp. 830–845.

HEINLEIN, W., and H. HOLMES, *Active Filters for Integrated Circuits*, Prentice-Hall International, 1974.

HUELSMAN, L. P., *Active RC Filters: Theory and Application*, Dowden, Hutchinson and Ross, 1976.

SENSITIVITY

BLOSTEIN, M. L., "Sensitivity Analysis of Parasitic Effects in Resistance Terminated LC Two-Ports," *IEEE Trans. Circuit Theory*, pp. 21–25, Mar. 1967.

BLOSTEIN, M. L., "Some Bounds on Sensitivity in RLC Networks," Proc. 1st Allerton Conf. Circuit System Theory, pp. 488–501, 1963.

BRUTON, L. T., "Sensitivity Comparison of High-Q Second-Order Active Filter Synthesis Techniques," *IEEE Trans. Circuits Syst.*, pp. 32–38, Jan. 1975.

DIRECTOR, S. W., and R. A. ROHRER, "Automated Network Design–The Frequency–Domain Case." *IEEE Trans. CT-16* (1969), pp. 330–337.

DIRECTOR, S. W., and R. A. ROHRER, "The Generalized Adjoint Network and Network Sensitivities." *IEEE Trans. CT-16* (1969), pp. 318–323.

GÉHER, K., *Theory of Network Tolerance*, Budapest, Akadémiai Kiado 1971.

HOLT, A. G. J., and M. R. LEE, "Sensitivity Comparison of Active-Cascade and Inductance-Simulation Schemes," *Proc. Inst. Elec. Eng.*, vol. 119, pp. 277–282, 1972.

MOSCHYTZ, G. S., "A Note on Pole Frequency and Q Sensitivity," *IEEE J. Solid-State Circuits*, vol. SC-6, Aug. 1971.

ORCHARD, J., "Inductorless Filters," *Electron. Lett.*, vol. 2, 224, Sept. 1966.

SCHOEFFLER, J. D., "The Synthesis of Minimum Sensitivity Networks," *IEEE Trans. Circuit Theory*, vol. CT-11, pp. 270–276, June 1964.

SWAMY, M. N. S., C. BUSHNAN, and K. THULASIRAMAN, "Bounds on the Sum of Element Sensitivity Magnitudes for Network Functions," Proc. IEEE Int. Symp. on Circuit Theory, IEEE Cat. No. 72CHO 594-2CT, pp. 21–25, Apr. 1972.

TEMES, G. C., and H. J. ORCHARD, "First-Order Sensitivity and Worstcase Analysis of Doubly Terminated Reactance Two-Ports," *IEEE Trans. Circuit Theory*, vol. CT-20, pp. 567–571, Sept. 1973.

BIQUADRATIC TWO-INTEGRATOR LOOP STRUCTURES

AKERBERG, D., and K. MOSSBERG, "Low-Sensitivity Easily Trimmed Standard Building Block for Active RC Filters," *Electron. Lett.*, vol. 5, no. 21, pp. 528–529, Oct. 1969.

AKERBERG, D., and K. MOSSBERG, "A Versatile Active RC Building Block with Inherent Compensation for the Finite Bandwidth of the Amplifier," *IEEE Trans. Circuits Syst.*, vol. CAS-21, pp. 75–78, Jan. 1974.

BUDAK, A., *Passive and Active Network Analysis and Synthesis*. Houghton Mifflin, 1974, chs. 10–12.

GIRLING, F. E. J., and E. F. GOOD, "Active Filters 7 and 8: The Two Integrator Loop," *Wireless World*, vol. 76, pp. 117–119, Mar. 1970, and pp. 134–139, Apr. 1970.

KERWIN, W. J., L. P. HUELSMAN, and R. W. NEWCOMB, "State-Variable Synthesis for Integrated Circuit Transfer Functions," *IEEE J. Solid-State Circuits*, vol. SC-2, pp. 87–92, Sept. 1967.

TARMI, R., and M. S. GHAUSI, "Very High-Q Insensitive Active RC Networks," *IEEE Trans. Circuit Theory*, vol. CT-17, pp. 358–366, Aug. 1970.

THOMAS, L. C., "The Biquad: Part 1—Some Practical Design Considerations," *IEEE Trans. Circuit Theory*, vol. CT-18, pp. 350–357, May 1971.

TOW, J., "Design Formulas for Active RC Filters Using Operational Amplifier Biquad," *Electronic Lett.*, pp. 339–341, July 1969.

TOW, J., "A Step-by-Step Active-Filter Design," *IEEE Spectrum*, vol. 6, pp. 64–68, Dec. 1969.

BIQUADRATIC SINGLE-AMPLIFIER CIRCUITS

FRIEND, J. J., C. A. HARRIS, and D. HILBERMAN, "Star: An Active Biquadratic Filter Section," *IEEE Trans. Circuit Syst.*, CAS-22 (2), 115–121, 1975.

MOSCHYTZ, G. S., "Gain-sensitivity Product—A Figure of Merit for Hybrid Integrated Filters Using Single Operational Amplifiers," *IEEE J. Solid-State Circuits*, SC-6 (3), 103–110, 1971.

MOSCHYTZ, G. S., "A Universal Low-Q Active-Filter Building Block Suitable for Hybrid-Integrated Circuit Implementation," *IEEE Trans. Circuit Theory*, CT-20 (1), 37–47, 1973.

WEYTEN, L., "Q- and ω_0-Sensitivities in Positive Gain Second-Order RC Active Filters," Proc. IEEE, 60 (11), 1462–1463, 1972.

BIQUADRATIC GENERALIZED IMMITTANCE CONVERTER CIRCUITS

BRUTON, L. T., "Biquadratic Sections Using Generalized Impedance Converters," *IREE J.*, vol. 41, no. 11, pp. 510–512, Nov. 1971.

FLIEGE, N., "A New Class of Second-Order RC-Active Filters with Two Operational Amplifiers," *Nachrichtentech. Z.*, 26(6), 279–282, 1973.

MIKHAEL, W. B., and B. B. BHATTACHARYYA, "A Practical Design for Insensitive RC-Active Filters." *IEEE Trans. Circ. Syst.*, CAS-22 (5), 407–415, 1975.

INDUCTANCE SIMULATION

ANTONIOU, A., "Realisation of Gyrators Using Operational Amplifiers, and Their Use in RC-Active-Network Synthesis," *Proc. IEEE*, 116(11), 1838–1850, 1969.

BRUTON, L. T., "Nonideal Performance of Two-Amplifier Positive-Impedance Converters," *IEEE Trans. Circuit Theory*, vol. CT-17, pp. 541–549, Nov. 1970.

BRUTON, L. T., and J. T. LIM, "High-Frequency Comparison of GIC-Simulated Inductance Circuits," *Int. J. Circuit Theory Appl.*, vol. 2, pp. 401–404, 1974.

HOLMES, W. H., W. E. HEINLEIN, and S. GRUTZMANN, "Sharp Cut-Off Low-Pass Filters Using Floating Gyrators," *IEEE J. Solid-State Circuits*, vol. SC, pp. 38–50, Feb. 1969.

HOLT, A. G. J., and M. R. LEE, "Sensitivity Comparison of Active-Cascade and Inductance-Simulation Schemes," *Proc. Inst. Elec. Eng.*, vol. 119, pp. 277–282, 1972.

ORCHARD, H. J., "Inductorless Filters," *Electron. Lett.*, 2(6), pp. 224–225, 1966.

ORCHARD, H. J., and D. F. SHEAHAN, "Inductorless Bandpass Filters," *IEEE J. Solid-State Circuits*, vol. SC-5, pp. 108–118, June 1970.

RIORDAN, R. H. S., "Simulated Inductors Using Differential Amplifiers," *Electron. Lett.*, 3(2), 50–51, 1967.

SHENOI, B. A., "Practical Realization of a Gyrator Circuit and RC-Gyrator Filters," *IEEE Trans. Circ. Theory*, CT-12(3), 374–380, 1965.

TRIMMEL H. R., and W. E. HEINLEIN, "Fully Floating Chain-Type Gyrator Circuit using Operational Transconductance Amplifiers," *IEEE Trans. Circuit Theory* (Corresp.), vol. CT-18, pp. 719–721, Nov. 1971.

FREQUENCY DEPENDENT NEGATIVE RESISTANCE

ANTONIOU, A., "Bandpass Transformation and Realization Using Frequency-Dependent Negative-Resistance Elements," *IEEE Trans. Circuit Theory*, (Corresp.), vol. CT-18, pp. 297–299, Mar. 1971.

BRUTON, L. T., "Network Transfer Functions Using the Concept of Frequency-Dependent Negative Resistance," *IEEE Trans. Circuit Theory*, vol. CT-16, pp. 406–408, Aug. 1969.

BRUTON, L. T., "Nonideal Performance of Two-Amplifier Positive-Impedance Converters," *IEEE Trans. Circuit Theory*, vol. CT-17, pp. 541–549, Nov. 1970.

SCHMIDT, C. E., and M. S. LEE, "Multipurpose Simulation Network with a Single Amplifier," *Electron. Lett.*, 11(1), 9–10, 1975.

TRIMMEL, H. R., "Realization of Canonical Bandpass Filters with Frequency-Dependent and Frequency-Independent Negative Resistances," *Proc. IEEE Int. Symp. on Circuit Theory*, pp. 134–137, 1973.

ACTIVE LADDER FILTERS: IMMITTANCE SIMULATION AND SFG SIMULATION

BRACKETT, P. O., and A. S. SEDRA, "Active Compensation for High-Frequency Effects in Op-Amp Circuits with Applications to Active RC Filters," *IEEE Trans. Circuits Syst.*, vol. CAS-23, pp. 68–72, Feb. 1976.

BRACKETT, P. O., and A. S. SEDRA, "Direct SFG Simulation of LC Ladder Networks with Applications to Active Filter Design," *IEEE Trans. Circuits Syst.*, vol. CAS-23, pp. 61–67, Feb. 1976.

BRUTON, L. T., "Network Transfer Functions Using the Concept of Frequency-Dependent Negative Resistance," *IEEE Trans. Circuit Theory*, vol. CT-16, pp. 406–408, Aug. 1969.

BRUTON, L. T., "Topological Equivalence of Inductorless Ladder Structures Using Integrators," *IEEE Trans. Circuit Theory*, vol. CT-20, pp. 434–437, July 1973.

BRUTON, L. T., and A. B. HAASE, "Sensitivity of Generalized Immittance Converter-Embedded Ladder Structures," *IEEE Trans. Circuits Syst.*, vol. CAS-21, pp. 245–250, Mar. 1974.

BRUTON, L. T., J. T. LIM, and J. VALIHORA, "The Feasibility of Active Filtering in Frequency Division Multiplex Systems," in Proc. Int. Symp. on Circuit and System Theory, San Francisco, Ca., April 1974.

BRUTON, L. T., and K. RAMAKRISHNA, "High Frequency Limitations of Active Ladder Networks," presented at the Asilomar Conf., Pacific Grove, Monterey, Ca., Dec. 1974.

BRUTON, L. T., and A. I. A. SALAMA, "Frequency Limitations of Coupled-Biquadratic Active Ladder Structures," *IEEE J. Solid-State Circuits*, vol. SC-9, pp. 69–71, Apr. 1974.

BRUTON, L. T., and D. TRELEAVEN, "Active Filter Design Using Generalized Impedance Converters, *EDN*, pp. 68–75, Feb. 1975.

GIRLING, F. E. J., and E. F. GOOD, "Active Filters 12 and 13: The Leapfrog or Active Ladder Synthesis; Applications of the Active Ladder Synthesis," *Wireless World*, vol. 76, pp. 341–345, July 1970, and pp. 445–450, Sept. 1970.

GORSKI-POPIEL, J., "RC-Active Synthesis Using PIC," *Electron. Lett.*, vol. 3, pp. 381–382, 1967.

HOLMES, W. H., W. E. HEINLEIN, and S. GRUTZMANN, "Sharp Cut-Off Low-Pass Filters Using Floating Gyrators," *IEEE J. Solid-State Circuits*, vol. SC-, pp. 38–50, Feb. 1969.

ORCHARD, H. J., and D. F. SHEAHAN, "Inductorless Bandpass Filters," *IEEE J. Solid-State Circuits*, vol. SC-5, pp. 108–118, June 1970.

SZENTIRMAI, G., "Multiple-Feedback Filters," *Proc. IEEE Int. Symp. Circuit Theory*, pp. 339, Apr. 1972.

INDEX

INDEX

T